TECHNOLOGY TRANSFER

NATO ADVANCED STUDY INSTITUTES SERIES

Proceedings of the Advanced Study Institute Programme, which aims at the dissemination of advanced knowledge and the formation of contacts among scientists from different countries.

The series is published by an international board of publishers in conjunction with NATO Scientific Affairs Division

A Life Sciences B Physics	Plenum Publishing Corporation London and New York
C Mathematical and Physical Sciences	D. Reidel Publishing Company Dordrecht and Boston
D Behavioural and Social Sciences	Sijthoff International Publishing Company Leiden
E Applied Sciences	Noordhoff International Publishing Leiden

Series E: Applied Sciences
Volume 6 – Technology Transfer

TECHNOLOGY TRANSFER

edited by

HAROLD F. DAVIDSON
MARVIN J. CETRON
JOEL D. GOLDHAR

NOORDHOFF – LEIDEN – 1974

Proceedings of the NATO Advanced Study Institute on Technology Transfer
Evry, France
June 24 – July 6, 1973

ISBN 90 286 0573 8

PREFACE

Several formerly independent trends including the increasing rate of technological change, the demand for greater efficiency and productivity in R&D and innovation, and the need for applying technology to the solution of pressing social and economic problems have recently begun to reinforce each other resulting in renewed interest in technology transfer activities and research. The proliferation of research efforts in the U.S. and Western Europe, increased corporate and governmental efforts to develop mechanisms and incentives for transfer and the growing closeness between transfer and innovation studies led to the decision to convene a meeting with representatives from both research and practice in transfer and innovation in order to establish a "state-of-the-art" baseline and to promote discussion between the researchers and practioners from all sectors in the hope that new research will better reflect the realities of current practice and new mechanisms created by practioners might be improved through the use of knowledge generated from research.

How close we came to the second goal is a question which must be left to the future. The proceedings which follow reflect the first goal. If the papers and conclusions occasionally appear to be contradictory or repetitive, or naive, or overly pessimistic or simply based upon too narrow a base or research or experience, they are an accurate reflection of both the state-of-the-art and the lack of communications among individuals and organizations involved in similar activities. Efficient and effective communications and transfer should begin at home and we hope that the personal contacts and increased awareness resulting from the two weeks of both formal and informal discussion will contribute to the further development of research and practice in technology transfer.

The proceedings which follow are not arranged in order of presentation, rather they are grouped in a sequence which seems logical as a result of the conference discussions. Part I includes the Chairman's introduction at the conference. Part II brings together all of the papers containing quantitative data based upon empirical research. Most of these papers were presented by university-based researchers who have been able to assemble data from a variety of sources. Their conclusions have both the advantages and disadvantages of a somewhat detached point of view. Part III on public sector experiences reflects a longer tradition of effort, particularly in the area of military and space activities. The successes and failures in public sector transfer efforts provide many useful insights into the process and guidelines for the increasing levels of international and corporate activity.

Part IV brings together the contributions of both public and private sector practioners active in the transfer of technology between nations. As technology becomes a major vehicle for international competition, trade and national development, the concerns and problems of both corporate and governmental transfer policies and activities will become identical. International technology transfer, embedded within a variety of national foreign policies is the area with the highest level of complexity, least efficiency and greatest potential for worldwide economic and social development. Industry experiences are collected in the final section.

Part V is somewhat smaller than it should be to represent the level of corporate technology transfer activities; but the confidential nature of such activities, especially within the large multinational corporation tends to reduce the number and variety of publicly presented discussions. The importance of the area, however, should not be underestimated. Corporations have a long history of technology transfer. Much like the man who one day discovered that all his life he had been speaking prose, corporations are becoming aware that all along they have been transferring technology. The new awareness of industry's role is very important as it opens up a large backlog of experience and a new arena for research; both of which can be expected to contribute to improved future transfer mechanisms and activities.

The proceedings are completed in Part VI with the keynote address which was delivered by United States Congressman J. Edward Roush midway through the meeting. By placing at the end of these proceedings, an address which describes the contributions of military and space R&D to the civilian economy, we hope to emphasize that technology transfer is not a sudden new innovation, but rather a time-tested, but little-understood activity, which, for the reasons cited at the beginning of this discussion, is growing in urgency and importance and which requires more basic understanding of its underlying processes and motivations.

Our most important debt is naturally to the many speakers who have contributed to this volume. We are grateful to them. We also owe a debt of gratitude to both NATO and the U.S. Army for their financial support and help in planning this conference. A special tribute must go to our wives--Gloria, Helene, and Laura--to whom this volume is dedicated; they have helped us with patience, encouragement and not a small amount of critical commentary. Naturally we assume the responsibility for all errors in the editing, both of commission and ommission, and hope that the users of this book will feel free to give us their reactions to the collection of material offered herein.

Washington, D.C. July 15, 1974

Harold F. Davidson
Marvin J. Cetron
Joel D. Goldhar

TABLE OF CONTENTS

PART I: An Overview of the Meeting

CHAIRMAN'S INTRODUCTION

TECHNOLOGY TRANSFER: WHERE WE STAND TODAY

Marvin J. Cetron

Forecasting International, Ltd.

Introduction

This paper surveys the state of the art in technology transfer. The purpose in presenting this survey is to catalyze discussion about the concepts and types of technology transfer, as well as to encourage thought about the barriers and stimuli to innovation, and the role they play in the licensing and the growth of the multinational firm. Each of these topics will be addressed in the paper. It is the hope of the committee that a stimulating and useful perspective may be created by focussing on some of the international aspects of technology transfer.

Concepts concerning the transfer of knowledge are really not new. Francis Bacon had the following thought on the subject some 350 years ago:

> The human mind is often so awkward and ill-regulated in the career of invention that it is first diffident and then it dispares itself, for it appears at first incredible that any such discovery should be made; and, when it has been made, it appears incredible that it should so long have escaped man's research, all of which afford good reason for the hope that a vast mass of inventions yet remains which may be deduced, not only from the investigation of new modes of operation, but also from transferring, comparing, and applying those already known by methods of what we have termed literate experiences. [1]

In the rapidly changing world of today, the desire to utilize our wealth of scientific and technological knowledge has increased considerably since Bacon's time. The desire, the need, or the necessity in some cases has created a demand to improve upon the natural 20 to 30 year rate or more of basic research reaching practical world use.

What is technology transfer today, then, with regard to research and development? It is many things to many people but some of the less complicated definitions are: the multilateral flow of information and techniques across the boundaries of the sciences, technology and the practical world; transferring research results to operations; accelerating the application of research and exploratory development results to industrial application; science and technology transfer to the would-be user at the earliest practicable date and in a language he can understand; the process of matching solutions in the form of existing science and engineering knowledge to problems in commerce or public programs.

The common theme that runs through all these definitions is getting the knowledge out of the laboratory and literature and into the hands of the users.

The concern for the transfer process has become increasingly important in the United States in the last ten years. The expenditures of billions of dollars every year on research and development at all levels has brought continual concern as to the return on the investment.

Types of Transfer

A brief review of technology transfer shows that there are three types of technology transfer:

1. Direct transfer utilizing technology for the same purpose, e.g., adaptation of a design concept.
2. Utilization of a technology for a new or different purpose, e.g., use of a sensor for medical purposes.
3. Application of a technology to different problems, e.g., use of a quality control instrument developed for aerospace hardware to measure pavement thickness.

There are many channels for transfer and many prerequisite conditions. However, the most effective transfer takes place on the basis of an exchange of personnel or interpersonal transmission.

A study of the U.S. technological position shows that the U.S. has been importing technology. Further, the long lead held immediately after World War II has diminished except in selected high technology fields and agriculture. Even here, the gap is narrowing. Evaluation of this trend leads to the conclusion that progress in technology is asymptotic. The gap is closing, but other countries will reach a levelling off in their growth also. The net effect will be a highly competitive environment with government support a prerequisite.

Evaluation of the U.S. position in world aviation, for instance, with regard to the need for government support contrasts strongly with foreign trends. If current trends continue, the U.S. will lose in the medium and short haul aircraft market due to outright subsidies of foreign industry and a decrease in technology transfer from the military sector within the U.S.

Further, the favorable European position is not due to infusions of American technology. Development trends have proceeded in parallel with little crossover, even within the American aviation industry. The American position has probably been more favorably affected than the European by any exchange due to a greater ability to apply technology.

Barriers and Stimuli to Innovation

While the debate over the factors which determine whether or not a firm is innovative has been waged for years, no consensus has been reached. Some experts maintain that the size of the firm is the critical determinant of an innovative posture with larger firms being more innovative than smaller ones. Others, however, see dynamic management as the key to innovation.

Insights as to the barriers to innovation may be gleaned by focusing on the development process. Some characteristic barriers are: laboratory barriers, developer/producer barriers, marketing/distribution channel barriers, state government user barriers, state barriers to local governments, local user blockages, public barriers and federal government barriers.

Licensing and the Multinational Firm

In the past decade the United States has been experiencing increasingly severe balance-of-payments problems as a result of the declining international competitiveness of American goods, particularly manufactured goods. Various explanations have been advanced to provide understanding and insight into the problem, but the most prevalent reason is based on the closing of the "technology gap," the decline of the superior American technological position. It is argued that various mechanisms have come into play in recent years, most prominently multinational corporations, that increase the transfer of technological knowledge from this country to others; the result being that the U.S. loses its technology edge.

Hypotheses concerning the patterns of behavior of the multinational firm deal with: the location of the parent firm; the firm's rank in the industry; the initial impetus to overseas investment; the demand for the most advanced technology; the effects of foreign investment on the domestic operations; and the type of transfer. These hypotheses help explain the positive influence the multinational firm has on the transfer of technology.

Summary

The components of the technology transfer process have been set forth. The remainder of the paper is devoted to the analysis of each separate component of technology transfer.

SECTION II: TECHNOLOGY TRANSFER -- A BRIEF REVIEW

Introduction

This section provides a brief review of the process of technology transfer. First, it defines technology transfer. Then it explains the process in terms of types of transfer, channels, and conditions and provides illustrative examples.

Definition

A dictionary definition of technology is "applied science." As used in this study, it means the application of knowledge to achieve practical ends. Consequently, technology transfer is a process which results in knowledge generated in one place being utilized elsewhere.[2,3,4]

Types of Transfer

There are basically three types of technology transfer processes.[5,6,7,8] The first and most simple occurs when the transfer is direct. For example, when one nation or industry utilizes the technology developed by another for the *same purpose*, e.g., an airframe manufacturer.[9] The second type of technology transfer is utilization of a technology for a new and different purpose, without basic change. For example, the remote environmental sensor which is adopted for thermographic medical diagnostic use. The third type of technology transfer is application of a technology to a different problem. An example here might be an instrument developed in the aerospace industry for quality control measurements of space hardware. The instrument later inspired the development of a non-destructive, pavement thickness measuring device based on similar principles.

The Type I transfer is primarily a hardware transfer process, while the Type III transfer is primarily a concept transfer process. The adaptive engineering requirements increase radically as we go from the Type I to the Type III transfer process.

Channels

There are many channels through which technology can be transferred.[10,11] A list of transfer channels follows:

1. Multinational corporations, internal personnel transfer or the acquisition of a foreign subsidiary.
2. Foreign student exchange program.
3. Foreign aid programs (economic).
4. Foreign aid programs (military).
5. Selling end items (with maintenance manuals, blueprints, etc.).
6. Industrial shows, exhibits and trade fairs.
7. Selling components (avionics packages, propulsion plants, etc.).
8. Technical meetings.
9. Patents.
10. Open literature.
11. Production licensing.
12. Immigration/emigration (the brain drain and reverse).
13. Technical literature.

Despite these many channels, the fact is that the most effective and efficient transfer of technology is the long-term transfer accomplished through the transfer of people. As a result, much valuable European (especially British) technology has been acquired cheaply by the U.S. through the simple process of hiring people away from such laboratories at higher wages.[12,13]

Although they may appear important, many of these channels apply to only a small number of items. Export licensing which has a direct impact on technology transfer is an example.[14,15]

On the other hand, multinational companies have speeded the diffusion of technology. The U.S. multinational company has become a technological innovator on a global scale. Undoubtedly, American companies in Europe have achieved most of the technology transfer from the U.S. to Europe in the 1960's. In 1964, the U.S. received $264 million for patents, licenses, and technological know-how from Great Britain, Germany, France and Italy, while paying only $46 million for technology imports from those countries. Significantly, 71% of U.S. technology receipts from Western Europe in 1964 came from transactions between American parent companies and their European subsidiaries.[16]

In the area of technology vendors, trade press, visits (specifically the large number of Japanese and French visiting the U.S. for technical observation), seminars, trade fairs, etc., play a great role especially in Type I transfer. Engineers get their information from the above sources, usually as a direct inquiry when faced with a specific problem. For example, our AID program has transferred a great deal of U.S. technology by sending large groups of foreign nationals to the U.S. to observe actual field operations. An American electronics engineer who migrated to Israel stated that it was impossible to get his Israeli colleagues to consider mobile transceivers for use on trucks until the AID people sent a group to the U.S. to observe the mobile equipment in action. Incidentally, this example illustrates one of the main findings of research on the diffusion of innovation--a "hands on" experience is very effective in securing adoption of new equipment. This is one of the main reasons for trade expositions and fairs.

Conditions for Transfer

There are many factors which affect the nature, rate and direction of technology transfer through these channels. Some of them can be associated directly with a given channel of transfer. Others are more general and apply to the process itself. The most important factors include:

1. National policies, laws and regulations
   a. Taxes and tax credits
   b. Tariffs
   c. Non-tariff barriers
   d. Standards requirements
   e. Immigration
   f. Trade union restrictions
2. Corporate policies
3. Market demand
4. Scientific "base" of the nation and industry
5. Chance

6. Level of R&D effort
7. Education
8. "Gatekeepers"[17]
9. Availability of translations
10. Individual personalities and roles
11. Availability of venture capital

Summary

This section has suggested a framework for consideration of technology transfer. However, technology transfer should not be viewed simply as a transfer of information. Successful technology transfer requires information *plus* an environment conducive to new innovations. The next section treats the technical standing of the United States relative to that of other countries.

SECTION III: RELATIVE POSITION OF THE U.S.

Introduction

This section examines briefly the overall position of U.S. technology in relation to the rest of the world. It explores the question of importing technology versus exporting technology and then evaluates this question.

Technology Imports

After World War II, the U.S., being the major world producer and user of technology, operated under the assumption of "invincibility." That is, its position was so strong that it felt it could afford to "give away" technology and still not be concerned about losing its competitive edge. We were not hampered by divisive pressures of nationalism, as were Japan and Europe, and were able to proceed with work on massive projects such as computers and nuclear power. Proportionally, the U.S. also had a much larger force of technical manpower. And U.S. companies moved fast to introduce new products. Nevertheless, as the process and implications of technology transfer were better understood, it became obvious that continued relatively unrestricted technology transfer policies should be considered as a potential threat to the U.S. position in world markets.[18,19]

In President Nixon's speech before the Congress in September 1971, he stated that "we need new programs to insure America's enormous wealth of scientific and technical talent is used to the fullest." The President could find many reasons for concern: the favorable U.S. trade balance in high-technology products has leveled off at about $9.6 billion. In recent years, U.S. imports of these products has been growing 2-1/2 times as fast as exports. And in its trade in high-technology products with Japan, the U.S. has been running a deficit since 1965, which has now reached the $1 billion level. New technology has not been contributing much to U.S. production efficiency. The rate of productivity growth in U.S. industry has dropped in the last few years from its postwar average of 2.6% a year to just 1.7%, a decline accelerated by the shift to a service economy. In contrast, Western Europe's productivity growth rate is running at 4.5% a year, and Japan's has zoomed to 10.6%.[20]

As these balance-of-payments problems, national unemployment problems, productivity problems and technical/scientific capability maintenance problems began to appear, they were associated with technology transfer.

Today, except in the field of aerospace/aviation, agriculture, electronics and computers, *the U.S. has reached the point where it can benefit in varying degrees by importing technology from other countries.*[21,22]

For example, of 13 major innovations in steelmaking, not one originated with U.S. steel producers.[23] The steel industry still spends only 0.35% of its sales dollar on R&D, and it generally has not been innovative in production. The latest innovation announced by U.S. Steel Corp.--a kind of cross between the basic oxygen process and the old Bessemer process--actually started with a West German steel producer. Although the automotive industry has always been ready to look at new materials and machines that would cut its manufacturing costs, front-wheel drive, disk brakes, radial tires, and the Wankel rotary engine all had to prove themselves in Europe first. U.S. automakers still are resisting electronic fuel injection, standard on several Volkswagens. Polypropylene, a plastic now widely used in U.S. cars and appliances, was invented in Italy. Extra high-voltage transmission, a more efficient way of conducting electric power for long distances, was pioneered in the U.S. by ASEA, the Swedish electrical equipment maker. When Colt Industries sought to rescue its faltering program for developing big diesel engines, it had to turn to Austrian consultants for help. Georgia-Pacific Corp. went to Britain's Imperial Chemical Industries when it wanted a more economical way to make methanol in its new Plaquemines (La.) plant. Some U.S. shipbuilders are finally re-adopting cost cutting methods that were originally developed in the U.S. during World War II, then adopted in Sweden and Japan as their use faded here. The American construction industry is just beginning to use "systems building" technology that is already widely applied in Europe.

Application of Technology

Those who see the U.S. losing its technological lead through the technology transfer process assume that technology and productive capability are available equally in the competitive country(s). Technology has continued strong in Europe, but the lower productive capabilities have held them back. Their failure to innovate, or innovate rapidly, and research discoveries, gave us the advantage. The U.S., because of our superior production capability, has been in a position to exploit, or exploit more effectively, technology imported from Europe. Many examples might be cited, but a few crucial examples follow:

> Britain had an early lead in nuclear energy for commercial use. But in the past few years, General Electric and Westinghouse have outsold the British in Europe with their light-water reactors, primarily because the capital cost of U.S. reactors is lower.
>
> Time and time again French and British airplane designs have been ahead of the American designs. The British and French were first with jet commercial aircraft (Comet, Viscount, and Caravelle), but unlike the Americans, they did not obtain orders for later models. Boeing first challenged European dominance of medium- and short-range aircraft in 1964 with the 727 and 737.
>
> Over a hundred years ago, an Englishman, Babbage, built an "analytical engine" that possessed all the major features of a computer. But it was the Americans who recognized the commercial applications of the computer.
>
> Europeans made early inventions and discoveries in lasers, cyrogenics, office copying machines, and fluidics. In all of these areas, American companies have made the most successful innovations.

What is indicated by these examples is that the U.S., with its superior production capability, is able to take ideas from Europe and develop them before the Europeans. The European industry's failure has been in marketing, not technical knowledge.

However, a portent of the future may be gained by looking at the historical position of the U.S. in relation to the four I's: Ideas, Inventions, Innovations and Imitations (Diffusion). If a country's importance in the generation of ideas is measured by the total number of Nobel prizes in Science that have been awarded to its citizens, then it is clear that until just before World War II, the U.S. was not able to compete scientifically with older countries. Tables 1 and 2 illustrate how the United States has gained the leading position.

Table 1 - Nobel Prizes in Science, 1901-30[24]

Germany	27	Denmark	4
England	15	Austria	3
France	11	Switzerland	3
Sweden	6	Italy	2
United States	6	U.S.S.R.	2
Holland	6		

Table 2 - Nobel Prizes in Science, 1931-60[25]

United States	33	Austria	4
England	18	Sweden	2
Germany	14	Italy	2
Switzerland	5	U.S.S.R.	2

This rather sudden reversal may be attributed in part to the substantial influx of scientists who were educated abroad and migrated to the U.S. because of the political turmoil of the 1930's. It is also a result of the great material resources which are available in the U.S. The areas where we lead are those which require expensive experimental equipment, which some nations cannot provide.

In contrast, U.S. contributions to technology have been important for many years. A crude documentation of this, using invention rates as an index of contributions, is presented in Table 3. Column A is a list of the nationalities of the authors of the major inventions from colonial times until the present. Such a tabulation can be attacked on the grounds that both chauvinism in the choice of inventors to whom credit is given and a lack of discrimination in the choice of inventions must be considered. As a corrective, Column B gives the average annual patenting rate for the years 1930-39 in the countries being considered. The leading role of the United States is evident, regardless of which measure of inventions is employed.

Table 3 - Invention Rates in Various Countries[26]

	A Total on Selected List 1600-present	B Average Annual Rate 1930-39
United States	203	38,300
Great Britain	58	9,050
Germany	32	14,600
France	29	9,550
Italy	14	3,900
Switzerland		3,130
Sweden	4	1,030

To reiterate, in the past, the U.S. has been weak in idea generation, better in invention and outstanding in innovation and imitation. However, the competitive edge that American products have always enjoyed due to the superiority of the American innovative process and the diffusion techniques appears to be slipping. For instance, the Organization for Economic Cooperation and Development states in their report, "Gaps in Technology" that except for the computer, aerospace, and heavy electronics industries, technology is primarily transferred into the United States from other countries.[27]

Even in those areas where the export/import ratio is not in the U.S. favor, we have had a comparative--but diminishing--advantage among nations because of our relatively superior production capability. "Where we had a lead," says Robert A. Winslow, President of Enjay Chemical Co., a Jersey Standard subsidiary, "was not in the development of technology itself, but in the application of that technology, the marketing and management of it." It now appears that we are in danger of losing that superior capability. This loss may be why the "gap" is disappearing.

Technology Exports

Even though the United States maintains a lead in high-technology fields, the gap is narrowing for aviation and electronics.

For example, in selected aviation areas, such as V/STOL,[28] European countries are starting to compete successfully. The value of the Franco-British Concorde, the European SST, is still widely disputed. But there are other signs of the new competition. An LTV Aerospace Corp. subsidiary is now assembling helicopters designed by France's Aerospatiale. Nippon Aeroplane Mfg. Co. is taking aim at U.S. feeder airlines with its YS-11 short-hop turboprop. The European builders of the Mercure short-haul and the A300-B airbus also are aiming at the U.S. market. Says Allen E. Puckett, Executive Vice President of Hughes Aircraft Co., "our R&D money just isn't keeping pace."

In the electronics industry, Japan is developing at such a great rate that it may soon be the prime world supplier of consumer electronics.[29,30]

To the contrary, the U.S. seems to retain a three to five year edge over other developed countries in the computer industry. This is in R&D and application.[31,32] There are three major reasons for this which are important to investigate if we are to be able to evaluate future aviation industry technology transfer requirements and policies:

1. Computers had their initial application in the area of national security, whereas developments in the aviation industry were stimulated somewhat equally by both military and civilian needs and markets. (This is not true of aviation research, which received major support from the military side of the house.)

2. During the formative stages, the number of companies competing in the software area of the computer market place was very large, compared to the size of the market, so that there was, and continues to be, a self-imposed security (proprietary) on technological developments. The competition is characteristically open and a function of the free market. In contrast, there are relatively fewer companies competing for the civilian aviation market. One of the major reasons for the survival of these companies and the development of new technologies is dependent primarily on government contracting policies for military and government systems and regulatory policies for the commercial aviation market.

3. The computer system development cycle is generally several years shorter than the aviation industry development cycle. The computer industry does not rely greatly on advanced sales as is characteristic of the aviation industry. Further, the computer industry does not require major tooling or retooling lead times, or elaborate and lengthy testing and qualification programs (although it does have debugging and compatibility problems) before it can market its products.

The similarities and differences discussed above are some of the reasons why it is becoming more difficult to maintain a sizable competitive lead in the aviation industry. Another major reason is that critical aeronautics/aviation "information" is generally more accessible than similar information in the computer industry. More specifically:

1. Aircraft and components are usually sold free and clear of the manufacturer to the purchaser, which usually leases them from a bank or equipment trust. Computers, on the other hand, are usually leased because the user does not want to make a large capital investment on an item which has yet to be "proven" in his specific application, and because he most often will still have to purchase hardware and software services. Agreements for such services are frequently of sufficient duration so that the hardware may have been "substituted" for by advanced hardware, and the "safeguarding" of the technology is no longer of importance. Computer technology is also safeguarded by the developer offering computer services as an alternative to leasing or purchasing the equipment.

2. The general aeronautics/aviation literature is frequently quite explicit in providing design and manufacturing details (e.g., stress analyses of wing sections, etc.). However, in the rare cases where some "blueprints" are available in the aviation industry, many actual modifications have been made and, therefore, the direct chanels of technology transfer discussed earlier does not take place. In the computer field even less detailed manufacturing data is available, especially for hardware which is currently marketable or for that which is being developed. Computer literature is generally concerned with either systems applications or *basic* technology. For example, the concepts supporting the design of a memory core element may be discussed in considerable detail, but rarely will the specific technical or manufacturing aspects of memory systems be presented.

3. Total systems performance of computer systems is generally known; this is one of the bases for competing in the market. However, the performance characteristics of critical subelements of computer systems are almost never found in the open literature.

4. In the aviation industry, a purchaser of an aircraft can specify a unique configuration by shopping the components that may come from many sources rather than solely from

> the producer of the aircraft, e.g., different jet engines on the same basic airframe. In the computer industry, the computer is most often built from scratch by the company and "tailored" to the specific requirements of the producer.

Evaluation

With some technologies costing more to develop than the entire GNP of many countries, the time is past when any nation can be entirely self-sufficient in developing its own technology; a worldwide view must be adopted. Each nation must decide which technologies it must and can emphasize as well as the best way to import others.[33]

Harvey Brooks, Dean of Engineering and Applied Physics at Harvard, sees this as a natural culmination of the events of the past two decades. Although other countries are closing the technology gap, he sees both their progress and our own as asymptotic. In other words, there is some sort of saturation point which the U.S. has reached ahead of the other industrialized countries. The other countries will probably reach but not surpass us on a steeply rising curve.[34] These remarks can be interpreted to mean much greater commercial competition in all fields of activity.

Faced with this developing situation, the United States appears to be electing the area of technology intensive products as the most deserving of emphasis in competition in world trade, for it is only in these products and agriculture that we have been able to maintain a lead or grow.[35] Nevertheless, with a lead time of ten years, the lead is slowly eroding due to a lack of funds and continuity. Programs have been cancelled and research teams disbanded. Due to lack of continuity in funding, proposals are submitted primarily for the short term to ensure funding to completion rather than responding to need. Lacking future scientific eminence and a crossover of scientific knowledge into technology, there will be an inability to support the desirable social goals which currently compete for the same funds.[36]

With the present problems in the balance of payments, the best hope for regaining a favorable balance and paying for future needs is through the export of technology intensive products. Aircraft appear to be one of the best exports. Development of resources in some type of partnership with the government would permit the exportation of products. Lacking this partnership, the aircraft companies will export technology through consortia with foreign companies.[37]

This states the need for government assistance. The need is heightened by consideration of the fact that foreign governments have major support programs for their technology. Faced with huge developmental outlays, it appears to be asking too much of private industry to compete on a toe-to-toe basis with heavily subsidized and supported foreign firms.

Summary

Immediately after the war, the U.S. was in the position of concentrating all its efforts on research and development, rather than on restructuring the economy and country, as were the European countries and Japan. As a result, technological developments were at an accelerated pace and the U.S. developed a design and production capability which far exceeded that of the rest of the world.

In the postwar years, the technology flow was basically in only one direction--out of the United States. In the 1960's this lead we held began to diminish as the other countries were able to concentrate more on their economic and technological developments. A two-way channel began to emerge as the U.S. slowly imported technology from Europe.

In the 1970's, and in the future, unless the U.S. continues to lend strong financial support and encouragement, the lead will close, and we will lag behind. John B. Dwyer, Vice President of Research and Engineering for M. W. Kellogg Co., expressed the situation well when he said "if we fall just one year behind in research, the loss . . . is much, much greater in the '70's than it was in the '20's."

SECTION IV: CHARACTERISTIC BARRIERS AND STIMULI TO INNOVATION

Introduction

It is important to remember that technology transfer not only means transferring knowledge but also refers to the application of that knowledge, which includes technological development, application, marketing and management of that technology. Barriers and stimuli to this innovative process may arise at any point in the sequence. This section lists some of the most common barriers and stimuli which should be noted and considered.

Characteristic Barriers to Innovation

Laboratory Barriers (Universities, Research Institutes, Industrial R&D Laboratories)

Conflicting motivations on the part of professional personnel

Inadequate communications with the user community (state and local governments)

Lack of initiative in approaching industry

Limited money (in particular, the relative lack of developer/producer funding

Developer/Producer Barriers

Informational
- Lack of university advisory participation
- Lack of R&D synthesis

Technological
- Appropriate reliability
- Technology turnover rate

Legal
- Patent
- Anti-trust
- Non-standard purchase conditions

Managerial
- Lack of familiarity with market
- Personnel retraining
- Established competition
- Pricing considerations

Institutional
- Nature of firm--size, stability, outlook, existing distribution channels, R&D orientation, organizational structure
- Competition with existing customers

Financial
- Accounting considerations
- Need for venture and investment capital
- Basis for investment decision

Marketing/Distribution Channel Barriers

Disaggregation of market

Lack of product definition

Lack of test markets

Lack of standard specifications
Lack of product acceptance testing procedures
Cost of sales
Inadequate distribution channels
Occasional requirement for leasing capability
Lack of needed and related product lines
Product life cycle

State Government User Barriers

Informational

- Flood of information which is uninterpreted and not user oriented
- Dissemination of information is irregular and unsystematic
- Sources of technology in many cases are unknown (technology disaggregation)

Institutional

- State agencies are fragmented
- Budget and planning procedure

Political

- Conflict between short and long range objectives
- Intergovernmental restraint
- Climate
- 2nd and 3rd order consequences
- Marketing practices of developer/producers

Economic

- Cost of technology
- Labor intensity
- Resource allocation
- Pricing practices of developer/producer

Legal

- Federal grant-in-aid process freezes technology
- Local codes inhibit statewide programs
- No State checkoff on R&D

Technological

- Lack of inhouse technology and systems analysis capability
- No R&D effort *per se*
- Lack of operational performance criteria
- Lack of trained operating personnel
- Lack of technology assessment and forecasting capability
- Lack of product acceptance testing capability

State Barriers to Local Governments

Constitutional

- Lack of Home Rule authority
- Appointment of some local officials by the State
- Police powers reside with the State

Legislative

- State governs taxing authority
- State governs bonding authority
- Certain legislative and policy making powers are withheld by the State

Administrative

- Regional cooperation must be approved by State legislature
- Policy and program standards are set at the State level (health, police, education, sanitation, etc.)
- Some personnel qualifications are set by the State
- Grants-in-aid
- Shared revenue

Local User Blockages

Informational

- Limited science and technology information available
- Information available is often not user oriented
- Information dissemination is irregular and unsystematic

Institutional

- Compacts
- Non-innovative climate
- Geographical dispersion

Political

- Conflict between short and long range objectives
- Intergovernmental restraint
- Political climate
- 2nd and 3rd order consequences
- Marketing practices of developer/producer
- Review process for Federal funds

Economic

- Cost of technology
- Labor intensity
- Resource allocation
- Pricing practices of developer/producers

Legal

Ordinances
Codes
Enforcement
Decentralized control
Regulation
Review process for Federal funds

Technological

Lack of inhouse technology and systems analysis capability
No R&D effort *per se*
Lack of operational performance criteria
Lack of trained operating personnel
Lack of technology assessment and forecasting capability
Lack of product acceptance testing capability

Public Barriers

Lack of recognition of opportunities for innovation
Inadequate understanding of technological benefits and/or consequences
Political impact of public positions

Federal Government Barriers

R&D priority setting
Inadequate information dissemination
Grant-in-aid process

Characteristic Stimuli For Innovation

Behavioral

1. Crises (local, regional, national, or international)
2. Focus and articulation of needs provided by media
3. Perceived attitude of electorate
4. Entrepreneurial efforts (in research, development, industry, or other)
5. Moral convictions of governmental leadership
6. Pressures from special interest groups
7. Desire for specific improvements by citizens
8. Desire for change from existing patterns and trends
9. Image enhancement efforts by industry
10. Desire for personal recognition

Economic

1. Profit motive
2. Availability of venture capital for new enterprises
3. Availability of excess capital in established firms
4. Competition
5. Stimulation resulting from loss of share of market
6. Tax-assisted investment opportunity
7. Unique combinations of resources (mergers, acquisitions, contract spin-off opportunities, special contract provisions, etc.)
8. Existence of available pools of specialized talents
9. Technological opportunity
10. Technological obsolescence
11. Need for job satisfaction and work-force stability

Legal

1. Statutory and/or regulatory constraints
2. Legal/economic framework (tax structure, accounting techniques, etc.)

SECTION V: HYPOTHESES CONCERNING THE ROLE OF LICENSING AND THE MULTINATIONAL FIRM IN TECHNOLOGY TRANSFER

Introduction

Licensing agreements and multinational firms are two of the most important channels which accelerate the transfer of technology. Licensing agreements provide a vehicle for "renting" technology from another country or organization. Subsidized licensing has often been cited as the preeminent reason for the explosive success of Japan in recent years. Since this practice is very prevalent, it is important to investigate the role that it plays in the transfer of technology.

Another topic which demands the attention of this conference is the contribution to technology transfer made by overseas manufacture and investment by domestic firms. This section lists some hypotheses about the multinational firm and its place in the transfer process. These hypotheses are presented in order to stimulate discussion and study about this means of technology transfer.

Hypotheses About Multinational Firms

1. That the overseas firm tends to be a leader in the industry and will be among the most profitable and

fastest growing firms. It possesses an agressive management posture that excels in introducing product and process improvements following the industry cycle theory.

2. That often the initial overseas investment efforts were defensive actions (such as market protection) that were successful. Because of the early successes, the firm began to consider itself and behave as a world operator. The firms which continue to go overseas do so because of continued successes. Conversely, the "non-goer" firms will be primarily characterized by their early failures at overseas investment and licensing, and the fear of continued failures precludes further international ventures.

3. That firms in advanced countries will use the most advanced technology available regardless of the source (country or firm) of that technology. Technology gives only a temporary advantage which may facilitate entry. The sustained competitive advantage, then, that a given firm accrues is likely to be a function more of the marketing and managerial capabilities in response to the utilization of the technological opportunity than the level of the technology *per se*. Of course, it is expected that this will vary considerably from industry to industry.

4. That the levels of technology in advanced countries within an industry tend to become equalized through the diffusion of technology. The technological modifications made in one plant of a firm are quickly adapted or improved upon and transferred regardless of the location of the firm. The transfer mechanism may be as formal as cross-licensing agreements or as informal as plant tours. The inevitable blending of foreign and domestic technology may make the determination of the original source of the technology difficult. However, it is expected that the net contributor/borrower relationship can be ascertained.

5. That initially the domestic operations of a firm may be stimulated by the overseas operations due to the incomplete product line of the preliminary foreign operation. This may serve to increase exports and domestic employment levels in the short run until the foreign operation becomes sufficiently large to supply its own markets. The domestic market then reverts to its pre-foreign operation level supplying its own markets while the foreign operation creates and supplies its own market network.

6. That most technology transfer flows to and from a country occurs in areas of rapid change due to shifts in demand.

SECTION VI: CONCLUSION

Scientists and technologists throughout the world have become increasingly aware of problems associated with technology transfer, problems which could be ameliorated with international cooperation. To date there is no surefire recipe for successfully transferring technology, but a large number of efforts are under way in an attempt to better understand and appraise current problems in the area. Technology transfer has reached a state of development where its discussion on a broad international scale would benefit a large number of interested parties and contribute substantially to an advancement of the state of the art. This international conference is both timely and pertinent. It was in this light that this NATO Advanced Study Institute of Technology Transfer was first proposed early in 1972.

There is a rapidly growing body of information relating to the overall process of technology transfer and innovation as illustrated by the attached comprehensive and annotated bibliography. This research represents the contributions of economists, behavioral scientists, lawyers, managers and innovators; ranging from carefully designed empirical research to nostalgic case studies. The literature falls under a variety of subject headings including: diffusion, invention, innovation, learning, communications, idea/information flow, technological change, new product marketing, economic growth, creative behavior, etc. Some of the best empirical work has been done by disciplinary specialists who look at technological innovation and transfer and the R&D environment in an effort to advance knowledge in their own area. The secondary and often unintended effect of this research is that R&D managers gain some useful and relevant insights into their jobs. But this method of obtaining new concepts is long and circuitous.

It is necessary to view the growing body of research on the various aspects of the process of technological innovation relating to R&D managers in the same manner that physics, chemistry, and mathematics is related to engineering. There has been little progress toward building the bridges between theory and practice (scholars and managers) which are needed for testing and evaluating the research and for the subsequent improvement of the practice. In this period of reduced R&D funding, inflation and increased worldwide competition in high technology products and processes, improvements in R&D productivity are imperative. The route to such improvements is better R&D management in large and small firms as well as public organizations. This improvement in the practice of R&D management will result from an understanding of the underlying process of innovation, as Dr. Joel D. Goldhar[38] has so aptly pointed out, but this will require three efforts:

A. Some additional research on various aspects of technological innovation and its subsidiary concerns of creativity, information flow and motivation.

B. A series of carefully designed experiments based upon the findings of past (and current) research which will yield tested hypotheses and more generalized and creditable concepts and techniques for managing R&D.

C. A technology transfer effort concentrating upon disseminating the new concepts and techniques of R&D management as widely as possible.

I charge this conference with the task of offering a beginning in the following areas:

Reviewing the findings and hypotheses suggested by previous research.

Suggesting further research and experiments designed to test previous findings and hypotheses (the criteria of success or goodness being the extent to which innovation and productivity is improved).

Providing thoughts on priorities and possible mechanisms for accomplishing both the experimental effort and the information effort required.

REFERENCES

[1]The National Industrial Conference Board, Inc. Transcript of The Conference Board Seminar, *Organizing Our Scientific Knowledge For Use* (New York: National Industrial Conference Board, Inc., 1967), p. 51.

[2]*Factors Affecting the International Transfer of Technology Among Developed Countries*. Report of the Panel on International Transfer of Technology to the Commerce Technical Advisory Board, U.S. Department of Commerce, February 1970.

[3]*Proceedings of a Conference on Technology Transfer and Innovation* held under the auspices of the National Planning Association and National Science Foundation, May 15-17, 1966. NSF Report 67-5.

[4]"Reflections on the Relation Between Science and Technology," by Hendrick W. Bode; "Scientific Research and the Innovative Process," by William J. Price and Lawrence W. Bass; "Performance of Information Channels on the Transfer of Technology," by Thomas J. Allen; "Technology Transfer," by Charles Kimball; "Coupling of Research and Development," by Harold Davidson; "TRACES-Technology in Retrospect and Critical Events in Science," by B. Bartocha, F. Narin, and C.A. Stone. Volume 3, Chapter 12, "Technology Transfer," pp. 1161-1268, *The Science of Managing Organized Technology* (ed. by Marvin J. Cetron and Joel D. Goldhar), Gordon and Breach Science Publishers, Inc., New York, 1970.

[5]*Technology, International Competition and Economic Growth: Some Lessons and Perspectives*, by Keith Pavitt, Visiting Lecturer, Woodrow Wilson School of Public and International Affairs, Princeton University, 1971. Draft Report.

[6]*Managerial Implications of Innovation Diffusion Generalizations Applied to International Manufacturing Technology Transfers*, by Ivan A. Valle. MBA Thesis (University of Texas at Austin, December 1970).

[7]"Technology Transfer by Multinational Companies," by James Brian Quinn, *Harvard Business Review*, November-December 1969, pp. 147-161.

[8]"Strategies for a Technology-Based Business," by H. Igor Ansoff and John M. Stewart, *Harvard Business Review*, November-December 1967, pp. 71-83.

[9]One of the famous examples was the case of Dr. H. S. Tsien who left (or was asked to leave) the U.S. for Red China and is often reputed to be the brains behind their most advanced weaponry development. Another example would be the Japanese "Zero," the World War II fighter plane, which most think was adapted from a Howard Hughes design.

[10]*Joint DOT-NASA Civil Aviation R&D Policy Study*, Report NASA SP-265/DOT TST-10-4, March 1971.

[11]Volume 3, Chapter 12, "Technology Transfer," *op.cit.*

[12]This form of technological transfer does not work internally in the United Kingdom because industrial research has relatively low prestige, and salaries in the public sector actually tend to be as high or higher than salaries in industry.

[13]"Government Policy and the Prospects for Technology-Based Industry," *Professional Engineer*, by Robert U. Ayres, December 1971, pp. 38-41.

[14]*The Japanese Aerospace Industry*, Research Report No. 408. Stanford Research Institute Long-Range Planning Service, June 1970.

[15]*Technology, International Competition and Economic Growth*, *op.cit.*

[16]*The OECD Observer*, April 1968, p. 27.

[17]"The International Technological Gatekeeper," by Thomas J. Allen, James M. Piermeirer, and S. Cooney, *Technology Review*, Vol. 73, No. 5, March 1971.

[18]*The Influence of Decreasing Costs on International Trade and Cooperation in Aerospace Products*, Doctoral Dissertation (Columbia University, 1968) by H. R. Biederman, Senior Economic Advisor, Lockheed Aircraft Corp., Burbank, California.

[19]"Reaching the World's Consumers," by J. Douglas McConnell, Economics Division, Stanford Research Institute. Presented by the SRI Long-Range Planning Service Client Conference, San Francisco, California, 1969.

[20]"Making U.S. Technology More Competitive," *Business Week*, January 15, 1973, pp. 44-49.

[21]*A Historical Study of the Benefits Derived From Application of Technical Advances to Civil Aviation*, Volume I/Summary and Report and Appendix A (Detailed Case Studies). Prepared by GARRING, Bethesda, Md., for the Joint DOT-NASA Civil Aviation R&D Policy Study. NASA CR-1809/DOT TST-10-2, February 1971.

[22] *Ibid.*, Volume II/Appendices B through I. NASA CR-1809/DOT TST-10-3.

[23] "Making U.S. Technology More Competitive," *Business Week*, January 15, 1972.

[24] Hendrik W. Bode, *Basic Research and National Goals*, A Report to the Committee on Science and Astronautics, U.S. House of Representatives by the National Academy of Sciences, March 1965, p. 52.

[25] *Ibid.*, p. 54.

[26] *Ibid.*, p. 53.

[27] *Gaps in Technology*, Organization for Economic Cooperation and Development, Paris, 1969.

[28] *Reader's Digest*, November 1971.

[29] *Western European Aerospace Industry*, *op. cit.*

[30] *Transfers of U.S. Aerospace Technology to Japan.* Report P-3875, by G. R. Hall and R. E. Johnson, The Rand Corporation, Santa Monica, California, July 1968.

[31] *Gaps in Technology Between Member Countries.* OECD Report, series presented at the Third Ministerial Meeting on Science of OECD Countries, March 1968.

[32] *The International Computer Industry: Innovation and Comparative Advantage*, by Alvin J. Harman, Harvard University Press, Cambridge, Mass., 1971.

[33] "Technology Transfer by Multinational Companies," by James Brian Quinn, *Harvard Business Review*, Nov.-Dec. 1969, p. 149.

[34] "What's Happening to the U.S. Lead in Technology?" by Harvey Brooks, *Harvard Business Review*, May-June 1972, pp. 110-118.

[35] "U.S. Needs a Strong Finish to Win World Trade Race," by Emilio Q. Daddario, *New York Times*, January 24, 1971.

[36] "U.S. Technology is Threatened," by Charles A. Anderson, *New York Times*, August 22, 1971.

[37] "Technology, Trade and Government," by Karl G. Harr, Jr., *The Wall Street Journal*, January 13, 1972.

[38] Joel D. Goldhar, "Research and Experiment Needs in Technology Transfer and Innovation." (Unpublished paper)

PART II: Fundamental Concepts and Empirical Evidence: Technology Transfer and Innovation

This section presents a substantial amount of quantitative data which readers will find of interest as well as the implications and conclusions drawn by the original investigations. The first paper, by Joel Goldhar, brings together the sources and channels of information transfer and the innovation process. The relationship between transfer and innovation is a key one which is emphasized by many of the succeeding papers. Pearson and Richards emphasize the human channels of transfer and note the social barriers which can limit such activity. Ron Johnson also emphasizes the social setting as it differs for Technology Transfer using experiences from university-industry, government-industry, and industry-industry efforts. These institutional differences and potential barriers to both transfer and innovation are also discussed in the paper by Bradbury and Rothwell. Rothwell, reporting on the SAPPHO studies, extends the discussion to the differences between successful and unsuccessful innovation (as defined by market acceptance) while Paul Jervis and T.C. Sinclair discuss SAPPHO from a public policy viewpoint.

More narrowly defined research efforts are reported

by George Foster and Jules Schwartz. These two papers offer the richness of data made possible by the smaller research population. Foster uses the experience of the petroleum refining industry to illustrate the need-pull motivation for technology transfer and the strategy of risk reduction which minimizes the opportunity for high risk, non-routine innovations in a mature industry. Conversely, in the context of a rapidly changing technology and a young organization, Jules Schwartz illustrates the "risk-sharing" mechanisms which makes possible radical innovation.

Moving from a microscopic to a macroscopic level, Tom Allen and Sean Conney discussed the impact upon technology transfer of the institutional mechanisms and public policies in a developing country, while Gerhard Mensch discusses the same problem in the context of a highly industrialized nation. Forecasting technological change is often a matter of anticipating the diffusion and transfer of technology. Robert Pry uses an understanding of technology transfer and diffusion processes to develop a methodology for forecasting which can predict the growth of a technology with considerable accuracy.

Completing this segment of the proceedings the paper by Albert Rubenstein summarizes much of the case study "micro-level" research and suggests several new lines of potentially fruitful research and experimentations. Finally, the reports of two small group workshops held during the concluding days of the institute are presented at this point. The workshop on research and experiment design raises many of the problems (and some of the possible solutions) which must be faced in new efforts to extend the state of our knowledge in technology transfer. The workshop on the process of innovation raises more questions that it answers and provides an important set of models and hypotheses for future research on the interrelationship between technology transfer and innovation.

INFORMATION, IDEA GENERATION AND TECHNOLOGICAL INNOVATION

Dr. Joel D. Goldhar

Office of Science Information Service,
National Science Foundation, U.S.A.

INTRODUCTION

Managing R&D is the most difficult and challenging management task in any organization. The rate of change in both the internal activity and the relevant external environment, the level of uncertainty, the skill mix of the work force and the key role R&D plays in organizational and economic growth and development all contribute to the problems facing R&D managers and policy makers. These difficulties are no excuse, however, for not managing R&D as efficiently and effectively as the state-of-the-art in "research-on-research" will allow. The recent changes in both governmental and corporate attitudes toward R&D expenditures makes this effort mandatory--the squeeze between reduced levels of funding and the increasing costs of technical effort leave no choice but to improve the productivity of each R&D dollar. The research described in this paper was undertaken in the faith that an improved understanding of the innovative process leading to *successful* new products is the first step in developing organizational structures, public policies and management styles appropriate to the task of improving the productivity of science and technology.

If, as the often quoted Booz-Allen and Hamilton study suggests,[1] only 2 out of 100 new product ideas become successful new products; the productivity of R&D is too low. In today's environment of declining R&D support, we must improve the return from each dollar of R&D investment and the place to gain the greatest leverage is at the beginning by improving the quality of new ideas and design concepts offered to management for possible R&D funding. That improvement will come only from better R&D management based upon an understanding of the innovative process and an appreciation of the role of scientific and technical information and communications. Both public and private sector policies and actions must be consistent with the underlying nature of the basic process which presents both constraints and opportunities.

The processes of technological innovation, technical and scientific communication and technology transfer have been the subjects of considerable research activity in recent years. All are very broad topics. My interest in this paper, however, will be limited to the consideration of the characteristics of information flows in the idea generation/design concept formulation phase of technological innovation in order to suggest means of improving and managing the process through manipulation, enhancement and control of information flows.

A brief discussion of the characteristics of the process of innovation will be used to set the stage for a presentation of research results from a three year study* on the information using habits of successful technological innovators using the "Industrial Research 100" award winners as a research population.

Other relevant research will also be discussed, but this paper will not attempt to offer a comprehensive review of the innovation and technology transfer literature.[2] The final part of the paper describes a new, information-handling oriented model of the idea generation/design concept formulation stage of technological innovation, draws a variety of conclusions and lists several possible management and policy actions implied by the new model.

In this paper *innovation* refers to technology actually brought into first use or application as distinguished from scientific discovery, or invention. As defined in the recent Battelle study, *Science, Technology, and Innovation.*[3]

> "It is a complex series of activities beginning at first conception when the original idea is conceived; proceeding through a succession of interwoven steps of research, development, engineering, design, market analysis, management decision making, etc; and ending at first realization; when an industrially successful product which may actually be a thing, a technique, or a process, is accepted in the marketplace."

---

* The author wishes to acknowledge his debts, both intellectual and personal, to Dr. Daniel D. Roman of the George Washington University, who provided guidance and incentive for the effort reported here.

The innovative act represents an "organization" of many diverse bits and pieces of both new and existing technical and market information. It is a behavioral as well as an economic activity representing both individual and organizational information handling and risk taking.

This paper has three main parts: a background discussion of the process of technological innovation with emphasis on idea stimulus and the behavioral characteristics of innovators; a description of the types, sources and channels of stimulus information reported by innovators; and a new model for the design concept formulation stage which emphasize information flows and implies a variety of possible managerial and policy actions.

THE PROCESS OF TECHNOLOGICAL INFORMATION

A number of recent retrospective investigations have focused on the process of Technological Innovation. Myers and Marquis[4] studied successful Technological Innovators paying particular attention to their information seeking behavior and the role of technology transfer. Their view of the innovative process is summed up in *Figure 1*. Utterback[5] studied the process in the instrument industry. Baker et. al.[6] studied it in a single large laboratory and Bragaw[7] used a version of the Myers and Marquis model to study innovative patterns in smaller firms. Langrish and others in England have made a similar study of the Queens Award Winners.[8] There is considerable consistency among the results of these studies. All tend to agree on some version of the model presented in *Figure 1* and report that between 2/3 and 3/4 of all successful innovations were stimulated by information about a market need (potential demand recognition) as shown in *Table 1*.*

Several other investigators have studied allied processes including creativity,[9] technology transfer,[10] and the information seeking and use habits of technical personnel.[11] These studies, however, rarely discriminate between successful and unsuccessful innovators or between innovators and non-innovators or between idea generation and problem solving stages and are usually of limited scope in terms of the demographic variables: industry, technology. number of labs, firm size, innovator background, etc.

---

*Note that these studies all focus upon the final innovation and data from them are not readily comparable to data from *TRACES I and II, Hindsight, The Materials Advisory Board Study on Scientific-Technical Interactions* and similar studies which focus upon *events*--of which there may be many in a single innovation.

Figure 1 shows the process of technological innovation as three distinct stages.[12] These can be simplified to:**

Idea Generation and Design Concept Formulation → Problem Solving and Engineering → Commercialization and Diffusion

In this topology an invention is defined as the first discovery of new knowledge and precedes the entire innovation process. An innovation then is the first economic use of an invention, and diffusion implies the widespread acceptance, use and limitation of an innovation. By these definitions the term "invention" in Figure 1 should really be "innovation." Two additional points must be noted relative to the definition of innovation. "Innovation" is subject to the observers' interpretation; some researchers insist that only the first use of an invention *anywhere* can be called an innovation, while others will credit as an innovation any use of an invention which is *new to the specific user*. Neither of these extremes seems entirely suitable and in the IR-100 research an "innovation" is defined as the first use of an invention *or unique combination* of inventions in a particular innovator-client/use combination. These may be either product or process innovations and although my study concentrated on new products, it must be recognized that nearly every product innovation eventually becomes part of someone else's process.[13] In addition, most modern technological innovations represent the integration of several inventions and areas of knowledge.[14]

From a behavioral point of view, technological change and innovation occur as the result of complex sets of human interactions, information flows and transfers, individual and organizational creativity, and invividual and organizational risk-taking and decision making. Each of these facets of the process involves human beings with their motivations, perceptions, attitudes, beliefs, abilities, ambitions, personalities and prior knowledge and experience--all of which contrain and help determine their information seeking and use habits and represent the variables which can be dealt with in the attempt to improve the probabilities of successful innovation coming out of R&D activities.

Consequently, technological innovation is, to some extent, a subset of the more general subject of human behavior in organizations which includes consumer behavior, managerial behavior,

---

**It is important to note that data and evidence developed by research on one stage may not be extendable to the other stages.

FIGURE 1
THE PROCESS OF TECHNICAL INNOVATION
(SOURCE: MYERS AND MARQUIS, SUCCESSFUL INDUSTRIAL INNOVATIONS, NSF 69-17, p. 4)

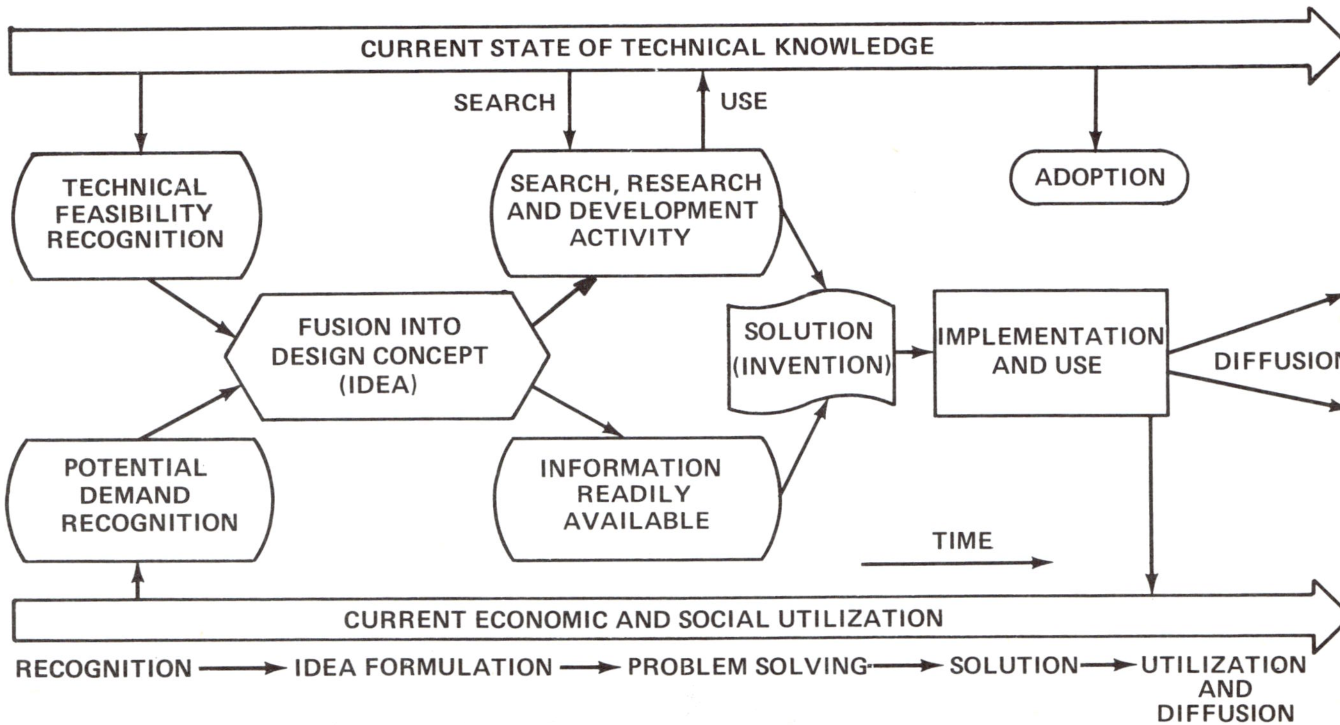

TABLE 1

Nature of Stimulus Information in Technological Innovation

Study	Nature of Stimulus Information Technology/Means		Market/Needs		Comment
Baker, et.al.	25%	(60)	75%	(212)	Ideas in a single lab.
Utterback -IR100 + control group	25	(8)	75	(24)	IR100 winners in the instruments industry plus control group of similar innovations in the same firm.
-IR100 sample only	40	(6)	60	(9)	
Bragaw	25	(11)	65	(20)	Smaller 1969 IR100 winners - diversified firms.
Goldhar	30	(57)	70	(135)	IR100, 1965-1972 winners. Generally diverse group of firms.
Goldhar	29	(12)	71	(32)	IR100, 1972. Diversified, slightly different questionnaire.
Myers Marquis	23	(130)	77	(937)	Railroads, housing and computer industries.
Langrish, et.al.	34	(29)	66	(55)	British Queen's Award winners for 1966-67.
Carter & Williams	27	(55)	73	(149)	116 British firms.

leadership, etc. Many of the concepts and research findings developed in these different areas are very similar and can be utilized to advance our understanding of technology transfer and innovative behavior. Particularly impressive to me is the "transferability" of consumer behavior concepts developed by marketing researchers to the analysis of technology transfer and innovative behavior; although no systematic work has yet been done in this area.

Marketing looks at "consumer innovators"[15] who are the first to try out or "adopt" a new product, process or service. This process (purchase behavior) is seen as one of *risk taking, problem solving* and *information handling*. Potential technological innovators (R&D workers) may be thought of as "consumers" of information. The adoption of new technological/scientific information or concept offers the risk of a loss of peer approval as well as the economic cost of a failure, if the idea does not work out. The acceptance of a newly identified economic or social need to which technology can be applied may prove to be equally risky for the innovator who misjudges the importance of the problem or wrongly combines particular sets of needs and technologies. The sanctions, both economic and social, which can result from such an error can be quite severe and are similar to the economic and social losses faced by the early consumer adopter of a new product or process which does not work out as expected.

The most general way of thinking about the information flows, human interactions, creativity and risk taking in the innovative process is in terms of the flow of knowledge from discovery to application as shown in Figure 2. The innovation stage is shown as the critical "valve" in the flow of technology into society. This is the process of bringing together scientific and technological "means" information with socio-economic "need" information. This is the activity which provides for the unique, creative *organization* of many diverse bits of knowledge resulting in the economic application of scientific knowledge and technological capabilities. It is the most uncertain and least manageable segment of the overall process; and therefore, the most important and interesting stage to study. The process of technological innovation as shown in Figure 1 begins with the recognition of *both* an economic/social need and a technical means. Having recognized the demand and the technical means of satisfying it, these elements are creatively fused into a design concept. This is the most abstract stage of the process; and it is also the point at which an evaluation must be made as to whether or not to continue the development of the idea. This evaluation, which may take place in either the private or public sector, must take into account the probability of technical success, size of estimated demand, cost of continuing, possible profit, long range impacts, etc. The cumulation of these decisions eventually determines the rate and direction of technological change.

FIGURE 2

THE FLOW OF KNOWLEDGE FROM DISCOVERY TO USE

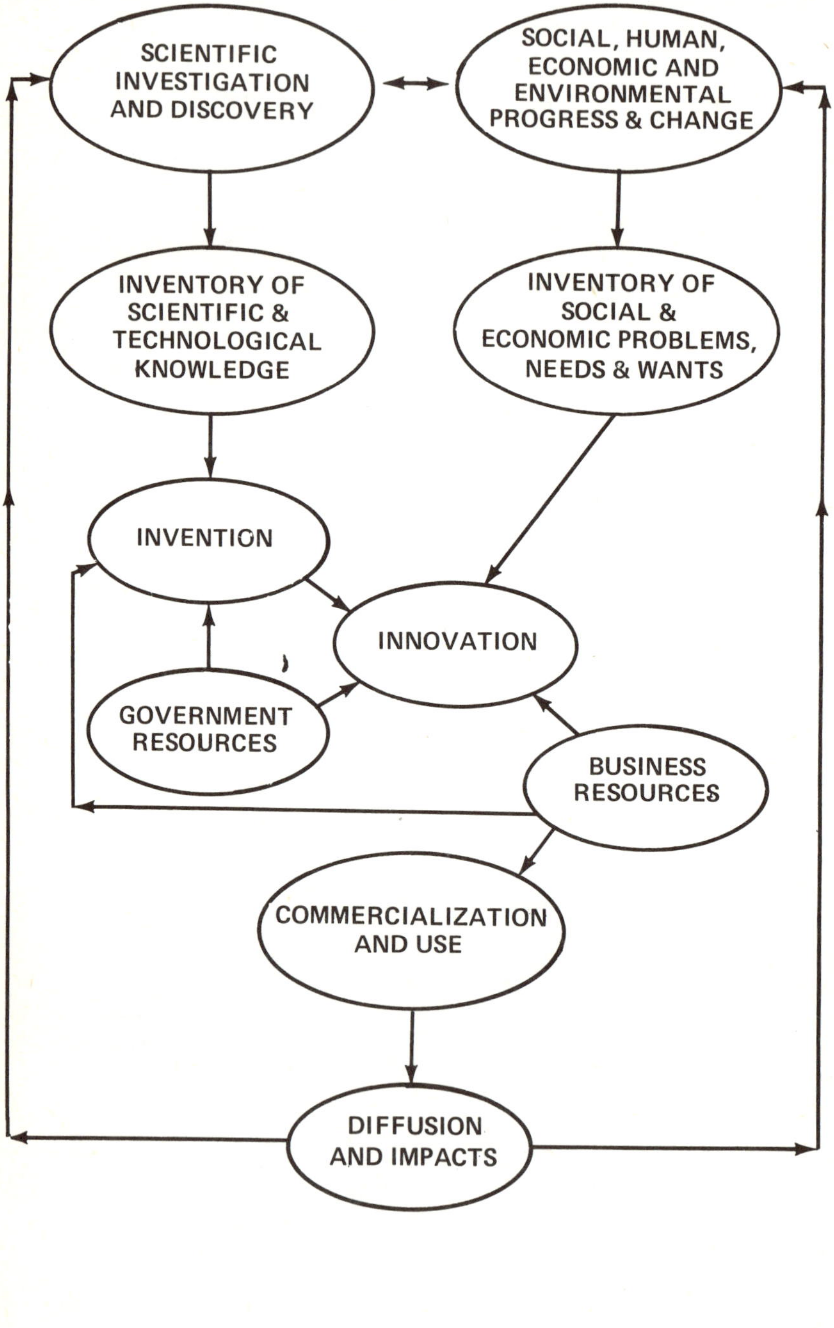

Figure 3 is an expanded model of the technological innovation process. It shows the parallel sequence of technical and market analysis and development and emphasizes the points of interaction between technological and marketing activities: Design Concept Formulation, Development Funding Decision, Commercialization Funding Decision and Manufacturing/Sales efforts. Each of these points represent a major decision and risk taking point for individuals and organizations. Each point also represents a combining and interpreting of data and information, usually by groups of people. Each individual in such a decision group will usually have a different point-of-view, background and mental "set" from all the others and the final result usually represents a negotiated compromise. In each case, technical information is being brought together with commercial/market information and at any stage new ideas may be stimulated and new design concepts formulated; making the process a more iterative and complex one than is shown in Figure 3.

STUDY POPULATION AND METHODOLOGY

The data and interpretations to be presented in this paper are based upon the analysis of a detailed questionnaire filled out by winners of the IR-100 award. One hundred and ten of the 500 winners between 1965 and 1969 responded to the first questionnaire (some answering for multiple awards) which was mailed to them in October 1969. As expected, only a few of the earlier winners felt they remembered enough about the events leading to their innovation to accurately answer the questions but roughly one-third of the most recent years' winners responded. Two replications of the survey (using about 2/3 of the first instrument's questions) in 1970 and 1971 achieved about a 40 percent response from those year's winners and the results of that survey have been combined with the initial data bank to yield a total population of 195 cases for over half of the original variables.

The IR-100 awards are given by *Industrial Research* Magazine which sponsors an annual competition "to select the 100 most significant new technical products of the year." The IR-100 winners are chosen by the magazine's Editorial Advisory Board (a group of about two dozen eminent scientists and engineers), "on the basis of their importance, uniqueness and usefulness from a technical standpoint." No claim is made, and for the purpose of this research effort none is needed, that the IR-100 population represents the "best" or top 100 innovations in any one year. However, the IR-100 designations verify the fact that the product or process at least fits the basic definition of an innovation.

FIGURE 3

THE PROCESS OF TECHNOLOGICAL INNOVATION

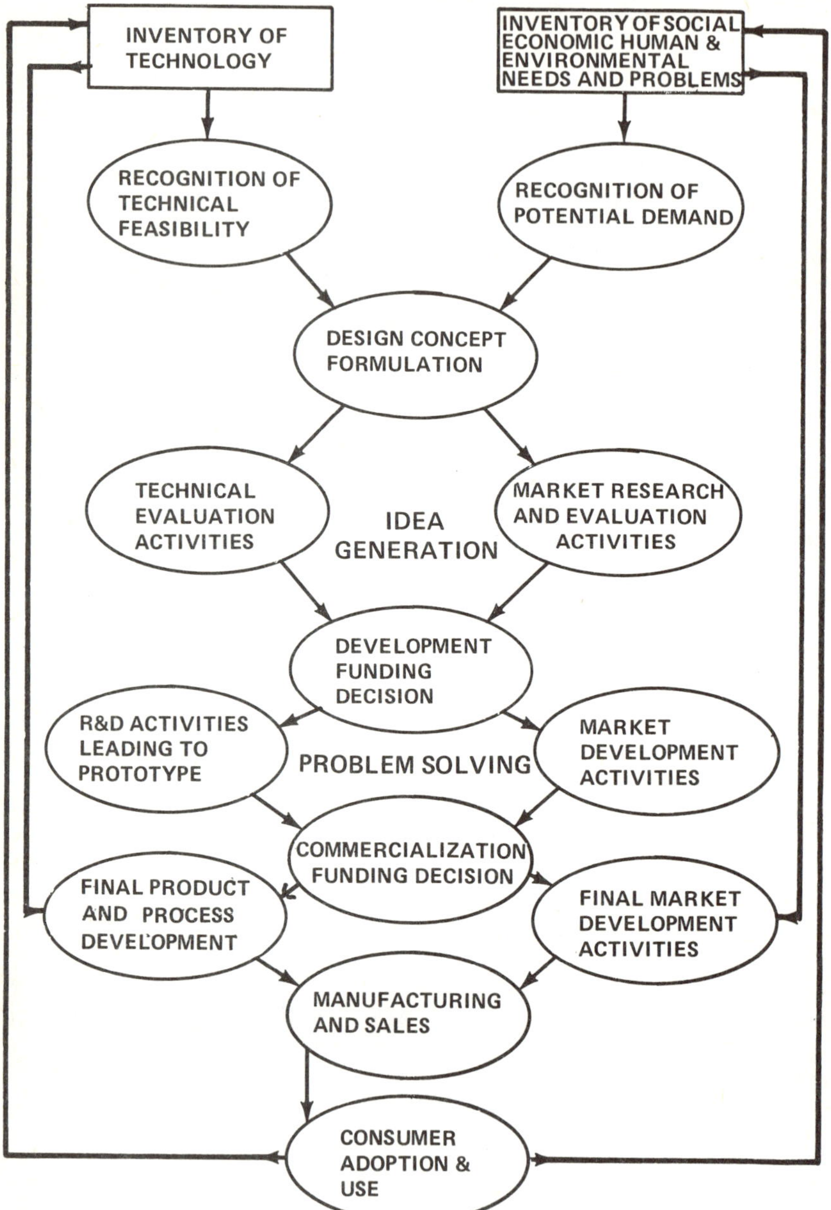

CHARACTERISTICS OF INNOVATORS AND INNOVATIONS

The original questionnaire provided a great deal of raw data on a variety of topics including: characteristics of innovators and innovations, the idea generation process, innovation stimulus patterns, and the role of various types, sources and channels of information. This section will attempt to present only a brief overview of the most general results in the hope that they will stimulate discussion and further research and set the stage for more detailed discussion of the information variables.

The first hypothesis of the original study was the existence of an identifiable set of individual characteristics unique to the population of *innovators*. This turned out to be a poor assumption. The "identity" variables including age, education, years with firm, salary, etc. were widely and evenly distributed.

Although one would hesitate to declare the IR100 award either a "representative" or a "random" sample of innovations, few studies have succeeded in gaining such a sample, and the results of this research might best be thought of as a study of the characteristics of almost 200 innovators and innovations representing a wide spectrum of demographic/identity variables.

Other studies referred to in this paper have generally used structured interviews. The general agreement between the stimulus (need vs. means) data collected by all of the surveys tends to lend additional credibility to the other data collected via the self-administered survey instrument. Of greatest interest was the large number of Ph. D. scientists in the sample (over one-third). In this sample, at least, the possession of a behavior and new product development -- a result somewhat at odds with the folklore which attributes a new product innovation to the more "practical" engineers. Moreover, the pattern of responses to the questions concerning information search and stimulus behavior showed no significant differences between scientists and engineers. Thus, in the innovative process both Ph. D.'s and non-Ph. D's appear to behave alike -- as innovators. There is, however, a significant tendency for the innovations of Ph. D.'s and scientists to be more often stimulated by information about new technology (means) rather than information concerning a market demand (need) and vice versa for engineers as shown in Table 2 below:

TABLE 2

Nature of Innovation Stimulus Information

Profession or Degree	Stimulus: Technical Means	Stimulus: Economic Need
Scientist	26 (36%)	47 (64%)
Engineer	31 (27%)	85 (73%)
Ph. D's	26 (36%)	47 (64%)
Non-Ph. D.'s	28 (24%)	87 (76%)

In summary, then, it seems reasonable to suggest that the *process* of innovation is not a function of the identity characteristics of the individual innovator. The initial stimulus, however, may differ and consequently the exact administrative and policy options for Ph.D. versus non-Ph.D.-oriented organizations (and for technology versus market-oriented organizations) may differ. Further research is required to establish the causal factors. Are Ph.D. scientists prone to technical means stimulation? Or are they mostly isolated from market information and responding to the only information available?

The survey also included a number of questions about the innovator's organization. The responses, for the most part, verified the "new product" and "development" emphasis of the sample population. Most innovations came from manufacturing organizations for whom product/process innovation is only one part of the firm's activities. The respondents' specific organizations were primarily development-oriented laboratories organized by product line. This is consistent with Seiler's findings[16] on R&D laboratories which suggests that laboratory organization bears no special relationship to successful innovation.

As noted above, the number of Ph.D. scientists in the sample is unexpectedly high. Most respondents also reported the presence of one or more Ph.D.'s at their labs suggesting (tenuously) a relationship between the presence of more basic scientific knowledge and individuals accustomed to keeping up with scientific information, and successful new product innovation. Firm size varied widely in the sample and the analysis showed no significance between size (in terms of annual sales) and any of the innovation process vairables. Ninety-two respondents worked for firms in the "Fortune 500" and the remainder were smaller ranging all the way down to eight firms with under $50,000 annual sales.

The innovators reported high levels of work autonomy having both personal freedom to plan and choose their work and considerable influence on any others who might have an input to their work planning and allocation of resources. This result is consistent with the Pelz and Andrews findings of a relationship between high levels of creativity and autonomy in scientific laboratories.[17] The match between the survey data on innovators and that of Pelz and Andrews on creative scientists may suggest that successful innovators are more like creative scientists than like business entrepeneurs; but that will take more research.[18] The innovators reported that they prefer work with long-term gains, great challenge and the development of basic new knowledge. It may be that each innovator is, as David Allison suggests, a "split personality" and each innovation is the result of a compromise between the individual as a "scientist" and as an "economic man."[19]

In general, the respondents supported the contention that the innovative process is continuous and cumulative,[20] something that Gilfillan told us 35 years ago. Seventy-eight percent of the respondents' innovations were related to a series of developments in the same area at the innovator's lab or firm and in 63 percent of the cases other innovations followed from the winning innovation. (This later figure is probably deflated due to the sample's bias toward the more recent IR-100 winners.) In addition, about 66 percent of the innovators had worked on similar innovations prior to the IR-100 winner and only 12 percent of the innovations occurred in a "totally unrecognized" technical area in the innovating organization. On the other hand only 23 percent of the innovations occurred in a specific area being given "high priority." Most innovations came from firms with *ongoing* but low priority programs in the general area of technology represented by the innovation. Although the continuity of the effort is important, it need not be large--60 percent of the 110 respondents in the initial survey reported that they devote less than five percent of sales to R&D. In turn, the proportion of basic and applied research to total R&D averaged 20 percent, somewhat lower than the industry average of 30 percent reported by the NSF. It will come as no surprise to industrial R&D managers that successful new product development is the result of a "development heavy" budget.

As shown below, the innovations caused changes in the other functional areas of the firm only in about 1/3 of the cases.

CHANGES CAUSED BY INNOVATION

Change	Frequency	
	Yes	No
Production Process/Capital Equipment	47	145
Marketing/Sales Activities	61	129
Organizational Structure	59	132

From the above data one might suspect that the choice of projects to fund is influenced by the expectation that the final new product result will not "rock the boat" too much. Such expectations may present a barrier to "radical" innovations which might drastically alter the future of the firm. The circumstances surrounding such "radical innovations" need to be studied in detail for these are the true "life-giving" inputs to aging organizations.

Support for the importance of technology-stimulated innovations comes from observation of the relationship between the innovation's initial stimulus ("need" versus "means" information) and the above organizational change variables. Respondents to the first survey reported that other innovations followed from 75 percent of technical means stimulated innovations but from only 58 percent of economic need stimulated innovations. In addition, 47 percent of means stimulated innovations caused changes in the firm's organizational structure as opposed to only 24 percent of need stimulated innovations having the same effect, while changes in marketing and sales activities were caused by 41 percent of the means-stimulated innovations and 28 percent of the need-stimulated cases. If the reader will accept the working hypothesis that technology-stimulated innovations are likely to be more "radical" than market-stimulated innovations, as indicated by the organizational changes initiated by the innovation; then the stimulus value of technological information must be given greater importance than that suggested by the initial observation that only 1/4 to 1/3 of new products are means stimulated.

This data suggests that the current fashion calling for market-oriented and directed R&D may be shortsighted.[21] Perhaps we need some technology-oriented/R&D-directed marketing efforts.

TYPES, SOURCES, AND CHANNELS OF INFORMATION IN THE IDEA GENERATION STAGE

In addition to asking the general nature of the innovator's stimulus (technical means or economic need), respondents were asked to detail the relative importance of a variety of types, sources and channels of information contributing to their design concept formulation. The data presented in Tables 3, 4 and 5 below are cumulations of three iterations of the questionnaire representing KR-100 winners from 1967-1971.

Generally, the types of stimulus information can be categorized as *Technological* (T) or *Economic* (E). The specific types chosen as having greatest value by respondents reported that economic information was responsible about 50% of the time for the stimulus to their innovation. This almost even split between technical and economic information and its inconsistency with the

30 percent means, 70 percent needs stimulus division reported earlier will be discussed in the next section of the paper. In addition, 110 respondents reported that the greatest value type of information stimulating their innovation was a type which they received regularly while 67 reported the opposite. Of these, 54 percent of means-stimulated innovators reported being stimulated by information of a type *not* regularly received while only 31 percent of need-stimulated innovators reported the stimulus as having come from an irregular type of information.

Turning to the sources of stimulus information (Table 4), 79 percent of respondents reported that the most important source was an internal (I) one while 21 percent reported external (E) "most important" sources as detailed in Table 4. Regularity of source was reported by 98 respondents while 80 indicated that the most important source of information stimulating their innovation was an irregular one. Fifty-three percent of the means-stimulated innovators reported an irregular source of information while 41 percent of need stimulated innovators reported similarly.

The channels of stimulus information can be categorized as either informal (I) or formal (F). Eighty-one percent of respondents reported informal channels as having the "greatest value" as detailed in Table 5. One hundred and twenty-eight respondents reported that the channel was one through which information was regularly received (the highest proportion of any of the information variables) while 60 indicated receipt of stimulus via an irregular channel. In addition, 39 percent of means-stimulated innovators reported irregular channels, as did only 29 percent of need-stimulated innovators (the lowest proportion for any of the information variables). One might suggest the information channel is the most difficult variable to manipulate and poses the largest barrier to innovation.

The frequency data tentatively confirms many of the results of prior research and some generally accepted managerial wisdom. Innovators appear to depend upon informal, internal, easy-to-access and low cost (economic *and* social) sources and channels of information.[22] Types of reported stimulus information appear to be weighted toward economic needs; again generally agreeing with other research.

For example, Utterback[23] suggests that need communication is often initiated by someone other than the person who generated the idea for the innovation, while communication about a technical means is initiated most often by the innovator himself. The role of work and contact diversing is also emphasized by Utterback[24] who reports that outside experts played a crucial role in the generation of ideas for 16 of 32 instrument innovations. Peters[25], in a study of idea generation among MIT faculty reports that 96 percent of faculty

TABLE 3

Type of Information Reported Having Greatest Values as a Stimulus to Innovation

	% Reported as Greatest Value
General Market Need (E)	22
A Specific Client's Need (E)	15
A Different Innovation (T)	6
Generally Available Technical Data (T)	2
Specific New Technical Information (T)	6
A Production Requirement (T)	3
Your Firm's Need to Offer a Competitive Product (E)	6
A Need to Reduce Product's Selling Price (E)	2
A Need to Reduce Production Costs (E)	1
Firm's Need for a New Product (E)	3
Your Own Recognition of New Technical Possibility (T)	28
Your Own Recognition of a New Economic Opportunity (E)	6

TABLE 4

Sources of Information Reported as Having "Greatest Stimulus" as a Stimulus to Innovation

Source	% Times Reported "Greatest Stimulus"
Innovator's Prior Experience and Training (I)	40
Technical Colleague within your Firm (I)	10
Technical Supervisor (I)	4
Client or Customer (E)	8
Technical Book (E)	0
Technical Article (E)	2
Technical Talk or Conference (E)	2
Company Production Group (I)	4
Company Marketing Group (I)	4
Non-Technical Book or Article (E)	0
Non-Technical Supervisor (I)	6
Non-Technical Colleagues in Innovator's Firm (I)	0
Firm's Long-Range Planning Group (I)	3
Competitor's Product (E)	0
Technical Colleague not in Innovator's Firm (E)	5
New Knowledge about Innovator's Technical Discipline (E)	7
Company Executive (I)	3
Patents (E)	2

TABLE 5

Channels of Information Reported as Having "Greatest Value" as a Stimulus to Innovation

Channel	% Reported as Greatest Value
Informal Discussion with Technical Colleagues within the Firm (I)	48
Informal Discussion with Non-Technical Colleagues within the Firm (I)	3
Informal Discussion with Technical Colleagues Outside Innovator's Firm (I)	14
Conferences at Company or Division Headquarters (F)	2
Written Memoranda (F)	2
Academic Courses (F)	1.5
Attendance at a Professional Meeting (F)	4
Bright Idea - No Outside Channel (I)	16
Professional Journal in Innovator's Field (F)	3
Non-Professional Magazine or Book (F)	1.5
Technical Book (F)	2
Professional Journal Outside Innovator's Field (F)	1.5
Industry Trade Magazine (F)	1.5

reporting innovative ideas engage in outside consulting as opposed to 55 percent of those not reporting ideas. Once again, the data confirms Gilfillan's original hypotheses[26]. Allen[27], in several studies, has developed data supporting the "Engineers can't write and don't read" dictum.

As shown in Tables 3, 4 and 5, information sources are predominantly internal and transmitted mostly by informal channels. Technical colleagues both in and out of the innovator's firm and informal discussion seem especially important. The unimportance of technical supervisors and company management in general and the innovator's heavy reliance upon his own knowledge, experience and creative abilities in the innovative process are two factors worthy of special note. What accounts for this "satisficing" mode of behavior? And how should we interpret it? Are the innovators insulated from the rest of the organization? If so, what accounted for the preponderance of "need"-stimulated innovations? Shall we accept the facts at face value and move completely to a market-oriented R&D effort with no "basic" work at all? Or should we interpret the data as an indictment of engineers *and* their managers for giving into laziness--and sloppy information-using habits--for becoming isolated and withdrawn from the external scientific world. This will lead us to the conclusion that innovation *can* be both managed and improved. The following discussion of the "mismatches" between information type and nature of stimulus will, I hope, serve to stimulate further research on the information handling and using phenomena of innovators.

The first sample of 110 pre-1970* innovators showed a 50-50 split between technical and economic information which appeared to be somewhat at odds with the original sample's 2/3 - 1/3 split between innovations reported as "need"-stimulated and those reported as "means"-stimulated. For this initial data an additional analysis was performed by combining the various classes of information variables and tracing the relationships among information type, source and channel and the reported nature of the innovation's stimulus.[26] Figure 4, the results of this analysis, shows 49 cases of "mismatch" between type of information and the stimulus it might normally be expected to generate-- that is, economic information and a reported means stimulus of technical information combined with a need stimulus. The implications of this data are discussed in the next section of the paper and used to develop a new model for understanding the innovative process and for developing management and policy options.

---

* This part of the analysis has not been performed in the larger sample due to a lack of computer capability but the 50-50 technical versus economic information split and the 70-30 need-means division are roughly the same moving from the smaller to the larger sample and the same analysis using the larger data base would probably yield similar results.

THE DESIGN CONCEPT FORMULATION PROCESS--A NEW MODEL

The most important interaction event (both human and informational) in the innovative process (Figure 3) is "Design Concept Formulation." This subprocess is the most conceptually creative, least predictable, least manageable, and least understood event in technological innovation. A design concept, as discussed earlier, requires the combination of a variety of diverse bits of information including evidence of an economic demand/social *need* and knowledge of a feasible technological *means* for satisfying that demand/need. The process of technological innovation, as modeled in Figure 1, suggests that the fusion of the needs and means recognitions happens as a single event--a sudden flash of insight.[29] However, other research suggests that the process is more likely a linear one.[30] The potential innovator begins with either information concerning an economic need followed by a search for the technical means to fulfill that need *or* he may begin with the recognition of a technical possibility and then searches for an economic need to justify the further development of the technology. Either way, the design concept formulation appears to be a linear pattern of an idea stimulus followed by information search as shown in Figure 5.

Empirical evidence (Table 1) shows that approximately 2/3 - 3/4 of the time the initial stimulus is reported as the knowledge of an economic need (shown as *Path A*). Here the design concept is completed by searching for the appropriate technology and results in a "demand-pull" innovation. The remaining 1/4 - 1/3 of the time, design concepts are, conversely, "technology-push" innovations (shown in *Path B*) in which the initial stimulus is the technical means and the information search centers upon identifying economic needs.

The relative importance of the 1/4 - 1/3 of technology-stimulated innovations should not be underrated. Although it is fashionable to point to these statistics as evidence of the need for a market responsive R&D effort, and the current trend is away from R&D completely; the technology-stimulated innovations are far more likely to offer "radical" new business opportunities to the organization and to further stimulate follow-on innovations (mostly need-oriented).[31] Therefore, arguments can be offered for the continued support of R&D in areas outside the constraints of the organization's existing manufacturing and marketing capabilities.

Figure 6 is an attempt at a representation of the complex interaction between the potential innovator and his "information environment." It presents a new hypothesis for Design Concept Formulation based upon previous empirical research which can be used as a framework for further study of the process. The R&D

FIGURE 4A

RELATIVE RELATIONSHIP AMONG INFORMATION SOURCES TYPES AND CHANNELS IN THE STIMULATION OF SUCCESSFUL TECHNOLOGICAL INNOVATION

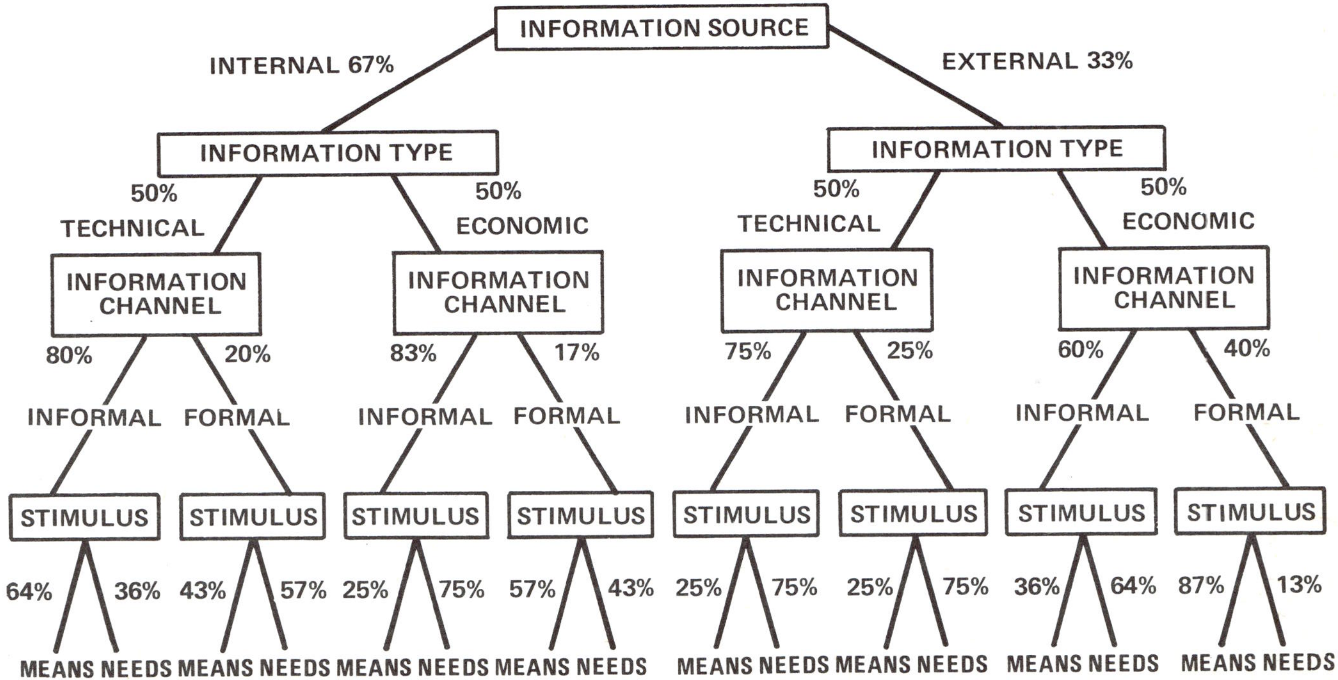

FIGURE 4B

Frequency Counts for Each Branch in Figure 3A

Source	Type	Channel	Stimulus	Frequency
Internal	Technical	Informal	Tech. Means	13
Internal	Technical	Informal	Market Need	11
Internal	Technical	Formal	Tech. Means	3
Internal	Technical	Formal	Market Need	4
Internal	Economic	Informal	Tech. Means	7
Internal	Economic	Informal	Market Need	21
Internal	Economic	Formal	Tech. Means	4
Internal	Economic	Formal	Market Need	3
External	Technical	Informal	Tech. Means	3
External	Technical	Informal	Market Need	9
External	Technical	Formal	Tech. Means	1
External	Technical	Formal	Market Need	3
External	Economic	Informal	Tech. Means	4
External	Economic	Informal	Market Need	7
External	Economic	Formal	Tech. Means	7
External	Economic	Formal	Market Need	1

worker exists in an environment of overabundant information and must develop mechanisms to cope with it. There are two main sequences through which the innovator chooses to absorb or discard available information: messages which *complement* information already stored in his mind and messages which complete a Design Concept Formulation (in the manner shown in Figure 1).

The model suggests that the innovator's thought process is a two-stage system with an internal as well as external "data bank." It seems plausible that in some cases the innovator's thought process follows the linear model (Figure 5), while in other cases he goes through an internal data bank in a rapid and almost unconscious fashion; but in retrospect (when questioned about his innovation) he reverses his perception of the innovation's actual stimulus. For example, an interesting bit of technical information gathered from the environment will pass through the innovator's "Internal Data Bank" where it may trigger a memory of an economic/social need. Having passed the "Quick Relevance Test," the knowledge returns to the innovator's mind and a search is begun for more detailed technical information to complete the design concept. Thus a "bit" of technological information triggered a "need" stimulus followed by a "means" search for more detailed technical data. Alternatively, the internal data bank may not be used: as for example, when the innovator is told by his boss, "We need a new product which will reduce the death toll from highway accidents." This is a linear sequence of economic information triggering a need stimulus followed by a search for a means without recourse to an internal relevance test.

It may also be hypothesized that the nature of the source determines whether or not the innovator goes through a "quick relevance test" loop or not. If we assume that the model is a representation of what happens when there is a reported "mismatch" between information type and stimulus (economic-means or technical need combinations) and that a more direct path bypassing the innovator's internal data bank represents the situation in which the innovator reports (as he perceives) a match between information type and stimulus (economic-need or technical-means combination), we can compare the results in terms of different sources of information as follows:

Information Type-Stimulus	Information Source	
	Internal	External
Match	40	12
Mismatch	26	23

$X^2 = 6.34$, DF =1

FIGURE 5

LINEAR FLOWS OF STIMULUS AND SEARCH PATTERNS IN DESIGN CONCEPT FORMULATION

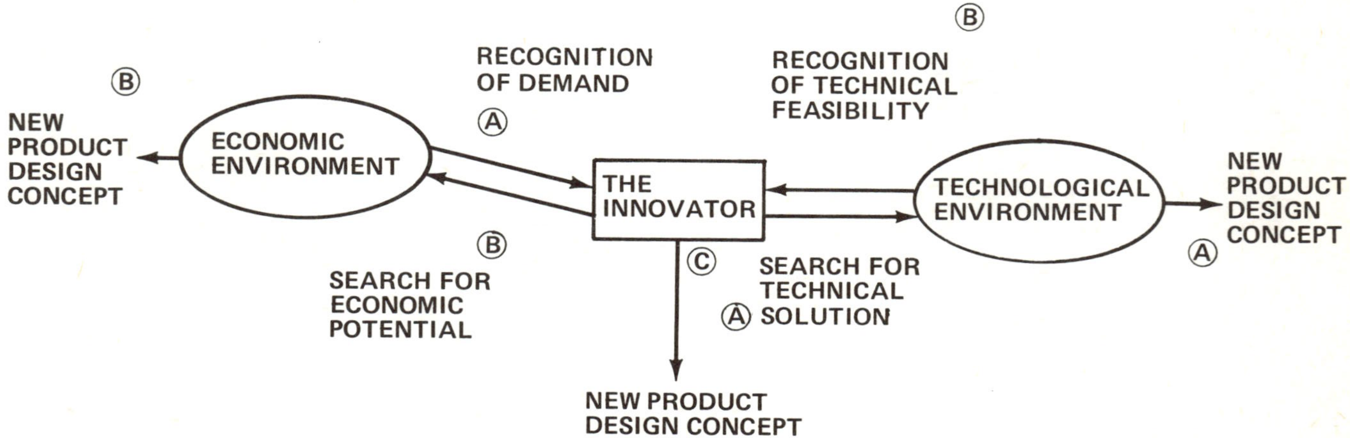

It would appear therefore that "mismatches" and the "quick relevance test" occur most often in response to external sources. A possible explanation for this phenomenon is that external sources generally will have less credibility and/or authority than internal sources. It may also be that internal sources have a hierarchy of creditability/authority and that cases of mismatches occur for the external sources which exhibit matching information type-stimulus patterns. Although statistically significant, this relationship is tenuous at best and needs to be tested through experiment and field observation. Again, the increased use of internal sources (from 65 percent in the pre-1970 group to 80 percent in the total group including 1970 respondents) appears to be an effect of the economic downturn which motivated a movement towards greater control and coordination of R&D activities with other areas of the firm and a tendency to find more short-term, quick payoff projects.

Consider the following scenario as an example of how the new model works:

> An engineer who works on radar is given an article from a journal describing a miniature radar for small airplane collision avoidance. The information enters his mind and coming from a low credibility/authority source is subjected to a subconscious relevance test to determine its value for retention in memory. The innovator's subconscious reports to him that there is a market need for an automobile collision avoidance product and asks if it could be met by a radar device. He then searches his external data banks for more detailed data on radar for the design of such a device. The technical data completes the design concept. The innovator then reports a "need" stimulus (auto collision device) generated by "technical" information (the journal article on radar).

Further research should attempt to correlate this process with the exact nature of the information source. Some additional support for this theory may be found in Gregory's[32] concepts of "feed foreword" and "oscillation" and in Haefele's proposition that "unusual creative ability is largely compounded of two things: (1) unusually skillful rapport with the unconscious and (2) unusual ability to deal with symbolization of a nonverbal character."[33] Perhaps some further research into innovation and the biological factors in idea generation will yield more understanding of the creative process. Ralph Gerard[34] also describes the process by which the brain creates ideas in a manner suggestive of the unconscious role of the innovator's internal data bank as described above.

One final note in support of the revised Design Concept Formulation Model. All of the studies on innovation report that most innovations are "economic need" or "market demand" stimulated. Yet casual observation of R&D activities suggests that the information environment is predominately technical with relatively few economic messages flowing through. Therefore, there must be some additional process and/or filter and a recognition probability function which results in economic need-stimulated innovations. This process may actually be the activities of several processes and in total, considerably more complex than the activities allowed for in the original model (Figure 1).

Clearly the role of information flows and technology transfer as well as the sources and channels of such information will vary considerably between the "technology-push" and "demand-pull" modes of technological innovation. Technology transfer is intimately related to the process of technological innovation and while it remains to be seen which is cause and which is effect, the important factor is the interrelationship between the two.

SUMMARY

Much of the discussion so far suggests the importance of the innovator's *environment(s)*. By designing and controlling these environments (organizational *and* informational), we can do much to stimulate and/or control the rate and direction of technological innovation. There appear to be at least six identifiable characteristics of environments which are conducive to technological innovation. These are:

1. Easy access to information by individuals.

2. Free flow of information both in and out of the organization.

3. Rewards for sharing, seeking, and utilizing "new" (externally developed) information.

4. Rewards for risk taking.

5. Rewards for accepting and adapting to change.

6. Encouragement of mobility and interpersonal contacts.

Therefore the "leverage points" at which individual, organizational or societal interests can effect change in the innovative process appear to be:

1. The interface between a scientific discovery and its technical implication,

2. The Design Concept Formulation stage.

3. The decision to fund a particular piece of technology through the development phase of the innovative process.

4. The decision to commercialize a particular piece of developed technology.

5. The impact analysis of the technological application on the economic, political, social and physical environments.

Combining the important characteristics of an innovative environment with the leverage points for management and policy options it appears that both government and private organizations can stimulate the flow of technology from the laboratory to the marketplace by focusing upon the following:

1. Efforts to increase the scope and rate of information flow via low cost (both economic and social) easily accessible channels.

2. Designing reward systems and operational environments which will legitimize and encourage the search for, and and use of, information from external sources.

3. Development of information services which reduce the cost of access and which require very little learning time for efficient use. A colateral effort to increase the awareness of information service availability is also needed.

4. Development of "information specialists" to interface between the potential innovator and available information services.

5. Reducing the level of both social and economic uncertainty and risk at the critical leverage points for individuals as well as organizations.

6. Improving feedbacks and impact analyses on potential legal restrictions as well as future market reactions and needs for new technology. This means Technological Forecasting, environmental (in the broadest sense of the term) monitoring and an overall interest in future research.

7. Efforts to increase the general level of understanding of the process of innovation and the role of scientific and technical information, leading to improvements in the management of R&D activities.

Finally, what does all this mean to managers of R&D activities? In addition to the ideas contained in the research conclusions above there appear to be two useful questions which R&D managers can ask about their organizations:

1. Is your R&D effort organized to promote informal interchange of both technical *and* economic/market information? In other words, are you making it easy for your R&D people to be creative?

2. Are your R&D people making enough use of external sources of economic and, especially, technological information or are they almost solely dependent upon internal data, thus limiting their ability to match technological means with economic need information in order to create a "design concept" which can be considered for support?

In summary, the scientist or engineer is a consumer of information. He is offered a surfeit of choice (just as he is in the supermarket) and develops (as do all consumers) short cuts and behavior patterns designed to help him cope with the overabundance of potential "purchase." And, like consumers, being an "early adopter" of a new scientific or technical idea or concept is a risky business with significant economic and social penalties for being wrong. Thinking of information as a product and the information user as a consumer is the basis for designing organizational and managerial policies which will motivate innovators to move from satisfying towards optimizing behavior by becoming more venturesome in their search for and acceptance of scientific, technological and economic information.

FIGURE 6

THE INNOVATOR'S THOUGHT PROCESSES DURING DESIGN CONCEPT FORMULATION

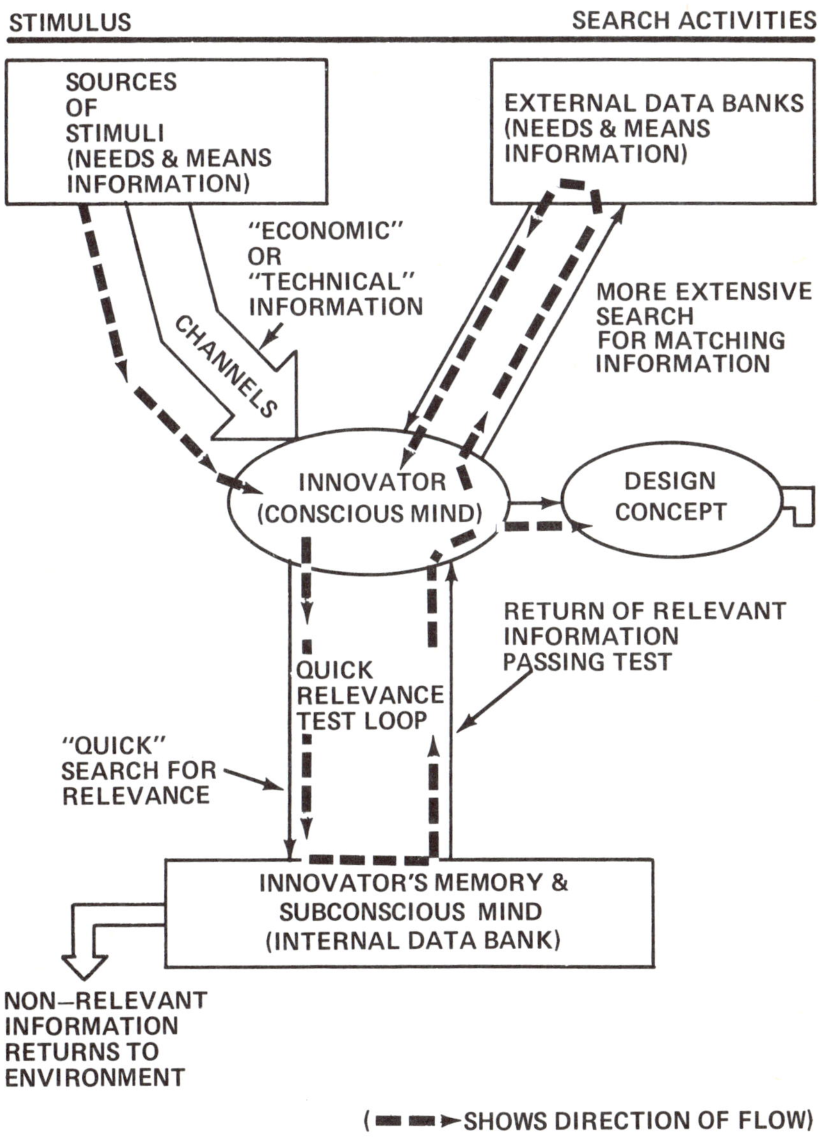

FOOTNOTES

1. Booz-Allen & Hamilton, Inc., *Management of New Products*, New York, Booz-Allen & Hamilton, Inc., Undated, P. 9.

2. For a review of the innovation literature see James Utterback, "The Process of Technological Innovation Within The Firm," *Academy of Management Journal*, March, 1971, p. 75-88. Scientific and Technical Communications are discussed at length by the papers in Carnot E. Nelson and Donald K. Pollock, *Communication Among Scientist and Engineers*, Lexington, Mass., D. C. Heath and Company, 1970. See also Richard S. Rosenbloom and Francis W. Wolek, *Technology and Information Transfer*, Boston, Mass., Division of Research, Harvard Business School, 1970 and Diana Crane, *Invisible Colleges*, Chicago, Ill.: The University of Chicago Press, 1972.

3. Battelle Columbus Laboratories, *Science, Technology and Innovation*, Washington, D.C.: National Science Foundation, February 1973.

4. Sumner Myers and Donald Marquis, *Successful Industrial Innovations*, Washington, D.C.: U.S. Government Printing Office, 1969.

5. James Utterback, *The Process of Technological Innovation in Instrument Firms*, Unpublished Ph.D. Dissertation, Sloan School of Management, M.I.T., 1969.

6. Norman R. Baker, Jack Siegman and Albert H. Rubenstein, "The Effects of Perceived Needs and Means of the Generation of Ideas for Industrial Research and Development Projects," *IEEE Transactions on Engineering Management*, Vol. EM-14, No. 4, December 1967, p. 156-163.

7. Louis K. Bragaw, *Some Characteristics of Successful Technological Innovations and Their Patterns*, Unpublished D.B.A. Dissertation, The George Washington University, June 1970.

8. J. J. Langrish, et.al., *Wealth From Knowledge*, London: Macmillan Press, 1972.

9. Donald C. Pelz and Frank M. Andrews, *Scientists in Organizations: Productive Climates for Research and Development*, New York: John Wiley & Sons, 1966.

10. Albert H. Rubenstein and Charles F. Douds, *A Real-Time Study of Technology Transfer in Industry*, Evanston, Illinois: Department of Industrial Engineering and Management Sciences, Northwestern University, January 1969. See also Richard S. Rosenbloom and Francis W. Wolek, *Technology and Information Transfer: A Survey of Practice in Industrial Organizations*, Boston, Mass.: Division of Research, Graduate School of Business Administration, Harvard University, 1970.

11. Thomas J. Allen, and Stephen I. Cohen, "Information Flow in Research and Development Laboratores," *Administrative Science Quarterly*, Vol. 14, No. 1, March 1969, p. 12-19 and Thomas J. Allen and Peter G. Gersberger, "Criteria for Selection of an Information Source," Cambridge, Mass.: Sloan School of Management, M.I.T., 1967.

12. See Bragaw, *op.cit.*, for a more detailed discussion of the differences in information seeking and use behavior as a function of each stage.

13. Jacob Schmookler, *Invention and Economic Growth*, Cambridge, Mass.: Harvard University Press, 1966, p. 101 ff.

14. IITRI, *Technology in Retrospect and Critical Events in Science* (TRACES), Washington, D.C.: National Science Foundation, December 15, 1968.

15. For a more complete discussion of consumer innovators see Thomas Robertson, *Innovative Behavior and Communication*, New York: Holt, Rinehard and Winston, Inc., 1971 and Everett Rogers and Floyd Shoemaker, *Communication of Innovations: A Cross-Cultural Approach*, New York: The Free Press, 1971.

16. Robert E. Seiler, *Improving the Effectiveness of Research and Development*, New York: McGraw-Hill, 1965, p. 40.

17. Pelz and Andrews, *op.cit.*, p. 881.

18. See Edward B. Roberts, "Entrepreneurship and Technology," in Cetron and Goldhar (eds.), *The Science of Managing Organized Technology*, New York: Gordon and Breach, 1970, p. 485 ff for an opposing point of view.

19. David Allison (ed.), *The R&D Game*, Cambridge, Mass.: MIT Press, 1969, p. 14 ff. See also Thomas Kuhn, "Scientific Research," in Trylor and Barron (eds.), *op.cit.*, p. 341 ff.

20. See S.C. Gilfillan, *The Sociology of Invention*, Cambridge, Mass.: MIT Press, 1970 (originally published in 1935) for one of the earliest and most complete discussions of this concept.

21. See for example, David Kiefer, "Fashions Change for Industrial R&D," *Chemical and Engineering News*, September 18, 1972, p. 8-9.

22. Allen and Cohen, *op.cit.*

23. James Utterback, "Process of Innovation: "A Study of the Origination and Development of Ideas for New Scientific Instruments," *IEEE Transactions on Engineering Management*, March, 1971, p. 75-88.

24. James Utterback in *Technological Change and Economic Growth*, G. W. Wilson, editor, Bloomington, Ind.: Division of Research, Indiana University, 1971.

25. D. H. Peters, *The Incidence and Exploitation of Commercial Ideas in University Departments and Laboratories*, Doctoral Dissertation, MIT, Sloan School of Management, Cambridge, Mass., 1968.

26. Gilfillan, *op. cit.*

27. For further discussion of this point see Thomas J. Allen, "Performance of Information Channels in the Transfer of Technology." *Industrial Management Review*, Vol. 8, 1966, pp. 879 ff. and Thomas J. Allen, "Studies of the Problem Solving Process in Engineering Design," *IEEE Transactions or Engineering Management*, Vol. EM-13, No. 2, June 1960, p. 72-82. See also, Thomas J. Allen, *Meeting the Technical Information Needs of Research and Development Projects*, Cambridge, Mass. : Sloan School of Management, M.I.T., 1969.

28. This analysis is possible using the variable stacking and transformation features of the U.C.L.A. Biomedical Program, BMD 08.

29. Sumner Myers and Donald G. Marquis, *Successful Industrial Innovations*. Washington, D.C.: U. S. Government Printing Office, 1969.

30. See Arthur Gersterfeld, *Effective Management of Research and Development*, Reading, Mass.: Addison-Wesley, 1970.

31. Keith Pavitt, *The Conditions for Success in Technological Innovation*, Paris, OECD, 1971.

32. Gregory, Carl E., *The Management of Intelligence*, New York: McGraw-Hill Book Company, 1967, p. 23.

33. Haefele, John W., *Creativity and Innovation,* New York: Reinhold Publishing Corporation, 1962, p. 16.

34. Gerard, Ralph, W., "How the Brain Creates Ideas" in Parnes and Harding (editors), *A Source Book for Creative Thinking,* New York: Chas. Scribner's Sons, 1962, p. 115 ff.

CURRENT PROBLEMS IN TRANSFERRING SCIENCE TO TECHNOLOGY

A.W. Pearson and T. Rickards

R & D Research Unit, Manchester Business School

ABSTRACT

It is generally accepted that 'most effective transfer takes place on the basis of an exchange of personnel or inter-personal transmission', but this is not enough. Two problems which can create barriers to the successful utilisation of science and technology are

1) the not-invented here (or NIH syndrome) which seriously affects the process of matching solutions in the form of existing science and engineering knowledge to problems in commerce or public programmes,

2) the difficulty experienced in communications between those with the knowledge of what is possible and those who may be able to put this knowledge to use, i.e. between the scientific and technological capability and the potential implementation. This barrier is often due to the inability of the former to communicate with the latter in a language he can understand.

Recent research suggests that some of these barriers can be overcome by the encouragement of group activity, making use of methods taken from the creative problem-solving area, such as brainstorming and synectics. This will be discussed in the light of experience of using these methods in our current research programme at Manchester.

INTRODUCTION

Much of the literature concerned with innovation gives the impression that science and technology to a large extent go their separate ways. The idea that one feeds directly onto the other seems to be inadequate. Unfortunately it is very difficult to really establish the influence of one on the other because of the time scales involved and the complexity of the innovation process. Hence the problems of attempting to establish the 'value' of scientific effort and the 'effectiveness' of research establishments. The net result of this appears to be a fairly wide acceptance that research can only be successful (in the sense of being exploited) where market needs have been clearly defined and ultimately where a specific customer exists for the output - preferably one who is prepared to pay for any effort which is expended.

In the U.K. this attitude, coming as it does at a time of economic stringency and a rethinking of expenditure in all areas of activity, has made R & D establishments more externally motivated. Both in the public and private sectors we are now seeing significant changes which are designed to encourage a closer relationship between the market needs and the technological capability or between the 'customers' and the 'contractors' as these are now commonly called.[1]

This is a significant trend, and it is having considerable impact on the method or organization of scientific and technological enterprises. It is helping to bring R & D effort closer to the needs of the marketplace in situations where these are readily identifiable. In such cases it is certainly having an effect on the rate of technology transfer, but it does not entirely eliminate the not-invented-here (or NIH) syndrome, and the communication barrier which exists between scientists and the potential users of their output. Current research suggests that these problems can be significantly reduced through the encouragement of group activity and the use of a number of creativity-spurring techniques.

The Not-Invented-Here Syndrome

This is a phenomenon often discussed in the post-war period. Its name is self-explanatory; it refers to the attitude displayed by people who tend to reject things which are not generated from within the group of which they form part - the boundaries of such groups often being very narrow. As a disease it is not confined to any one area of activity. Many new and potentially profitable products have been rejected by existing divisions of a company, but it has been found among scientists and technologists who have often felt it necessary to reject ideas coming from outside their immediate environment, perhaps seeing them as a threat to their own reputations. These two groups of people have also rejected ideas which have come from each other, as many studies of innovation have shown, and (just as important) they have rejected ideas that have come from other parts of the company, e.g. the marketing function, and also from outside the organization altogether.

This attitude is now changing, often as a consequence of the creation of new structures within the organization. For example, in the U.K. there has been a significant increase in the number of new venture operations which have been established in the last few years, most of them with terms of reference which encourage them to seek out ideas from any source which can be developed into new businesses. In looking outwards some have found in-house R & D not to be the most profitable source of ideas, particularly in the short-term. At least one company has replaced its central corporate research by a new venture structure[2]. In addition many companies have recently looked carefully at their objectives and at the alternative ways of achieving them, and in-house R & D now finds itself competing more vigorously with acquisition and the buying-in of know-how. This has resulted in R & D becoming more market oriented and taking positive steps to actively display its skills to all those who might make use of them, in some cases developing a 'marketing' type activity for itself[3].

Only recently the central research establishment of a divisionalized company has established a small 'new project group'. This is located within the research laboratory and is directly responsible to the research director. Its terms of reference are to identify areas of activity with which the company should be concerned and in which R & D has a major part to play. Its method of operation, at least in the first instance, will be to go out into the divisions seeking out ideas, looking for possibilities for transferring know-how between divisions and for developing new and modifying existing technology for the benefit of the organization as a whole. There is no question of "not invented here," but a real attempt to draw on the resources of people from all functional areas. Unfortunately it is not quite as simple as this, because some very basic problems of communication can arise which prevent people, however willing, from contributing to the best of their ability.

Communication barriers

As so often pointed out in the past, there has been little if any actual communication between the different groups within an organization. Our own research suggests that production and marketing people contribute little to the pool of ideas from which the R & D department draws its projects. The final customer, 'the man in the street' makes very little contribution except indirectly, via market surveys, perhaps conducted by questionnaire - and of course through the complaints department and the newspapers. Surely these people have a lot to offer - the question is whether it is the N.I.H. syndrome which prevents many of them from being allowed to make their contribution or whether the problem is more fundamental. The evidence suggests that it is the former. For example, attempts to invite such people to discussions on topics which appear relevant to all parties often lead to a polarization, with each group defending its own interests and/or omitting to listen to any of the points raised by the other. This is a natural reaction and one which it is not easy to overcome and is often made worse by a complicating factor which is not immediately obvious - the difficulty experienced by individuals understanding what others really mean. We have encountered a number of examples of this which are interesting and illustrative of the point.

In one case during the course of an idea generation session run at the Manchester Business School the whole group was concentrating on a problem posed by one of the members. This was concerned with the possibilities for utilizing 'knots' which arose as a by-product of the processing of wood. The session went on for about ten minutes before the person whose problem it was burst in and said "the knots are in the form of a slurry and are not solid". Clearly the rest of the group had not appreciated this, and the ideas they produced reflected their lack of understanding of the problem.

A second example occurred in a session taking place in an industrial organization in which both research and marketing people were involved. The problem was to identify possible advertising material for a new product and the group had been got together because neither party had been able to agree the other's suggestions. This was not surprising when after the group had been in session for only a short time a marketing man said in amazement "do you mean that one of the materials involved is a <u>liquid</u>?" Again a demonstration of how lack of understanding led to an inability to produce a solution which was acceptable enough to be adopted.

Coupling through Group Activity

We have recently set up a research programme to examine the way in which the innovative activity is influenced by the encouragement of group activity and the introduction of creativity-spurring techniques such as brainstorming and synectics. The results of some of this work have been reported[4]. In the last year over 100 managers have participated in 10 small-group seminars, many others have been introduced to the techniques through the graduate, management and executive programmes at the Manchester Business School, and in one case a session was conducted with over 100 postgraduate students from a variety of scientific and technological disciplines. In addition a number of exercises have been carried out working with groups of people from within the same company. These sessions start with a discussion of the techniques, illustrated with the help of some simple problem situations. The group then examines in detail one or more problems which are directly relevant to the participants, and it would be fair to say that in most cases new insights have been gained, particularly by those involved in the problem. The participant whose problem has been tackled has often gone away with ideas and potential solutions which he was prepared to admit were interesting and had not previously been thought of by his colleagues or himself. This is not to say 'miracles' have been worked or even that on further evaluation they will be found to be worthwhile. Sufficient to say they were novel to the people concerned. This is important, particularly as they were often generated by a group which at first sight would appear to have little to contribute to the area, that is, by a group which was not necessarily familiar with the technology and whose members in some cases were not even technologists. The following example illustrates this:

In a seminar at the Manchester Business School this Spring a group of approximately a dozen managers was examining the concept of problem redefinition as a source of new insights. (This concept is central to the structured aids to creativity[5]). A technical director provided a real problem from his own environment. He worked for a company in the textile industry, and considered that one of his main products had not adapted to the technological changes of the last fifteen years. He presented his problem as "how to make a better X". (He described 'X' in detail for a few minutes only). The group was asked by the discussion leader to redefine the problem in their own terms, and then to suggest 'wishful' starting points to solutions. The other participants came from quite different backgrounds - and included managers from the private sector, an operations research specialist, and a secretary. Several of the starting points caused the client to think hard about the ideas, but the one of greatest impact on him came from the secretary (herself a user of his product). In itself the idea was unworkable, but the client took it as a starting point. He literally withdrew from the exercise and began scribbling on a pad. The rest of the group continued working on the problem. After a few minutes he looked up and announced that there was no need to continue - he had found just what he was looking for. He had taken the secretary's idea and linked it with one of his own to produce a valuable new product concept. He is currently running validation experiments in his laboratory.

It is interesting to note that the person concerned was an innovator in his own right, with a successful track record and now a research director. In short, a 'product champion'. He was therefore willing and able to accept other peoples' ideas, although technically very deficient, and build on them using his own expertise. This characteristic may be very important and necessary to have in order to promote and develop technology at a faster rate.

General Discussion

It is not easy to generalize about the effectiveness or otherwise of a particular technique or method of organization on the basis of a selectively small number of specific examples. It could even be dangerous. We are therefore setting up a research programme which we hope will provide some information about the potential value of group activity and creativity-spurring aids in the information process. Measurement is difficult in this area, but some data have already been collected.

Our preliminary work with U.K. managers seems to confirm earlier work[6,7] and suggests that

1) the general atmosphere/attitude of problem solvers can influence both quantity and quality of ideas;

2) simple mechanistic devices can jog individuals out of set ways of thinking about problems (e.g. encouragement to indulge in wishful thinking, introduction of random stimuli, use of metaphor and analogy);

3) when individuals are replaced by a group, the other members of the group can supply the stimuli, but special procedures are needed to deal with normal interpersonal behaviour which reduces chances of ideas surviving.

An example of the effect of mechanistic devices on idea generation is shown in the table. This pattern is repeated when any of a series of model idea-generation tasks is presented. We are hoping to make a careful analysis of data from individuals and groups shortly.

Number of ideas produced in two minutes with and without encouragement to indulge in wishful thinking
(Each individual from a group of 21 managers from various industrial organizations, in a training seminar, took part in the experiment).

Subject Number	Total ideas unpromoted (after 30 minute lecture)	Total ideas promoted (after a further 30 minutes)
1	8	6
2	5	7
3	2	5
4	1	5
5	3	5
6	7	6
7	3	7
8	3	5
9	6	4
10	2	8
11	3	8
12	9	4
13	3	6
14	3	4
15	1	9
16	4	7
17	1	6
18	3	5
19	3	6
20	4	6
21	3	0

Conclusions

The rate at which new technology is introduced depends upon the acceptance by people that it will be useful to them in achieving their goals. Ready acceptance comes when the parties concerned understand fully the implications for them of any changes and in particular the problems that may be encountered. In order to promote a better understanding of these problems, in advance if possible, and also to throw more light on how they might be solved as and when they arise, it is suggested that contributions should be solicited from a wide variety of people. This is only possible if they are able to see the sort of contributions they might make and that others will accept this as being potentially useful, even if it might appear to be a threat to their own competence.

This is unlikely to happen if the more usual channels of communication are used and if "ideas" meetings are run on the same lines as decision taking, information gathering, or general discussion meetings. The alternative is to change the format sufficiently to allow more participation by diverse elements in idea-generation meetings. This is often helped by the introduction of a number of creativity-spurring techniques. Results to date suggest that (i) such meetings can produce many more ideas, of a wider-ranging nature and that (ii) people from a variety of backgrounds are able to contribute. However, it is clear that the way in which the meeting is run is important. At the present time we know something about this, and in particular about the need for a good leader, but are a little short of factual information about the way in which such leaders might be identified in advance, as distinct from being recognised on their performances in a whole series of meetings. In addition we have little information about how the membership of such a meeting should be made up. It is too easy for a leader to choose such people from those he has previously worked with and/or from those whose personal characteristics are acceptable to him. These are important points on which we need more research and on which we expect to undertake work in the near future. In the meantime we intend to continue with our seminar and in-company training programmes, as the response we have had to date has been extremely encouraging and suggests that the people both from within and outside an organisation can make significant contributions to innovation activity and can help to couple science and technology not only to each other but also to other parts of the organization.

Addendum

The following example illustrates spin-off which can be achieved by bringing together two different technologies through the intermediary of a neutral third party. In a recent seminar demonstrating problem-solving techniques one participant brought the following problem: he represented an engineering firm which had been designing machinery which came into contact with metalliferous ores. Precipitation of inorganic salts was seriously affecting the performance of the machines, and the company's research approach (based on metallurgical and engineering know-how) had not solved the problem satisfactorily.

One or two leads to new approaches were suggested in a demonstration session of the technique known as synectics. The leader of the group refrained from introducing his own ideas, as this is one of the principles of the technique--the leader acts as an organizer, not an idea-generator. However, his own background included work in a completely different industry that had committed a lot of applied research resources into 'wet' inorganic chemistry.

After the meeting the group leader was able to bring together the group development manager of a company with the know-how ('means') and the engineering manager with the technological need. A joint project between the companies is currently being negotiated.

This approach seems to be a more effective one than if in the synectics session the group leader had attempted to solve the problem. The danger of N.I.H. was reduced because the offer was clearly made without any vested interests likely to conflict with the engineering organization. The third party role seems not unlike that of a technological gatekeeper.

References

1. The Organization and Management of Government R & D: A report by Lord Rothschild, Head of the Central Policy Review Staff, in "A Framework for Government Research and Development", Command 4814.

2. Gardner, J.B. "Innovation through New Ventures: new venture concept in B.O.C." R & D Management, Vol.3, No.2, February 1973

3. Griffiths, D and Pearson, A.W. "The Marketing of Research and Development", Industrial Marketing Management, Vol.2, 1973.

4. Rickards, T. "Brainstorming in an R & D Environment", R & D Management, Vol.3, No.3, June 1973.

5. Rickards, T. "Open-ended Problem-Solving" to be published by Gower Press, 1974.

6. Osborn, A. "Applied Imagination"

COUPLING PATTERNS IN SCIENCE AND TECHNOLOGY

Dr. Ron Johnston , Manchester University

Technology transfer is a term which has been applied to a wide range of activities associated with the movement of knowledge--in the crude state as ideas or embodied in literature or in hardware, across institutional and organizational interfaces. This wide compass has the advantage, appropriate to such a subject, of promoting cross-fertilization between studies of the various forms of technology transfer. It has the disadvantage that the lack of consistent definitions can hinder the processes of conceptualization and analysis.

One particular class of technology transfer, which has been the subject of a great deal of study, has been concerned with the interface between science and technology. The motivations for the study of the relationship between science and technology have varied. In recent years, governments concerned at the size and rate of growth of the science budget, and the limited degree of accountability for this expenditure, have been demanding a clear demonstration of the mechanism and the amount by which scientific research contributes to industrial growth and economic prosperity. In many countries industry has shown a concern with increasing the efficiency of its own investments in research as well as expanding access to science and technology developed outside the firm.

The most cited studies of the link between fundamental scientific research and industrial technology have been Project Hindsight (1) and TRACES (2) which are too well known to require detailed elaboration here. The methodological assumptions common to these studies are that innovations can be considered as composed of a series of R & D events and that the contribution of science to technology can be assessed by tracing the origin of these events. A more recent TRACES study (3) has overcome some of the objections to this approach (4), but has clung to the 'critical events' methodology.

Similar assumptions are evident in the arguments of scientists engaged in the defence of the science budget (e.g. see Perutz (5)) by tracing the technological consequences of specific scientific discoveries. However, science is only one of many inputs to successful innovation and hence it cannot claim for itself complete responsibility for such technology, or demand that it be judged only on the basis of its successes.

The other major field of investigation of the interface between science and technology has been in the form of empirical studies of the innovation process which have incidentally examined the contribution of science. (6,7,8) These studies have produced a considerable array of findings but there has been general agreement that the previously accepted linear model of the interaction between science and technology, i.e., that technology is the direct application of basic science, is not valid; in fact, science plays little part, at least directly, in the vast majority of technological innovations.

It is apparent that the results of any study of the science-technology interface and the flow of information between the two sectors will depend critically on what is accepted as the demarcation between them. Science and technology can be considered distinct in the sense that it is possible to identify two professional groups whose activities are describable in terms of the attitudes, values and norms of scientists and technologists, respectively. However they cannot be distinguished in terms of method, as recent studies of the form of progress in science (9) and in technology (10) suggest that within somewhat different constraints, both the academic scientist and the innovator in industry approach the definition and resolution of problems in very similar ways, each relying on an extensive background of craft knowledge and using a combination of understanding and heuristic to achieve the desired goal.

On this subject of definitions, both the 'tracing' and the 'innovation' types of studies are unified in their acceptance of fundamental knowledge as the only, or at least the most important, output of science. This view of science is based, however, on an outdated conception of it as a systematised body of knowledge. A cursory examination of the activities of scientists today shows that the majority are not engaged solely, or even primarily, in the pursuit of fundamental knowledge, even within universities. The pressures exerted on science by its very success, its growth, its professionalization and its politicization have combined to alter its nature substantially.

Hence any examination of the transfer of knowledge between science and technology, and vice-versa--which is based on the equation of science with knowledge production and technology with knowledge application -- is clearly innappropriate for the contemporary situation. Rather, knowledge is both produced and applied in a variety of institutions such as universities, independent research institutes, industry and government. Each of these has a different function, operates under different constraints and within different environments. Nevertheless for the effective achievement of national goals they must interact positively as elements of a research-production system. On this basis the major problem may be seen as that of identifying means of reducing institutional barriers and assuring effective coupling between the various elements.

A recent study of the information used in the resolution of technical problems arising during the course of development of a wide range of innovations in British industry has allowed the examination of a variety of factors affecting the transfer of information between these various institutions.

DESCRIPTION OF THE STUDY

This study was oriented towards answering the question, "What types of information are used in the resolution of technical problems arising during the development of an innovation in British industry and how is it transferred from source to user?" The form of the question exposes some of the assumptions underlying the investigation. Firstly, it is assumed that in its pursuit of technological innovation industry encounters technical problems the solutions of which are describable in terms of certain information inputs. No attempt was made to synthesise these inputs into the creative insight which, in fact, is the solution. Secondly, by focussing on the resolution of technical problems _per se_, one is clearly, and from the outset, avoiding the difficult question of whether the process of problem-resolution will subsequently lead to a "successful" technological" innovation - that is, one which is commercially successful for the firm in question. The solution of technical problems is a necessary but not sufficient condition for successful technological innovation. Many other factors, financial and organizational, must be fused with technological capability and market need before one can begin to analyse the mercurial notion of success in this context. In fact, most of the products whose development provided the data for this study now appear to have at least been successfully marketed. But their subsequent fate will not influence its conclusions.

Unlike many of the recent empirical studies which have set out to determine the relative contributions of science and technology to technological innovation, we have attempted primarily, to describe the characteristics of the information which have contributed to the resolution of technical problems and secondarily to determine, if any of these are further describable as scientific or technological information.

The study is structured around the notion of problem-solving. It draws primarily on that view of organizational activity which regards members of organizations as decision makers and problem-solvers and this, in turn, suggests that the individual, or team, in the process of problem-solving, might be the most suitable unit to focus on. (11) However, the type of technical problem which arises in industry is characteristically different from those conventionally used in the study of problem-solving behaviour. "For a technical problem there is often no correct, or even best, solution in the long run. In fact, there is frequently no terminal state : both problems and solutions are themselves often dynamic. (12) Also, the criteria for evaluation of a potential solution are not absolute - they are ultimately a matter of judgment and different individuals evaluate them differently. Nevertheless, the resolution of technical problems can be viewed as the assemblage and evaluation of information and we have chosen this perspective for the examination of the roles of science and technology in technological innovation. The problem-solver is regarded as operating within an informational matrix which comprises knowledge and skills with respect to the technology in which he is currently engaged as well as the backlog of previous experience, educational as well as industrial. In the event of a technical problem, the problem-solver can be considered, ideally, to approach the solution by moving from the familiar to the unfamliiar and from the routine to the novel. In the course of this process he may need to extend his informational resources, first of all by consulting sources readily available to him - such sources include colleagues, books, periodicals, among others. If these sources prove inadequate the problem-solver will begin to consider sources less familiar or less readily available. It should be emphasized that the problem-solver can, and in practice frequently does, solve, or more appropriately, resolve the problem by redefining it in accord with the information he possesses.

A number of variables were developed to describe the characteristics of the information inputs. The first of these was the general source of the information, by which the information was sub-divided according to whether the problem-solver obtained the information from sources outside the company, from sources within the company, or possessed the information himself at the start of the innovation. The general source was further categorised into the specific source which described where the problem-solver obtained the information. No attempt at further tracing to ultimate origins was made. Other variables, adapted from Marquis and Myers (13), described the nature or content of the information, the channels by which it reached the problem-solver, and the impact the information made on the particular technical problem. Measures of various characteristics of the organization, the innovation, and the education and experience of the problem-solver were also obtained.

A feature of this study concerns the choice of recent or on-going developments. Using this criterion of selection many of the problems involved in the reconstruction of past events are avoided. In most cases, it was possible to interview the individuals who were directly concerned in the problem solution about events which were relatively fresh in their memories. Also, data gathered in this way are more likely to be descriptive of the contemporary institutional situation of science and technology and so may provide more relevant information for policy makers who seek to influence the environment in which science and technology are produced, and the transfer of technology occurs.

COUPLING PATTERNS IN SCIENCE AND TECHNOLOGY

The three categories of general source contributed approximately equal numbers of information inputs to the resolution of technical problems. In terms of examining coupling between science and technology, it is the information from outside the company which is of greatest interest. The various categories of specific source together with the distribution of information inputs among them are shown in Table 1.

Recent studies of technological innovation have made use of the concept of technology transfer to describe how information, often crucial to the success of an innovation, is acquired by the innovating firm. The general consensus has been that technology transfer occurs most efficiently through person-to-person contact : technology transfer 'on the hoof' (8,14) By contrast, this study concludes that information from outside the firm is transferred equally by inspection of the literature and person-to-person contact. More specifically, the ratio of scientific information transferred by the literature to that transferred by contact with individuals was found to be 54 : 53 while that for technological information was 95 : 98. However the personal contact was often necessary in order to interpret or 'translate' the literature.

Table 1

Specific Source of Information with Respect to General Source

Number of Information Inputs

Outside Company	No.	%	Inside Company	No.	%
Trade Literature	43	14	Analysis and Experiment	183	69
Technical Lit.	22	7	Superior	25	9
Scientific Lit.	36	12	Colleague	17	6
Text Books	8	2	Other Dept.	17	6
Handbooks	20	7	Other Division	7	3
Reps. & Customers	20	7	Company Manual	9	4
Supplier Cos.	39	13	Company Report	6	2
Universities	30	1	Internal Library or Reference System	3	1
Professional Institutions	3	1			
British Standards	12	4			
Exhibitions	2	1			
Consultants (commercial)	12	4			
Government Agencies	11	4			
Field Tests	8	3			
Conferences	4	1			
Research Assns.	24	8			
Patents	6	2			

In the following paragraphs, the discussion will center on the use of and contribution made by information inputs transferred along these two main routes : literature and personal contact. After this, some attempt will be made to relate the use of information sources to the educational background of the problem-solver and to characteristics of the organization.

Literature

The major form of printed matter that contributed scientific information to the resolution of technical problems was the primary and secondary scientific literature. This literature consists of publications of original research and reviews of particular fields of scientific interest. This type of information made up about 30% of the total scientific inputs and was also the third most commonly used source from outside the company (see Table 1). Examination of the use of this information in individual innovations, however, revealed that in over half of the innovations no use was made of the scientific literature at all (Table 2). This phenomenon in which the scientific literature was used frequently or not at all has been denoted by the phrase 'zero or multiple coupling' and will be discussed more fully below.

The information inputs obtained from the scientific literature mainly described either the properties, composition or characteristics of materials and components or provided theories and general principles. It is of particular significance that of the information from ten major external sources, that provided by the scientific literature had the greatest impact on the resolution of technical problems; that is, it contributed most directly to the solution adopted. These results which were derived by correlating characteristics of information inputs from outside the company with the specific source of the information are presented in Table 3.

Table 2

Distribution of Use of Outside Company Sources

Source	Number of Innovations having specified number of Information Inputs								
	0	1	2	3	4	5	6	>6	
Trade Literature	7	10	9	2	1	1	0	0	
Technical Literature	18	5	4	3	0	0	0	0	
Scientific Literature	20	5	0	1	0	0	3	1	(10)
Textbooks	26	2	0	2	0	0	0	0	
Handbooks	18	6	4	2	0	0	0	0	
Reps. & Customers	18	7	2	3	0	0	0	0	
Supplier Companies	13	7	7	1	1	0	0	1	(11)
Universities	18	2	6	2	1	0	1	0	
Consultants	24	3	1	1	1	0	0	0	
Research Assoc.	25	2	0	0	0	1	1	1	(11)

The majority of scientific literature used was in the form of reports of original research work, mostly carried out in universities and was not directed toward achieving anything related to the innovation in which it was subsequently used. The average age of the literature (from the date of publication) was 12.2 years with a standard deviation of 9.1 years. Moreover, 26 of the 36 inputs from the scientific literature had been produced within the decade previous to the innovation. As the average time elapsed since the problem-solver left the education system was 17 years (with a standard deviation of 7.7 years) it is apparent that the problem-solver <u>does</u> rely on information generated by scientific research, and that he continues to obtain such new information after he completes his formal education. These results are in direct contrast to the widely accepted "education cycle" hypothesis (15), which states that there is a time delay of the order of twenty to thirty years between most basic research and its technological application.

Table 3

Comparison of Characteristics of Outside Information Inputs with the Specific Source of the Information

Characteristics	Number of Information Inputs Source								
	A	B	C	D	E	F	G	H	I
Nature									
Existence or availability of equipment or materials with particular capabilities	20	5	1	9	7	9	2	1	4
Properties, composition, characteristics of materials or components	14	7	17	7	5	6	4	3	7
Test procedures and techniques	1	3	1	5	1	3	1	3	1
Operating guidelines, required specifications, technical limitations	1	1	2	0	2	2	0	2	0
Location of information	1	1	0	0	1	4	7	1	3
Theoretical basis, explanatory general principle	3	2	13	4	1	2	5	0	3
Design-based information	3	3	1	3	1	9	4	2	4
Existence of specialist facilities/services	0	1	1	0	2	4	7	0	2
Mechanism									
Search expt.	39	20	35	28	0	1	6	1	5
Sought help	0	1	1	0	16	31	20	11	11
Other's initiative	3	0	0	0	3	6	4	0	6
Serendipity	1	0	0	0	1	0	0	0	0
Non-specific search	0	0	0	0	0	1	0	0	2

A = Trade Lit., B = Tech.Lit., C = Sci.Lit., D = Textbooks, E = Reps. & Customers, F = Supplier Cos., G = University, H = Consultants, I = Research Associations

The other printed sources of scientific information inputs were textbooks describing scientific theories and principles, and scientific handbooks providing data and characteristics of fundamental or natural materials. The use of these sources also displayed the characteristic 'zero or multiple' coupling.

Another major vehicle of information transfer by means of printed matter was the technical literature. A technical journal was defined as one containing articles of a technical nature, advertisements of products and processes and general news of relevance to a single or a group of industries. Its aims are different from those of the scientific literature; it attempts, not only to provide new knowledge, but, more than the scientific literature, to provide a focus for groups, which may reside in diverse industries, facing common technical problems. The technical literature was used in a large proportion of the innovations studied and provided a wide range of information which was, of course, generally available throughout the industry.

By comparison the trade literature - that disseminated by firms to provide information about their products - in the form of catalogues was the major single source of technical information from outside the company (see Table 1). It was used in almost all innovations (see Table 2) and, not unexpectedly, provided in particular information about the existence or availability of materials or equipment with particular properties (Table 3). It was apparent that many of the technical problems encountered during the innovation process involve finding a material or a component with suitable properties to fulfil a specific purpose and the search strategy is to determine, firstly, if there is a suitable product already on the market.

Personal Contact

The other major form of coupling of science to technology occurred through the contact of the problem-solver with members of the scientific community. The scientists themselves were located in a variety of institutions including universities, government laboratories and research associations.

Of particular interest here is the close coupling of the problem-solver with scientists working in universities. As can be seen from Table 1, contact with these scientists provided an important source of information from outside the firm. The pattern of person-to-person contact in the transfer of scientific information was similar to that which occurred in the use of scientific literature. That is, less than one half of the innovations depended for their completion on information acquired by contact with scientists but those that did obtained a number of information inputs from the same or a related source (see Table 2).

Information was obtained from universities through a number of modes of interaction : occasionally by employing academic scientists directly as consultants and by supporting research in a university which was relevant to company interests, but more frequently by requesting advice and assistance, and by the use of specialist facilities such as instruments or libraries. As we have remarked, these modes of coupling did not occur as single isolated events in the course of problem-solving. Rather, they tended to occur frequently or not at all. The establishment of coupling between a firm and a university characteristically followed a 'cascading' pattern; originating perhaps with the firm hiring a graduate from the university or sponsoring research there through the gradual establishment of closer relations, culminating in a real mutual understanding of interests and needs. Scientists at universities were used largely in a supportive role-- attesting to the feasibility of a proposed solution or providing details of the location of specific information or specialist facilities and services of direct relevance to the problem under consideration. Such a pattern corresponds very closely to that uncovered in Project Sappho (16) : that successful innovations exhibited a closer coupling with the scientific community regarding the specific projects than did unsuccessful innovations. Of course, some relationship must exist prior to the innovation in order to provide the opportunity for this more specific form of coupling.

The essentially supportive role of person-to-person contact with scientists in universities is illustrated by the finding that the impact of the information transferred was not in the provision of basic ideas for the innovation. Rather, the scientist tended to respond to a previously posed problem and while he was in some cases able to provide the solution directly, he was more frequently able to suggest alternative ways in which the problem might be tackled or to narrow the range of considerations or to provide equipment and procedures with which to test the feasibility of an industrially-generated solution. It frequently happened that once started on the path suggested by the scientists the solution followed with few major stumbling blocks.

The research associations provided an extreme sample of the 'zero or multiple use' of information sources which we have described previously. This source provided the fifth highest number of information inputs (Table 1) but was, in fact, used in only five of the innovations - three of these five firms made extensive use of this source (Table 3). While the number of innovations involved with this source is too small to admit any generalizations to be made, it may be relevant that only medium-and low-research intensity industries made use of the research associations. This would suggest, and such comments were made more than once during the interviews, that firms, lacking an 'in-house' capability, rely on expertise and the facilities of the research associations when they do decide to develop a new product.

The type of information which this source provided was a mixture of the scientific and the technological and involved the properties or characteristics of materials, design information and the location of specific pieces of information. Thus, the research associations appear to play a role complementary to that of the universities and, because they are situated within both the scientific and technological communities, would appear to be in a unique position to promote coupling between the scientific community and industry.

Contacts in industry such as sales representatives, customers, and the technical sales staff in supplier companies were one of the two major sources of technological information from outside the company and were used by the vast majority of problem-solvers during the process of innovation though in most cases only once or twice. It comes as no surprise to learn that many of the information inputs are acquired from trade sources, but the pattern of use so different from that for scientific information is of interest. It would appear that because of the structure of the industrial community the problem-solver can initiate short-term coupling with other members to obtain specific information or assistance but that little or no long-term benefits ensue from such a relationship.

The Role of Education

It might be expected that the education of the problem-solver would considerably affect both the type and the source of information he characteristically seeks. Examination of the use of general sources of information in terms of the qualifications of the problem-solvers revealed that problem-solvers with all standards of educational qualification contributed approximately the same total number of information inputs except for the three problem-solvers with O-levels. However, comparison of the use of the three general sources shows that those with a university education and particularly those with a Ph.D. relied much more on external sources of information while those with an industrial education depended on information they already possessed (see Table 4). In particular, the problem-solver with a university education used the scientific literature, commercial consultants and scientists in universities to obtain information to solve technical problems.

The only external source used significantly more by those with an industrial education were the experts in companies supplying them with raw materials. Hence the reliance of the problem-solver with a university education on sources of information outside the company would appear to be closely related to the sources of information he learns about during his education, i.e., the scientific literature and the availability of expertise in universities.

The sources of personal expertise relied on by the problem-solver with an industrial education were identified as education, experience in industry, trade contacts and regular browsing of trade literature. Hence those with an industrial education rely more on their education and experience to solve technical problems whereas those with a university education use it to extend their resources.

Table 4

Comparison of Use of the Various Sources of Information Inputs with the Education of the Problem-Solver

	Number of Information Inputs Educational Qualifications					
	O-level	Batchelor degree	Ph.D.	ONC	HNC	Grad. Inst.
a. General Source						
Outside Company						
Total	17	195	86	34	103	39
Mean	5.7	9.8	14.3	8.5	10.3	7.8
Inside Company						
Total	15	228	49	43	78	45
Mean	5.0	11.4	8.2	10.8	7.8	9.0
Personal						
Total	13	187	56	56	144	78
Mean	4.3	9.4	9.3	14.0	14.4	15.6
Totals						
Number	45	610	191	133	325	162
Mean	15.0	30.5	31.8	33.3	32.5	32.4
b. Specific Source						
Outside Company						
Trade Literature	0	18	4	5	13	3
Technical Lit.	2	7	5	0	7	1
Scientific Lit.	0	31	2	0	1	2
Textbooks	0	5	0	0	3	0
Handbooks	0	7	2	1	8	3
Reps & Customers	0	8	2	3	6	0
Supplier Customers	3	13	4	3	16	0
Universities	2	15	3	0	6	4
Consultants	0	13	8	1	1	0
Research Ass.	0	22	12	11	11	0

Organizational Characteristics

High research-intensity industries, i.e., science-based, could be expected to show different patterns of coupling and information use than low research-intensity, i.e., craft-based industries. Reference to Table 5a shows that use of the three general sources does not vary significantly from a random distribution for the high and medium research-intensity industries, but the low research-intensity industries relied far more on external than on internal sources. Having only a small R D capability, they are forced to use external sources of information in order to carry through an innovation. Table 5b shows that these industries show a distinct bias towards the use of trade contacts such as sales representatives and customers, and also consultants and research associations compared with the more research intensive industries.

Table 5

Comparison of Use of the Various Sources of Information Inputs with the Research Intensity of the Industry

	Number of Information Inputs Research Intensity		
	High	Medium	Low
a. General Source			
Outside company	163	74	63
Inside company	156	74	37
Personal	208	75	37
b. Specific Source			
Outside company			
Trade literature	26	8	9
Technical literature	12	9	1
Scientific literature	25	7	4
Textbooks	4	1	3
Handbooks	12	5	3
Reps. Customers	7	8	5
Supplier Cos.	22	7	10
Universities	20	5	5
Consultants	4	2	6
Research Assocs.	0	16	8

Other significant findings were that the more active the strategy of innovation adopted, the more the problem-solver relied on scientific literature to resolve technical problems; the larger the scale of the innovation, the more sources of information from outside the company were used. Also the added literature, both scientific and technical, rather than personal contact, provided the major sources of information.

CONCLUSIONS

The mode of coupling of the industrial community with scientific sources has been shown to be distinctly different from that with technological sources. Scientific sources were used either not at all or to a considerable extent, i.e., zero or multiple interaction; whereas the pattern of use for technological sources was distributed more evenly at a medium level throughout the innovations studied. This pattern of 'zero or multiple' use for scientific sources has been found in the use of scientific literature reporting original research, scientific handbooks and textbooks, and in the contact of the problem-solvers with the scientific community in universities and in research associations. Similarly, the pattern of "even" use of technological sources was found for both the use of technological literature and contact with the technological community. It would appear that the cost of establishing coupling with science is, or is perceived to be, higher than that required for technology, but that once this barrier is overcome, a cascading expansion of coupling occurs leading to a considerable contribution of information resolving technical difficulties and promoting effective innovation. Hence it is apparent that different strategies for coupling need to be adopted depending on the nature of the information sought and its likely location. One particular key to successful coupling with both the scientific community and its literature would appear to be the possession of a university education. A number of mechanisms to promote this coupling have been described in recent publications. (17, 18, 19)

Effective performance of a national research-production system depends not only on each element efficiently executing its appropriate functions but also on the coupling between these elements at both policy and execution levels. In most countries there would appear to be the potential for a considerable return by investing greater resources in improving this coupling. Hence efforts must be directed to developing environments in the various institutions conducive to coupling without, however, unduly interfering in the internal mechanisms of them. However, it must be remembered that coupling is most effective at the informal and personal levels. While opportunities can be provided for this kind of personal interaction to be initiated, it cannot be legislated into existence.

In recent years considerable effort has been devoted to the development of sophisticated information systems. The assumption underlying this approach to information problems is that information is an essentially passive element; thus improving technology transfer is merely the problem of delivering the appropriate units of information when required to the researcher or innovator. However, as this study has shown, the most effective coupling occurs when an active approach is adopted on both sides of the interface. Even when the ultimate source of particular information is in literature, the coupling agent often has played a vital role in suggesting that literature because of his familiarity with it and his awareness of the problem, or he has translated a particular piece of information contained in the literature into a form in which it is directly applicable to the industrial innovator's problem. It is the promotion of an active rather than a passive conception of coupling in all the institutions of the research-production system that holds the greatest prospect for the improvement of technology transfer.

REFERENCES

1. Office of the Director of Defense Research and Engineering, "Project Hindsight Final Report", Washington, D.C., 1969

2. Illinois Institute of Technology Research Institute, "Technology in Retrospect and Critical Events in Science", Chicago, 1968

3. Battelle Institute, "Interactions of Science and Technology in the Innovative Process : Some Case Studies", NSF-C667, 1973

4. Kreilkamp, K., "Hindsight and the Real World of Science Policy", Science Studies 1, 43, 1971

5. Perutz, M.F., "Health and the Medical Research Council", Nature 235, 191, 1972

6. Organization for Economic Co-operation and Development, The Conditions for Success in Technological Innovation, Paris, 1971

7. Cahn, R.W., "Case Histories of Innovations", Nature 225, 693, 1970

8. Price, W.J. and Bass, L.W., "Scientific Research and the Innovation Process", Science 164, 802, 1969

9. Ravetz, J.R., Scientific Knowledge and its Social Relations, Clarendon Press, Oxford, 1971

10. Johnston, R.D., "The Internal Structure of Technology", Monographs of the Sociological Review 18, 117, 1972

11. March, J.G. and Simon, H.A., Organizations, John Wiley, New York, 1967

12. Frischmuth, D.S. and Allen, T.J., "A Model for the Description and Evaluation of Technical Problem-Solving", IEEE Transactions on Engineering Management, EM-16, 58, 1969

13. Myers, S. and Marquis, D.G., "Successful Industrial Innovation", US Government Printing Office, Washington, NSF 69-17, 1969

14. Langrish, J., Evans, W.G., Gibbons, M. and Jevons, F.R., Wealth from Knowledge, Macmillan, London, 1972

15. de Solla Price, D.J., "Is Technology Historically Independent of Science? A Study in Statistical Historiography", Technology and Culture 6, 533, 1965

16. "Success and Failure in Industrial Innovation : Report on Project SAPPHO by the Science Policy Research Unit, University of Sussex", Centre for the Study of Industrial Innovation, London, 1972

17. OECD, "The Research System", Vol. 1, Paris, 1972

18. Confederation of British Industry, "Industry, Science and Universities : Report of a Working Party on Universities and Industry Joint Committee", London, 1970

19. Roy, R., "University - Industry Interaction Patterns", Science 178, 995, 1972

20. Gibbons, M. and Johnston, R.D., "The Interaction of Science and Technology", Report to the Council for Scientific Policy, Department of Liberal Studies in Science, Manchester University, 1972

COUPLING INVENTION to INNOVATION

F.R. BRADBURY

Professor of Industrial Science, University of Stirling, STIRLING, Scotland.

The title of this lecture 'Coupling Invention to Innovation' was presumably coined by Marvin Cetron, certainly not by me. Nonetheless, with certain reservations which I will mention later, it will serve my purpose well enough.

The letter of invitation to lecture to this gathering refers to 'practitioners and developers' of technology. The nature of this paper is the extraction of patterns from my experience as a practitioner and developer of technology in the chemical industry, reviewed against the academic frameworks subsequently acquired at the University of Stirling.

I must begin by defining the terms 'invention' and 'innovation'. By the former I wish to indicate a discovery which is relevant to innovation; by 'innovation' I denote the whole process of bringing new or modified technology to the users on a full commercial scale. In contrast with innovation, invention is something of an event, albeit an event which may be preceded by lengthy periods of gestation. The distinction between invention and innovation, which we owe to Schumpeter (1), is important for the organisation and practice of research and development. If R & D aimed at innovation is overweighted towards invention the vastly longer and more expensive innovation may fail for want of adequate and appropriate resources, and the inventions thereby remain unfruitful. An uncritical acceptance of loose dubbing of the innovative process as 'invention', not uncommon in the literature (2), predisposes technologists to such errors of resource allocation. My experience in the chemical industry tells me that the recruitment of top class research scientists into R & D departments imbued with the view that the coupling of their inventions to innovation is largely a routine and readily attainable outcome can result in a very meagre return to the investment. In other words invention is all too readily uncoupled from innovation. The inadequacy of natural science alone or engineering alone as a training for innovation is becoming increasingly evident as we move through the seventies. The society we work in is not any longer eager to take from our innovators whatever they may create; there is so much social accounting now demanded of both products and processes that the social sciences must now contribute to the process of innovation and to the education of innovators.

My view of innovation is of a process in which a complex existing technological-economic system is modified by the introduction of new or modified technology: (like Schon (3) I include in technology products processes and organisation). This view is in accord with Schumpeter's characterisation of innovation as "putting productive resources to uses hitherto untried in practice, and withdrawing them from the uses they have served so far." (1). Thus although the technology introduced may be new, the resources used in its deployment are not. This leads to the very important, if elementary canon of systems theory, that the particular innovation under consideration must be viewed in its full context; 'look out before you look in' might be an appropriate maxim.

For the innovation process the outward look requires that the new should be checked at intervals against company strategy and objectives as well as being related to current technology and assets. This may most conveniently be done by including in project evaluation, check lists which ask questions about the interaction of the proposed innovation with company plans for expansion and diversification, short term and long term to uncover mismatches as well as good fits. This includes, of course, some degree of technology assessment. The changes which the innovation may make on the deployment of company resources may be discussed in terms of Ansoff's concepts of synergy (4). Check lists for strategy and synergy were developed by one of my students, Dr. W.M. Gallagher, working in I.C.I. and these have been published in R & D Management (5). Again the contextual mapping of innovation should not, of course, be limited to exploring its place within the company. This internal appraisal, as it is termed by Ansoff (4), must be complemented by an external appraisal with particular reference to potential competition. So, in assessing the project for synergy, the advantages claimed for positive gains to the project from existing company assets, such as distribution systems or technological proficiency, must be weighed against the advantages that might be expected to accrue to a similar project in the hands of the most likely competitors. In other words synergy is to be taken relatively, not absolutely.

When we turn from analysis of invention and innovation to the anatomy of the process by which the existing technological system is to be changed a conceptual model (Figure 1) is of assistance (6). This diagram attempts to identify the phases of product innovation in a process industry but, with minor modifications and changes in terminology, it can also describe the process of innovation in engineering and construction industries.

The model is not entirely appropriate for a discussion of the coupling of innovation with invention because it both exalts and understates the role of invention: invention is exalted by showing it as a prime source of innovation, the starting point, whereas innovation need not be led by invention; the model underrates invention by failing to indicate its frequent contribution to all phases of the innovation process. The inventiveness of technologists can evade obstacles to innovation in process definition, in design of plant, in start up and production as well as in market research and market development. A better representation of the innovation mechanism is one in which invention is shown as a potential contributor to all phases of the process. Invention may do more than circumvent obstacles; it may change the product and/or lead to an alternative technology, a contribution I designate as shaping and to which I will return.

Coupling implies the existence of a disjunction. These are found between phases of the innovation process (Figure 1) as well as between inventions and the innovation process (Figure 2).

These disjunctions are more evident in large establishments and it is commonplace to point to the advantages of the very small firm in the matter of easier communications and barrier avoidance. My view from experience of innovation in the large firm is that in the development of a major system restructuring (innovation) barriers or disjunctions may be a source of strength rather than of weakness. (Although I do not propose to enter the territory of psychology and creativity my observations tell me that the disjunction of inventor from innovator is usual, a factor which must be taken into account in any attempt to couple invention with innovation).

The positive contribution of barriers to the coupling process, be this between invention and innovation or be it between phases of innovation, lies in the synergy resulting from grouping or concentration of skills and other resources.

In discussing the progress of innovation stages for the weedkillers diquat and paraquat, Bradbury, McCarthy & Suckling identified contributions from various and distinct parts of

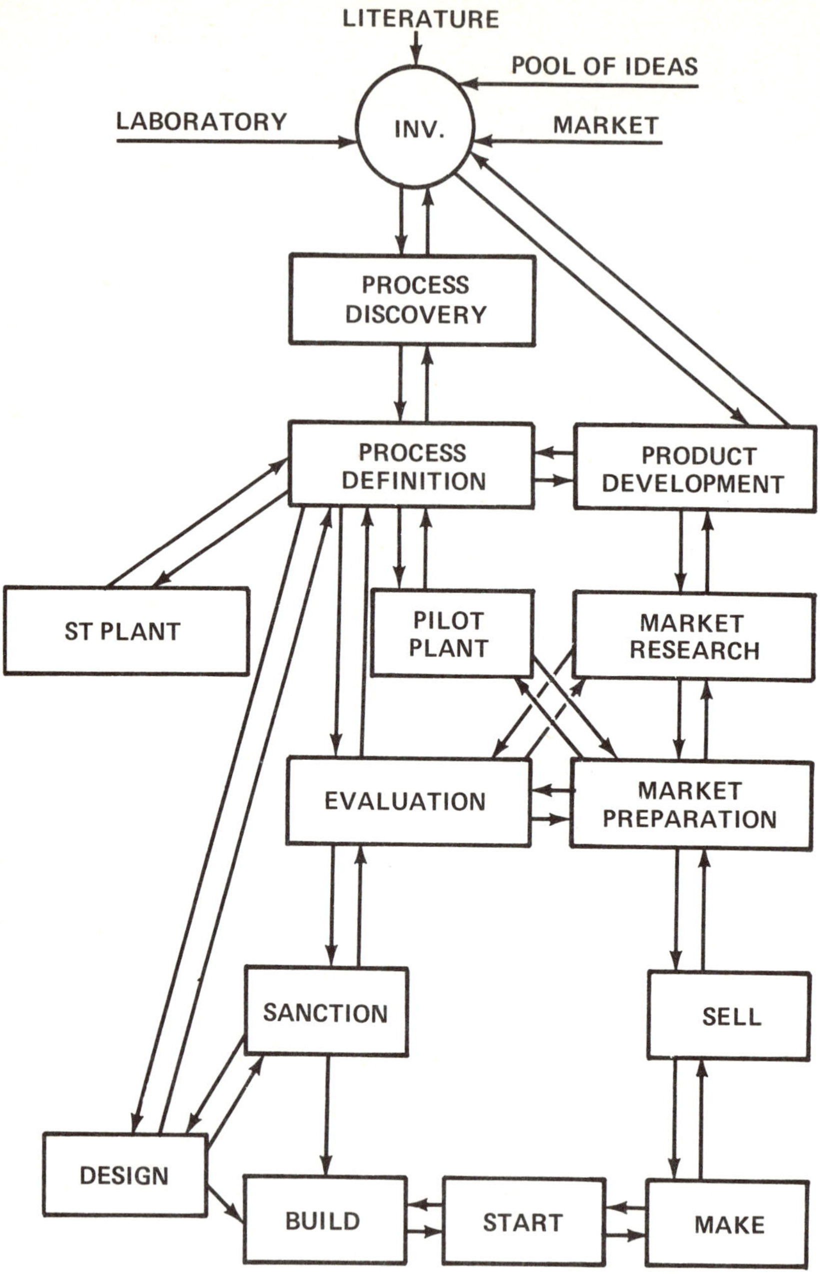

FIGURE 1 INNOVATION PROCESS STAGES

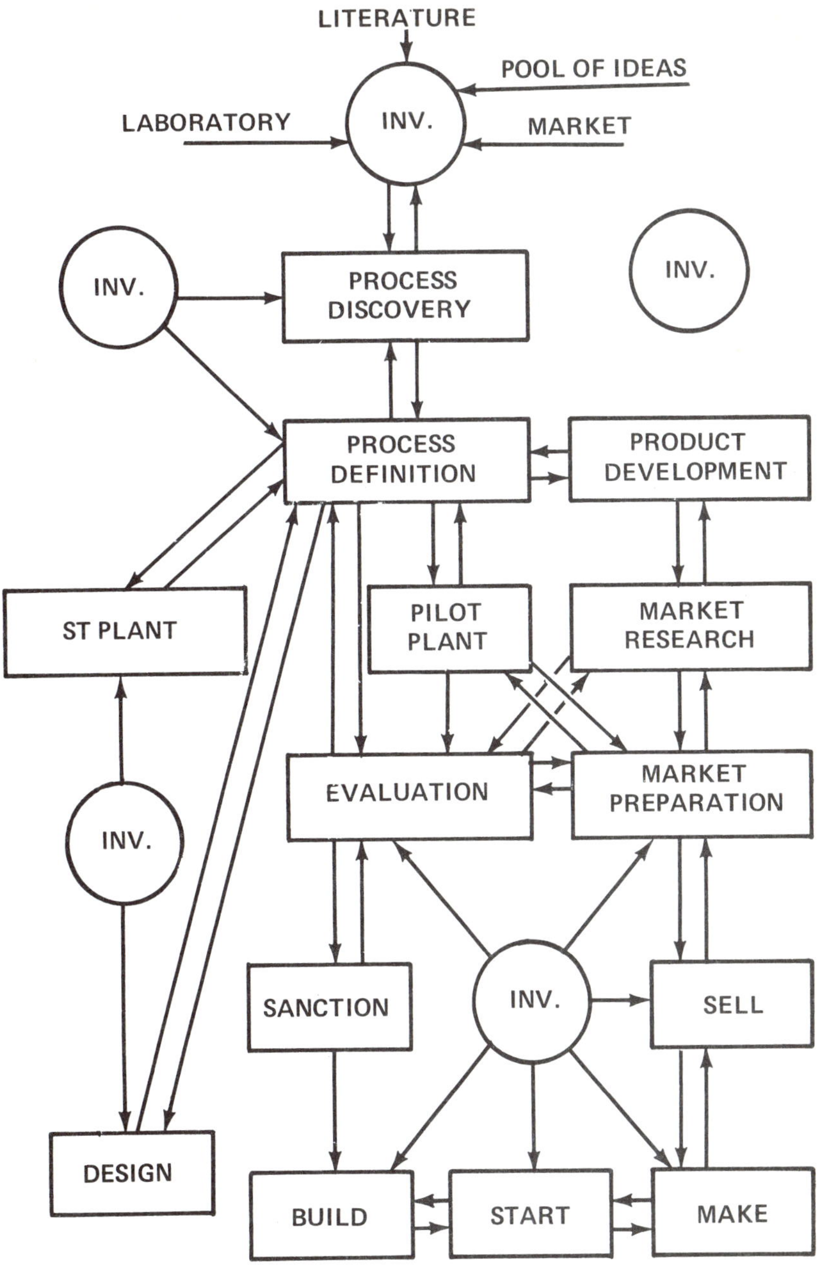

FIGURE 2 INNOVATION PROCESS SHOWING ROLE OF INVENTIONS.

the company (I.C.I.). In the search process by which the products were invented, in the search for viable manufacturing routes for the two compounds, in the definition of processes, scale-up, design and manufacture, and in the market research the main effort at any one time was directed by the division or other part of the company best equipped for that phase of the task. We wrote: "each part of the organisation was asked to make decisions in a business area in which it was already expert. If, following Ansoff (4), diversification is defined as the combining of a new product and a new mission - a new mission being a customer need not previously satisfied by the firm - bipyridyls did not involve any individual group in I.C.I. in diversification. Because of this, each group was able to place the decisions required in diquat's and paraquat's development in the context of previous decisions. Neither Organics nor Mond Division - the manufacturers - had to take decisions about the merits of rival herbicides, but were concerned with assessing investments in chemical processes, a decision familiar to each. Similarly, Plant Protection - the marketers - was able to relate its decisions on the level of applications research to experience previously gained with herbicides. For all groups, familiarity with the type of decision that was required informed judgement. A second reply, again connected with the industrial organisation, is to be found in the spreading risk among the various parts of I.C.I. involved. There were two consequences of this. The large research investment in devising manufacturing routes was shared by Organics and Mond, so that the burden was not carried by Plant Protection alone. And each group, although responsible for its own profits, felt that the risk that it individually was incurring was commensurate with the benefits offered. The division of the overall objective between the different groups within I.C.I. was effective in that each contributed an expertise and on a scale, appropriate to its resources.

The difficulties of managing innovation within large organisations are often publicised. It is a useful counterweight to consider an example where a large organisation by effectively using its component strengths and skills, can achieve a complex innovation probably too large, too risky and too difficult to be completed by a smaller company." There is strength in tension between subsystems.

The positive contribution of divisions and departments to the innovation process cannot be realised, of course, if the barriers are impermeable. How is the desired permeability to be achieved? The matrix systems of organisation can greatly assist this process (and it may be noted that any move from hierarchical structures to matrix structures does not diminish but increases the number of sub-units and barriers), as can the various forms of project team and project management. These organistic organisation forms (to use Tom Burns' terminology) (8) are effective in so far as they promote the movement of people across boundaries. The very quotable tag from Burns' contribution to the 1966 M.I.T. Conference on transfer of technology (9) that 'transfer is by agents not agencies' is an important piece of wisdom that would-be couplers should observe. I have direct experience of the benefits of this movement of people during the emergence of innovation. The barriers between process defining in R & D, and scale up in experimental plants, and design in an engineering department, and production in a works, to say nothing of the barriers between all of these and marketing and selling organisations, are a severe constraint on innovation when they are departmentalised and geographically separated, even by small distances. We found in I.C.I. that an effective way to overcome these departmental constraints was by the movement of key individuals across the fences. Whilst the concentrations of expertise in departments was a valuable asset for invention and innovation effectiveness and whilst the leadership in the various phases could pass with advantage from one department to another as the innovation entered its various phases, it was necessary to identify in advance key men for subsequent stages and to involve them in the planning and execution of preceding stages.

Insularity of objectives and suboptimisation, with concomitant inefficient use of resources, must result if means are not found to make barriers permeable. My experience of scale up experiments in intermediate size plants in the chemical industry has led me to identify a number of heresies. These are described elsewhere (10). They include the follcwing: (i)that the object of the scale-up is a well functioning miniature plant; (ii) that successful operation at semi-scale predicates successful operation at full scale; (iii) that substantial cost savings may be made by economies in semi-scale operations; (iv) that mistakes can be rectified later; (v) that emotional attitudes are of no account in fault diagnosis ... and so on. All of these can be traced to departmental boundaries acting as blinkers which prevent the custodians of this phase of innovation (in this case the R & D department) from visualising the total objective. The result is optimising on inadequate departmental criteria.

The serious interference of impermeable barriers with coupling is greatly reinforced if the barriers are not interdepartmental ones but inter-institutional ones. One of our research projects has been to study the transfer of technology from a Government Research Establishment to industrial establishments. We have found, by joining with G.R.E. teams attempting to 'sell' new technology to industrial firms, frequent failure to achieve the dialogue necessary to effect the transfer. A recent enquiry into the functioning of British Research Associations, the Bessborough Report (11), describes the transfer of information from the Associations to their member firms as a crucial aspect of the work of the Associations. In urging more effective communications across the institutional/firm barriers the report stresses the value of word-of-mouth transfer whilst recognising its relatively high cost.

Earlier it was commented that 'coupling' is not a good description of the interaction of innovation and invention. It is not helpful to describe this interaction in a way suggestive of a docking operation between invention modules and the innovation ship, the linking of two hitherto disconnected entities. There are in fact numerous and repeated probings of the barriers and, in the absence of these, transfer of technology from stage to stage or from user to user, is unlikely to be achieved. In the case study referred to above, on transfer of technology from a G.R.E. (the inventors and early developers) to potential users (who would be expected to carry the innovation forward to first commercial exploitation) it was never a simple coupling; the usual pattern encountered in the discourse between inventors and possible developers was one of mild disbelief in the claims concerning the benefits that would follow the adoption, leading to suggestions for further work on the project by the G.R.E. before any adoption by the firm concerned would be contemplated further. If the G.R.E. subsequently found its technologists unwilling or lacking the facilities for such further work the attempted coupling failed. In other words, the interfaces must be dynamic. Again one is reminded of Burns' contribution to the 1966 symposium (9) in which he evoked a picture of technology transfer through the coming together of groups into 'postures of consensual alliance' having a transient dynamic character.

Moreover, if we accept the Schmooklerian view (12), which I am prepared to do, that demand - if not necessity - is at least the midwife of invention, then the coupling is likely to be influenced by a strong pull from existing technologies and earlier innovations to invention, rather than by a push in the opposite direction. Some of the most effective inventors I know are well aware of this and deliberately expose themselves to the influence of existing technologies to help match their embryo ideas to innovatory opportunities in the existing systems of production.

The transfer from invention to innovation and through the various phases of the innovation process is indeed better characterised by the repeated presentation of ideas and designs, resulting in rejection more often than acceptance. That is, the transfer is a series of attempts to make a fit of one phase to another by trial and error, by a process of screening; many starters and few finishers is the characteristic not only of innovation of new 'effects', as is encountered in the chemical and pharmaceutical industries, but in engineering too. Milner has recently shown (13) that the progress of an innovation in shipbuilding is marked by rejection of misfits at every stage, at a rate decreasing as the project proceeds.

If coupling turns out to be a search for fit, it is appropriate to consider the operation in the terminology of screens. It is therefore germane to look at the so-called coupling process in relation to the parameters of screening recorded by Baines Bradbury & Suckling (6). That analysis was written with the early stages of the invention/innovation process as its substrate, but one may, if my diagnosis of the coupling process is sound, analyse all transfer phases of the innovation process as a series of screening operations.

We recognised six important parameters of screening. They fall into three pairs and are: target definition and width of spectrum; mesh size and quantitative content; abstraction and sufficiency. Each pair of parameters demarks the limits of a range of possible screening patterns. Tight target definition in which the search is sharply focussed on a perceived opportunity is at the opposite pole to a very wide spectrum screen which will accept a range of possible effects for further attention. Large mesh size will reject few candidates from the selection process whilst high quantitative content may make passage through the screen dependent upon a close fit to defined characteristics. High abstraction in screening designates the use of proxy criteria relying heavily on engineering and science theory, but which may thereby lack in sufficiency, a term used to denote isomorphism or goodness of fit to the real world complexities of the system into which the successful candidate is to be injected.

All of these parameters are clearly relevant to the process of coupling or internal transfers within the innovation process, and it may not be necessary to comment further on the first two pairs, target definition, width of spectrum, mesh size and quantitative content; sufficiency and abstraction may bear a closer look, however. In general the invention/innovation coupling, the subject of this lecture, is low on sufficiency as well as being light on theoretical guide lines, leading to a high rejection rate (fine mesh size) in the coupling attempts. It is a territory characterised by the sampling and observational problems of the early search studies of operational researchers. Viewed as a screening process the whole innovation procedure is an exercise in stepwise approach to the degree of sufficiency which will be demanded by those who eventually buy the innovation or its products; and the role of abstraction, the use of theory, or maybe simple heuristics, is to reduce the consumption of resources consumed in the progress of the new product or process towards an acceptable level of sufficiency. This level will be determined by the expected returns from the venture. Reduction in resource consumption is an outcome of abstractions in screening in so far as many of the rejections can be made on the basis of response to inexpensive tests which range from rejection on theory without any experimentation to the use of simple proxy variables such as physical and chemical properties readily determined by laboratory measurements. An extreme case of this is in the screening of chemicals as inhalant anaesthetics which can be identified with a fair degree of sufficiency by gas-liquid chromatography data (14).

There is in this matter of abstraction and sufficiency an important time trend to be observed. If, as now, one is fitting an innovation into a tight set of technological and social constraints there may be great difficulty in achieving the needed degree of sufficiency in screening and testing. Tighter specifications which include social costs and values are not easily abstracted and this sharply puts up the costs of innovation.

In characterising innovation progress in terms of screening there is danger of undervaluing the contribution of technologists in assisting candidates through the screen meshes. Abstraction or theory will make these meshes more or less precise in shape; the technologist faced by a rejection of a candidate by the screen may adopt one or both of two tactics, the first being to alter the shape of the candidate and the second to alter the shape of the mesh, the outcome being a pass to a following stage. The second tactic, adjusting theory or heuristic to accept something it would normally reject is not the criminal offence against the canons of science that it might appear. The best of theory is a comparatively crude abstraction from the complexity of the real world, heuristics much cruder. Rejection by such proxy criteria may well be erroneous therefore and challengeable. However, there is risk involved in changing the rules or in moving the goal posts during the game and role of the product champion who adopts these tactics is therefore a hazardous one.

To sum up this section, effective progressing of innovation at acceptable cost requires couplings or transfers to be guided by screening operations which combine abstraction with high sufficiency. This implies that those managing the coupling operations must look forward along the various possible pathways of the innovation (an operation which may be greatly assisted by the discreet use of decision trees) and devise acceptance/rejection procedures which realistically mimic the barriers that will be encountered ahead. This brings us full circle to the transfer of technology by agents, the movement of people up the innovation chain to positions nearer the origin than those they will subsequently man; in this way many errors of abstraction may be avoided and the cost of others reduced in so far as they are detected in advance of heavy commitment of time and effort to them.

I have so far avoided discussing the diffusion of technology although this is inextricably bound up with the innovation process. However it is important to consider the diffusion operation during the innovation period, since this too should affect the design of screening procedures. The acceptance of a breakpoint between innovation and diffusion will lead to errors in the conduct of the innovation process. I have referred to the need for anticipation of barriers by using them as a template for the construction of screens; barriers to diffusion of the innovation should also be anticipated by the screens.

Two cases studied by Bradbury, McCarthy and Suckling are relevant, halothane and paraquat (7). The first is an anaesthetic and the second a weedkiller. In the case of halothane one barrier to diffusion, that of compatibility of the new anaesthetic with existing anaesthetic administering equipment was successfully anticipated by the devising of a machine giving tighter control of dosage and allowing recycle of the expensive anaesthetic. A second one, difficulties of communication with the community of practising anaesthetics, was not adequately anticipated with the result that diffusion of the new anaesthetic $CF_3CHClBr$ was delayed by those potential adopters who assumed, wrongly, that it carried the same set of undesirable toxic properties as chloroform, $CHCl_3$. Again in the case of the weedkiller there was an equipment barrier, the absence of suitable agricultural machinery for its application, which was successfully anticipated.

It will be apparent from the terminology used in the discussion of innovation diffusion that I am referring to the paradigm of Rogers (15). His characterisation of the diffusion process, and especially his recognition in an innovation of the key characteristics for diffusion, are important. They are relative advantage, complexity, communicability, compatibility and devisibility. In the devising of screens to effect the couplings of innovation development each of Rogers' characteristics requires specific attention. It may require substantial amounts of time and effort to prepare the innovation for its subsequent diffusion to users but the pay-off for so doing may be great. The rate of sales growth may be directly affected by slow diffusion and delays in take off of product from newly completed plant, occurring as they will at the nadir of cash flow of the investment, will be paid for dearly.

My final comments on the invention/innovation coupling problem are these. Having redefined coupling as screening for fit, and having extended the concept to interphase transfers in the innovation process, I have made frequent reference to the economic use of resources in these screening operations. As in the extraction of work from chemical reaction systems (16), the most efficient forward movement of the emerging innovation is likely to occur at near-equilibrium conditions, avoiding large stepwise changes or transfers. I refer to the concept of balanced attack. This may be described in the following terms.

The innovation development is characterised by the attack on many centres of doubt and potential error residing in the total system, all demanding to a greater or lesser degree, R & D effort to reduce them. The most efficient attack on innovation is one which progresses this reduction of uncertainty in a balanced way. This means that no centre or centres of doubt lag behind the others in their potential error costs. This contention is based on the view that work which runs ahead of substantial resolved uncertainties elsewhere in the total programme is at risk of being made redundant by later developments involving a change of direction. It is useful to look at this in terms of opportunity costs. Opportunity costs of scientists in innovation can vary from very low (when there is not enough work for them to do) to very high (when they are not used to the best advantage). Opportunity costs of scientists therefore can vary with the workload of the department and the relative importance of different aspects of the innovation process. As the programme progresses the relative importance of different facets will change. The raising of raw material efficiencies in the process from 80 to 90% may be relatively unimportant whereas, at an earlier stage the lift from 40 to 80% may have been critical for the survival of the project. If the increase to 90% is achieved at the cost of neglecting serious effluent problems, for example, the consequences may be serious. It may turn out, for example, that the problem of effluent from the process under study is intractible and make it necessary to switch to another manufacturing route. This will make nonsense of the premature attempt to optimise yield in the original candidate process. A more extreme case is the overrunning of the whole technical side of innovation in relation to the development of the potential market. The outcome of this may be that people are working vigorously at a process for making a product which no-one will buy. These examples underline the importance of opportunity costs and balanced attack in the innovation process. In order to achieve this balance it is necessary to keep the total system under review by screening progress on all fronts in order to give early warning of neglect of critical aspects.

In making these comments on the desirability of maintaining a balanced attack in an R & D programme I am not unaware of the doctrine of the unbalanced attack and its virtues as argued by Hirschman and Lindblom in 1962 (17). Their view, which is that there are self-correcting mechanisms for one-sided surges in R & D problem solving and that such a progression towards the goal is more rapid, even though less efficient, than proceeding in a balanced way, does not conflict with the views I have advanced here. Hirschman and Lindblom concede that the unilateral tacks must not be allowed to go on for too long or progress will be lost.

The Hirschman and Lindblom statement is an arresting and engaging form of the cyclic view of progress in problem solving postulated by my colleague, Professor B.J. Loasby and others, a theme which is reflected in Herbert Simon's likening of the progression towards goals to the zig-zag tacking of the movement of ants! (18).

I would like to give the last word on the coupling of invention to innovation to Schumpeter (1): "Successful innovation is..........a feat not of the intellect but of the will. It is a special case of the social phenomenon of leadership."

REFERENCES

1. SCHUMPETER J. "The Instability of Capitalism" reprinted in "The Economics of Technical Change", edited by Nathan Rosenberg, Penguin Economic Readings, London, 1971.

2. JEWKES, SAWERS & STILLERMAN: "The Sources of Invention", MacMillan, London, 1969.

3. SCHON D. "Technology and Change", Pergamon, London 1967.

4. ANSOFF H.I. "Corporate Strategy", McGraw-Hill, New York, 1965.

5. BRADBURY, GALLAGHER & SUCKLING: "R & D Management" 1973, Vol. 3, No. 2, p.49.

6. BAINES, BRADBURY & SUCKLING: "Research in the Chemical Industry",Elsevier, London, 1969.

7. BRADBURY, McCARTHY & SUCKLING: "Chemistry & Industry", 1972, p. 195.

8. BURNS T. & STALKER G: "The Management of Innovation", Tavistock Publications, London, 1961.

9. BURNS T: "Models, Images and Myths" in"Factors in the Transfer of Technology", Gruber & Marquis M.I.T. Press, 1969.

10. BRADBURY F.R: "Chemical Technology" (in the press).

11. BESSBOROUGH: "Industrial Research and Development", White Crescent Press, Luton, 1973.

12. SCHMOOKLER: "Economic Sources of Inventive Activity" reprinted in "The Economics of Technical Change", edited by Nathan Rosenberg, Penguin Economic Readings, London, 1971.

13. MILNER C.G: "R & D Management", 1971, Vol. 2, p. 41.

14. BRADBURY & SUCKLING: (unpublished work).

15. ROGERS & SHOEMAKER: "Communication of Innovations", MacMillan, London, 1971.

16. VON BERTALANNFY: "Thermodynamics of Open Systems" reprinted in "Systems Thinking" edited by Emery F.E., Penguin Books, London, 1969.

17. HIRSCHMAN & LINDBLOM: "Economic Development, R & D Policy Making" ibid. p. 351.

18. SIMON: "The Sciences of the Artificial", M.I.T. Press, London, 1970.

FROM INVENTION TO THE MARKETPLACE - SUCCESSFULLY!

Dr. Roy Rothwell

Research Fellow in Industrial Innovation,
Science Policy Research Unit,
University of Sussex.

INTRODUCTION

It is becoming increasingly accepted by economists that technical change plays a major role in causing economic growth. But technical change, of itself, is not sufficient to guarantee such growth: economic growth derives from the commercialization of new technical know-how, i.e. through innovation. Innovation, and in particular successful innovation is, therefore, of prime importance to economic progress, both at the national level and at the level of the individual industrial organisation.

The title above, "From Invention to the Marketplace - Successfully", both defines the limits to the innovation process and indicates the theme of this paper, which is a discussion of the conditions for success in industrial innovation. The innovation process consists of a series of functionally separate, but interdependent phases, and Figure 1 gives a simple schematic diagram of the process. This shows the innovation sub-process and also some of the intra-firm and extra-firm interactions which combine them to form the overall process.

The various links represent communication channels, or at least demonstrate where communication channels ought to exist, and the successful conclusion of innovation will, presumably, be to a great extent dependent upon the establishment and maintenance of good communications throughout. This paper does not focus specifically on the role which communications and technology transfer play in successful innovation (this is achieved in a later paper), but rather covers the broader spectrum of activities associated with innovation. Emphasis is, however,

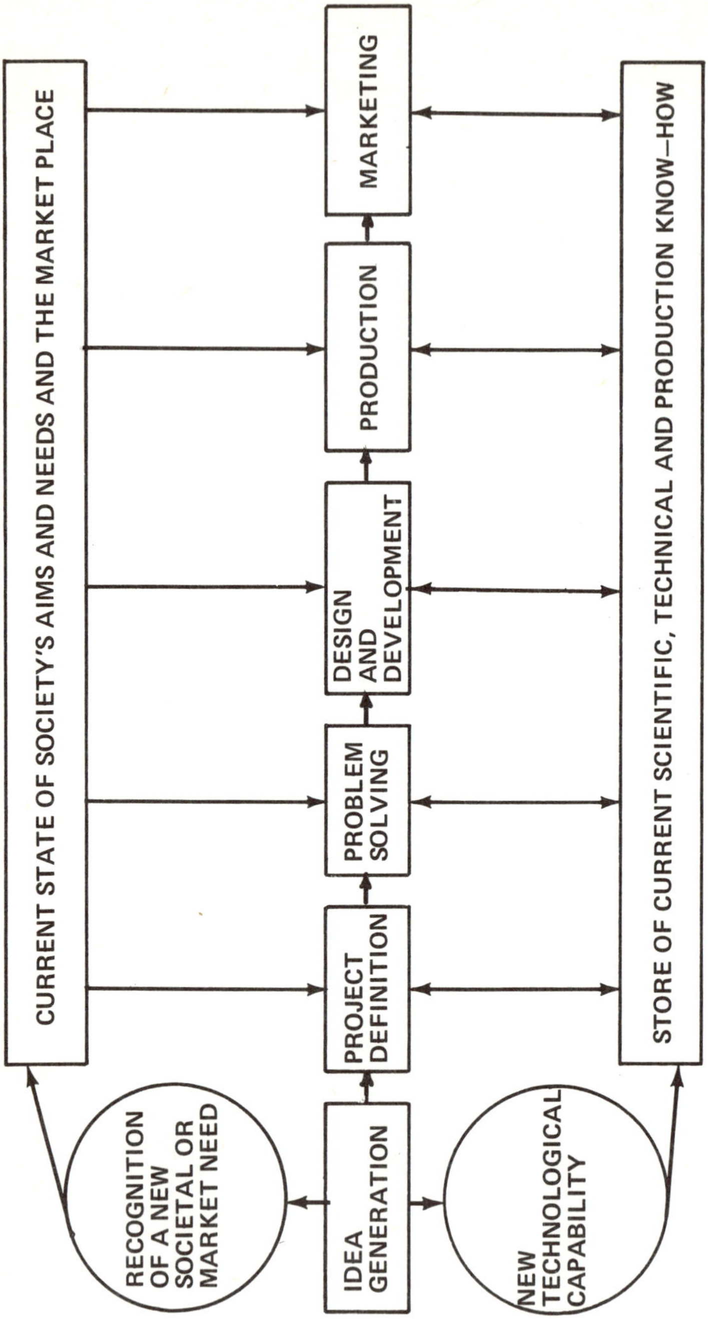

FIGURE 1 MODEL FOR THE INNOVATION PROCESS

placed on discussing the significance of penetrating the interface which exists between the innovator and his market. The definition of innovation adopted here corresponds to that used by the Central Advisory Council on Science and Technology, which is as follows:

> "The technical, industrial and commercial steps which lead to the marketing of new manufactured products and to the commercial use of new technical processes and equipment." (1)

Considering the enormous sums expended on innovating, it is surprising how little effort has gone into studying the innovation process, how few researchers have investigated and analysed past attempts at innovation to discover what lessons might thus be learned. Of the studies which have been made, the great majority have related to the experiences of a single individual or organization, and have consisted of purely descriptive case studies. Consequently any conclusions reached for these particular innovations are not suitable for generalization. During the last few decades, however, a number of studies have been undertaken which have been both comprehensive and analytical, among the most notable being those of Carter & Williams (2), Myers and Marquis (3), Burns and Stalker (4) and Langrish et al.(5). They have, though, all been concerned primarily with successful innovation, making little or no reference to failure. Project SAPPHO*, on the other hand, considers failures equally with successes and lays emphasis on seeking the differences between the two: this paper describes project SAPPHO, its aims, methodology, results and conclusions and discusses the significance of these results to the problem of overcoming the innovator/customer interfacial barrier. The results of some other workers in this area are listed briefly for comparison.

PROJECT SAPPHO - METHODOLOGY

Project SAPPHO was designed as a systematic attempt to discover differences between successful and unsuccessful innovations. The technique used was one of paired comparison, where a successful innovation was compared with an unsuccessful innovation, any differences between the two being noted. The criterion for the formation of a pair was commercial, being that the two innovations competed for the same market. The criterion for success was also commercial, a success being defined as an innovation which obtained a worthwhile profit and/or market share, and a failure being defined as an innovation which failed

* Scientific Activity Predictor from Patterns with Heuristic Origins.

to achieve this. Comparison was made using a large number (122) of 'measures', which between them provide the essential information concerning the innovation: the innovating organisation, its R & D, marketing, management, environment and production, etc. Comparison between measures is made purely on a greater than, equal to, or less than basis. The innovation attempts all occurred during the post-Second World War period, about one-third between 1945 and 1960, and two-thirds since 1960. The gap between launching the success and failure onto the market was normally only a few years or less. If the time between launches was such that the success arose simply because of extraneous scientific or technical developments, then the pair would be excluded from the analysis. One of the advantages inherent in this type of comparative analysis is that external influences are minimised.

The data collection technique was one of open-ended interview, and the investigator interviewed as many people as was practicable who were concerned with the innovation, often returning two or three times to interview a single individual. Where face to face interviewing was not possible, extensive effort was put into telephone interviews and written communication. In all cases vigorous attempts were made to cross-check the data.

It was recognised at the outset that generalizations which might be valid for one branch of manufacturing industry might not be valid for others, which have differing characteristics. Accordingly, two branches of industry were chosen as being suitable for study, the chemical <u>process</u> industry and the scientific instrument industry: in the latter sector, all the innovations were <u>products</u>. As expected, there are notable differences in the results for the two industries. The sample of innovations, which is listed in Table 1, is international, being taken from a total of eight countries.

In all, a total of 43 pairs have been investigated, 22 in chemical processes and 21 in scientific instruments. The results have been analysed using a variety of techniques: univariate analysis, principal components analysis and factor analysis, and ten 'index variables' were constructed, which are simple linear combinations of some of the 122 individual measures.

RESULTS

(i) Univariate Analysis

Depending on whether an individual measure, compared within a particular pair, is judged to weigh in favour of success (S > F)

Table 1

List of 43 pairs

Chemicals:

1. Accelerated Freeze-Drying of Food (Solid)
2. Acetic Acid Preparation
3. Acetylcne from Natural Gas
4. Acrylonitrile I
5. Acrylonitrile II
6. Acrylonitrile III
7. Acrylonitrile IV
8. Ammonia Synthesis
9. Caprolactam I
10. Caprolactam II
11. Caprolactam III
12. Caprolactam IV
13. Ductile Titanium
14. Extraction of Aromatics
15. Extraction of n. Paraffins
16. Glass Ceramics
17. Hydrogenation of Benzene to Cyclohexane
18. Methanol Synthesis
19. Oxidation of Cyclohexane
20. Phenol Synthesis
21. Steam Naphtha Reforming
22. Urea Manufacture

Scientific Instruments:

1. Amlec Eddy-Current Crack Detector
2. Atomic Absorption Spectrophotometer
3. Automatic clinical analysers I
4. Automatic clinical analysers II
5. Automatic clinical analysers III
6. Automatic clinical analysers IV
7. Digital Voltmeters
8. Electromagnetic Bloodflowmeter
9. Electronic Checkweighing I
10. Electronic Checkweighing II
11. Electronic Checkweighing III
12. Electronic Checkweighing IV
13. Foreign Bodies in Bottles Detector
14. Milk Analysers
15. Nuclear Magnetic Resonance Spectrometers
16. Nucleonic Thickness Gauges
17. Optical Character Recognition
18. Particle Counters
19. Roundness Measurement
20. Scanning Electron Microscope
21. X-Ray Microanalyser

in favour of failure ($S < F$), or neither ($S = F$), it is accorded a +1, -1, or 0 respectively. The zero, $S = F$ codings contain no information of use, and many of these codings arose simply because insufficient data was available on which to form the comparison. It is the distribution of +1s and -1s found to occur for a particular variable, taken from the total sample of pairs, which provides information concerning whether or not the variable differentiates in favour of success or failure. The statistical significance of a distribution is determined using the binomial test, which gives the probability of the observed pattern of +1s and -1s, occurring naturally. The smaller this probability, the greater is the statistical significance of the measure for success or failure.

The results of the binomial test are given in Tables 2, 3 and 4, (right-hand column), for the most significant variables, the measures being ranked in order of statistical significance (left-hand column). The first table, Table 2, gives the aggregate results of the combined data, and lists those variables significant to 1% or better. Tables 3 and 4 show the results for the chemical and instrument industries separately, and again lists only those variables significant to better than 1%.

Some inter-industry differences have emerged which, generally speaking, relate to the underlying environmental and structural differences between the two industries. For example, in the chemical industry, the power, status, responsibility, experience, commitment and length of service of the business innovator all differentiate in favour of success.* In the instrument industry, on the other hand, the experience, commitment and responsibility of the business innovator distinguish for success, but the measures of formal status or degree of power failed to distinguish success from failure. In the highly structured, hierarchical and bureaucratic environment of the chemical industry, the key individual requires to have high status and power, otherwise he is unable to perform effectively. In the often unstructured, small firm environment of the instrument industry however, with consequent absence of bureaucracy, the enthusiasm, experience and commitment of the individual entrepreneur emerge as important.

The individual variables listed in Table 2, 3 and 4 are not all, of course, unrelated, and they can be grouped together to define a number of 'areas of competence': since it is difficult to envisage the innovation process in terms of a large number of single variables, such a grouping gives a more realistic view of innovation. It is found that many of the statistically most significant variables can be grouped under five general headings,

* For definition of the business innovator, and other 'key individuals, see Appendix I.

Table 2

Aggregate Results for Both Industries

(a) Variables for which the probability of chance occurrence is less than 0.1%.

Ranking	Variable	S>F	S=F	S<F	Binomial Test *
1	Successful firms understand user needs better	33	10	-	1.2E-10
2	Successful innovations fewer after sales problems	1	13	31	7.7E-09
3	Successful firms employ larger sales efforts	22	21	-	2.4E-07
4	Successful innovations have fewer 'bugs' in production	1	17	25	4.0E-07
5	Successful firms have better coupling in specialised areas	23	19	1	1.5E-06
6	Successful firms pay more attention to educating users	23	19	1	1.5E-06
7	Successful innovations need less adaptation by users	0	28	15	3.1E-05
8	Successful innovations need fewer unexpected adjustments in production	2	20	21	3.3E-05
9	In successful firms there is less opposition to the innovation on commercial grounds	3	20	20	2.4E-04
10	Successful firms give more publicity to the innovation	17	24	2	3.6E-04
11	Successful firms benefit from outside technology during production	25	12	6	4.4E-04
12	Successful firms drop products as a result of the innovation	11	32	2	4.9E-04

* For convenience, 10^{-x} is written as E-x.

Table 2 continued

13	Successful firms seek the innovation more deliberately	18	22	3	7.4E-04
14	Successful firms have better external communications	18	22	3	7.4E-04
15	The executive in charge of the successful innovation has more power	20	19	4	7.7E-04

(b) Variables for which the probability of chance occurrence is less than 1%.

16	The executive in charge of success has more responsibility	18	20	4	1.3E-03
17	The innovation is made as a result of marketing decisions rather than production decisions	9	34	-	2.0E-03
18	The executive in charge of success has more diverse experience	20	18	5	2.0E-03
19	The executive in charge of success has more enthusiasm	14	27	2	2.1E-03
20	The executive in charge of success has higher status	18	21	4	2.2E-03
21	In success firms there is less opposition to the innovation on technical grounds	3	24	16	2.2E-03
22	Successful firms have better internal communications	13	28	2	3.7E-03
24	Successful innovations are more radical for world technology	16	23	4	5.9E-03

A further 17 variables emerged as significant at the 5% level, giving in all a total of 41 measures which are significant to better than 5%.

Table 3

Results for the Chemical Industry

Variables for which the probability of chance occurrence is less than 1%.

Ranking	Variable	S>F	S=F	S<F	Binomial Test
1	Successful firms understand user needs better	17	5	-	7.6E-06
2	Successful innovations have fewer aftersales problems	-	8	14	6.1E-05
3	The executive in charge of success has more responsibility	12	10	-	2.4E-04
4	Successful firms have better coupling in the specialised areas	11	11	-	4.9E-04
5	Successful firms drop processes as a result of the innovation	11	11	-	4.9E-04
6	Successful firms pay more attention to user education	10	12	-	9.8E-04
7	Successful innovations need less adaptation by users	-	13	9	2.0E-03
8	Successful innovations are less modified in the course of development	2	6	14	2.1E-03
9	Successful firms benefit from outside technology during production	14	6	2	2.1E-03
10	The executive in charge of successes has more power	11	10	1	3.2E-03
11	Successful innovations have fewer unexpected adjustments in production	1	10	11	3.2E-03
12	The executive in charge of successes has more diverse experience	10	11	1	5.9E-03

Table 3 continued

13	The executive in charge of successes has higher status	10	11	1	5.9E-03
14	Successful firms initially rate the prospects of technical lower	2	8	12	6.5E-03
15	Successful firms employ greater sales efforts	7	15	-	7.8E-03
16	Successful firms employ larger teams at the beginning of the project	15	3	4	9.6E-03

A further 9 variables emerged as significant to 5%.

Table 4

Results for the Instrument Industry

Variables for which the probability of chance occurrence is less than 1%.

Ranking	Variable	S>F	S=F	S<F	Binomial Test
1	Successful firms understand user needs better	16	5	-	1.5E-05
2	Successful innovations have fewer bugs in production	-	5	16	1.5E-05
3	Successful firms employ greater sales efforts	15	6	-	3.1E-05
4	Successful innovations have fewer after-sales problems	1	3	17	7.2E-05
5	Successful firms have better external communications	11	10	-	4.9E-04
6	Successful firms have a single technical innovator in the organisation	11	10	-	4.9E-04
7	Successful firms pay more attention to user education	13	7	1	9.2E-04
8	In successful firms there is less opposition to the innovation on technical grounds	-	11	10	9.7E-04
9	In successful firms there is less opposition to the innovation on commercial grounds	1	8	12	1.7E-03
10	Successful firms have better coupling in the specialised areas concerned	12	8	1	1.7E-03
11	In successful firms there is no parent involvement in financial matters	-	12	9	2.0E-03

Table 4 continued

12	Successful innovations have less unexpected adjustments after commercial sales	1	10	10	5.9E-03
13	Successful firms give more publicity to the innovation	10	10	1	5.9E-03
14	Successful innovations are aimed at markets with which the firm is familiar	2	7	12	6.5E-03
15	In successful firms there is no parent involvement with decision-making	-	14	7	7.8E-03
16	Successful firms have better internal communications	7	14	-	7.8E-03

A further 10 variables emerged as significant to 5%.

which are: (1) understanding user needs, (2) marketing effort, (3) efficiency of development, (4) communications, (5) characteristics of key individuals. Those factors which systematically emerge as distinguishing between successful and unsuccessful innovations can be described, using the aggregate data for both industries, by the following statements:

1. Successful innovators have a much fuller and more imaginative understanding of user needs:- successful innovations need less adaption by users and need fewer modifications as a result of user experience. Successful innovators see user problems earlier and understand user needs better, than do failures.

2. Successful innovators pay more attention to marketing and publicity:- successful firms employ greater sales efforts, devote more effort to educating users and give more publicity to the innovation, than do failures.

3. Successful innovators make more use of outside technology and scientific advice:- successful firms have better coupling with the scientific and technological community in the specialised areas concerned, benefit more from dependence on outside technology during production and have better internal and external communications, than do failures.

4. The responsible individuals in the successful attempts are usually more senior, and have greater authority than their counterparts who fail:- in successes the R & D chief has greater authority, and the business innovator has more responsibility, more diverse experience (including overseas experience), higher status and more enthusiasm for the innovation. Successful firms also contain someone who plays the role of 'product champion', as opposed to failures who do not.

5. Successful innovators perform their development work more efficiently than failures but not necessarily more quickly:- successful innovations have fewer bugs and need fewer unexpected adjustments in production, have fewer aftersales problems and are less modified during the course of production than failures, and they do not require modifications after commercial sales. Successful innovators benefit from outside technology during production, make more use of development engineers in planning and costing for production, do not order new tools or equipment for commercial production before the decision to launch full-scale production has been made, and do not underestimate production costs.

(ii) Index Variables Analysis

In order to further investigate those various 'areas of competence' mentioned previously, and others, ten 'index variables' were formed. These are simple linear combinations of some of the individual variables, each variable being used only once, and being accorded an equal weighting of +1 or -1, depending on whether it differentiates in favour of, or against success. The definitions of the ten index variables and the structure of variables Marketing and Userneeds are given in Appendix I.

The following table lists the ten index variables in rank order of importance for the aggregate data, and for chemicals and instruments separately. Ranking is achieved according to the number of points 'correctly classified', i.e. with a positive sign, and is denoted by the number in brackets.

Index Variable	% of pairs for which S > F		
	Aggregate	Chemicals	Instruments
Marketing	(1) 83.7	(2) 72.7	(1) 95.2
RDStrong	(2) 76.7	(1) 81.8	(4) 71.4
Userneeds	(3) 74.4	(2) 72.7	(2) 76.2
Communication	(4) 69.8	(5) 63.7	(2) 76.2
Management Strength	(5) 65.1	(4) 68.2	(5) 61.9
Familiarity	(6) 58.1	(8) 54.5	(5) 61.9
Techniques	(7) 55.8	(8) 54.5	(7) 57.1
Pressure	(8) 51.2	(7) 59.1	(9) 42.9
Organic	(9) 44.2	(10) 31.8	(7) 57.1
Risk	(10) 39.5	(5) 63.6	(10) 19.0

Significance rankings of Index Variables

Perhaps the most notable feature of the above data is the difference in ranking of the index variable RISK, ranked fifth for the chemical process sector and tenth for scientific instruments. Successful instrument innovators operated in markets and with technologies with which they were more familiar, and tended to be second to market, whereas successful chemical process innovations were often first to market and tended to be more radical for world technology. The very large firms which dominate the chemical process industry will, of course, have more resources available to develop radical technology, while in the small firm, low risk capital atmosphere of the scientific instruments sector, the emphasis is directed towards the minimization of risk.

Other important differences in ranking between the two industries occur for the index variables RDSTRONG, COMMUNICATION and ORGANIC, and again these can be explained in terms of inter-industry differences. In the first case the relatively large scale of the chemical process projects demands a greater allocation of resources (larger R & D team size, continuous effort) and the hierarchical company structure requires R & D chiefs of higher seniority (often on the main board). Also unexpected production adjustments and after sales problems tend to be more catastrophic for chemical process innovations, simply because of their larger scale with consequent greater difficulty of introducing fundamental changes when completed. In the second case, chemical process firms, because of their considerable resources, have a relatively greater in-house capability, and consequently rely less on the import of technology than do the smaller instrument firms. In the third case, the same differences in scale can be used to explain the relatively greater importance of flexibility and adaptability in the instrument sector.

(iii) Multivariate Analysis

In section i, the measures which consistently differentiated between success and failure were grouped under five general headings, i.e. understanding user needs, marketing and sales performance, efficiency of development, communications and characteristics of management. In section ii, these five areas were further investigated using, among others, the index variables USERNEEDS, MARKETING, RDSTRONG, COMMCATN and MGSTRONG, and again they emerged as describing much of the difference between success and failure. But both of these procedures, while perfectly valid, are highly subjective, and in order to take a more rigorous and objective look at the data, two sophisticated analytical techniques were employed, namely factor analysis and principal components analysis.

Using both of these techniques, the underlying areas of competence contained within the individual variables data were investigated. The earlier work was confirmed, in that the following five fundamental areas of difference between successful and unsuccessful innovations emerged from the multivariate analysis:

Marketing performance
Understanding user needs
R & D efficiency of development
Communications
Strength of management and characteristics of managers.

Furthermore, the analysis showed that strong correlations exist between the scores on these five factors, and that on average the successful innovator out-performed the unsuccessful innovator in all five areas of competence. This is an important result since it shows that successful innovation cannot, generally speaking, be explained in terms of a single factor, but that innovators should look to the complete range of activities if they would wish for a successful outcome to their endeavours.

QUALITATIVE REVIEW OF FAILURE CASES

The statistical analysis described above revealed a pattern of characteristics which distinguished between successful and unsuccessful innovations. It did not, however, weigh the relative importance of each characteristic of the success-failure pattern to a particular success or failure, nor did it provide an interpretation of some of the underlying reasons for the poor performance of the failures: nor does the statistical analysis reveal which variables may be amenable to deliberate policy manipulation by technically progressive management. Thus, for example, the analysis revealed that one characteristic which frequently differentiated between success and failure was the variable 'understanding user needs': what the analysis did not show is <u>why</u> the failure did not understand user needs.

Accordingly, a review was undertaken of 34 SAPPHO failure cases in an attempt to say, in the subjective opinion of the researcher, why the failure performed in such a manner as to result in failure, and to identify those factors which are amenable to deliberate management manipulation.

The results were analysed under the headings of the four index variables USERNEEDS, MARKETING, RDSTRONG and MANAGEMENT STRENGTH. The review concentrated exclusively on those factors which, in the subjective judgment of the researcher, dominated the failure, and omitted everything else.

1. USERNEEDS

The neglect of user needs or failure to interpret such needs in design.

(a)	Made no enquiries of users at all:	3 - instruments 1 - chemicals
(b)	Made enquiries, but of too few, or a-typical users:	4 - instruments 2 - chemicals
(c)	Made enquiries but ignored the answers:	2 - instruments
(d)	Mis-interpreted or mis-understood answers to enquiries:	2 - instruments
(e)	Committed to a pre-conceived design:	2 - instruments 4 - chemicals
(f)	Made user enquiries but no on-the-spot investigations:	3 - instrument

Total instruments: 16
Total chemicals : 7

Comment

Of the above, (a), (b), (c) and (f) are all clearly "manipulable", i.e. subject to policy control by good management practice. (d) may be regarded as errors of judgement arising from the uncertainty inherent in innovation. To some extent (e) may also be regarded in this light, although amenable to modification through better communications with the outside community, and better management intelligence. Although subject to management control (b) may present very serious problems of identification of potential users in the case of radically new products and processes.

2. MARKETING

(a)	Market research neglected or ignored:	8 - instruments 3 - chemicals
(b)	Neglect of publicity or under-investment in marketing effort:	6 - instruments 1 - chemicals
(c)	Failure to educate users:	4 - instruments 1 - chemicals
(d)	Unforeseen changes in the market (e.g. price falls, demand shifts, new legislation, low cost substitutes):	2 - instruments 5 - chemicals

Total instruments: 20
Total chemicals : 10

Comment

Of the above (a), (b) and (c) are all amenable to good management control, but (d) may be regarded as in part true market uncertainty, beyond the control of even the best management.

3. R & D STRENGTH

(a) Poor or incomplete development work: 9 - instruments; 5 - chemicals

(b) Over-dependence on outside technology, i.e. relative lack of in-house capacity: 2 - instruments; 1 - chemicals

(c) Insufficient resources for development work: 5 - instruments

(d) Unexpected superior competitive technology: 1 - instruments; 8 - chemicals

Total instruments: 17
Total chemicals : 13

<u>Comment</u>

Of the above, (a), (b) and (c) may be regarded as amenable to a large extent to management control. Although ill-luck plays its part in development, good management screening and resting techniques could at least prevent the launch of a thoroughly unsatisfactory product or process. In the case of (d) this may be regarded as largely beyond management manipulation, although good intelligence and communications might avoid some of these failures.

4. MANAGEMENT STRENGTH

Failure to assess or control the project satisfactorily during its development and launch, failure to establish effective internal and external communication networks, failure to integrate the project into overall company strategy.

(a) Project not taken seriously by top management or not integrated into company strategy: 8 - instruments; 2 - chemicals

(b) Inadequate project evaluation or control: 2 - instruments; 5 - chemicals

(c) Failure to communicate with critically important outside interests: 6 - instruments; 5 - chemicals

(d) Business innovator too junior, too weak, inexperienced or non-existent: 5 - instruments; 6 - chemicals

(e) Powerful, committed but mistaken product champion: 1 - instruments 2 - chemicals

Total instruments: 22
Total chemicals : 20

Comment

(a) and (d) are clearly susceptible to good management practice, but (e) may be regarded either as a matter of "luck", or ill-judgement on the part of a single individual.

It is clear that the areas covered by the index variables "Userneeds", "Marketing", and "R & D Strength", particularly where subject to management control, are more significant to failures in scientific instruments than to chemical process failures. In the latter case "ill-luck" seems to play a more significant role (2 (d), 3 (d)). "Management Strength" appears to play an equally important role in failure in both industries. The results clearly demonstrate that improvement of the quality of innovation management is clearly both possible and desirable, even though risk and uncertainty cannot be eradicated.

THE INNOVATOR/MARKET INTERFACE

Of all the various hypotheses which have been forwarded as affording an explanation for innovative success, the hypothesis finding most confirmation from the SAPPHO results is that concerning "strong marketing". The variables relating to the area of 'marketing' and 'user needs' are shown in Table 5: clearly, they support the contention that success attends those who pay close attention to the requirements of the marketplace.

Successful innovations tend to arise in response to a market need, and successful innovations pay more attention to marketing and publicity than do failures. In fact, it is this very positive attitude to the marketplace which strongly characterises successes. Failures, on the other hand, are content to sit back and wait, (generally in vain) for the world to beat a path to their door!

Another important characteristic of successful innovators is that they involve potential users with the innovation, often as early as the design and development stage, and thus discover user problems while the innovation is still being developed. This means that fewer modifications are required after the innovation has passed the prototype stage and, more important, that there are fewer after-sales problems. Successful innovators also adapt the innovation to suit the customer,

TABLE 5

ATTENTION TO MARKET EXPLANATION

	Both Industries			Chemicals			Instruments		
	S>F	S=F	S<F	S>F	S=F	S<F	S>F	S=F	S<F
Marketing decision	9	34	0	4	18	0	5	16	0
Innovation part of a general marketing policy	8	17	8	6	11	5	2	16	3
Opposition within firm on commercial grounds	3	20	20	2	12	8	1	8	12
Market forecast by sales organization	14	21	8	6	10	6	8	11	2
Market forecast by others in firm	7	30	6	4	15	3	3	15	3
Market forecast by outsiders	4	37	2	3	17	2	1	20	0
Market forecasting taken earlier	4	32	7	4	16	5	0	16	2
Attention to publicity	17	24	4	7	14	1	10	10	3
Sales effort	22	21	0	7	15	0	15	6	0
Re-structuring of marketing organization	12	24	7	4	14	4	8	10	3
Customers involved at development stage	16	19	8	6	11	5	10	8	3
Sophisticated technological problems for users	6	32	5	6	15	1	0	17	4
Steps to educate users	23	19	1	10	12	0	13	7	1
Users obliged to adapt	0	28	15	0	13	9	0	15	6
User problems became apparent later in developt.	3	27	13	1	16	5	2	11	8
Modifications after prototype	7	25	11	3	17	2	4	8	9
After sales problems	1	11	31	0	8	14	1	3	17
User needs more fully understood	33	10	0	17	5	0	16	5	0
Going into an unrelated market	9	17	17	7	10	5	2	7	12
First to market	17	8	18	14	5	6	7	3	11

rather than the reverse. The fact that the innovation presents sophisticated acceptance problems to users is not, of itself, necessarily a barrier to success: failure to educate users in the proper use of the innovation, and thus help them to overcome problems does, however, result very definitely in failure.

The single variable 'understanding user needs' differentiates most strongly, of all the individual variables, in favour of success, and in no case does it differentiate for failure. The importance of strong marketing and efficient customer liaison are further underlined by the high rankings of the index variables USERNEEDS and MARKETING, being ranked (3) and (1), (2) and (2), and (2) and (1) for the aggregate data, for chemicals, and for instruments respectively.

The smaller emphasis on marketing in the chemical process industry can be explained, to a great extent, by the fact that in chemicals the innovators will often be the first users of their own new processes and, since there are relatively few chemical process users, will know their markets better and generally have well established contacts with potential customers. The small firm environment of the instrument industry on the other hand, is more conducive to open market competition. However, when the chemical process innovations were made by contractors, their relationship with the chemical firms, particularly as regards to understanding 'user needs', was of crucial importance.

THE RESULTS OF OTHER WORKERS

The results of other researchers pertaining to the importance of efficient marketing to successful innovation are tabulated briefly below.

Carter and Williams[2]	75% of successful projects arise in response to a need. Vigorous and effective selling policy characterises technically progressive firms. Efficient aftersales service and user education are important. (200 manufacturing innovations studied in the U.K.).
Myers and Marquis[3]	75% of successful innovations arose in response to a recognised market potential or a recognised need in the production process. (567 innovations studied in the U.S.).
Burns and Stalker[4]	Firms which fared badly did not match their technical expansion with a

	comparable expansion in sales activities. Unsuccessful firms expected customers to act as a 'sink' for their innovations rather than as a 'source' of ideas for exploitation (about 20 electronics firms studied in the U.K.).
Langrish et al.(5)	"Clear identification of a need" was a factor of importance in a firm's success. "No market or need" a factor of importance in causing the delay of an innovation. (Studied 84 firms which won the Queen's Award to Industry).
Utterback(6)	75% of award winning instrument innovations arose in response to a recognised need. (30 instrument innovations studied in U.S.).
U.S. Materials Advisory Board (7)	In 90% of cases studied, the recognition of a need was an important factor (10 materials innovations studied in the U.S.).
Peplow(8)	35 out of 48 successful R & D projects arose in response to a perceived need. Most of the 36 failures lay with jobs started by R & D initiative. Extensive interaction with customers is important to success.
Baker et al.(9)	Studied 300 ideas arising in an R & D laboratory. Found that both a 'need-event' and a 'means-event' were connected with each idea. However, 85% of the "best" ideas and 75% of "good" ideas were accounted for by a need-event.
<u>On The Shelf</u>(10)	Studied 53 projects which were 'shelved' before market launch. Projects were shelved for unattractively small market, uncertainty with monopsonistic buyers, unattractive level of competition, lack of marketing capacity or expertise (10 reasons given in all).

It can be seen that there is a concensus of opinion concerning the importance of market awareness to innovative success, and that the above results strongly support those of project SAPPHO.

DISCUSSION

The results of project SAPPHO describe the differences between successful and unsuccessful innovations in terms of five fundamental areas of competence, namely: understanding user needs, marketing and sales performance, management structure and characteristics of management, communications and R & D efficiency of development. These same five factors emerged using a variety of analytical techniques.

In the specific area of market performance, both the comparative analysis study and the qualitative review clearly demonstrate that failure attends those who neglect the market-place: that close and continuing interaction with potential customers is essential throughout the course of the innovation and that successful innovators mount an effective and comprehensive marketing and sales campaign and train customers in the right use of their innovations. These results are strongly reinforced by the results of other workers.

Of the interfacial barriers which exist between the various functional phases of the innovation process (e.g. R & D/production interface, production/marketing interface) perhaps the most important, and the most difficult to surmount, is the barrier between the innovating organisation and its intended market. It would seem that the greatest strength of successful innovators lies in their ability to successfully overcome this barrier.

REFERENCES

(1) Technological Innovation in Britain, H.M.S.O., 1968.

(2) C.F. Carter & B.R. Williams, Industry & Technical Progress, O.U.P., 1958.

(3) S. Myers & D.G. Marquis, Successful Industrial Innovation, N.S.F., 69-17, 1969.

(4) Burns & Stalker, Management of Innovation, Tavistock Publications, 1961.

(5) Langrish et al., Wealth from Knowledge, Macmillan, 1972.

(6) J.M. Utterback, "The Process of Innovation - A Study of the Origination and Development of Ideas for New Scientific Instruments", I.E.E. Transactions on Engineering Management, Vol. EM-18, No.4, Nov. 1971.

(7) Materials Advisory Board, "Report on the Ad Hoc Committee on Principles of Research Engineering Interaction", Washington, D.C., Nat. Acad. Sci., National Res. Council, 1966.

(8) "Design Acceptance" in Gregory, S.A., Ed., The Design Method, Butterworth, 1960.

(9) N.R. Baker, J. Siegman and A.H. Rubenstein, "The effects of perceived means and needs on the generation of ideas for R & D", I.E.E. Trans. Eng. Management, Vol.EM-14, Dec. 1967, pp.156-163.

(10) C.S.I.I., "On The Shelf", A Survey of Industrial R & D Projects Abandoned for Non-Technical Reasons, C.S.I.I. 162 Regent Street, London, 1971.

APPENDIX I

Index Variables

RISK : A measure of the degree of risk taken by the innovating organisation.

TECHNQES : A measure of the extent to which management techniques explained the success/failure differences.

MGTSTRONG : A measure of the strength of the management of the innovating organization.

ORGANIC : An attempt to classify the structure of the innovating organisation as organic or mechanistic.

RDSTRONG : A measure of the performance of the development work concerned with the innovation.

COMMCATN : A measure of the effectiveness of the innovating organizations' communication network with the outside scientific and technical community.

PRESSURE : A measure of the competitive situation facing the innovating organization.

FAMILIARITY : A measure of the extent to which the innovating organization was familiar with technical problems posed by the innovation, and with the market.

USERNEED : A measure of the efficiency with which market research or other procedures have established the precise requirements of the customer.
= (-USERADPT + MKTFRCT + OTHRFRCT - WHENSEEN + MFCNPRTO + NDSUNDSD + USERINVD) ÷ 7

MARKTING : A measure of the marketing effort deployed by the innovating organization.
= (GENPOLCY + PUBLCITY + USEREDUC + SLESEFRT + TYPDECSN) ÷ 5

Mnemonics used in Index Variables USERNEED and MARKETING

USERADPT : Did the innovation have to be adapted by users?

MKTFRCT : Was any systematic forecasting by the marketing (or sales) department involved in the decision to add the innovation to product lines or to existing processes?

OTHRFRCT : Was any systematic forecasting by others in the firm involved in the decision to add the innovation to product lines or to existing processes?

WHENSEEN : If there were user problems, at what stage did they become apparent?

MFCNPRTO : Were any modifications introduced after the prototype stage as a result of user experience?

NDSUNDSD : Were user needs more fully understood by the innovators in one case than in the other?

USERINVD : Were potential users involved in development at any stage?

GENPOLCY : Was the innovation part of a general marketing policy?

PUBLCITY : How much attention was given to publicity and advertising?

USEREDUC : Were any steps taken to educate users?

SLESEFRT : Was the sales effort a major factor in the success or failure of the innovation?

TYPDECSN : In each case was it a marketing decision (the observation of a 'need') or a production decision (addition to product lines created by new technology)?

Definitions of 'key-individuals'

(i) 'Technical innovator'. The 'inventor' or individual who made the major contribution on the technical side to the development and/or design of the innovation. He would normally, but not necessarily, be a member of the innovating organization. He would sometimes, but not always, be the 'inventor' of the new product or process.

(ii) 'Business innovator'. The individual who was actually responsible within the management structure for the overall progress of this project. He might sometimes be the technical director or the research director. He might sometimes be the same man as the 'technical innovator'. He could be the sales director, or chief engineer. Occasionally, especially in smaller firms, he could be the chief executive for the organization as a whole.

(iii) 'Chief executive'. The individual who is formally the head of the executive structure of the innovating organization, usually but not necessarily with the job title of 'managing director'.

(iv) 'Product champion'. Any individual who made a decisive contribution to the innovation by actively and enthusiastically promoting its progress through critical stages.

CONDITIONS FOR SUCCESSFUL T.T. AND INNOVATION IN THE U.K.

P. Jervis

Research Fellow, Sussex University
Now at the AERE, Harwell

T.C. Sinclair

Senior Research Fellow,
University of Sussex,
Now Head of Innovation
Management, IIMT, Milan

The paper will outline some of the attitudes to, and expectations of, the process of technology transfer illustrated by comments from government and industry in the United Kingdom. Some of the research findings on technology transfer will then be briefly discussed, and used to illustrate an apparent disparity between those areas selected for policy formation and the factors which have been hypothesised to be of importance in successful transfer. Some published information about the industrial programme at the United Kingdom Atomic Energy Authority's Harwell Laboratory, which has now gained considerable experience in the reorientation of its work, will then be examined. Finally, it is hoped to show that some interesting facts about technology transfer can be elicited from an examination of some of the case studies carried out during the Science Policy Research Unit's study of innovation, Project SAPPHO. While the subset of data that can be used for this purpose is small, it indicates that a modified version of the SAPPHO methodology might be adopted for further studies of technology transfer.

Perhaps the most rigorous definition of technology transfer has been attempted by Brooks(1), and it will be adopted for the purposes of this discussion:

> 'Technology transfer is the process by which science and technology are diffused throughout human activity. Wherever systematic rational knowledge developed by one group or institution is embodied in a way of doing

> things by other institutions or groups we have technology transfer. This can be either transfer from more basic scientific knowledge into technology, or adaptation of an existing technology to a new use. Technology transfer differs from ordinary scientific information transfer in the fact that to be really transferred it must be embodied in an actual operation of some kind'.

Brooks extends this discussion to recognise two types of transfer, which he designates as 'vertical' and 'horizontal'. Vertical transfer is the process by which technology is transferred from the more general to the more specific, for instance by the incorporation of new scientific knowledge into technology and by the embodiment of a "state of the art" into a system. Horizontal transfer involves the adaptation of a technology from one application to another, as in the adaptation of a laboratory analytical instrument for on-line process control.

Policies towards technology transfer

In most countries governments are responsible for funding a significant part of total national expenditure on science and technology, and much of such work is carried out in the public sector, in universities, government laboratories, etc. It is frequently assumed or implied that much of the benefit to the nation of this work is not realised until its results have, in some way, been transferred into industry. This raises questions about national priorities, mechanisms for funding science and technology and, of course, about the process of technology transfer itself. However, at this level discussions of technology transfer are often inextricably involved with wider discussions of national science and technology policy.

Consider the example of the United Kingdom. In the Zuckerman report(2), it was stated that "Britain is responding vigorously, by a variety of measures, to the challenge of world technological and commercial competition". The measures itemised included

- special government support for critical industries such as computers and nuclear power
- encouragement, through the National Research Development Corporation (NRDC) and the Industrial Research Associations, of small firms which work in fields where technological innovation does not demand vast industrial resources
- the promotion of industrial mergers, particularly in fields where individual projects are large
- sharing aircraft projects with other countries

- the avoidance of large space projects and the consequent conservation of resources for other fields
- the search for ways to promote the development of industries on a continental scale throughout Europe.

After outlining the factors considered necessary for success in technological innovation and the increase of productivity, the Report proposed a number of actions. These included the suggestion that the many highly-trained and experienced research people working in government research establishments should be brought into closer relation to industry. Recognising that this still left the R & D man outside the control of "dynamic and commercially motivated management" the Report suggested that:

> "There is a growing need to consider other measures, such as the short or long-term transfer of specialist staffs to industry for particular projects, and even other resources, from government establishments".

The discussions in the Zuckerman report led to a number of recommendations, among which is clearly implied a commitment to the transfer of technology:

> "The encouragement of Government research establishments to act on a commercial basis as agencies for private manufacturing industry not only in the R & D phase, but also in the later stages of the innovative chain".

This is a typical example of many of the statements found in such reports, where technology transfer is seen as a desirable aim without however discussion of the factors involved in achieving it, or the conditions necessary for success.

Some two years after the publication of the controversial Zuckerman report the Ministry of Technology produced a discussion paper on the organization for industrial research and development in government laboratories.(3) In this the problems that occur when technology has to be transferred were implicitly recognised. It was pointed out that

> "If the firms capable of applying new inventions or processes develop them in their own laboratories, early exploitation is obviously more likely and the larger units which are a feature of those sectors of British industry which have recently been restructured are obviously better equipped to undertake the R & D they need and are more likely to appreciate its potentialities".

This consideration was not sufficient, however, to modify with fuller explanatory justification the recommendation that the Reactor and Research Groups of the Atomic Energy Authority and the Ministry of Technology Industrial Research Establishments (IREs) (plus the NRDC) be merged into a British Research and Development

Corporation which would have as one of its aims:

> "to encourage and support the development and application of innovation and technological improvement in industry for the benefit of the U.K. economy, and to carry out research and development for this purpose, both itself and in collaboration with industry and on repayment".

The Customer/Contractor principle

In 1971 Lord Rothschild, head of the Central Policy Review Staff, published his report on the new mechanisms for government funding of applied research, in which he advanced the customer/contractor principle. That is "the customer says what he wants; the contractor does it (if he can); and the customer pays".(4) While the Rothschild report was concerned with the organisation and management of government R & D, and concentrated on the mechanisms for funding and reviewing such work, the customer/contractor principle has, potentially, very significant implications for industrially oriented R & D, performed outside industry.

In the White Paper announcing the Government's intention to implement the customer/contractor principle(5) attention was similarly focussed on mechanisms and administrative structures. The setting up of a series of Requirement Boards to commission work was announced:

> "The third main block of research and development (costing about £26 million in 1972-73) includes the work of the Department's Industrial Research Establishments, the non-nuclear work carried out by the UKAEA extra-mural contracts which the Department places directly with industry and universities and the funds provided to the Industrial Research Associations. In future responsibility for commissioning much of this work will be channelled through a series of Requirements Boards, on which customer interests in the Department, in industry and elsewhere will be represented and, where appropriate, the contractors. The new Boards will cover a number of fields such as standards the metrology; ship and marine technology; mechanical engineering and machine tools; and chemical and mineral processes and plant".

The White Paper advanced the new organization for commissioning government applied research and development as a way of determining priorities for R & D programmes and also for deciding how the results of such programmes could be fully exploited to the benefit of the community as a whole. It concludes

> "The new framework provides a partnership within which

science will have more influence on the Government's central policy-making activities than before, and which will contribute more directly and more effectively to the task of making the best use of science and technology for the needs of the community as a whole".

Once again it seems clear that of the success of this new organization for applied R & D must remain dependent on effective transfer of technology, yet the words 'technology transfer' appear neither in the Rothschild Report nor the White Paper.

The role of the Industrial Research Establishments

In 1972 the Parliamentary Select Committee on Science and Technology examined the work of the IREs(6). This revealed some differences of approach to work for customers. The National Engineering Laboratory stated in its annual report that the resources of the Laboratory were "directly related to the 'real needs of industry' ". The Minister for Aerospace thought that this was an important target, though unsure of the meaning of such a generalization. The problem seemed never to have been closely analysed either by the IREs or by the government. The Director of the National Physical Laboratory said that work which was done under contract <u>for industry</u> was "merely spin-off" and that the Laboratory's main task area would be where there was a customer-contractor relationship with the Government. The Select Committee was led to comment that it seemed clear to them that the industrial research establishments and industry had a great deal to learn about each others' respective needs and potential services if the objective of transferring the technology to industry was to be achieved to the advantage of the country. The Select Committee stated that

> "we also recommend . . . that the Government should state on what basis they propose that the IREs should transfer their technology to industry".

The Secretary of State for Trade and Industry replied to this report(7) and his comments included a description of methods by which technology is transferred.

The brief outline might seem somewhat over-simplified. While the headings under which technology transfer is discussed include those which recognize something of its personal and esoteric nature, the overall impression given is of a mechanistic process. The statement is made that technology is transferred by

- publication and dissemination of information
- personal contact
- advisory and consultancy services

- licensing
- R & D contracts
- interchange of scientists and engineers

This brief review of some of the recent statements which have been made about government policies towards science and technology should serve to illustrate the point that many of the policies being employed or advocated are partly dependent for their success on an effective transfer of technology from research laboratories of one kind or another to industry. At the same time it is implicitly recognised that technology transfer is a difficult process, and there are strong advantages in avoiding the necessity for it, by carrying out R & D within industry. This view is advanced in statements made by the Confederation of British Industries. For example:

> "Integration of applied R & D . . . with an undertaking's other activities (launching of a development onto the market) is, therefore, the key to its success, and the relatively poor national economic performance in relation to the expenditure on R & D can most probably be ascribed to failures in this respect. Integration is easier in theory than in practice, but industrial experience shows that R & D intended for ultimate commercial exploitation should be done as close as possible, both physically and organisationally, to the point of ultimate application. The CBI believes this to be most important: R & D intended to help industry should be done within industry".[(8)]

To summarize the implication of all these comments, it appears that while exploitation is expected to be more rapid if R & D is carried out in industry this does not prevent recommendations to carry out R & D for industry in government laboratories. There is apparently a belief that it is preferable that the customer (industry) pays for such R & D. Further, the importance of individuals in the process is recognized in comments about their role in communications and the importance of mobility. However, the whole debate is dominated by thoughts about doing the technology, not about using it. There seems to be insufficient consideration of the mechanisms by which technology, at the micro-level, is transferred, and little attention to the ways in which the process can be made more efficient.

The Research View

There is, of course, much literature on technology transfer. A large part of this concentrates on the international transfer of technology, and particularly from industrialized to developing countries. But a significant part of the information concentrates on the transfer from research laboratories to industry. Douds[9] has commented that there are three categories of literature dealing with technology transfer. Much of it is discursive in nature, and indicates significant concern with the problem in industry and government. A smaller section consists of case studies which illustrate particular aspects of the process, but the number of empirically based field studies is very small. There has been little in the way of theory building accompanied by field tests of hypotheses.

One of the more recent studies of government involvement in technology transfer has been produced by Doctors[10]. In this he concentrates on examining the horizontal transfer of technology, across mission lines to areas outside the ambit of agencies and their industrial bases, in the United States. He suggests that possible barriers are:

- the mission orientation of most agency technical personnel
- the vertically integrated nature of agencies
- conflicting policies concerning legal rights to patentable innovations and other proprietory data among the agencies
- institutional barriers to information flow in the aerospace/ weapon-system industry
- the low rate of technological mobility from the aerospace/ weapon-system industry to the commercial sector
- the low value placed on the transfer function by scientific and technological personnel engaged in federally sponsored R&D
- the political nature of institutions for transfer
- overzealous security restrictions
- poor and antiquated methods of information retrieval and evaluation
- poor understanding of the transfer process
- the power structure of the agencies themselves

Doctors has suggested a number of important concepts for policy formation. He believes that since space/defence technology is more likely to be quality-oriented than cost-oriented, and nearer the "science" end of the science-technology-application spectrum, transfer should concentrate on organisations whose participants can easily understand the sophisticated technology. He suggests that the industry structure must also be such as to allow the introduction of the higher-cost technology. If these criteria are not met then more effort will be required by the donor to adapt the innovation for use by the recipient. In addition, since innovations originating in government funded information appear more likely to require more adaptation for application, transfer of this type is more likely to occur where the recipients' standards of performance are high, seed money is available for initial development, and exclusive rights to the invention are available to provide protection for the larger investment required.

Belief in the importance of mobility of technical personnel in the process of technology transfer, coupled with the key role of spin-off companies from parent organizations, led Doctors to suggest that both should be assisted. He also recognized the importance of the entrepreneurial role in the transfer of specific innovations:

> "The methods appropriate to the transfer of generalized areas of technology are not necessarily appropriate to the transfer of specific innovations and vice-versa. The former can be accomplished by a variety of interpersonal contacts and probably require longer periods of time; the latter appear to be more dependent on entrepreneurial activities and careful need-oriented preparation by the transferor".

Doctors examines in detail the NASA Technology Utilization programme in the U.S. and states that since it is attempting to transfer technology horizontally the "champion" must be found in the transferor organization. Just as industrial organizations must provide champions to promote the development of an innovation or new area of technology, so a Regional Dissemination Centre must provide a champion until one can be identified within the transferee organization.

Doctors concerns himself mainly with the mechanisms which have been established to promote the transfer of technology, but recognizes the important role that people play in individual cases. Strong support for this importance of people is found in many other studies, and has been expressed in some detail by Price, Schon and Shimshoni, among many others. Price argues that:

> "it is crucial that on the basis of this model both types of transfer (i.e. transfer that occurs purely within the structure of technology, creating new results on the basis of old, and that which occurs in a more complicated way as part of the weak interaction that controls the relationship between science and technology) must depend on the research front being contained in people and transferred by the movement of people".(11)

Schon has paid considerable attention to the "people" aspects of innovation and technology transfer. He indicates that the process of transfer can be identified with all the problems of technological innovation. He suggests that when people migrate from one culture or from one industrial sub-system to another, they carry with them the technologies and ways of operating that they learned in the old situation. These migrations are principle sources of technological change in our society, and the process of technology transfer falls within this framework. Schon states that

> "Certain processes of change - movement, migration, invasion, are already in process, and our real question becomes "How can we create an environment favourable to this sort of change?" In answering such a question, I believe that the movement of people, organizations and institutions - not information - is the issue. "Information", "transfer" and "documentation" are merely luggage."(12)

Already, from a brief review of policy statements at the government level and an even more selective examination of the comments of some of those who have investigated the transfer process there appears to be a discrepancy between the variables manipulated for policy reasons and the factors which seem to affect the success of attempts to transfer. Burns[13] emphasises this disparity in perhaps the clearest and most critical way. He points to the important clarifying notion that the mechanism of technological transfer is one of agents, not agencies, and of the movement of people among establishments, rather than of the routing of information through communications systems. Burns points out that the set of assumptions which govern the greater part of the NASA-sponsored programmes on technology utilization are less than satisfactory. The assumptions derive from the set of beliefs which prevail in Europe and America about the nature of what it is which is transferred and of the process of transfer. In essence they are that technology transfer is a matter of either the unplanned percolation, or the planned transmission, of ideas, technical routines, or information from research/(scientific and technological) to manufacturing industry and so into commercial exploitation and eventual use. The conclusion from the argument is vital to the understanding of technology transfer.

> "The point is that all the thinking which has gone into attempts to expedite and widen the range of communication processes is implicit with the notion of some real or reified object which is transferred from one set of individuals or institutions to a different set of individuals or institutions. The jargon of space-age technology itself contains in terms like "spin-off" and "fall-out" the basic assumption of technology as an assembage of pieces of information which can be extracted or expelled from one sector of organised creativity and transposed

> to another to produce different outputs. Such thinking derives from an image of society, of the pathways for the significant traffic of social organisation, and of the relationship between men, ideas and institutional forms, which belong to the past."

From a long and detailed development of this analysis Burns concludes:

> "First, although technological transfer is now accomplished by agents rather than agencies, institutional means have to be devised to facilitate, expedite, and multiply the movement through the institutional systems between which transfer has to take place. Second, the institutional means so devised will inevitably, in the long run, inhibit the self-same processes".

A Practitioner's View

So far we have considered the views of those who have been associated with the process of transfer largely at second-hand. In the United Kingdom one of the laboratories which has had most experience of attempting to transfer technology to industry is the Harwell laboratory of the United Kingdom Atomic Energy Authority's Research Group. Some comments made by Harwell's Director, Marshall, add an extra dimension to consideration of the transfer problem. He describes how Harwell tried to see if it could work so closely with industry that the laboratory's research was strongly market oriented, and that there was a close link between all the parts of an innovative chain. In view of Doctor's comments about exclusive rights to an invention it is interesting to note that Marshall has stated:

> "For each specific idea, if it involves substantial risk and a considerable research and development programme, we usually search for a single industrial partner with whom we collaborate on an exclusive basis. This has been called the application of 'maximum unfairness' and in a sense this description is true; but in practice we discover that industrial companies are quite happy about the logic of it because it follows from the realisation that it is the final market opportunities, and the problem of penetrating that market, that dictates what industrial research is worth attempting".(14)

Marshall also commented on a practice designed to make the process of industrial collaboration efficient.

> "the very simple idea, of placing the Harwell research programme directly in the control of our industrial partner, is another way of insuring that the research

> is properly oriented, that there is an efficient two-way flow of information, and that the NIH factor (not invented here) is minimized".

Harwell has decided to place the full responsibility for research projects on the individual scientists leading them. This is believed to be the correct approach because it means that each scientist is directly concerned with the market and final application of his work: this gets him thoroughly involved, committed, and enthusiastic to make his work a success.

In a later discussion Marshall revealed some interesting facts about Harwell's approach to licensing.

> "We have also learned that it is very difficult to license a patent at the "ideas stage" and we now think it is right to develop the idea further before we become involved with an industrial partner . we can supply the "products" to allow them to test the market without the company being faced with the management and technical problems of small-scale production. When we come to the licensing stage both parties are then in a position of having a fairly fully developed product which has been tested on the market and its commercial value more clearly identified."(15)

That an industrial programme at a laboratory like Harwell can involve more than one type of "product" is indicated by a more recent description of the Laboratory's work.(16) . This states that Harwell offers three different approaches for sale.

The first stems from its special facilities. For instance, it sells neutrons worth £200,000 a year by giving outside scientists access to its research reactors . . .. The second category is contract research specifically for a single customer, where the "customer/contractor" principle applies. The third category of research service is what it calls "exploitation", where Harwell itself participates in the financial risk of exploiting some product, process or technique.

These comments about the Harwell experience indicate that they have encountered the same problem with non-exclusive licenses that Doctors commented on. In addition, there has been a decision to continue the Laboratory's involvement with product development beyond the initial idea stage through to prototype production and test marketing. This is apparently in conflict with the evidence of some studies,(17) that successful firms transfer in ideas as early as possible in development. Finally, it appears that there are different types of transfer mechanism, including both "contract research" and "exploitation", a concept apparently not explicitly recognised in any of the technology transfer literature.

A new research approach

Some of the data gathered during Project SAPPHO, a study of success and failure in industrial innovation, can be used to examine the conditions which appear to lead to successful transfer of technology. The methodology adopted in Project SAPPHO involved collecting case histories of pairs of attempts at innovation where, within each pair, one member had achieved commercial success and the other had been commercially unsuccessful, and where both members of the pair had been aimed at the same market outlet. The results of the Project were obtained by identifying factors which consistently appeared to show a difference in performance between success and failure in the collection of pairs examined. The Project initially concentrated on two industry sectors, chemicals and scientific instruments.

The significance of Project SAPPHO for discussions of technology transfer lies in the data from the instrument industry. This is a sector in which many of the most significant innovations have originated in research laboratories before being transferred to industry. White[18] and Shimshoni[19] have commented on this process at length, and more recently Utterback[20] has studied some aspects of this subject. Of SAPPHO's twelve instrument pairs, four involved repeated attempts to licence the same device, and in three more pairs an in-house development was compared with devices brought in from research laboratories. (Of these three in-house developments two succeeded and one failed). Finally, one pair consisted of two firms which had attempted to import technology, but from different sources. Thus of the twenty-four individual attempts at innovation in the instrument industry documented during the first stage of SAPPHO, thirteen involved the transfer of technology. It is instructive to examine the events involved in some of these individual attempts at innovation.

Electron-optical instruments

Two types of instrument in the general field of electron-optics were examined during SAPPHO, the scanning electron microscope and the X-ray microanalyser. The early history of electron optics has been extensively documented, and work on both the scanning microscope and the microanalyser began quite early in the history of the technology. However, the commercial development of both these instruments came some time after that of the transmission electron microscope. Work on scanning microscopy had been carried out in Germany, America and France before it commenced in the United Kingdom.

The scanning electron microscope

The construction of a scanning electron microscope was started in 1948 as a Ph.D. project by D. McMullan, a student of C.W. Oatley's at the Engineering Department of Cambridge University. Oatley had been given the responsibility for reorienting the Department's research after the war, since previously it had concentrated purely on power engineering. Oatley had become interested in scanning microscopy after reading the book written by Zurovykin and his colleagues describing their work on electron optics. This book had become something of a "bible" for researchers, and contained a description of a prototype instrument built in America. Oatley also realised that by incorporating a new type of electron multiplier developed in the nearby Cavendish Laboratory some of the signal to noise problems of the earlier scanning microscopes might be solved.

McMullan built a prototype instrument and described it at public meetings in 1952. At the same time a second research student, K.C.A. Smith, began a project which, by 1956, had resulted in an improved instrument. At this time the research prototype was seen by a member of the Canadian Pulp and Paper Research Institute who was spending a year at the Cavendish Laboratory. The performance of the scanning microscope in examining fibres was so impressive that the Canadian Institute commissioned Smith to build them an instrument. Smith did this, basing the construction on a conventional transmission microscope sold by AEI, the only British company active in this technology. Oatley had very good contacts with the middle and senior management in AEI, a result of his war-time work on radar. Through these he arranged for the firm to act as agents in the sale and shipping of the instrument to Canada. The Canadian instrument was completed and shipped in 1958 and Smith went to Canada for two years to help with the research. He returned to Cambridge in 1960.

As early as 1956 Oatley had asked AEI to undertake the manufacture of the scanning microscope. Although they agreed to this, in nearly five years they sold only one instrument. This sale was to a colleague of Oatley's, Professor Bowden of the Cavendish Laboratory. The AEI instrument was delivered late, was technically unsatisfactory, and was eventually returned to the manufacturer. There seemed to be very little communication between the AEI development and production personnel and Oatley's group, which by this time included several postgraduate researchers and had accumulated considerable expertise in the area. In addition the market for transmission microscopes was extremely healthy, and AEI appeared not to give the development of scanning microscopy a very high priority.

Oatley was disappointed with AEI's slow progress and by 1960 there was another British company which had entered the electron-optical field. This was the local Cambridge company, Cambridge Scientific Instruments, and their appearance in the technology, a result of the development at the Cavendish Laboratory of an X-ray microanalyser, will be described later. In 1960 Oatley requested to AEI that they relinquish interest in scanning microscopy so that he could work with other firms. Once this had been agreed, Cambridge Scientific started the construction of a prototype scanning microscope, and recruited one of Oatley's Ph.D. students, A.G.D. Stewart, to assist in the work. The prototype was exhibited in conjunction with the Institute of Physics exhibition in January 1962.

Late in 1962 a representative of du Pont, who had heard about the scanning microscope through their Canadian company's contacts with the Pulp and Paper Research Institute, visited Cambridge, saw a second prototype and placed an order. Delivery to du Pont was made in the autumn of 1964, and the company committed itself to a production batch of five instruments. These were delivered to customers in the summer and autumn of 1965. Although at this time Cambridge Scientific Instruments did not envisage a large market for the scanning microscope they nevertheless launched an intensive publicity campaign which made much use of applications studies done on their prototypes and the series of research instruments which had been built at the Engineering Department. The success of this sales drive exceeded all the company's expectations, and within three years sales of scanning microscopes had risen to over one hundred per annum, representing perhaps 40% of total sales of the parent, the Cambridge Instrument Company.

Cambridge Scientific Instruments received Queen's Awards for technological innovation (1967) and export performance (1970) for the scanning electron microscope, and the instrument was also selected as one of the "Industrial Research" 100 most significant new products. The company had placed great emphasis on engineering the instrument for ease of use by non-specialist staff. Stewart made a major contribution to this work, and was in charge of development from the middle of 1963. His work was recognized by a New Scientist Award in 1968.

The X-ray microanalyser

When R. Castaing published the results of his work on X-ray microanalysis in Paris in 1951 the potential importance of the technique for metallurgists was quickly recognized. In Britain the British Iron and Steel Research Association's Metallurgy Division had an Electron Microscope Sub-committee which included Dr. V.E. Cosslett of the Cavendish Laboratory and a representative from AEI. Discussions in this committee of Castaing's work led to the parallel development of two British instruments. A team at AEI's Aldermaston Research Laboratory was advised and funded by a consortium of four metallurgical research associations. Independently, Cosslett decided to make the construction of a microanalyser the subject of a research project for one of his Ph.D. students, P. Duncumb. Duncumb started work in 1953 and by 1956 had built a prototype instrument which incorporated the technique of scanning the electron beam across the specimen, thus enabling a two-dimensional display of elemental distribution to be obtained. In doing this Duncumb obtained assistance from Smith who was working on the scanning microscope at the Engineering Department. Neither Castaing's original instrument nor the AEI version under development at the same time as Duncumb's used scanning, but relied on mechanical techniques for positioning the electron beam on different parts of the specimen.

When Duncumb had completed his prototype it was used to examine some specimens of steel for D.A. Melford, a research worker at Tube Investments Research Laboratories at Hinxton Hall, near Cambridge. Melford had heard of Duncumb's work through the Head of the Metal Physics Group at Hinxton Hall, who had previously worked at the Cavendish. Melford was so impressed with the potential of Duncumb's scanning microanalyser that he obtained resources to build an instrument at Hinxton Hall. Duncumb was used as a consultant during this exercise, and later joined Tube Investments' Research Laboratories. The instrument developed by Duncumb and Melford was designed with the particular requirement of metallurgists in mind, and was seen by a number of metallurgists at Hinxton Hall, where there was a high throughput of visiting scientists.

In 1958 the Cambridge Instrument Company appointed a new managing director, H.C. Pritchard. One of Pritchard's tasks was to rejuvenate the company's product range, which had remained virtually unchanged for a number of years. Pritchard wanted to introduce an advanced technology to the company, and knew Tube Investments' Research Director. From these contacts the Company was offered, and accepted, the chance to licence the Duncumb and Melford microanalyser. This instrument was put into production by Cambridge Scientific Instruments and came to market early in 1960, at almost the same time as the AEI-developed static probe microanalyser. The scanning microanalyser immediately attracted more enthusiasm from metallurgists than the non-scanning version, and sold more successfully. It was this entry into electron-optical technology that enabled Cambridge Scientific Instruments to undertake development of Oatley's scanning microscope.(21)

Cambridge Scientific Instruments were undoubtedly extremely successful in introducing the new technology of scanning microscopy and X-ray microanalysis to the market, but financially the parent company did not do so well. The development expenditure on a whole range of new instruments, together with the production problems of introducing sophisticated new products, meant that the Company's record was not impressive. For example, from 1955 to 1967 sales increased from £1.4m. to £5.3m., but the profit before taxation in 1967 was £0.54m., almost exactly the same as the 1955 figure of £0.53m. Return on net assets employed was 34.6% in 1955 and 15.7% in 1967. The Cambridge Instrument Company was taken over by the Kent Group in 1967.

Atomic absorption spectrometry

Although the basic science behind the principle of atomic absorption spectroscopy was discussed by Kirchhoff and Bunsen in 1860, practical application of the knowledge was not attempted till the 1950's. The Commonwealth Scientific and Industrial Research Organisation (CSIRO) in Australia has as one of its remits to investigate science which might have implications for Australian industry. Dr. A. Walsh of the Division of Industrial Chemistry in the CSIRO started work on atomic absorption in 1952. He was an Englishman who had worked for the British Non-Ferrous Metals Research Association before going to Australia. During his career he had designed two components which had been licenced to firms in the instrument industry, one to the British firm of Hilger & Watts and the other to Perkin-Elmer in the United States.

Walsh had demonstrated the basic usefulness of atomic absorption as an analytical tool by 1954 and CSIRO had patented the results of his work. Because of his earlier contacts with the firm, Walsh discussed his atomic absorption research with the Research Director of Hilger and Watts. As a result Hilgers licenced the technique from CSIRO in 1954. Walsh's first publication, in 1955, coincided with that of two researchers at the State University of Utrecht, C.T.J. Alkemade and J.M.W. Milatz. It was the work of these latter researchers that first gained attention in the United States, and which eventually prompted Perkin-Elmer to design an atomic absorption spectrometer to Alkemade's design.

Hilger and Watts decided that the best way to introduce atomic absorption to the market was to produce an attachment to one of their existing ultraviolet instruments. For a number of reasons this device did not give satisfactory results, and in addition Hilgers gave little sales support to the technique. Only a few atomic absorption attachments were sold and since Walsh's work was not well known in the United Kingdom, the poor performance of the device started to give the technique a bad name.

In Australia and New Zealand a number of researchers started using atomic absorption, and between 1955 and 1958 some significant papers were published. Walsh was disappointed in the slow diffusion of the technique, and started to build instruments for local use using three small firms to help with electronics, glassware and burners. This resulted in over 50 kits being sold in Australia. The beginnings of commercial development in Australia made Walsh even more dissatisfied with Hilgers' progress, and on a visit to America he encouraged Perkin-Elmer to take up the CSIRO instrument. This visit coincided with a proposal from within the company, from W. Slavin, to develop a new atomic absorption instrument, the one based on Alkemade's design having never been put into commercial production. CSIRO made Hilgers' licence non-exclusive, and in 1959 Perkin-Elmer licenced the technique. Slavin championed the development of a new instrument which was given intensive marketing support and accompanied by a programme of user education which rapidly led to a considerable commercial success. In Australia the three small firms which Walsh had used merged, and the resulting organisation, Techtron, enjoyed considerable success in the home market.

Technology transfer and the SAPPHO results

The aggregate results obtained from the collection of success/ failure pairs have been published widely,(22)(23) are also discussed elsewhere at this conference, and will not be repeated at any length here. However, the essential difference between successes and failures can be summarised in five factors, in all of which successes appeared to perform better:

- marketing and sales performance
- efficiency of development
- understanding user needs
- strength of management and characteristics of managers
- effectiveness of communications

The pattern of success which emerged from an analysis of those pairs involving technology transfer was the same as for other pairs. When firms were successful with "imported" devices they were seen to have understood user needs better, developed the product more effectively, carried out the marketing and sales more successfully, and so on. The firms which failed after importing technology were found to have failed to anticipate user needs, to adapt products, and so forth. If the pattern of commercial success is virtually the same in cases involving technology transfer as for those concerning in-house innovations then examination must concentrate on the mechanisms by which barriers to success have been overcome in some attempts but not in others.

The three case studies outlined above illustrate a number of the lessons which seem important for successful technology transfer. The success of Cambridge Scientific Instruments' scanning microscope and microanalyser developments was aided by a complex, and largely informal, communications network. Undoubtedly the geographical proximity of two "centres of excellence" in electron-optics, the Engineering Department and the Cavendish Laboratory, and a large industrial research laboratory, Hinxton Hall, was of great benefit to the firm. The early stages of development also show the importance of good communications. Oatley's use of the Cavendish electron-multiplier and Smith's assistance to Duncumb are examples. And it was as a result of contacts made at Cambridge by a man from the Pulp and Paper Research Institute that led both to Smith's work in Canada and, eventually, to du Pont's interest which led to the first commercial sale by Cambridge Scientific. There is also no doubt that Melford benefitted considerably from Duncumb's help in producing the microanalyser which was licenced to Cambridge Scientific.

In Cambridge's successful development, however, they capitalised on the advantages of these communications networks, and took steps to maximise the effectiveness of the technology transfer. The recruitment of Stewart demonstrated the effectiveness of the method of transfer commended by so many.(24) In contrast, although AEI might have capitalised on external expertise they did not do so. They made little use of Oatley or his colleagues in attempting to develop the scanning microscope, and the microanalyser development was complicated by a need to transfer the work from the Central Research Laboratory at Aldermaston to the production department at Manchester. It was suggested that communications between the two departments were not good, and no transfer of personnel accompanied the move from research to production. AEI failed to appreciate the full potential of scanning on the microanalyser, as indeed did their Advisory Committee of Metallurgists, and some people have seen in this evidence for the "not invented here" hypothesis.

Atomic absorption spectrometry illustrates another facet of the process. Walsh had good contacts with both firms involved, but it was only when one member of Perkin-Elmer, Slavin, adopted the technique and became its "product champion" that success was achieved. Walsh himself was geographically isolated from most of the commercial developments in England and the United States, but achieved considerable success in Australia by working closely with the three firms which produced instruments to his design.

In all three successful cases a great deal of effort was put into adapting the innovations concerned to the requirements of the users, and also in educating potential users about the possibilities of the new techniques. Cambridge Scientific, aided considerably by the recruitment of Stewart, did this for the scanning microscope, Melford at Hinxton Hall developed the microanalyser for use by metallurgists, and Perkin-Elmer put considerable effort behind the dissemination of information about atomic absorption.

An analysis of all the cases involving transfer confirms the pattern suggested by the three discussed here.

The role of individuals appears to be very important. In gathering the case studies the SAPPHO team had made provision for collecting information about individuals who might occupy a number of key roles. Three, in particular, were found to be important in technology transfer:

> 'Technical Innovator'
>
> The 'inventor' or single individual who made the major technical contribution to the development and/or design of the innovation. He would often, but not always, be a member of the innovating organization.
>
> 'Innovation Manager'
>
> The individual who was actually responsible within the management structure for the overall progress of the innovation project.
>
> 'Product Champion'
>
> Any individual who made a decisive contribution to the innovation by actively and enthusiastically promoting its progress through critical stages, whether he had management responsibility for all or part of the project or not.

The Sussex team did <u>not</u> expect to find individuals who fitted these roles in <u>every</u> case. Where they were identified, information was collected about their age, educational background and previous employment, position and status in the innovating organization, and involvement with the particular innovation project. In some cases one person played more than one role. Thus, in a small instrument firm set up to exploit a particular idea the Technical Innovator was also the Innovation Manager and Chief Executive, and because of his commitment and enthusiasm for the particular idea was also considered a Product Champion. This case was, of course, exceptional, but there were others where, for instance, Technical Innovators or Innovation Managers behaved as Product Champions.

When technology was being transferred it was usually possible to identify a single technical innovator, a member of a government laboratory or university department, although in one or two cases it would be unfair to single out one person in this role. As illustrated by the example of Walsh, the technical innovator usually showed an enthusiasm for the commercial exploitation of his product, and in several cases it was largely as a result of his championship of it that industry began to take an interest. However, there is a limit to the amount that any outsider can influence the managerial decision-making of a firm. However much the Technical Innovator in these cases tried to promote his product, he was not able to produce the effect that Schon attributed to his Product Champions, since he had no formal status in the organisation concerned. In many cases the Technical Innovators felt that their advice had been ignored, or that they had been insufficiently consulted by the firms concerned in the unsuccessful attempts. That this happens in many attempts to transfer technology is supported by a recent study of the activities of NRDC.(25)

> "it was frequently said (by university inventors) that the inventor was not kept properly in touch with the development by NRDC and that his technical abilities to help in the development were not properly used. . . .
>
> The major criticism (by private inventors) was that after having been requested to assign the rights to NRDC the inventor was unable to influence or effectively participate in the development and exploitation of his invention. . . .
>
> The dislike on the part of certain Industrial Research Establishments of having exploitation carried out at second hand arises partly from . . . technical arguments resting on the desirability of maintaining the closest links between the inventor, the development team and exploitation".

These comments relate to a situation when a third party is interposed between the donor and recipient, which was encountered in some of the SAPPHO transfer cases. Even when only donor and recipient were involved the same problems were frequently identified.

When products were transferred successfully it was possible to identify an Innovation Manager within the receiving organisation who enthusiastically supported the project, and had sufficient experience and status within the firm to obtain the necessary resources and commitment. In some cases it was found that another member of the organization receiving the technology would adopt the project and act as its Product Champion. In some of the successful cases firms recruited staff from the laboratory where the innovation had originated, to aid in the process of technology transfer. In many cases these men would then assume the role of Product Champion in their new organizations, and in some cases were given managerial responsibility and became Innovation Managers.

In the chemical industry, with its largely in-house developments, a pattern was found that the successful innovations were less modified during development. However, the opposite was true of the pairs which involved transfer. In these cases the successful firms made more extensive modifications, and these were largely concerned with adapting the device from a form suitable to a research environment to that which would be acceptable to the (commercial) users. The importance of understanding user needs, and directing the development of a prototype product towards the requirements of the market place are emphasised in SAPPHO, and the individual cases involving technology transfer also show this clearly. Thus, to summarise the conclusion which can be drawn from the Sussex project, apart from confirming the importance of people in the transfer process the case studies indicate that adaptation of the products being transferred to suit the needs of the market is usually necessary for success, and that this function has normally to be performed by the recipient rather than the donor.

Postscript - implications for further studies of technology transfer

The methodology used in SAPPHO may also provide a basis for a more structured study of technology transfer. Inevitably, in carrying out the individual case studies, both donor and recipient were interviewed, and thus the two sides of the transfer mechanism were examined. Bar-Zakay[26] has stressed the importance of considering the roles of both donor and recipient in technology transfer, and the SAPPHO case study method achieves this. In addition the success/failure pairing means that a framework can be developed within which the important factors for success can be isolated. The direct extension of the success/failure pairing is hard to imagine, because only rarely will the convenient pattern of repeated attempts to transfer the same product be found. However, if success is taken to mean, in Brooks' words, 'the embodiment in an actual operation of some kind' it may be possible to embark on a series of success and failure comparisons using similar products, or technologies, rather than identical ones. But, of course, care would have to be taken to include safeguards against the possible influence of external factors which might make comparison invalid. This was relatively easy to achieve in SAPPHO. In a technology transfer study it might be more difficult.

Some other dimensions could be added in future studies. Projects such as SAPPHO show that investigations at the micro-level, that of the particular innovation or project, can be very revealing. They also appear to indicate barriers to success, and methods for achieving success, which are often ignored at the macro-level, when organizations or mechanisms for encouraging transfer are discussed.

Nearly all the case studies of technology transfer are concerned with specific product innovations, and this is certainly true of the SAPPHO examples. Yet comments indicate that there is another mode of transfer, which may not deal so much with product development but with process improvement. As a recent enquiry into the Research Associations commented:

> 'RAs must not be thought of as attempting cooperatively to 'invent' products. The fact that RAs are concerned mainly with processes or production efficiency rather than with invention of new products for the market is not widely appreciated. Nevertheless they do in this way contribute to product development in their industries, and, moreover, it is often the apparently elementary developments, which may lack scientific glamour, which make the greatest impact on industrial productivity.'(27)

Further it has often been pointed out that the sectors of industry which contribute most to the UK's GNP are usually not those which spend most on research and technology.(28) But they are those in which the pressure for increased productivity, and thus process improvement, may be both greatest, and the results most beneficial. This has been advanced as an argument for transferring some of the national R & D effort from the usual 'advanced technology sectors' to the 'high performers' on the GNP ladder. If this argument is to be extended to imply that government laboratories should try to transfer technology to the non R & D-intensive sectors of industry then it may be that they would have to concentrate on process improvement, rather than product innovation.

Marshall, Director of Harwell, has indicated that there is more than one mode in which work for industry can be carried out in a government laboratory. There is "contract research", done under the customer/contractor principle, and there is "exploitation", where the laboratory takes some financial risk, and by implication develops a piece of technology in advance of the identification of a definite customer.

One is led to ask if the ingredients of successful technology transfer are the same for product innovation and process improvement, and for "contract research" and "exploitation". It may well be that it is, but this assumption should not be made unless there is research evidence to support it. There is considerable scope for research in this area.

References

(1) H. Brooks, "National Science Policy and Technology Transfer" in Proceedings of a Conference on Technology Transfer and Innovation, National Science Foundation, NSF67-5, Washington, 1967.

(2) "Technological Innovation in Britain", HMSO, London, 1968.

(3) "Industrial Research and Development in Government Laboratories", Ministry of Technology, HMSO, London, 1970.

(4) "A Framework for Government Research and Development", Cmnd 4814, HMSO, London, 1971.

(5) "A Framework for Government Research and Development", Cmnd. 5046, HMSO, London, 1972.

(6) Third Report from the Select Committee on Science and Technology, Session 1971-72, HC302, HMSO, London, 1972.

(7) Department of Trade and Industry, "The Non-Reactor Research and Development Activities of the Atomic Energy Authority and the Industrial Research Establishments of the Department of Trade and Industry", Cmnd. 5176, HMSO, London, 1972.

(8) D.W. Budworth, "Government Role in R & D", CBI Review,

(9) C.F. Douds, "The State of the Art in the Study of Technology Transfer - a brief survey", R & D Management 1, 3, 1971.

(10) S.I. Doctors, "The Role of Federal Agencies in Technology Transfer", MIT Press, Cambridge, Mass., 1969.

(11) D.J. de S. Price, "The Structures of Publication in Science and Technology", in Factors in the Transfer of Technology, eds. W.H. Gruber and D.G. Marquis, MIT Press, Cambridge, Mass., 1969.

(12) D.A. Schon, "Innovation: the development and utilization of technology", in Factors in the Transfer of Technology, op.cit.

(13) T. Burns, "Models, images and myths" in Factors in the Transfer of Technology, op.cit.

(14) W. Marshall, "Harwell and Industrial Research", Sixth Maurice Lubbock Memorial Lecture, 9 May 1969, OUP, 1969.

(15) W. Marshall, "The Industrial Research Programme at Harwell", Atom, No.185, March 1972.

(16) D. Fishlock, "Harwell goes commercial", Financial Times, 12 January, 1973.

(17) B. Achilladelis, P. Jervis, A. Robertson, "Project SAPPHO: A Study of Success and Failure in Innovation", Science Policy Research Unit, September 1971.

(18) F.A. White, "American Industrial Research Laboratories", Public Affairs Press, Washington, 1961.

(19) D. Shimshoni, "The Mobile Scientist in the American Instrument Industry", Minerva, 3, 1, January 1970.

(20) J.M. Utterback, "The Process of Innovation: A Study of the Origination and Development of Ideas for New Scientific Instruments", IEEE, Transactions on Engineering Management, Vol.EM-18, No.4, November 1971.

(21) For a more detailed account of these events see P. Jervis, "Innovation in electron-optical instruments: two British case histories", Research Policy, 1, 2, 1971-72.

(22) B. Achilladelis, P. Jervis, A. Robertson, op.cit.

(23) "Success and Failure in Industrial Innovation", Centre for the Study of Industrial Innovation, London, March 1972.

(24) e.g. T. Burns, op.cit.

(25) First Special Report from the Select Committee on Science and Technology, Session 1972-73, HC43, HMSO, London, 1972.

(26) S.N. Bar-Zakay, "Technology Transfer Model", Ind. Res. and Development News, 6, (3) 1972.

(27) "Industrial Research and Development", Conference of Industrial Research Associations, London, April 1973.

(28) I. Maddock, "Low growth rate in the midst of high technology", Proc. Inst. Electronic and Radio Engineers, Vol.7, No.3, May/June, 1969.

AN EXAMINATION OF PROCESS INNOVATION IN PETROLEUM REFINING

George Foster

Forecasting International, Ltd.

This paper presents the results of an inquiry[1] into the circumstances surrounding the introduction of new processes into the petroleum refining industry. The findings of this study are then compared with those of Myers and Marquis[2] in their studies of industrial innovation and those of Project Hindsight.[3]

The original concept of this study was to determine if large and small corporations differed in their methods of introducing new technology into the firm. A second goal was to determine and describe the factors which led a firm to search for and employ new technology.

For this analysis, the processes of gasoline refining in the United States were chosen, and the first implementation of an innovation was emphasized. This area of analysis was chosen as the processes of refining are well-defined and known so that it is possible to determine whether a process innovation is, in fact, relevant to the gasoline refining industry.

The first implementation was emphasized because it is a clear departure from previous practice and begins a sequence of imitation and improvement. The decision to install the first unit is both more critical and difficult than the decision to implement after its success has been demonstrated or to make major improvements in existing processes.

In the early stages of the study it was anticipated that the study could concentrate on process innovations introduced since World War II as this was a critical period of activity; railroads became dieselized; jet aircraft became commercial; home heating with oil became widespread; high performance automobiles became popular.

And it was the age of research.[4] Contrary to expectations, these changes did not lead to the introduction of new processes except for catalytic reforming in 1949, hence, it was necessary to examine process innovations of a much earlier period of change. The first of the major processes -- continuous distillation -- was introduced in 1911.

A fair question is whether the analysis of events of so long ago will contribute any useful insight into the present. There are two factors which do indeed contribute to the relevance of this effort. First, the phenomena of innovation does not seem to have changed much in the past century and the situation which occurred fifty years ago will seem very familiar to present-day workers. Secondly, the real question is how a large technically sophisticated industry reacts to significant change in regulatory environment and demand structure.

In 1909 the American petroleum industry was fifty years old. The most important product was kerosene for lamps as well as some gasoline for stove fuel; the market for motor fuels was growing rapidly. The industry was dominated by the Standard Oil Company which was under attack for its remarkably aggressive business tactics and its monopolistic position. It was technologically aggressive as well. Standard Oil of Indiana, for example, had been established in anticipation of a major improvement in treating high sulfur crude oil. Atlantic Refining, another Standard company, was doing excellent work on developing new refining processes. Vacuum Oil was a leader in the production of lubricants. Standard Oil as a whole was well aware of foreign developments and was quick to apply them domestically. Standard Oil was the IBM of 1909.

By 1911 the Standard Oil was broken up under the anti-trust laws and the sales of motor gasoline exceeded sales of kerosene. Dissolution of the Standard Oil trust may have made it easier for new firms to enter the industry, although this effect was not very great. The change in demand patterns, however, clearly created the opportunity and incentive for process innovation.

Kerosene technology is not adequate for production of gasoline as the major product. Shift in demand created a need for new technology.

Three processes -- distillation, cracking and reforming -- are the core of the modern refinery. Distillation separates crude oil into fractions boiling at different temperature ranges. Cracking converts high boiling to lower boiling material, especially gasoline. Reforming increases the octane number of gasoline stocks. (Octane number is an important index of gasoline quality; it measures the tendency of gasoline to knock in the engine).

There are several less important processes used in a typical refinery. Polymerization, alkylation and isomerization were introduced to convert waste gases into gasoline. With the advent of catalytic cracking and catalytic reforming the original need has become less important but the products are very good gasoline stocks, so the processes are still used. Solvent extraction which was introduced to improve kerosene from aromatic crudes is now used primarily to improve lubricating stocks and to isolate aromatics for chemicals and high-test gasoline. Hydrogenation which was introduced as an alternative to cracking is used to remove sulfur from various stocks and to only a slight (but growing) extent for its original purpose.

Table 1 lists the refining processes considered in this study. The first groups are the core processes which make up the heart of a refinery. The remainder are secondary processes of somewhat lesser importance. About half the processes are now obsolete but are included because they were important in the development of current processes. The other columns give the inventing organization, the innovator and the date of the first installation. The column headed nature of innovation classifies the innovation as to whether the technology used was a departure from contemporary practice (D) or an extension of contemporary practices.

Tables 2 and 3 are "score cards" which show the number of processes invented and innovated by individuals and companies. The data show that petroleum refineries have not been particularly inventive.

Individuals invented five of the six processes introduced before 1930. Of the remaining nine, individuals invented two and Universal Oil Products (UOP) invented three. No refiner invented more than one. Individuals invented five of the nine processes which were substantial departures from contemporary practice and UOP invented three. Shell invented the remaining one.

It must not be thought that the individual inventors were the wild eyed men of legend. Most had very good technical educations. The exceptions were Milo Trumble who invented the pipe still and Jesse Dubbs who is credited with the invention of the Dubbs Process. In the latter case, however, the actual reduction to practice probably should be credited to others, especially Shell engineers.

Shell was the most innovative company. American Shell probably still is, although the arena has shifted from refinery to discovery and production of crude oil. Jersey is second. An important element in their performance was corporate management's interest in the firm's technology.

TABLE 1

	Inventor	Innovator	Date	Nature of Innovation
Core Processes				
Pipe Still	IND	Small	1911	D
Thermal Cracking				
Burton	SOInd	SOInd	1913	E
Tube and Tank	IND	Small	1921	E
Dubbs	IND	Shell	1922	E
Cat. Cracking		SONY	1936	D
Fixed Bed	IND	SUN	1937	
Moving Bed	SONY	SONY	1943	E
Fluid Bed	IND	SONJ	1942	D
Cat. Reforming	UOP	Small	1949	D
Secondary Processes				
Solvent Extraction	IND	Shell	1915	D
Hydrogenation	IND	SONJ	1929	D
Polymerization	Pure	Pure		
Thermal	Phillips	Phillips	1934	E
Catalytic	UOP	Shell	1935	D
Alkylation				
Sulfuric Acid	Anglo Iranian	SONJ Shell	1939	E
HF	UOP	Phillips	1943	D
Isomerization	Shell	Shell	1941	D

TABLE 2

Invention Record

Individuals	7
UOP	3
Shell	1
SOInd	1
SONY	1
Others	2
	15

TABLE 3

Innovation Record

Shell	5
SONJ	3
SOInd	1
SONY	2
Small	3
Others	3
	17*

* Three innovations were introduced simultaneously by two firms.

In Shell's case they were engaged in bitter competition with the Standard Oil Trust before its dissolution, and the Standard Oil Companies thereafter, and had to use whatever weapons they could. One they exploited was new technology.

In the case of Standard Oil of New Jersey, the original impetus seemed to be the high royalty payments Jersey was making to Indiana for the rights to the Burton Process which was not well suited to Jersey's needs. Walter Teagle, President of Jersey, deliberately set out to raise his company's level of technical competence. This led to the development of the tube and tank process. Strangely enough, Jersey did not innovate the process even though they developed it. Actually, Jersey's record is somewhat better than indicated here. They innovated olefin hydration, an important petro-chemical process, the production of butyl rubber, the Krouse process for tetraethyl lead and the first viscosity additive for motor oils. In general, they were cautious about innovation. They developed their own catalytic cracking process (jointly with Kellog UOP and Shell) only when they were faced with paying huge royalties to Houdry for use of the fixed bed process.

The development of petroleum hydrogenation was a bold step. (At the time it was thought to be the most costly industrial development program ever undertaken). The process stretched the state of the art technically and was ahead of its time economically. The company hedged its bets however and brought in other refiners to share in the costs.

A contrary example is provided by Standard of Indiana. It was founded in anticipation of a technological development and until about 1920 it was one of the more innovative oil companies having, for example, developed the first thermal cracking process. However, when Burton, the inventor of thermal cracking, became president of the company he blocked further development of the process and this helped drive away his best technical man. Thereafter, the company paid more attention to acquiring crude reserves than to technology and the company was no longer a successful innovator.

All of the innovations described here were invented and innovated in response to perceived needs. However, a problem which is common to most firms within an industry probably is not perceived as a need but as a fact of life. The firm which does recognize it and which responds by innovating is likely to be one which has a competitive disadvantage and feels the need with particular intensity. Consider catalytic cracking, for example. The first process was invented in 1929 by Eugene Houdry, an independent inventor, who wanted to produce a superior gasoline. When Houdry ran out of money, his work was supported by Vacuum Oil and when they lost interest, by Sun Oil. Sun was interested because the Houdry process produced a high quality gasoline and Sun's marketing strategy was to sell a single high-grade lead-free grade of gasoline. Eventually Sun needed more money too and Vacuum (then Socony-Vacuum) contributed to the final development. The first units were installed by Socony in 1936 and Sun in 1937.

Note that Vacuum became interested in the process at a time when it had inadequate crude reserves, marketing outlets and refining capacity. When Vacuum merged with Socony these deficiencies were alleviated and the company lost enthusiasm for the development. Then it was taken up by Sun which had a special need for the process because of their strategy of marketing a single grade of gasoline.

Perhaps the Houdry success would have aroused greater interest if it had promised increased yield as well as quality improvement. (As it happened, the process did give increased yields but this was not apparent at the time). A refiner will feel the need for improved quality when his competitors produce better quality than he does and his customers recognize it. When his products are competitive, quality improvement is not perceived as a need, and, in fact, it is not a need.

This is only slightly less true of yields and costs. As long as a refiner's costs and yields are about the same as his competitors, improving them is perceived as nice to do but not as a pressing need. A refiner operating on purchased crude is likely to be very conscious of yields and be interested in introducing processes which promise to conserve crude. He would probably try to acquire crude reserves rather than to innovate new processes, however.

This suggests ways to reduce process innovation. One is to make the industry more homogeneous. As the several companies come to resemble each other the likelihood is reduced that any one company will feel a particular need strongly enough to innovate. A second way is to increase technical cooperation between companies and reduce rivalry. This could be accomplished by relatively open exchange of technical information and making advances available to all comers at modest royalties.

The refining industry has been behaving in precisely these ways since the twenties. Petroleum companies have experienced vertical integration into all aspects of the business--crude production, transportation, refining, bulk distribution and retail marketing. Also, they have tended to operate over larger geographic areas. As the integration process continues, the differences between them disappear.

Furthermore, petroleum firms have become more cooperative after the ferocious legal battles which accompanied development of thermal cracking. Jersey made hydrogenation available to any refiner that wished to participate in developing it. Fluid catalytic cracking was a joint development of several refiners as well as Kellogg and UOP. The result has been to reduce both the incentives for innovation and the penalties for non-innovation.

In general, the results of this study parallel and reinforce the conclusions reached by Myers and Marquis in their study of industrial innovations. Their investigation was much broader than the present study and they were able to discuss incremental innovations which they showed were important in overall technological progress.

The present study does not bear on this point except inferentially. Only fairly large advances are discussed but it must be recognized each has been considerably improved since its first introduction.

Also Myers and Marquis collected and analyzed data on ideation and information sources. The present study did not consider this area, nevertheless, it supports one of their conclusions. They found that major information inputs were general in nature and widely accessible. In the case of petroleum processes only fluid cracking depends on a principle which was discovered during the course of development. The Houdry process depended on finding a particular catalyst formulation but the fact of catalytic cracking, e.g. with aluminium chloride was known and, in fact, had been considered by Burton and Humphrey.

Myers and Marquis showed the effectiveness of demand pull versus technology push, that is, the recognition of demand or need is a more frequent factor in innovation than recognition of technical potential. Each of the innovations described in this study was purposely developed in response to a widely recognized economic need.

Myers and Marquis also found that adopted innovations; that is, innovations developed by another firm, were a significant portion (23%) of the innovations they studied. Two of the more important petroleum innovations--catalytic reforming and catalytic polymerization--were developed by UOP and innovated by refiners. Hydrogenation was adapted from a process developed to treat coal and strictly speaking the Tube and Tank process was an adaption since it was developed by Jersey and first installed by an independent refiner but this is a rather special case. In addition the Dubbs process, sulfuric acid alkylation and olefin hydration were at least partly developed outside of the innovating firm.

Finally, in what may be their most important conclusion, Myers and Marquis say that innovation is more than R&D, a conclusion which is abundantly substantiated by the present study.

Another, and probably the best known retrospective study of the development of specific technologies, was Project Hindsight. It was conducted by the Department of Defense under the direction of Colonel Raymond Isenson and undertook to determine the source of scientific and technological event which essentially contributes to development of twenty weapons systems.

Briefly, Project Hindsight showed that development of a new weapons system involves the skilful combination of a large number of events, e.g. the development of a new material. The large majority (88%) of these events were motivated by DOD needs. About 60% were motivated by specific system needs; 20% of the events were responses to the needs of the specific systems being studied so that 40% were in response to requirements of other systems. In the average, about a third of the events occurred after the system was committed to engineering design or production.

Isenson concluded that a high rate of technology utilization requires recognition of need, an educated talent pool to provide ideas, capital resources and an adequate communications path to potential users.

Isenson's conclusions seem reasonable when applied to the processes of innovation discussed here if the analogy between a complex process and a weapons system is accepted. Like the weapons systems, they were quite complex and were at the forefront of the state of the art. As a matter of fact, most of them required several significant inventions to become operational which were invented in response to the needs of the system.

The major difference between the two areas seems to lie in the way needs were identified. In petroleum process, the identification of needs was and, for the most part still is, a haphazard process. The case of the weapons systems needs were sought out in a systematic manner and about 85% of the events treated in the Project Hindsight were responses to needs identified by applications groups are only 15% were responses to needs recognized by research performers. In other words, in weapons systems, at least, researchers are not particularly good at identifying needs. On the other hand, the users were not particularly good at devising technical approaches. A clear division of labor.

The implication is that an R&D group is unlikely on its own in isolation to establish useful goals. This may be the reason for the inadequate performance of petroleum R&D in the post war period. Much of the high R&D productivity of Universal Oil Products in its heyday was probably due to its special relationship in the petroleum industry. The firm functioned as the engineering department for the smaller refiners and as a result had to know their client's problems and respond to them in an inventive fashion to stay alive.

REFERENCES

1 George FOSTER, *Process Innovation in Petroleum Refining,* MIT Working Paper 490-10, Sloan School of Management, 1970.

2 Sumner MYERS and Donald G. MARQUIS, *Successful Industrial Innovations: A Study of Factors Underlying Innovation in Selected Firms.* (National Science Foundation NSF 69-17: Washington, D.C.) May, 1969.

3 Office of the Director of Defense, Research and Engineering, *Project Hindsight: Final Report Task 1* (DOD: Washington, D.C.) July 1, 1967.

4 The number of research scientists and engineers in the petroleum refining industry increased from 1,800 in 1950 to 3,800 in 1964 (3,600 in 1966) from *Employment of Scientists and Engineers in the United States 1950-1966,* (National Science Foundation: Washington, D.C. NSF 68-30).

HOW AN ORGANIZATION DECIDES TO INNOVATE

Jules J. Schwartz

University of Pennsylvania

Introduction

This paper reports some of the insights gained in studying a number of product and process innovation decisions in both an integrated, dominant product and a diversified firm. It is my conclusion that innovation, whether based on internal ideas or transferred technology, is a manageable process. The failure of innovative efforts is more often influenced by bad decisions than by technical or informational deficiencies.

To better understand the decision process in a large organiztion it is useful to turn to the work of J. L. Bower who studied the resource allocation process in a diversified firm and reported his findings in his book, Managing the Resource Allocation Process. Bower suggests that the process is best viewed as two related subprocesses mediated by the structural context of the organization. The first subprocess he terms "definition." Definition is the means by which a proposal gains its substantive content. The second subprocess he call "impetus," the means by which the proposal gains advocacy and support and progresses toward approval. Corporate structural context includes the articulated strategy of the firm, its formal organization and its control, incentive and resource allocation systems.

Definition, Bower found, is chiefly determined at the lower levels of the organization, where technical specialists confront problems and propose solutions.

Impetus, on the other hand, is chiefly the product of the activities of middle managers, who filter the many proposals to determine what will go forward to compete for scarce resources, serve as translators between senior managers and specialists, and absorb risk for senior management by staking their own repuation where there is uncertainty.

Bower concluded that it was principally the role of senior level managers to establish the context that governs the definition and impetus subprocesses.

Bower's study concerned itself with routine resource allocation decisions that usually involved responses to discrepancies between manufacturing capacity and market demands. My study sought to develop a description of the decision process where technological innovation was involved. Successful firms in high-technology industries were studied in order to develop prescriptions regarding the decision process. This paper presents some of my findings and is divided into six sections: a brief summary of the peculiar problems of high-technology industries; the implications of an innovative strategy in terms of demands on the firm; top-management problems; possible responses to these problems; the key tasks for middle management; and finally, a set of general conclusions.

Special Problems in High-Technology Industries

Three special problems are faced by a firm choosing to compete in high-technology areas: (1) a high rate of technical obsolescence, (2) high risks and (3) large development investments. These problems are discussed in turn.

The nature of competition in high-technology industries is a rapid and continuous creation of new products, many of which render older products technically obsolete. Process innovation is eschewed because of fear that the new process may produce a product that has become obsolete.

Firms operating in high-technology industries face both technical and marketing risks. The risk of technical failure in a rapid development effort is high. The threat of technical obsolescence caused by competitive products is also severe. In addition, the introduction of a new product exposes the firm to the uncertainties of the marketplace. A significant effort may be required to convince potential customers that the new product is superior enough to justify their investment in an untried item.

If a firm is to operate in a high-technology area, it must be prepared to invest substantial portions of its discretionary funds in research and development and to forego alternative investments. The stakes tend to be big because entire businesses are threatened by obsolescence.

Implications of Innovative Strategies

What are the implications of a high obsolescence rate, great risk, and big bets to a firm? My research suggests that four requirements are imposed.

First, technical and marketing intelligence must quickly come to the attention of those who need it for proposals and for decision making. Proposals that do not have the benefit of the latest ideas available are likely to be inferior.

Second, there must exist in the firm an ethos of risk-taking. Firms are made up of people who must assume an individual risk of failure in order to advocate change.

Third, there must be a high order coordination of the activities of specialists. This coordination is necessary to facilitate a rapid exchange of ideas. But equally important, it is necessary for effective business planning. The specialist who can design and propose a technological innovation is rarely equipped to provide the market plan necessary to justify a commitment to his idea or the manufacturing plan to produce it economically, if approved. Even if he were, the urgency of change dictates that these tasks be performed simultaneously.

Finally, recognizing the constant press of urgent operational problems, there is a special need to focus the attention of senior managers on strategic matters such as innovation. Selectivity is the key. The firm will generally have more proposals than it can fund.

A striking result of my research is the critical need to reduce the time required to develop alternatives and to choose one, then to enlist commitment and generate impetus. A high-technology market must often be recognized _and_ exploited in the time required to take many routine decisions. Management's contribution in a decision to innovate includes finding ways to speed up the decision process.

My study also finds that management must concern itself that proposals be properly scaled. The changes proposed are rarely global in scale; that is, they are seldom significant relative to the entire activity of the corporation. Senior managers are obliged to seek big opportunities in order to allocate resources appropriately.

Parochialism is another problem. Subunits tend to shape their proposals to fit them into their present business charters. In answer to market threats, they seek solutions that the subunit can control.

Development specialists prefer to work in a loose, unstructured environment, without close supervision. Management is faced with the need to gain commitment from these specialists in order to produce consistently good results. The question management must then confront is how to keep people satisfied when a program is rejected after commitment develops. Management must be selective, but proponents of new projects must also be encouraged.

To control the decision process, management must have good, timely data on which to base its choices. There is concern that proposals will be biased by denigrating the alternative in order to sell the option selected by subordinates.

Management Responses

The responses I see successful managers making to these problems fall into four categories: (1) contextual -- that is the formal organization and the established control, resource allocation and reward systems; (2) personal style of the managers; (3) specific management techniques; and (4) particular roles assigned to the various levels of management.

In each situation studied, *ad hoc* groups were established to prepare the innovation proposal and to manage the effort when it was funded. Such groupings serve several purposes. They can be established on a part-time basis within operational units without being disruptive. They provide a means of assembling and supervising the activities of essential specialists. They permit their members to make commitments to a project, while giving them reassurance that they have a "home" to return to if the effort fails or is cancelled.

The incentive systems in the firms studied were structured to provide big rewards to successful managers of innovative activities. The rewards in the companies took the form of salary increases, promotions, stock options and grants of greater authority, autonomy and responsibility. Employees seemed content to accept the delayed payoffs that were attached to the longer-term strategic tasks. This may have been in part because they were many highly visible real models to attest to the certainty and size of the rewards they could expect. Perhaps as important, the penalties for failure did not seem great to the people interviewed.

Management style is still another factor which may affect the ability of a firm to innovate. A "shirt sleeves" style for senior managers shows they are involved and are sharing the risks. They are also approachable, demonstrating the openness of communication desired for their organizations.

The study confirmed previous findings that managers impose schedules, and more specifically, deadlines as control measures on innovative activities. Several advantages attach to such controls:

1. A necessary sense of urgency is imparted to subordinates. They know not only that there will be a reckoning, but when.

2. Management can order priorities for employees who have several jobs to do. This also makes coordination of activities simpler. The schedule also serves to budget manpower.

3. Meeting intermediate goals builds confidence and morale in individuals and organizations. Risk can be distributed among the managers responsible for the individual tasks.

4. Perhaps most important, the workers are forced to make tradeoffs. They just don't have the time to find a non-existent "best" solution.

Managers also employ another technique that is closely related to establishing schedules and that we shall term "factoring."

Managers are constantly faced with proposals that would require them to make large-scale commitments of resources very early in a program. Understandably, subordinates wish to avoid uncertainty in funding, so they package their entire request to get all the money needed as quickly as possible. Managers sought to factor such proposals by dividing them into separate, measurable work packages.

Factoring accomplishes several functions that managers will appreciate but that subordinates may miss in their own planning. Factoring breaks big decisions up and allows management to make a series of small, more evolutionary ones. This means that resources can be committed as information is accumulated and the absolute level of initial risk is reduced. In each of the innovations studied, the proposals made slow progress toward approval until they were factored.

Factoring a proposal is also one way that an intermediate manager can demonstrate that he grasps the problem and the means to implement its solution. This approach will often allow the evaluator to judge a plan better; if he has had experience in any of the work packages, he can at least determine the merit of the proposal in those areas.

To an experienced manager, none of this should seem new. Senior managers are always faced with the pressure of schedules. Opportunities are fleeting and must be exploited promptly. Time is one dimension of the risks they confront, and so they seek to include it in the planning of those subordinates who do not seem to sense the same urgency and thus do not volunteer a schedule.

Certain elements of corporate policy can act to drive the innovation process. For example, if a firm decides to try to seize a market with a new product, it can often maintain a dominant share by passing along to its customers the cost-savings realized as it gains production experience. This strategy places great pressure on the manufacturing group to make continuous incremental improvements in its processes.

Having discussed some of the responses and techniques available to managers in high-technology firms, it is now possible to consider how the specific tasks to be accomplished are divided among the various management levels.

It is important to understand that this work is being done in an organization, so the system concepts of feedback and iteration apply. Further, to make the system work, many of the tasks must be accomplished simultaneously, rather than sequentially. Finally, although the tasks may be performed during different phases of a proposal effort, the same manager may serve several roles, and so perform several of the jobs.

In the situation studied, the tasks of coordinating specialists and business planning were performed by first level managers. Second-level, or middle managers, accomplished the functions of negotiating criteria, factoring and scheduling. Senior, or third-level managers designed and implemented the structural context that mediated the decision process. These tasks can be viewed as bargaining, in which the managers shape the proposal definition by withholding or offering their advocacy in return for modifications to the substance of the plan.

The managers and specialists participate in a series of exchanges and negotiations that result in contributions to the proposal definition, apportionment of risk and impetus toward approval. Senior managers are understandably reluctant to commit themselves to a new proposal when they sense that there is still dissension among the specialists. Such discord casts doubt on the technical choices made.

Business planning is also best performed by a first level manager who can bring together specialists' technical, production and marketing inputs.

A critical problem faced in developing a new product or process is securing general agreement on what criteria will be used in design. Different levels in the organization may have different ideas regarding what makes a product good. Middle managers usually serve this role of vertical integration. They negotiate to develop a common perception of the problem.

The task of factoring proposals and adding the time dimension of schedules and deadlines also seems to fall to second-level managers. This activity is analogous to risk brokerage in that it provides for factoring work and assigning measurable performance goals. Thus, individual responsibility(or risk) is imposed on subordinates in return for the support (risk-taking) of the manager. The work packages, schedules and deadlines then become a part of the proposal definition.

The elements of context have been discussed before. It is reasonable to ask why the managements fostered the elaborate system of negotiated plans and goals rather than a more directive system of managing the decision process. This is a consistent approach when one considers that the firm wants to encourage risky (innovative) proposals from its subordinates. First, managers realize that subordinates must have some direction in the process in order to develop good ideas and that participation is also a form of reward. Equally important, the sharing of risk legitimizes the large rewards conferred to encourage innovation.

Implicit in the function of setting structural context is the need to articulate corporate strategy clearly. If the contributions in the initiating phase do not reflect this strategy, they are doomed from their inception.

Summary

In summary:

If a firm wishes to assume high risks, it must structure its systems to provide big rewards.

If it wishes to manage these risks, it must negotiate rather than direct the decision process. Impetus develops when intermediate managers are satisfied that advocating the proposal, as defined, results in an acceptable level of risk both to themselves and the firm.

Just as an innovative firm seeks to factor proposals, it also tends to factor the decision process among various levels of management.

Shared problem conception is a principle requirement in timely innovation decisions. A clearly understood corporate strategy is a first essential in developing this conception.

As Shumpeter suggested almost fifty years ago, successful innovation is a feat of will, not intellect . "It is a special case of the special phenomenon of leadership."

INSTITUTIONAL ROLES IN TECHNOLOGY TRANSFER: A DIAGNOSIS OF THE SITUATION IN ONE SMALL COUNTRY

Thomas J. Allen

Massachusetts Institute of Technology

Sean Cooney

The Agricultural Institute

Ireland

Introduction

The rapid expansion of the world's store of scientific and technological knowledge presents the small nation with a particularly vexing problem in maintaining its scientific and technological communities abreast of foreign developments. To assure its survival and growth the small country must not only acquire foreign scientific and technological information, but it must also solve the more difficult problem of seeing that this information is disseminated to those points where it can be utilized.

This problem has been discussed extensively in recent years and numerous solutions have been proposed. The majority of these, however, are concerned only with the first problem, that of bringing information into the country (acquisition), and they either ignore or give only cursory treatment to dissemination. As many research laboratories have sadly learned, the mere possession of information does not guarantee its utlization by those who require it.

The problem of dissemination is a particularly difficult one. Most of the recent work in the field of technical information has gone into the development of the dissemination problem. National dissemination of scientific and technological information is highly dependent upon the ways in which science and technology are organized in the country, upon the existing institutional forms, and upon relations among both institutions and people. This is true whether the source of information is domestic or foreign. To advocate reorganizing or establishing new or additional institutions without first understanding the roles in the dissemination process of existing ones would indeed be foolhardy. As a first step, a thorough investigation must be undertaken to ascertain the respective roles and the effectiveness of the various types of institutions which exist in the country.

Similarly, acquisition mechanisms must be subjected to scrutiny. The allocation of resources to different methods of information procurement should be determined on the basis of their relative cost-effectiveness. Although considerable effort has been devoted to evaluating the effectiveness of information acquisition mechanisms, this effort has been, for the most part, restricted to the evaluation of hardware and software systems. Little is known about the human element in the acquisition process. The effectiveness of mechanisms such as foreign training of scientists and engineers, employment of foreign nationals, attendance at international conferences, foreign sabbaticals, etc., must be determined. Since research into the dissemination process has shown the overwhelming importance of personal contact, such approaches to acquisition will have a natural kinship with the dissemination system. In fact, they may prove to be more effective than all of the hardware, software and print-oriented devices combined.

Any country, whatever its size, would certainly benefit from a better understanding and evaluation of its strategies for acquiring and disseminating technical information. For the small, growing nation, however, the need is at once both more acute and more capable of fulfillment. Fortunately, because of its size much more can be learned about the functioning of science in a small country. The entire country can be made a unit of analysis, and the entire scientific and technological community can be reached and studied in considerable detail, at reasonable cost.

The rationale for this approach to the problems of the small country grows out of a program of research, which was initiated at the Massachusetts Institute of Technology in 1963. The implications of this work have been widely disseminated (see, for example, Allen, 1970), and are well known in the context of individual research laboratories in government and industry, where much of the earlier studies were carried out. More recent work suggests, however, that the relevance of these questions, and the answers provided by communication research, goes beyond the individual laboratory and deserves the attention of national governments, international organizations and other agencies concerned with the relation of information transfer or dissemination to economic development.

A pilot study was carried out (Allen, et. al., 1971) in one Research Institute, and clearly demonstrated the existence of the international technological gatekeeper as an intermediary in the transfer of technology between nations. The international gatekeepers display similar characteristics to their U. S. counterparts, being, on average, more technically competent and more productive, more highly qualified than their colleagues and reading the literature more extensively. The pilot study has now been extended to include the entire research and development community of the Republic of Ireland.

Examination of the entire R&D community of a country affords the opportunity, for the first time, of viewing the process of information acquisition and dissemination at a more inclusive system level. Individuals can now be grouped according to their organizational affiliation (or other relevant aggregations) and the roles of, and relations between institutions can be studied. More particularly, one can ask whether any of the phenomena which have been observed to operate at an individual level, within single organizations, also function at an organizational level. Do, for example, gatekeeper organizations exist, to which other organizations turn when in need of information from outside of the country? What factors influence the development of communication between organizations of various types in a country? It is to questions such as these that the present paper is addressed.

The Survey

Ireland has a population of some three million in an area of 70,000 sq. km. Compared to many countries it is sparsely populated, at a density of 42 sq. km. One of the chief causes of the sparseness of population has been heavy emigration, which continued for over a hundred years. Encouraging emigration has been the chronic high volume of unemployment or undermeployment.

The Irish government in recent years has sought to stem excessive emigration and create more and better employment opportunities by a series of development programs. These have used grants to agriculture and industry, investment in state-sponsored (or "semi-state") companies and improved services to agriculture and industry. One of the objectives of policy is to improve the contribution of science to economic and social development.

A survey was conducted in 1970 and 1971 of the communication patterns of 1,500 research and development scientists and engineers then working in Ireland. The techniques used in studying industrial firms were adapted to suit the conditions within a country. Data collection was in two stages. In the first, covering one major research institute, responses were received from almost 200 scientists. These were analyzed and results published (Allen, et. al., 1971). In the national survey, responses were received from more than 1,200, or 80 percent of the population responded. Of these, more than one hundred either disqualified themselves or had to be disqualified for various reasons. The principle reason for disqualification was that the respondent had become completely involved in research administration, and was no longer an active researcher. A number of other respondents either answered questions in ways that were uninterpretable or gave answers that were unusable, so for any given question the size of the sample was often further reduced, resulting in about 1,000 questionnaires, plus or minus 25, that were analyzed on any given question. This is about two-thirds of the estimated population size, and should be large enough to provide results that are valid for the population as a whole.

The questionnaire, itself, requested information on demographic variables, such as age, education, field of research activity and years of technical experience, but in addition, information was obtained on each respondent's present and former employers, both within and outside of Ireland, and on each respondent's communication activity at three levels: within his organization; within the country; outside of his organization; and outside of the country. Data such as these can be analyzed in many ways. The present paper is concerned only with the flow of information into and among the various research institutions in the country. Once communication measurements as these have been made at an individual level, it is a relatively straightforward task to aggregate them by specific organization, and to examine the relative strengths of communication bonds between organizations. This provides a measure of the extent to which technical and scientific information flows, via personal contact, from organization to organization or from sector to sector, within the country.

Each respondent was asked to report the frequency (once a month, once every six months, etc.) with which he maintained contact with those outside his organization, and the data were analyzed in terms of a high frequency network (once a month or more frequently) and a low frequency network (less than once a month, but at least once a year).

The Communication Bond

In order to compare the amount of communication among organizations, an index had to be developed. There are many possibilities for such an index, none of which are completely satisfactory. The one which has been chosen is one which should allow valid comparisons to be made with a minimum of distortion to the data. The index is based on the number of individuals in any organization, who are reported as communication partners by those in another organization.

$$C_{ij} = K \frac{n_{ij} + n_{ji}}{N_i N_j} 1-e^{-\frac{N_j}{10^2}}$$

where:

C_{ij} = strength of the communication bond between organization i and organization j

K = a scale factor, in this case $K = 2 \times 10^2$

n_{ij}; n_{ji} = number of individuals in organization j or i who are named as communication partners by their counterparts in organization i or j

N_i = number of respondents in the larger of the two organizations

N_j = number of respondents in the smaller organization

The constant term is simply a scale factor to avoid the use of very small decimal fractions, while the exponential term is a correction factor to offset the effect of wide differences in size of organization.

Results

As mentioned above, the existence of the "internatioaal technological gatekeeper" was established in the first phase, and carried out in one research institute in 1970. Although these results have already been published (Allen, et. al., 1971), it may help to mention them here, since they illustrate the implications of the national study very clearly.

The research personnel of this institute are overwhelmingly Irish by nationality (95% are citizens of the Republic or of Northern Ireland). Of the sample population, 30% had Ph.D. degrees; most of the remainder held qualifications higher than bachelor level.

The structure of the internal communication network was modelled by examining responses to the questions regarding each scientist's "most frequent" discussion partners. This gives an index to the number of colleagues who turn to a given individual for information, or the number who can (potentially) be reached by information held by that person. Someone who is consulted by many colleagues is likely to be an important information source for his colleagues; he may be called a "communication star."

For such a "communication star" to also be a "technological

gatekeeper" he must be well integrated into an external network of foreign information sources; he must use the network regularly, and his network must be broad enough to include a diversity of such foreign sources. Two criteria -- the frequency of foreign technical correspondence and the frequency of foreign scientific and professional meeting attendance -- were used. Each of these channels can be used frequently and each allows contact with a large number of overseas colleagues. The degree to which they are used is therefore judged to be a good measure of integration into the external network.

The data from this phase of the survey did, in fact, show that technical discussion stars do receive significantly more international technical information than nonstars, as measured by frequency of foreign correspondence and attendance at foreign scientific and professional society conferences. They also read a significantly greater number of foreign journals. The gatekeeper hypothesis is quite strongly supported. A total of 26 scientists were identified in the organization whose communication activities fit the operational definition of a "gatekeeper."

In addition to his communication activity, the international gatekeeper displays somewhat greater technical competence than his non-gatekeeper colleagues. Those in the institute are, on the average, significantly superior to their colleagues in both publications and patents, and a higher proportion have a Ph.D. degree. Technical performance is extremely difficult to measure, but to the extent that publication of scientific and professional papers and the acquisition of patents serve as an indicator, the international gatekeeper is a high performer indeed.

Sector to Sector Flow

There are three principal classes of organizations performing research in Ireland. They are: the Universities, the Research Institutes and Industry. In addition, there are a number of smaller organizations including Research Sections in several government departments. There are four major universities in which some of the staff perform research in conjunction with their teaching responsibilities while others are engaged solely in research. There are three major research institutes, which are supported primarily by government funds and which perform research for both industry and agriculture. Finally, there are a number of industrial firms, both private and state-sponsored, which support research and/or development work within their own organizations. These are but a very small proportion of Irish industrial firms, but it is important to see how they acquire technology, since they tend to be either the largest firms, or those in industries whose technology changes rapidly, or both. They are an important indicator of the country's capacity to maintain a competitive position in world trade. A further study involving firms which conduct little or no research is in progress.

Broadly, the main communications are provided by the universities and the research institutes. There is relatively little contact between the other organizations. The relative strengths of the bonds (Table I) show communication to be heaviest between research institutes and industry and weakest between universities and industry. If research institutes are, as it seems, the only available intermediary between more fundamental university research and the needs of industry, the communication bonds would indicate that they are performing one part of this function somewhat better than the other. The bond extending forward to industry is nearly one and one-half times that extending back to the university base. This linking role may be clarified when communication between research departments in different organizations has been analyzed.

Table I

The Strength of Sector to Sector Communication Bonds (Monthly or More Frequent Contact)

and between	Communication Bond (C_{ij}) research institutes	 industry
universities	0.285	0.098
research institutes	--	0.422

Of course, the universities are not the only source of new knowledge. In fact, the research institutes themselves may often be the leaders in creating knowledge, particularly of the sort that would be relevant to industrial innovation so, to some extent, the research institutes may be able to operate independently of the university. The communication with which the institutes should really be concerned is the one with industry. In this sense, the situation presented in Table I is probably not a bad one.

In addition to creating knowledge through their own efforts, the research institutes have developed a network of contacts with foreign universities, research institutes and firms. This network allows them to tap information sources elsewhere in the world, as well as from the universities.

Foreign Contacts by Sector

The three main research sectors differ both in the extent and amount of foreign contact by their research staff (Table II). Surprisingly, the three research institutes show the lowest frequency of foreign communication. It is not surprising that the universities have strong foreign ties, particularly to the U. K., but it is surprising to see a higher degree of direct foreign contact by industry than by the research institutes. The temptation is to explain this as a result of foreign ownership of industrial firms, a common enough situation under the programs for industrial development which have been pursued for more than a decade. The fact that slightly more than half of the research staff in the industrial sector are employed by semi-state firms however, discounts this explanation. The greater number of foreign contacts are maintained by private (rather than state sponsored) firms. Not all of these are foreign owned however, while nearly half of the foreign contacts, themselves, are employed by organizations other than industrial firms.

Table II

Foreign Contact by Research Sector

	Foreign Communication Index (C_{if})*
universities	0.851
research institutes	0.356
industry	0.449

$$*C_{if} = \frac{N_{if}}{N_i}$$

where: n_{if} = number of individuals in foreign countries with whom contact is reported by individuals in sector i

N_i - number of individuals reporting in sector i

University contacts, in addition to being greater in number, are more evenly spread around the world (Table III). The research institutes and industry turn more frequently to British sources for their information. This situation will undoubtedly alter somewhat with the entry of Ireland into the European Economic Community. Links by all three sectors to the North of Ireland are uniformly weak. Of course, there are fewer contacts available in the North, than in the U. S., Britain or Continental Europe. Even so, one would expect far more contact with a community of scientists who are culturally, linguistically and geographically so close.

Table III

Location of Foreign Contacts by Sector

research sector	Great Britain	Continental Europe	United States	other countries	North of Ireland*
universities	0.382	0.204	0.174	0.063	0.053
research institutes	0.209	0.090	0.027	0.032	0.047
industry	0.291	0.120	0.019	0.019	0.006

*Not included as foreign in the totals

The Flow from Institution to Institution

Within each research sector are a number of institutions. These institutions are connected among themselves to varying degrees, and the coupling to other sectors varies considerably from one institution to the next.

Communication Among the Universtiies

The two major universities have between them a total of five campuses, one of which was too small to be distinguished in this analysis. Of the remaining four (which will be referred to as A, B, C, and D), two are in Dublin, the capital, and two in other cities. The Dublin campuses account for about two-thirds of the research staffs in the sector. The level of communication between university personnel is given in Table IV.

Communication between A and D is the strongest of the six possible pairings. Campus C is, as regards communication, the most isolated of the four. The bond between A and B is moderately strong. Generally, the more physically isolated campuses had the weakest communication bond.

Table IV

Communication Among the Four Universities
(Monthly or More Frequent Contact)

	Communication Bond (C_{ij})		
	Univ A	Univ B	Univ C
University D	0.412	0.067	0.096
University A	–	0.236	0.079
University B	–	–	0.040

The Position of the Universities

The universities are more closely linked to the public sector than to any of the other segments of the research community in Ireland. Three of the four have very strong communication bonds (Figure 1). Only C, possibly because of its geographic location, has failed to develop strong contact. A is responsible for most of the university communication with research institutes. D and B have a moderate level of contact, while C's is light.

Only D and A have even a moderate level of communication with industry, and then only with state-sponsored firms. None of the universities have even moderate contact with private industry.

The colleges of technology, medical research institutions and other academic institutions show, at best, a moderate degree of interaction with the universities. In some of these, however, one would expect communication to take place with only one part of a university (e.g., a medical school). Final judgement on this situation therefore must await analysis of university communication at the departmental level (to be reported in a later paper).

Foreign Contact by the Universities

All four universities show similar levels of foreign Communication (Table V). C, somewhat surprisingly, is the leader, showing an even higher level (proportional to its size) of contact with Britain than D. Campus A has established the strongest ties to the continent, and with leads in communication with the United States. D has far more contact with the North of Ireland than the other three.

Foreign contact is primarily with university personnel, although a high proportion is with the staff of "research institutes" abroad.

This is something of a "catch all" category for all laboratories that are not clearly universities or industrial. It includes personnel from foreign government departments and universities and some quasi-public research organizations, as well as such governmental research institutions as the U. K. Atomic Energy Authority laboratories at Harwell and the Naval Research Laboratory in the United States.

The Research Institutes

There are three major research institutes in Ireland. Their spheres of activity are broadly speaking, Industry, Planning and Environment and Agriculture and Food.

Table V

Location of Foreign Contacts by the Four Universities

location of contacts	Communication Bond (C_{if}) Univ D	Univ A	Univ B	Univ C
Great Britain	0.459	0.346	0.264	0.478
university	0.294	0.231	0.181	0.366
research institute	0.156	0.093	0.055	0.116
industry	0.009	0.022	0.028	0
Continental Europe	0.147	0.275	0.195	0.116
university	0.101	0.110	0.139	0.087
research institute	0.046	0.110	0.042	0.029
industry	0	0.055	0.014	0
United States	0.138	0.187	0.167	0.188
university	0.092	0.143	0.139	0.159
research institute	0.018	0.033	0.014	0
industry	0.028	0.011	0.014	0.029
Other Foreign	0.046	0.044	0.097	0.101
university	0.037	0.033	0.069	0.029
research institute	0.009	0.011	0.028	0.072
industry	0	0	0	0
Total Foreign	0.790	0.852	0.723	0.883
North of Ireland	0.101	0.039	0.069	0
university	0.101	0.033	0.055	0
research institute	0	0.006	0.014	0
industry	0	0	0	0

Communication Among the Research Institutes

Communication among the three research institutes is at a weak to moderate level (Table IV). It is neither as low in frequency as the lowest of the university pairings, nor as high as the highest. Institutes I and III have stronger communication bonds with II than they have with each other, but the amount of communication could probably be increased generally in this sector, with beneficial results.

Table VI

Communication Among the Three Research Institutes (Monthly or More Frequent Contact)

	Communication Bond (C_{ij})	
	Inst. I	Inst. II
Institute III	0.074	0.104
Institute I	-	0.210

The Position of the Research Institutes

Institute II clearly dominated the institute sector (Figure 2). It is the only institute to show a high level of interaction with other parts of the research community. It is responsible for the majority of contact with the universities, and is the only research institute to show a high level of contact with private industry. It must be remembered, however, that the only industrial firms surveyed were those which themselves carried out research and development. That necessarily excludes many firms; perhaps it excludes a higher proportion of the firms served by the I and III than of those served by II. On the other hand, the survey in no way prevented individuals in any of the institutes from reporting contacts they had with people in any industrial firm, whether or not it is supported R&D.

So the results are, at best, a little dismaying. This is especially true for state-sponsored industry, which is not linked by strong bonds with any of the three institutes.

When contacts that occur less often than once a month are included in the analysis (Table VII), there is little change in the results. There is apparently much that can be done to strengthen communication bonds between the research institutes and industry.

Foreign Contact by the Research Institutes

Institute III has developed the greatest number of British contacts, but reported no contacts outside of Europe (Table VIII). Institute I has developed contact in a large number of countries around the world, but the actual number of individuals is still small. It has the lowest index of foreign contact of the three. Institute II has well developed contacts throughout the world, and maintains its contacts across a wide variety of institutions. It is the only one of the three that comes even close to the universities in the extent of its foreign contact.

Institute III could certainly stand to develop stronger communications with American firms, universities and research institutes. It will probably develop more communication with the continent now that Ireland has joined the European Community. Its overwhelming contact with British sources probably indicates too great reliance on these as compared to other foreign sources. A broader base of contact might stimulate innovative ideas, and this as we have seen, is the purpose of R&D aimed mainly at economic development. Institute I is a relatively new organization, so perhaps it has not had time to properly develop a range of foreign information sources. It should certainly pay attention to the critical effort a good foreign network of sources can have on a small country's information intake.

Communication Among Other Parts of the Research and Development Community

The network for the remaining portions of the R&D community is a somewhat sparse one (Figure 3). This category includes local authorities; government departments and agencies concerned with meteorology, ordinance survey, geological survey, and forestry plus a number of quasi-public agencies.

With the exception of the block called "other public agencies and departments," which has strong communication with the medical field and with semi-state industry, there is just very little communication among other agencies.

This network better than any other displays the central role played by the research institutes and universities in integrating the flow of communication in Ireland. Not only do they directly support most of the research and development work performed in the country, but without the communication links which they provide, other organizations supporting research would have little or no chance of learning of developments outside their own organizations.

Much of this report could be taken to be critical of the performance of the universities and research institutes in communicating among themselves, with other organizations in Ireland and with scientists abroad. But without their contribution to integrating information, the universities with foreign science, and the research institutes within the country, the communication network would be very poorly connected.

A More General View of the Network and Discussion

We entered the present study, if not with tightly structured hypotheses, then at least with some preconceptions of what a communication network in a small country like Ireland might look like. These preconceptions have not been completely supported, but neither have they been completely denied.

One might expect, for example, that the research institutes would form a buffer stage between less directed research of the universities and the needs of industry. And, in fact, we find that they do play this role, to some degree. Communication between research institutes and universities and between research institutes and industry is much stronger than direct university-industry communication. But then, not all research institutes are equally effective in establishing these bonds (Figure 4). State-sponsored industry, in particular, fails to receive the communication it must be assumed to need, either directly or through this buffer stage.

In a similar way, one might expect the universities to be the principal link to foreign sources, while research institutes play a secondary role, with both passing the information along to industry. In fact, the universities do have the strongest foreign contact, but they communicate little of this information directly to industry. Moreover, the research institutes generally have neither the amount nor the diversity of foreign sources that industry itself has developed. Partly through the mechanisms of foreign ownerships, but largely through sheer need (Cf., Allen and Reilly, 1973) industry appears to have developed its own foreign information sources. This is not entirely unsatisfactory, but it does have several discomforting features. Judging from the extent of their foreign contacts, both the research institutes and the universities should be able to perform a "gatekeeping" role for industry. The gatekeeper, however, must both have extensive outside contact and also be highly sought for consultation by colleagues within his organization. The four universities and three research institutes, to some degree, meet the first of these conditions. What most of them lack, though, is a strong communication bond with industry. With the possible exception of Institute II, none of these seven organizations has assumed a gatekeeper role, between Irish industry and foreign technology.

Industry has apparently found it more useful to develop its own foreign sources, directly. There is certainly nothing wrong with an industrial firm keeping abreast of foreign technology in this way, but at a system level it appears terribly inefficient. Moreover, if an industrial firm learns of a foreign development, competitive pressures and the general isolation of the individual firm tend to severely limit the dissemination of that information. The research institutes or universitites, being free of such competitive pressure,should make such information more generally available. This is particularly critical in the case of the small domestically controlled firm, which often has neither the business associations nor the resources to allow it to cultivate foreign sources of technology.

The research institutes were established to bring new technology to industry, and it is a stated policy of the National Science Council to encourage cooperation between universities and industry. It is therefore surprising and disturbing to find so little communication between these elements of the R&D community.

The Development of Communication Bonds

It is one thing to say that communication between any two organizations or between sectors should be improved, but it is quite another thing to specify just how one would go about it. Fortunately, participants in the survey were asked (by means of coded categories) to indicate how they first met each of the individuals, outside their organization, with whom they regularly discuss scientific and technical subjects. There was a total of 1,282 such contacts for which respondents indicated the way in which the contact first came about. Of these, 495, or 38.5 percent, were with scientists or technologists in other countries. This statistic, in itself, is impressive, showing the extent of foreign contact by Irish scientists.

Domestic Contacts

Of the remaining domestic contacts, 36 percent were the result of working relationships between individuals in different organizations that were in effect at the time of the study. Work relationships, or project membership are a very strong determinant of communication patterns (Walsh & Baker, 1972). For the present purposes, however, it is more important to determine how those contacts, which were not required by the nature of the work, came about. These, it seems, are caused most often by people becoming acquainted through working together in the same organization (Table IX). In these cases, the contact had either worked in the respondent's organization and then changed jobs; the respondent had formerly worked in the contact's organization; or both had worked in a third organization. Job mobility, to the extent that it exists, appears to be a very important determinant of interorganizational communication.

Acquaintances made in the university are next in line of

importance. These include professor-student relations, as well as relations among former students. In a small country, such as Ireland, with a limited number of university departments in any field, one might expect that a reasonable number of scientists would, in any specialty, have known one another from university days. Still, to find such a strong effect on inter-organizational communication patterns is somewhat surprising.

Professional society conferences and meetings rank fourth in importance. Most of these were, of course, held within the country. Nevertheless, a surprising number of Irish scientists meet one another for the first time while attending a conference in another country.

Table IX

Sources of Domestic Communication Contacts (787 Instances)

Way in which contact was established	proportion (percent)
participation in current working relationship	36.1
previously worked together in same organization	19.1
met in university	13.6
met through professional society membership or conference	11.9
introduced by mutual acquaintance	6.2
formerly had working relationship	3.0
other	10.0
	100.0

Introductions and former working relationships account for less than 10 percent of the total. An important point, to be made here, concerns the difference between the number of communication contacts resulting from <u>current</u> work relations and the number from <u>past</u> work relations. The work-induced force to communicate is apparently very strong, but not terribly persistent.

Finally, the category "other" in Table IX includes a vast range of reasons, from the use of common facilities to family relationships. The most commonly cited reason in this group though is that of having common research interests. Apparently, one member of the pair learns of the other's research, probably through the published literature, and then seeks more direct and continuing contact with the second party.

Sector to Sector Contacts

The experience of having worked together in the same organization is the dominant underlying cause of communication among the three sectors of the R&D community in Ireland (Table X). This is especially true in the case of the universities, in which former students and former staff members maintain contact with their old research teams for some period after joining research institutes or industrial firms. Just how long these relationships endure is not known at this point, but will be the subject of furter investigation. At any rate, the existence of people, particularly in industry, who have recently left a university, provides a potential for improving contact between universities and the other two sectors. Such people should be encouraged to keep up with the work of their old university research groups, and be given some form of inducement to maintain contact with their old colleagues.

Next to job mobility, professional society meetings and conferences provide an effective mechanism for promoting contact among sectors. This is especially true in the case of research institute and industry relations. The professional conference provides an excellent vehicle for forging new acquaintanceships and renewing old ones among scientists. To stimulate inter-sector communication, agenda should be set in such a way that industry personnel are attracted, and then research institute and university staff should be encouraged to attend. While it may be a bit more difficult to induce university attendance, attendance by the staff of the research institutes should not present so much of a problem.

Table X

Sources of Communication Contacts
Between Individuals in Different Sectors
(Excluding Current Working Relationships)

way in which contact was established	contact between universities and research institutes	research institutes and industry	universities and industry
previously worked together	28.4%	22.8%	36.1%
met in university	28.4	13.5	22.2
prof. society or conference	21.0	34.0	19.5
mutual acquaintance	16.0	6.8	2.8
former working relationship	3.7	0	8.4
other	2.5	22.8	11.1

Contacts Within Sectors

Job mobility once again is a critical factor im promoting communication among university faculty and staff (Table XI). It is less important in the case of research institutes and of industry. In the former, there were very few contacts for which information was obtained concerning how the two individuals first met,[1] so the data in the second column of Table XI have little meaning. In the case of industry, associations among firms (belonging to the same group of firms; contractual relationships; etc.) are the overwhelming determinant of communication contacts. In other words, formal organizational relationships are most important in that sector. In general, however, communicaton among industrial firms, particularly semi-state firms,is very poor. Perhaps greater job mobility among semi-state firms might improve this situation. At any rate, it is certainly something, along with greater conference attendance, that could be seriously considered.

1 This question was not included in the questionnaire administered to An Foras Taluntais, the largest of the three research institutes.

Table XI

Sources of Communication Contacts
Among Individuals in the Same Sector
(Excluding Current Working Relationships)

way in which contact was established	universities	research institutes	industry
previously worked together	30.8%	0%	11.7%
met in university	23.4	0	20.6
prof. society of conference	13.9	0	11.7
mutual acquaintance	12.7	0	5.8
former working relationship	2.2	100	5.8
other	17.0	0	44.2

Foreign Contacts

As stated earlier, nearly 40 percent of communication contacts are with individuals outside of the country. It is interesting to note once again that job mobility is an important factor in this, at least in the case of universities and research institutes (Table XII). This is very interesting, since it presents a positive aspect of the tremendous brain drain that Ireland has suffered. In recent years, Irish universities have produced trained personnel at a rate far exceeding the economy's capacity to absorb them. Consequently, a very high proportion of Irish graduates are forced to seek employment in other countries. The present data would indicate that there is a reverse flow as well, and that the combination of the two will provide very definite long term benefits to the Irish economy. It is not known, at this point, whether job mobility produced few foreign contacts by the industrial sector, because returning emigrants did not go to work for industrial firms or for other reasons, but the fact remains that it is extremely important for the other sectors.

Table XII

Sources of Foreign Communication Contacts
(495 Instances)

way in which contact was established	contacts by individuals in			
	universities	institutes	industry	other
previously worked together	38.3%	26.4%	8.1%	21.0%
met in university	10.3	26.4	6.8	13.6
prof. society or conference	20.6	36.8	35.1	22.2
current working relationship	12.5	0	23.0	16.0
former working relationship	3.8	0	6.8	7.4
mutual acquaintance	5.6	5.2	6.8	11.1
other	9.0	5.2	13.5	8.6

The returning emigrant scientist brings back with him not only a range of technical skills developed during his foreign employment, but a cadre of foreign personal contacts to help him maintain those skills, as well. The data here further support the finding by Allen, et. al. (1971) that foreign employment is the critical factor in developing international gatekeepers. The emigrant who has spent some time working in an R&D organization outside the country is a potential gatekeeper. He must first be attracted back home, then properly integrated into a research organization, and finally provided with the wherewithal to maintain his contact with foreign colleagues. But to begin with, he must be attracted into returning home. A strong program to attract back emigrants in critical skill areas would have an enormous potential for technology transfer. The form for such a program will be treated in subsequent papers in this series, following further analysis of the data, and more intensive consideration of the emigration problem.

Before leaving the subject of foreign contacts, the reader should note the importance of professional society conferences. It is a bit surprising to see that these are more important for research institutes and industry than they are for university staff. Nevertheless, they remain important. Since most of these were foreign conferences, the data would provide an argument for increased travel budgets to promote conference attendance. In addition, since the returned emigrant is more likely to know some of the foreigners attending a conference (Cf., Allen, et. al., 1971), he should be given some consideration when selecting the people to be sent. Again, this is a subject to be given more careful consideration in a subsequent report.

Conclusions

It appears that the scientist's ability to keep abreast of foreign technological developments through personal contact is an effective way for a country to import technological information. It also appears to be a relatively inexpensive way; the cost of allowing an individual to maintain his foreign contacts through periodic travel is quite small. The present evidence indicates that his external contacts are most effectively developed through active research experience in a foreign country.

The implications of these results are unambiguous. If international transfer of technology is to be fostered, nations should seek to open those channels through which information flows most effectively. They should assist able, domestically educated research personnel to do research abroad. They should encourage researchers to have foreign sabbaticals, fellowships, and other forms of extended foreign technical experiences. They should help gatekeepers to perform their role as an efficient mechanism for transfer by providing funds to maintain existing contacts through foreign travel. They should stimulate contact among organizations by the mechanisms described in this paper. They should assist organizations in properly developing their specific roles vis-a-vis technology transfer. The development of interpersonal and inter-organizational contacts is not in itself the complete answer to international technology transfer. It is, however, an essential part of the process. Where it works, it is doubtless the least expensive of the available mechanisms -- and we are now learning ways to further improve its effectiveness.

REFERENCES

Allen, T. J. (1970), Communication Networks in R&D Laboratories R&D Management, 1, 14-21.

Allen, T. J., Piepmeier, J. M. and Cooney, S. (1971), The International Technological Gatekeeper, Technology Review, 73, 37-44.

Allen, T. J. and Reilly, V. (1973), Getting the Word Around: Report of a Pilot Study on Technology Transfer to Irish Industry, Cambridge, Mass.: MIT Sloan School of Management, Working Paper No. 650-73.

Walsh, V. M. and Baker, A. G. (1972), Project Management And Communication Patterns in Industrial Research, R&D Management, 2 103-109.

INSTITUTIONAL BARRIERS TO THE SCIENCE AND TECHNOLOGY INTERACTION

Prof. Dr. Gerhard Mensch

International Institute of Management
Science Center Berlin

INTRODUCTION

Following is the menu of this paper:

Apéritif: Motivation

Deux Menus à la Choix

1. The Technology - Science - Interaction
2. The Science - Technology - Interaction

Le Repas: Do Barriers Exist?

1. Hors d'oeuvre varié:
   - basic innovations
   - inventions
2. Le Vin: Line of Reasoning
3. Entrée:
   - Blocking of Basic Innovations
   - Crowding of Basic Innovations
4. Plat de Résistence:
   - Delaying of Basic Innovations
   - Rushing of Basic Innovations
5. Dessert:

   Inefficiencies in the Knowledge Supply?

Café: Policy Problem

Aperitif

Man is a tool making animal (Benjamin Franklin, 1778). Societies of modern man are built on a highly diversified technology base. In Western civilization, we live with and feed from several technology systems that interest one another.

The interplay within such a conglomerate of technologies can be illustrated by recalling the Le Châtelier principle. Our technological world may be thought of as being in a state of dynamical equilibrium - whatever that is - if the rate of quantity change and quality change is somewhat modest. Specifically, to perpetuate this Nirvana none of the existing technology systems should break loose and expand so vigorously in size as to overpower the other systems, thereby conquering the lion's share in capital, manpower and brainpower. Moreover, none of the preexisting technology systems, which are still in their incubation phase though ready for being innovated as brand-new, hyper-modern technology systems, none of these potential intruders into the existing technological world should offer new goods and services that are so radically different from the available ones as to grossly upset the balance between all technologies. Conclusion: Modesty in quantity and quality changes is the general rule in the interplay of technology systems.

Radical quantity and quality changes invade the way of life of everybody in many direct ways and via secondary effects. Many people have a low capability for absorbing the shocks of involuntary changes. They suffer from it. Therefore, it is in the interest of many people and parties that the general rule of modesty is being observed by innovators. This rule is a built-in barrier against a high rate of radical change in any social system. It can take various institutional forms. We should focus on some.

Deux Menus à la Choix

In the science-and-technology interaction, we may distinguish two areas.

1. The Technology - Science - Interaction

In the interplay between the existing technology systems of a given society (technology - technology - interaction), scientists play several roles. For example, they provide the knowledge base for modest quality changes and for quantity changes. For the product and service industries, scientific manpower is an input factor. The input factor markets are closely connected with the product and service markets of these industries; and it is in these output markets where the need for further scientific knowledge originates. Thus, the technology - technology interaction necessitates a specific labor market for scientists. This demand-pull of existing technologies for scientist's time I call the technology-science-interaction.

We have seen that modern society at large, and the ruling powers in particular, have a need for control over quantity and quality changes in the technology base. If such controls actually exist, then they must have the form of barriers against certain kinds or certain numbers of outputs of technologies, or, much more subtly, against an efficient technology-science transfer, and these barriers must - for the purposes of the controlling powers - possess a NO GO - GO switch. Note that one can not easily control the thought processes of scientists. But it is relatively easy to control scientific activities in terms of working conditions. Scientists have found some answers to the Damocles sword of control: they diversified the types of scientific institutions.

Thus, the demand-pull of existing technologies for scientist's services, i.e., the technology-science-interaction, takes place in a network of related institutions. Whether this market compound is efficient is subject to serious doubt. However, in the technology-science interplay, it might be difficult to distinguish between institutional barriers which A has erected against a potentially rich flow from B to C, and between lack of supply from B or lack of demand from C. In any case, market economists will find that the technology-science interaction is just another example of an imperfect market.

2. The Science - Technology - Interaction

Therefore, let us focus on what we may call the knowledge transfer in the science-technology-interaction. In a given society, this is the interplay between existing technology systems and those scientists and engineers who are suggesting some new goods or services for which there is no market as yet. There is a knowledge supply push toward innovation.

R & D economists have often argued that such a knowledge supply push does not exist sui generis. That is, they say, whenever there has been evidence for a successful knowledge push for innovation, then there always has been demand for the new good or service in the first place. Therefore, they say, we need not bother to look at the knowledge transfer. Note that this argument diverts the attention from possibly existing barriers against an efficient knowledge transfer.

Moreover, the argument is misleading. If there was always demand for the novelty in cases of successful knowledge supply push, and there was no demand in cases of unsuccessful knowledge supply push, we must ask: was there no demand by the general public, or was there no demand by industries with vested interests in competing goods and services? Do we not have the electrical car because we did not want it so far? Or is it because of alleged direct and indirect opposition of the oil and automobile industry, who are said to have their foundations, news media, lobbyists, and other beneficiaries frustrate the research and the demand articulation for the electric car?

Institutional barriers to the knowledge supply push - if such obstacles actually exist - would deserve our primary interest. Such barriers could discourage both knowledge accumulation and demand articulation for certain innovations, and they could erect institutional walls between the rider and the horse, i.e., between those people who have the knowledge and those people who have command over the means to innovate.

But do such obstacles really exist? It is this question which we shall pursue now.

Le Repas

1. Hors d'oeuvre varié

"The phenomenon of human knowledge is no doubt the greatest miracle in our universe (Karl Popper, 1971). Therefore, to start with, I offer fruits from the tree of scientific knowledge: Basic innovations. In Tables 1 and 2 the "basic innovations" of modern chemistry and electro-technology are given. An innovation, in my work[1], is the doing of a new thing or the doing of a thing in a new way. That is Schumpeter's definition. I distinguish between basic and improvement innovations[2]. Basic innovations in a field like chemo-technology or electro-technology are the first instances of commercial application, marketing, or serious production of the most successful products or services. These are outstanding achievements of risk-taking entrepreneurs.

In Tables 1 and 2, the innovation dates are listed in the first columns. For the purposes of timing the basic innovations, we may have taken the date of opening the first factory for producing the new product (pharmaceuticals, for example), or the data of setting up a market for the new service (telephone, for example).

In the second columns of Tables 1 and 2, the invention dates are given. To find these dates, one has to go back through the branches of the tree of knowledge of the corresponding science. The branches correspond to the various specialties of the science. The method of finding "the" invention date shares many features with the tracing algorithm. A full account of the information condensed in Tables 1 and 2 is given elsewhere[3].

The next two Tables, 3 and 4, contain dates for basic innovations and inventions which are the product of several trees of scientific knowledge that came to bear fruits at the same time: the basic innovations in Table 3 are from the first half of the 19th century, those in Table 4 from the first half of the 20th century.

With these four sets of data from Mensch,[3] we shall **establish our claim:** there are barriers to the knowledge transfer.

2. Le Vin

And this is the spirit in which we shall proceed in answering the existence question. This line of thought we shall follow:

The transfer of scientific knowledge into practical use will be dealt with as the process from invention to innovation. For specific classes of technological innovations we shall observe how they came about.

The existence of barriers against innovations can be taken for granted if the **band wagon** effect in the emergence of innovations exists. By that behavior of innovations we mean

(a) the early innovations in a specific class are unduly retarded. But once they break through,

(b) the subsequent innovations show a crowding and rushing as if a dam is broken.

Conversely, if the timing of basic innovations is compatible with a random walk behavior, then the existence of barriers against innovations is highly in doubt. For the beforementioned data I have tested the randomness of occurrences with nonparametric statistics (with the runs test). The inference is: the basic innovations emerged in such a grape clustering fashion that the pattern is unlikely (with significance $\alpha > 95\%$) to be the result of a ran-

Basisinnovationen Chemie 2. Haelfte 19. Jahrhundert

	Innovation	Invention	Dauer	100% / Dauer
Thomas-Stahl	1878	1855	23	4,34
Sicherheitszuendhoelzer	1866	1805	61	1,63
Anilin-Farben	1860	1771	89	1,12
Speisefett	1882	1811	71	1,26
Indigo-Synthese	1897	1880	17	5,88
Soda	1861	1791	70	1,42
Aluminium	1887	1827	60	1,66
Kaeltetechnik	1895	1873	22	4,54
Kunsteseide	1890	1857	33	3,03
Kupferkunstseide	1898	1890	8	12,50
Heiz-Gas	1875	1780	95	1,05
Acetylen-Schweissen	1892	1862	30	3,33
Dynamit	1867	1844	23	4,34
Kunstduenger	1885	1840	45	2,22
Konservierungsmittel	1873	1839	44	2,27
Elektrolyse	1887	1789	98	1,02
Antitoxine	1894	1877	17	5,88
Chloroform (Narkosemittel)	1884	1831	53	1,88
Jodoform (Septikum)	1880	1822	58	1,72
Veronal (Barbiturate)	1882	1863	19	5,26
Aspirin	1898	1853	45	2,22
Antipyrin (Synthet. Schmerzmittel)	1883	1828	55	1,81
Backpulver	1856	1764	92	1,08
Gipsverband	1852	1750	102	0,98
Schwefelsaeueregross-fabrikation	1875	1819	56	1,78
Synthet. Alkaloide (Kokain)	1885	1844	41	2,43
Synthet. Alkaloide (Chinolin)	1880	1834	46	2,17
Edelstaehle	1856	1771	85	1,17

Table 1

Basisinnovationen Elektrotechnik 2. Haelfte 19. Jahrhundert

	Innovation	Invention	Dauer	100% / Dauer
Elektrodynamische Messung	1846	1745	101	0,99
Bleiplatten-Akkumulator	1859	1780	79	1,26
Doppelanker-Dynamo	1867	1820	47	2,12
Kummutator	1869	1833	36	2,77
Trommelankermotoren	1872	1785	87	1,14
Bogenlampe	1873	1802	71	1,40
Gluehlampe	1879	1800	79	1,26
E-Lokomotive	1879	1841	38	2,63
Elektro-Heizung	1882	1859	23	4,34
Kabelkonstruktion	1882	1820	62	1,61
Telefon	1881	1854	27	3,70
Dampfturbinen	1884	1842	42	2,38
Wasserturbinen	1880	1824	56	1,78
Transformatoren	1885	1831	54	1,85
Widerstandsschweissen	1886	1841	45	2,22
Lichtobogenschweissen	1898	1849	49	2,04
Induktions-Schmelzen	1891	1860	31	3,22
Zaehler	1888	1844	44	2,27
E-Eisenbahn	1895	1879	16	6,25
Entdaempfung von Fernsprechleitungen	1910	1893	17	5,88
Hochspannungs-isolatoren	1910	1897	13	7,69
Benzinmotor	1886	1860	26	3,84

Table 2

Basisinnovationen 1. Haelfte 19. Jahrhundert

	Innovation	Invention	Dauer	100% / Dauer
Starkstromerzeuger	1849	1820	29	3,44
Reizstrominduktor	1846	1831	15	6,66
Tiefseekabel	1866	1847	19	5,26
Herstellung von Elektrizitaet	1800	1708	92	1,08
Isolierte Stromleitung	1820	1744	76	1,31
Bogenlampe	1844	1810	34	2,94
Pedalrad	1839	1818	21	4,76
Gewalzte Schienen	1835	1773	62	1,61
Gewalzter Draht	1820	1773	47	2,12
Puddelofen	1824	1783	41	2,43
Kokshochofen	1796	1713	83	1,20
Tiegelgussstahl	1811	1740	71	1,40
Erste deutsche Eisenbahnstrecke	1834			
Lokomotiven	1824	1769	55	1,81
Telegraphie	1833	1793	40	2,50
Deutscher Zollverein	1834			
Bleikammerverfahren	1819	1740	79	1,26
Pharma-Fabrikation	1827	1771	56	1,78
Chininfabrikation	1820	1790	30	3,33
Hartkautschuk	1852	1832	20	5,00
Portlandzement	1824	1756	68	1,47
Kaliumchlorat	1831	1777	54	1,85
Photographie	1838	1727	111	0,90

Table 3

Moderne technologische Basis-Innovationen
(extrahiert aus Jewkes, Sawers und Stillermann [11])

Nr. i	Bezeichnung der Innovation	Erfindung t_{1i}	Anwendung t_{2i}	Dauer d_i	Tempo $S_i = \frac{100\%}{d_i}$
1	Hydramatisches Getriebe	1904	1939	35	2,86
2	Hudraulische Kupplung	1904	1937	33	3,03
3	Kugelschreiber	1888	1938	50	2,00
4	Katalytische Petrol-Trennung	1915	1935	20	5,00
5	Wasserfestes Zellophan	1900	1926	26	3,85
6	Cinerama	1937	1953	16	6,25
7	Kontinuierlicher Stahlguss	1927	1948	21	4,76
8	Kontinuierliches Heiss-Walzen	1892	1923	31	3,23
9	Baumwoll-Pfluecker (Campbell)	1920	1942	42	2,38
10	Baumwoll-Pfluecker (Rust)	1924	1941	38	2,63
11	Faltenlose Stoffe	1906	1932	26	3,85
12	Zyklotron	1929	1937*	12	8,33
13	DDT	1939	1942*	3	33,33
14	Diesel-Lokomotive	1895	1934	39	2,56
15	Elektrische Praezipation	1884	1909*	25	4,00
16	Fluoreszierende Beleuchtung	1852	1934	82	1,22
17	Freon Kuehler	1930	1931*	1	100,00
18	Gyro-Kampass	1827	1909	82	1,22
19	Fett-Haertung	1900	1909*	9	11,11
20	Helikopter	1904	1936	32	3,13
21	Insulin	1889	1922	33	3,03
22	Duesenmaschine	1928	1941	13	7,69
23	Kodachrome	1910	1935	25	4,00
24	Magnetophon	1898	1937	39	2,56
25	Plexiglas	1877	1935	58	1,72
26	Neopren	1906	1931	25	4,00
27	Nylon. Perlon	1927	1938	11	9,09
28	Penizillin	1922	1941	19	5,26
29	Polyethylen	1933	1953	20	5,00
30	Kraft-Lenkung	1900	1930	30	3,33
31	Radar	1887	1934	47	2,13
32	Radio	1887	1922	35	2,86
33	Raketen	1903	1935	32	3,13
34	Sicherheits-Rasierer	1895	1904*	9	11,11
35	Automatische Uhr	1922	1928*	6	16,67
36	Cronings Gussform	1941	1948*	7	14,29
37	Silikon	1904	1946	42	2,38
38	Rostfreier Stahl	1904	1912*	8	12,50
39	Streptomycin	1921	1944	23	4,35
40	Sulzer Webstuhl	1928	1945	17	5,88
41	Synthetische Waschmittel	1886	1928	42	2,38
42	Synthetischer Polarisierer	1857	1932	75	1,33
43	Fernsehen	1907	1936	29	3,45
44	Terylen Fasern	1941	1955	14	7,14
45	Klopffestes Benzin	1912	1935	23	4,35
46	Titanium	1885	1937	52	1,92
47	Transistor	1940	1950	10	10,00
48	Tungsten Karbid	1900	1926	26	3,85
49	Xerographie	1934	1950	16	6,25
50	Reissverschluss	1891	1923	32	3,13

*) Dauer der kommerziellen Reifung

Table 4

dom process[4]. This pattern I call the band wagon effect in the emergence of basic innovations.

3. Entrée

The bandwaggon effect in group behavior denotes the particularity that the individuals - instead of passing by with a somewhat uniform frequency rate - first do not come at all, but shortly after some forerunners do show up, the whole party marches by together.

Evidence for the existence of the band wagon effect is presented in Figure 1, where the frequency of basic innovations from Tables 1-4 is plotted as a time series. The distribution exhibits dramatic swings in the frequency of basic innovations.

A swing from "very few at a time" to "very many at a time" is typical for a horde that presses through a widening gap in the fence.

4. Plat de Résistence

Let us substantiate this phenomenon also in another way. The band wagon effect implies a coincidence in blocking and delaying and in crowding and rushing of innovations:

- There is a low frequency of basic innovations when the barriers are high. Then the advancement of innovative change is slow.

- And there is a high frequency of basic innovations when the barriers are down. Then the advancement of innovative change is fast.

- In any case, the cumulative frequency of a group of basic innovations should be correlated with some measure of the speed of innovative change.

As a measure of the average speed of innovative change we take the reciprocal of the time span between invention and innovation, expressed as a percentage value[5]. For an innovation i, occurring at t_{2i}, that time span is

$$d_i = t_{2i} - t_{1i}$$

FIG. 1

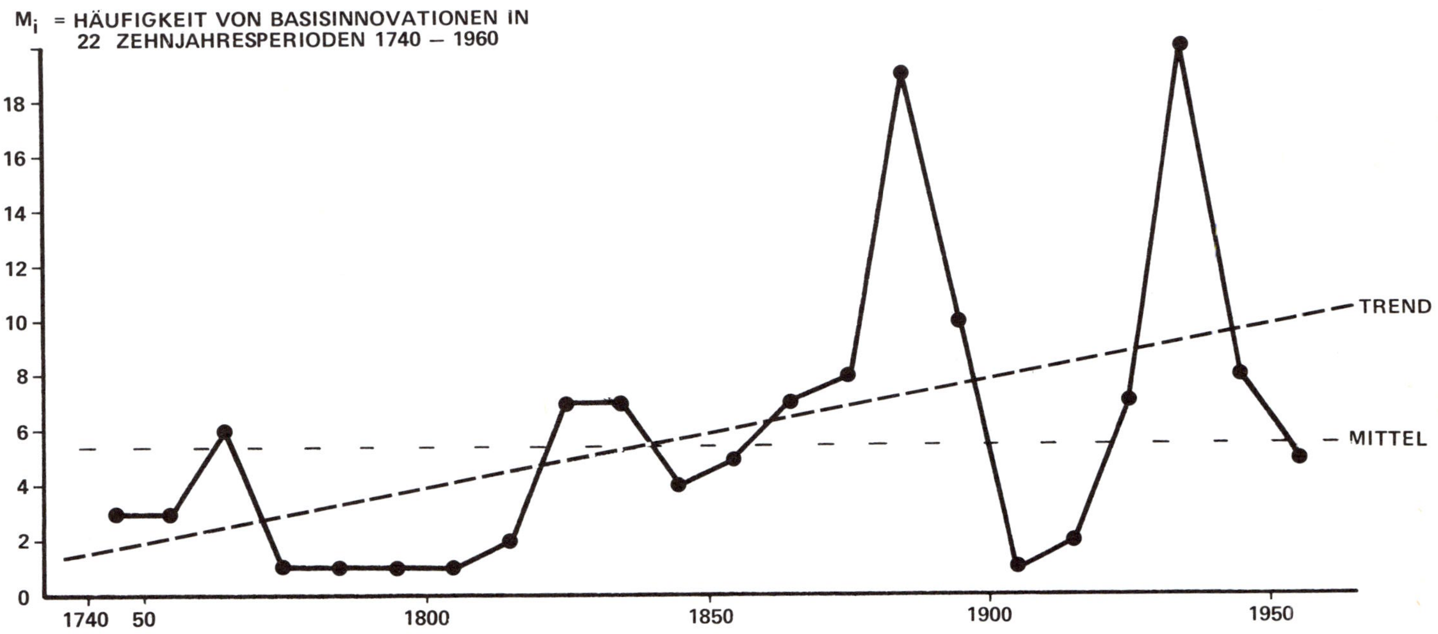
M_i = HÄUFIGKEIT VON BASISINNOVATIONEN IN
22 ZEHNJAHRESPERIODEN 1740 – 1960
TREND
MITTEL
0
2
4
6
8
10
12
14
16
18
1740
50
1800
1850
1900
1950

if t_{1i} is the date of the invention as given in the data tables. And

$$S_i = \frac{100\%}{d_i}$$

is the average annual advancement of the innovative process that connects the two events.

Let D_i denote the regression function through the cloud of points (d_i, t_{1i}), and let S_i denote the regression function through the cloud of points (S_i, t_{1i}). Then

$$S_1 = 100\% \frac{1}{D_1}$$

Similarly defined are D_2 and S_2. These functions, for the set of basic innovations of Table 4, are given in Figure 2 and 3. In Figure 4, the average speed of change function D_2 is plotted together with F(t), the cumulation distribution function of the basic innovation in Table 4.

Note that the speed measure S_2 in Figure 4 exhibits exactly that retardation - acceleration pattern which barriers against innovation do produce before and after they crumble: first there is a standstill and then a sudden spurt. Curve S_2 describes a GO SLOW - GO FAST switch.

We are now demonstrating that this switching pattern of blocked and freed advance of change holds not only for the bulk of basic innovations in the first half of this century (those in Table 4) but also for those in the 19th century (Tables 1-3). As a preliminary step observe that the beforementioned formulas link S_2 to D_1. If D_1 has the shape as in Figure 2 (the middle piece slopes down with -45°), then S_2 has the shape as in Figure 3 and 4 (the middle piece jumps up with 90°). Thus, by inspection of the Figures 5-7, depicting the D_1-curves for the sets of basic innovations in Tables 1-3, the observer will see that these D_1-curves have a center piece with -45° slope. Therefore, the corresponding S_2 curves for the average speed of change have the predicted shape: they represent GO SLOW - GO FAST switches.

Conclusion: Barriers to the science transfer into basic innovations do exist. And sometimes they break down.

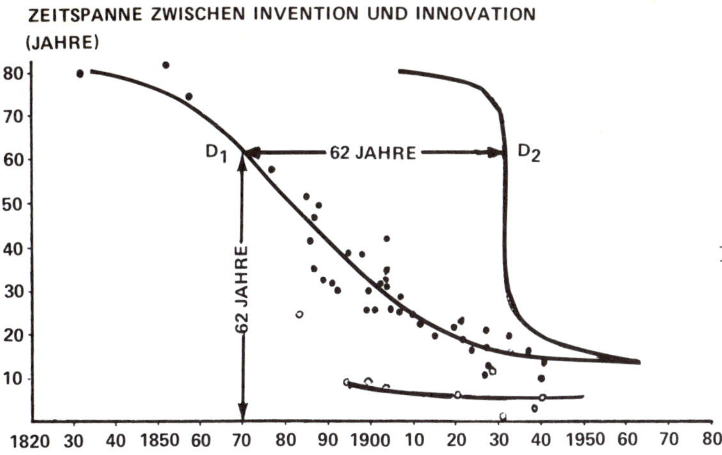

FIGURE 2

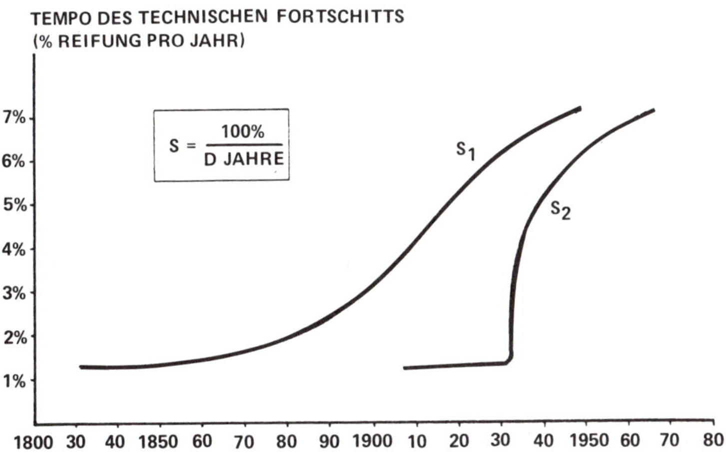

FIGURE 3

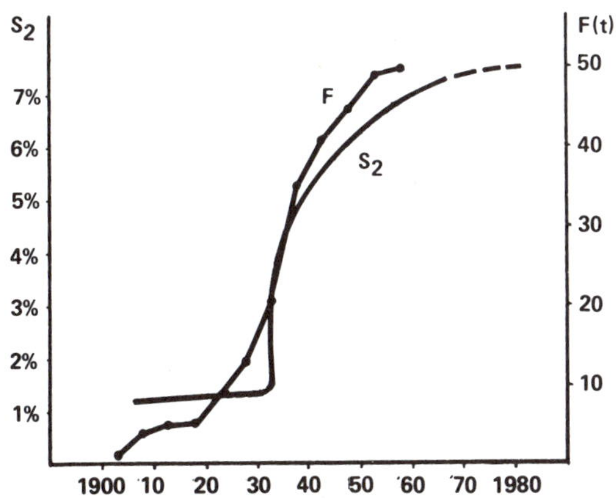

FIGURE 4

INVENTIONEN – INNOVATIONEN : TECHNOLOGISCHE ERRUNGENSCHAFTEN IN DER CHEMIE

(ZWEITE HALFTE DES 19. JAHRHUNDERTS)

FIGURE 5

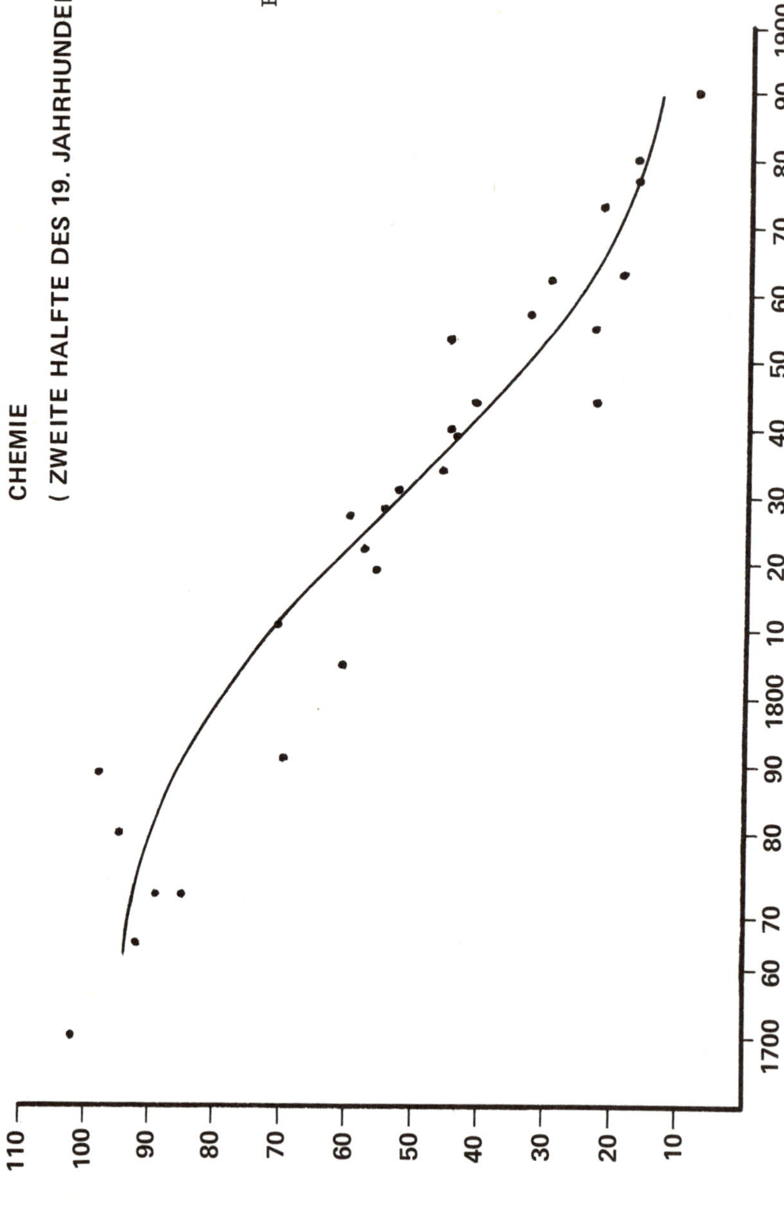

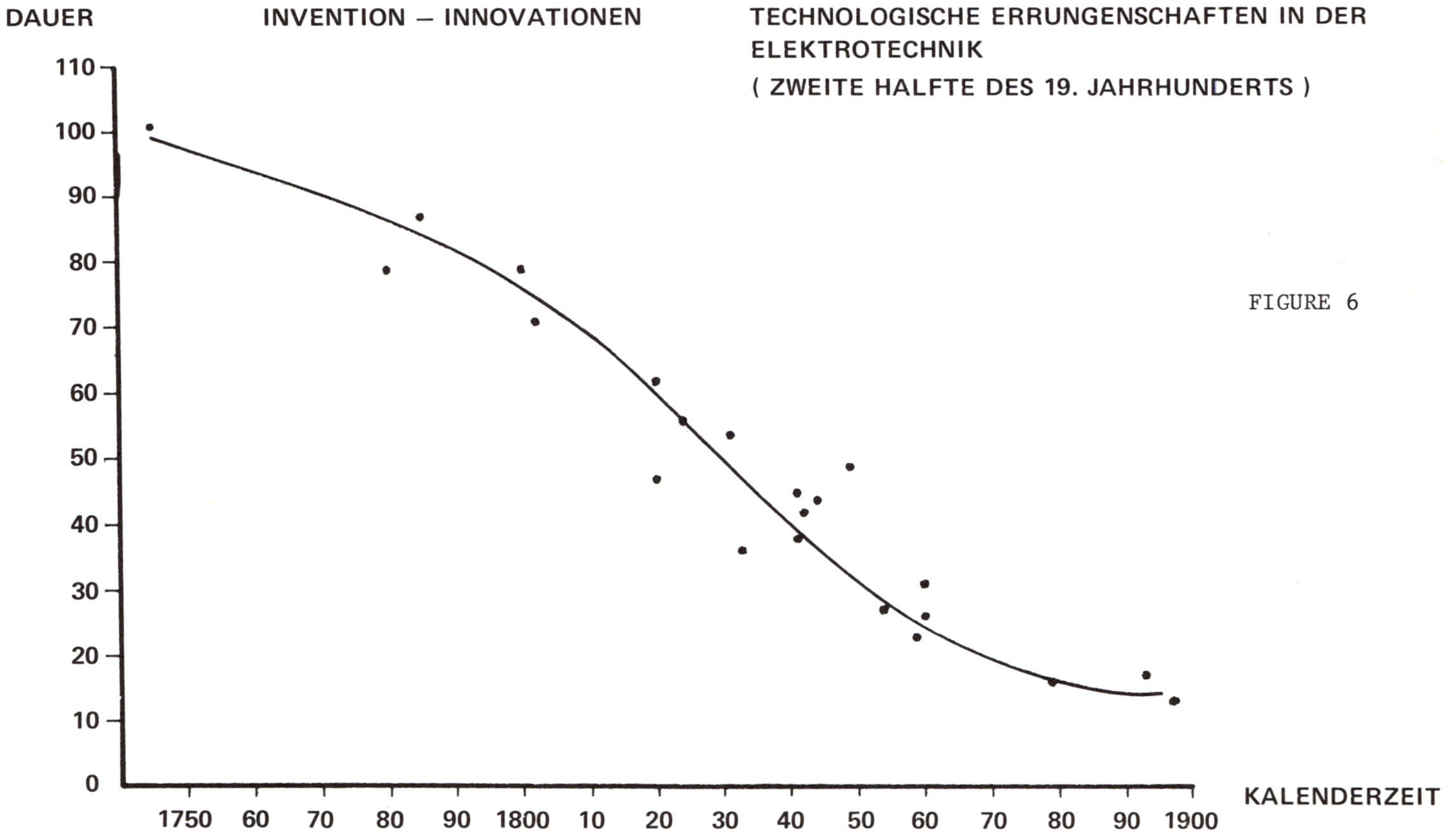
DAUER
INVENTION – INNOVATIONEN
TECHNOLOGISCHE ERRUNGENSCHAFTEN IN DER ELEKTROTECHNIK
(ZWEITE HALFTE DES 19. JAHRHUNDERTS)
110
100
90
80
70
60
50
40
30
20
10
0
1750
60
70
80
90
1800
10
20
30
40
50
60
70
80
90
1900
KALENDERZEIT

FIGURE 6

INVENTIONEN – INNOVATIONEN: TECHNOLOGISCHE ERRUNGENSCHAFTEN IN DER ERSTEN HALFTE DES 19. JAHRHUDERTS

FIGURE 7

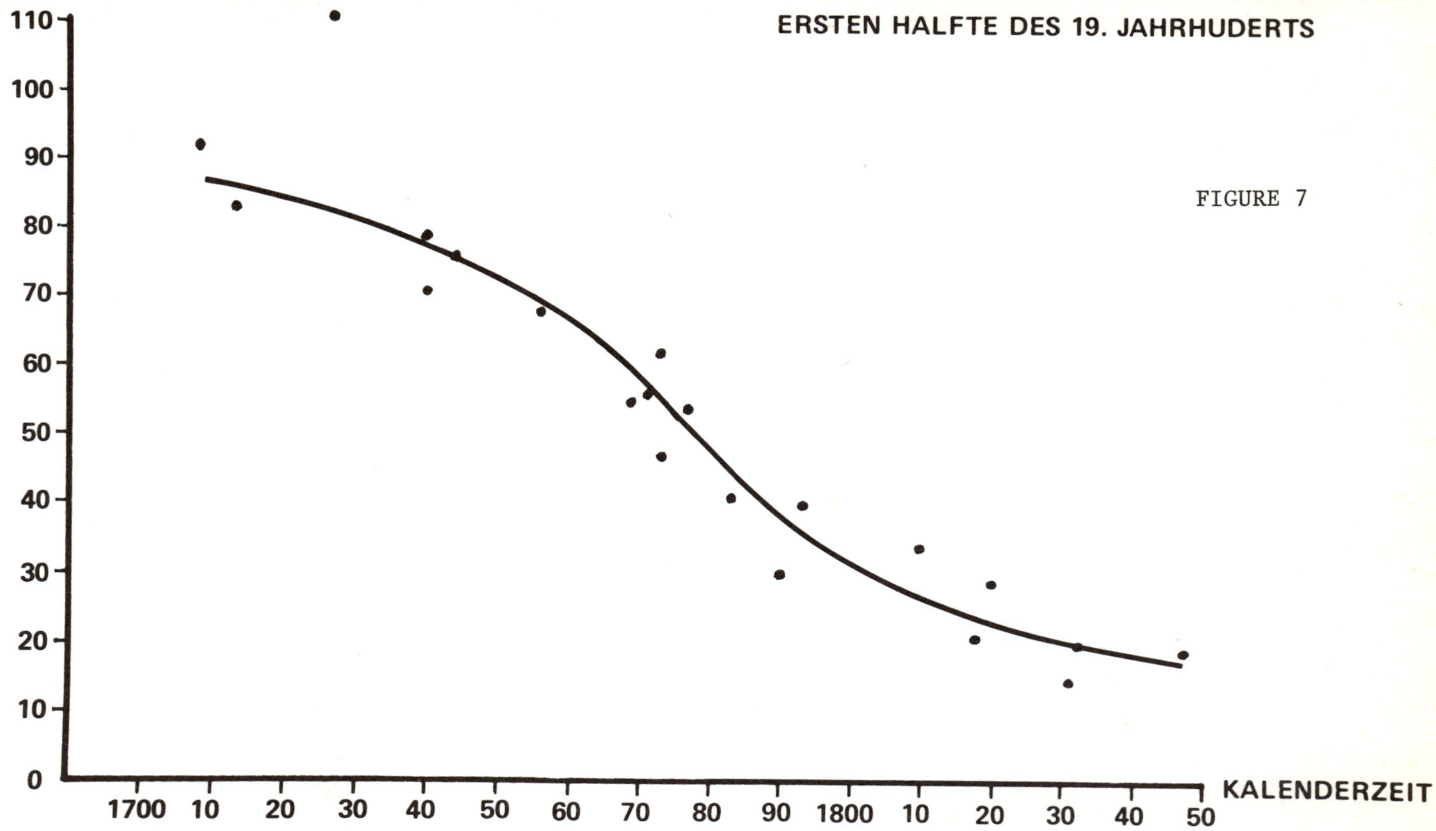

5. Dessert

To round up our point, after having shown the existence of barriers against basic innovations, some characterizations of these barriers are in place.

Admittedly, we know very little about these barriers. We know more about the circumstance of their crumbling; innovations research has given much more attention to cases of successful innovation than to cases of delay or failure. And the latter cases may contain the hidden message.

Who puts these barriers up, so that only a few basic innovations occur at some time, and who breaks them down, so that many basic innovations occur at another time? Can we hold fluctuations in the knowledge supply responsible for these discontinuities in the rate of innovation? In order to check this hypothesis, we look at the distribution of invention and innovation dates from Tables 1-4. These are depicted in Figures 8-10.

Inspection of these graphs reveals: the inventions occurred over a much longer time span than the corresponding innovations. While the frequency distributions of inventions over time are rather flat and stretched out, the frequency distributions of basic innovations are peaked up while squeezed into a shorter time period. Thus, continuity in accumulation of scientific knowledge is empirically compatible with the discontinuity in making use of this knowledge.

A rich flow of scientific discoveries does not necessarily imply a rich flow of usable innovation projects at all times. Thus, one can imagine a constant propensity of the economy to innovate, but a lack of innovation projects does not allow for a continuous allocation of a significant portion of investment capital to basic innovations. This hypothesis, again, ascribes the responsibility for the discontinuity to the failure of scientists. The hypothesis can be checked by means of the distributions of basic innovations in Figure 10. The peaks in innovative activities of both key industries, in the chemical and the electrical technology, coincide. That means: the barriers against different types of innovations have broken down at the same time. I would not know how to explain this fact as a sudden removal of inefficiencies in the knowledge generation process, namely, to say it bluntly, as a sudden enlightenment of generally ignorant scientists.

Rather, this timing of basic innovations is easily explicable by remembering that in all cases of peaks in the frequency of basic innovations, Western countries had just gone through

FIGURE 8

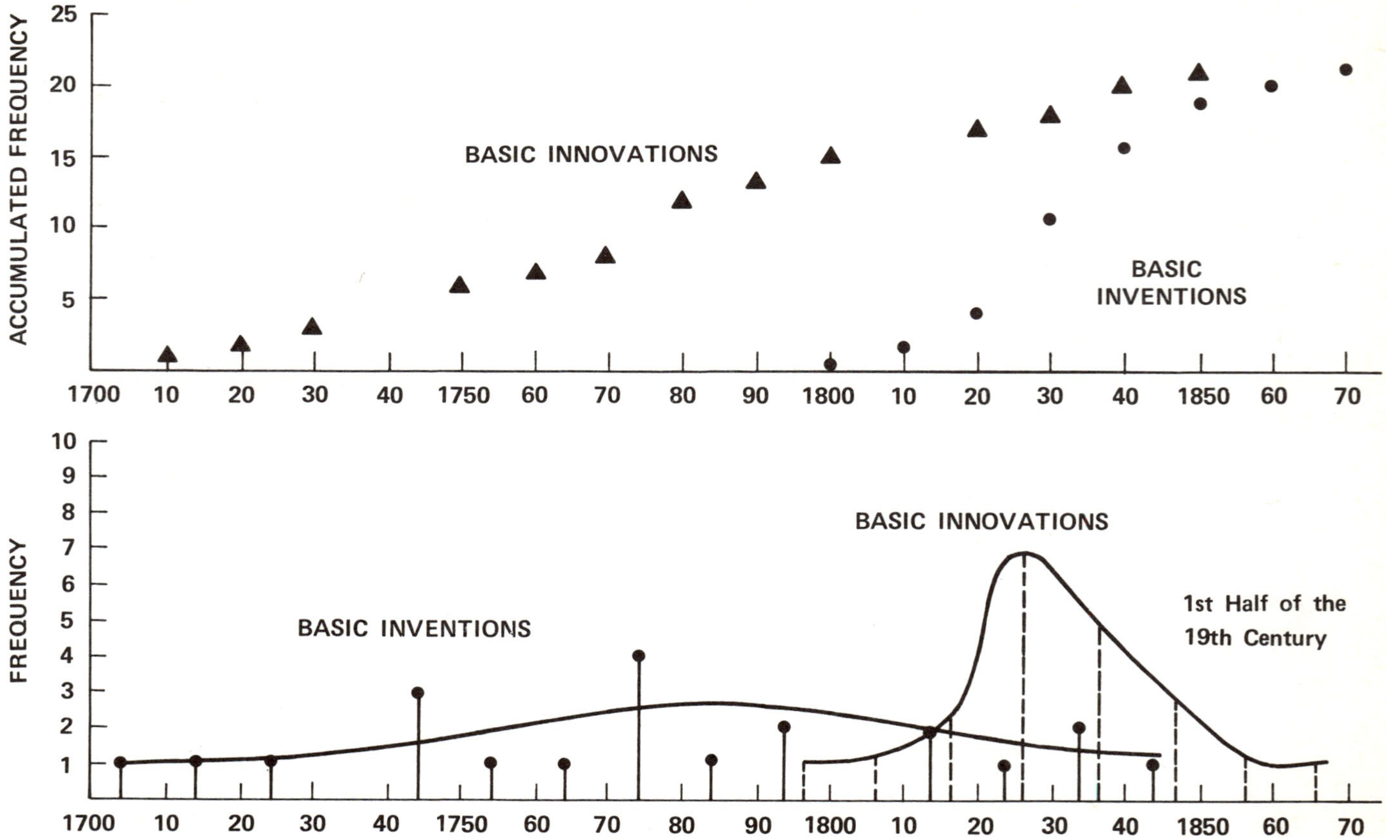

ACCUMULATED FREQUENCY
25
20
15
10
5
BASIC INNOVATIONS
BASIC
INVENTIONS
1700 10 20 30 40 1750 60 70 80 90 1800 10 20 30 40 1850 60 70
FREQUENCY
10
9
8
7
6
5
4
3
2
1
BASIC INNOVATIONS
1st Half of the
19th Century
BASIC INVENTIONS
1700 10 20 30 40 1750 60 70 80 90 1800 10 20 30 40 1850 60 70

FIGURE 9

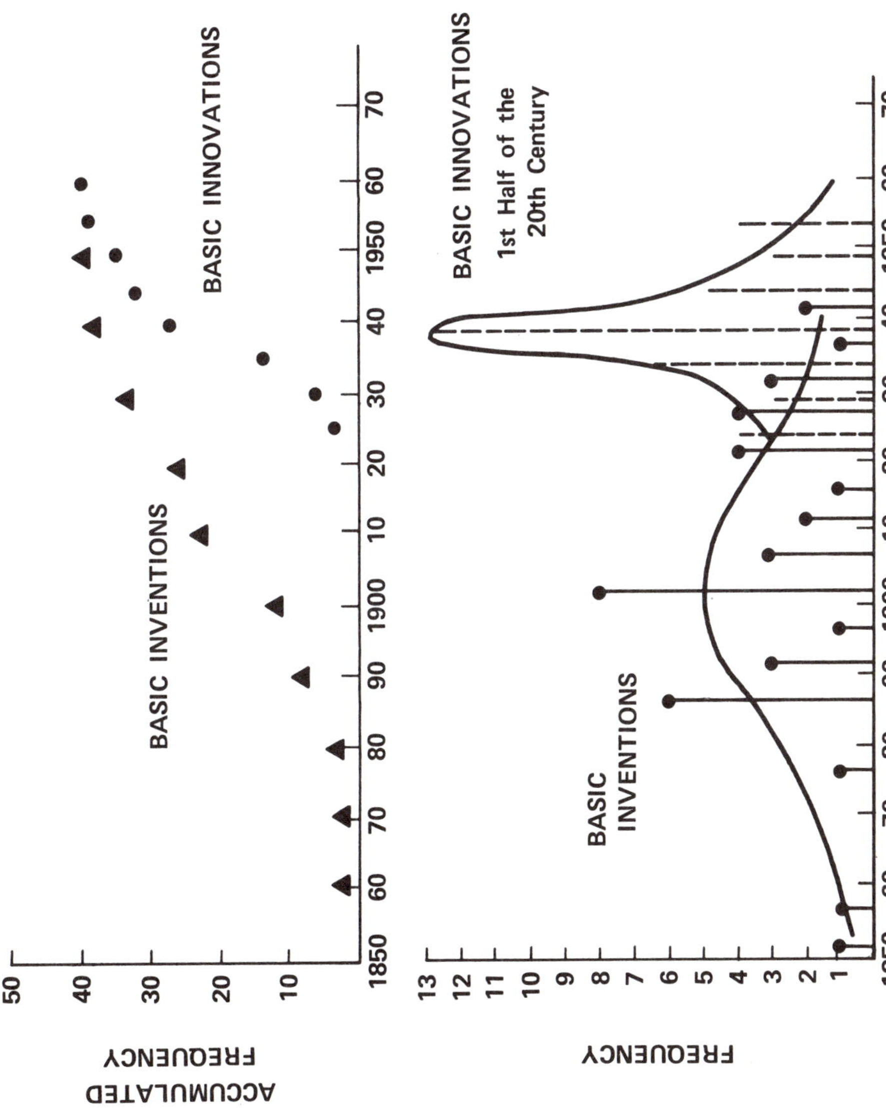

BASIC INVENTIONS
BASIC INNOVATIONS
ACCUMULATED FREQUENCY
50
40
30
20
10
1850
60
70
80
90
1900
10
20
30
40
1950
60
70
BASIC INNOVATIONS
1st Half of the 20th Century
BASIC INVENTIONS
FREQUENCY
13
12
11
10
9
8
7
6
5
4
3
2
1

FIGURE 10

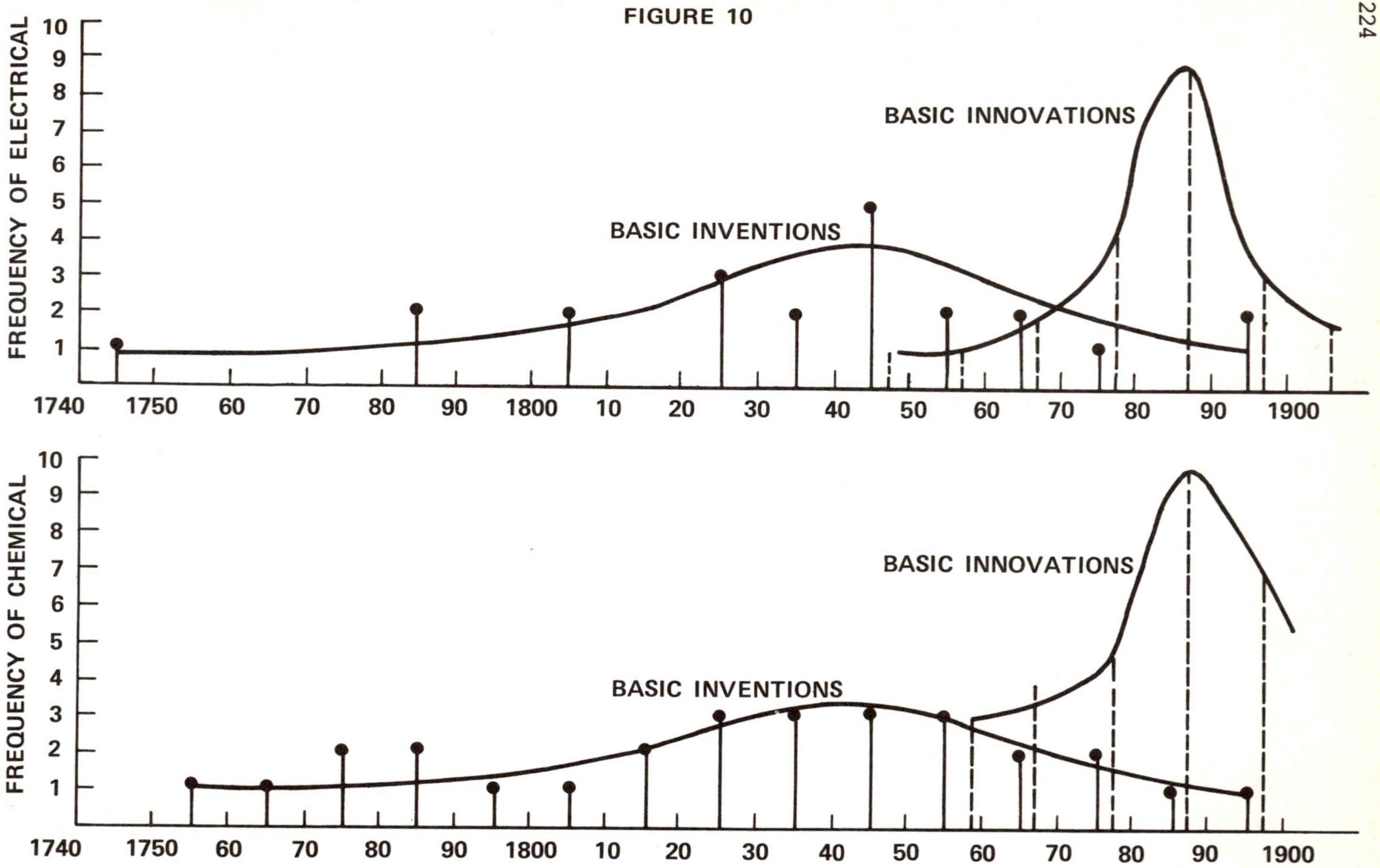
FREQUENCY OF ELECTRICAL
BASIC INVENTIONS
BASIC INNOVATIONS
FREQUENCY OF CHEMICAL
BASIC INVENTIONS
BASIC INNOVATIONS
1740
1750
60
70
80
90
1800
10
20
30
40
50
60
70
80
90
1900

deep international monetary and structural crises. During these crises, old commitments of capital and labor to stagnant technologies had gone overboard while firms went bankrupt by the thousands and people lost their work by the millions. In the past, it was extremely costly to the public that basic innovations had been blocked for too long by barriers within the economic system. It is a burning policy problem to make the reallocation of capital and labor resources less crises ridden.

Café

After a substantial meal, coffee is the reminder of action. Appropriate measures in the area of science policy and economic policy must be taken if we should come to enjoy the fruits from the trees of scientific knowledge without first suffering from structural crises. The decision making process in the market economy has some weaknesses if it comes to decisions about basic innovations.

In judging the pros and cons of basic innovations, a firm who may think of investing in an innovation prospect is bound to be fooled by the cost-benefit-ratio. The evaluation of costs and benefits is done by employment of methods that are good for standard investment planning only. Applied to innovations, the costs are overrated in order to compensate for "uncertainty" - lack of knowledge, and the benefits are underrated, again, because of lack of knowledge about the secondary effects. The cost-benefit criterion, notably the way it is evaluated, brings individual firms to shy away from basic innovations.

This criterion, and the way it is applied in the institutional framework of the techno-structure, is the pivot element in the barriers to the science-technology interaction.

REFERENCES

[1]G. Mensch, *Theory of Innovation*, Berlin, 1973

[2]G. Mensch, "Basisinnovationen und Verbesserungsinnovationen," *Zeitschrift für Betriebswirtschaft*, 42 (1972), 291-97.

[3]G. Mensch, *Innovationen und Industrielle Evolution*, Working Paper I/73-29, International Institute of Management, Berlin, 1973.

[4]Mensch, *Innovationen und Industrielle Evolution*, *op.cit.*

[5]G. Mensch, "Zur Dynamik des technischen Fortschritts," *Zeitschrift für Betriebswirtschaft*, 41 (1971, 295-314.

FORECASTING THE DIFFUSION OF TECHNOLOGY

Robert H. Pry

General Electric Company
Research and Development Center

INTRODUCTION

Three distinct stages are apparent in the birth, growth and diffusion of a science-technology area. First, the stage from scientific discovery of a new physical phenomena or a new theoretical concept to the broad acceptance of its applicability to technology. This gestation period involves a broadening of the concept, probing the depth and breadth of the experimental phenomenon and subsequently, assimilating the findings into a common technical-knowledge base. The time period for this process varies greatly, has been shortening considerably over the last century, but still may take 40 to 50 years to complete (e.g., electromagnetism in the 19th Century; superconductivity in the Twentieth).

The second stage consists of the time between the recognition of a scientific solution to a technological need and the time a first viable product emerges utilizing the solution. Several studies undertaken in the U.S. in the last 20 years show this time to be between six and thirty years, with the average time between 10 and 15 years.

The third stage consists of the time required for the new product or service to grow and diffuse through the technology and society. As was pointed out previously,[1] the span of time required for this process of technological and/or product substitution to occur has not shortened in any discernible way over the last 60 or 70 years. The examination of a number of cases of product substitution led to the intuitive conclusion that the time required for the completion of this stage did vary with the breadth of the product impact, the capital requirements, social changes required, marketing and distribution patterns, as well as product superiority or other technically related factors, but not with the specific dates of the substitution.

The purpose of this paper is to examine this third stage of technology growth and diffusion a little further utilizing the substitution model. Some emphasis will be placed on comparing like substitutions in different countries, in comparing related substitutions, and in substitution in consumer goods.

The substitution model supposes technical advance to be a process by which one product, service or behavior is substituted for another, such as cars for horses, service worker for manufacturing worker, synthetic fibers for cotton, wool and silk, or plastic for leather.

The rate at which the substitution proceeds is given as $F/(1-F) = \exp \alpha(T-T_0)$ where F is the fraction of the new product or service that has substituted at a time, T, T_0 is the time when the substitution is half complete and α is a rate constant that can be defined in terms of the time, T, required for the substitute fraction,

$$F = \frac{\text{substitute}}{\text{Total}},$$

to go from 0.1 to 0.9; that is, $\alpha = 2 \ln 9/ \Delta T$.

Thus, the total competitive substitution process can be characterized by two parameters, T_0, the substitution midpoint, and at ΔT, the time for the substitution to go from 10 to 90 percent. It was shown in the paper cited above that a number of diverse technical advances follow this simple model.

INNOVATION IN STEEL

In the late 1940's, a new process for refining steel was introduced in Austria, called a Basic Oxygen Process, referred to below as BOP steel. This new process appeared on the scene at a time when the world steel refining capacity was either largely exhausted or destroyed by the ravages of a major war.

At that time, four process choices were available for the production of crude steel; the old Bessemer process which was displaced in the U.S. by the more versatile open hearth process during the early part of this century; the electric arc process, a superior process for alloy and specialty steels; and the new basic oxygen furnace.

For the major steel-producing countries, the timing could not have been worse. Faced with an immediate need to replace damaged or worn-out equipment at a time when the experimental BOF process had not yet reached a commercial scale, few real alternatives existed to a heavy investment in the well-developed open hearth, and where possible, electric furnace processes.

Thus, faced with a world investment in new open hearth plants, the new BOF innovation made little real progress in substituting for other processes until 1960. The real competitive takeover of BOF steel from other forms in various countries of the world all started seriously in 1960 and should be nearly complete, worldwide, by 1980.

Figure 1 shows the tonnage of crude steel production in the four countries, Japan, U.S.S.R., U.S.A., and West Germany, up to the present.[2] Notice that each has had a somewhat different absolute trend of production during recent years. Japan shows the most dramatic growth in BOF steel, starting as it did from a base of little steel production in 1960, but with a strong policy to become a world leader. Figure 2 shows this same data as a percentage of the country total. The similarities in the takeover trends in these four countries now begin to emerge. Data from each country now can be seen to follow similar trend lines displaced in time.

When the empirically determined potential production of electric arc steel is subtracted from the total steel production for each case (i.e., approximately 20% in Japan, 10% in the U.S.S.R. 10% in West Germany, and 20% in the U.S.A.) and substitution curves plotted for the takeover of BOF steel for the remaining steel is plotted, the results are as shown in Figure 3. These curves suggest a number of rather striking simple conclusions.

First, notice that the takeover time (the time from 10% to 90% production) for Japan is 9 years with a one-half substitution time of 1963. The U.S. and West Germany substitution curves are nearly identical, with takeover times of 12 years and a one-half time of 1969, about 5 1/2 years later than Japan.

Since Japan had little steel capacity in 1960, but a real commitment to increase production, its curve could be considered to be an experimental determination of the rate at which capacity could be installed in an advanced country, limited only by construction constraints and industry learning curves.

Using Japan as a base, one might say that the U.S. and West Germany, whose behavior was and is nearly identical, demonstrate the effect of the delay in technology diffusion caused by heavy investment in an existing technology and a slightly more conservative investment policy. Several articles (e.g., 3,4) have been written in the U.S. debating the issue of "if and why the U.S. lagged Germany in utilizing this invention." Looking at the data presented in this manner, one is led to wonder what the debate was about, unless redirected at the constraints existing in both of

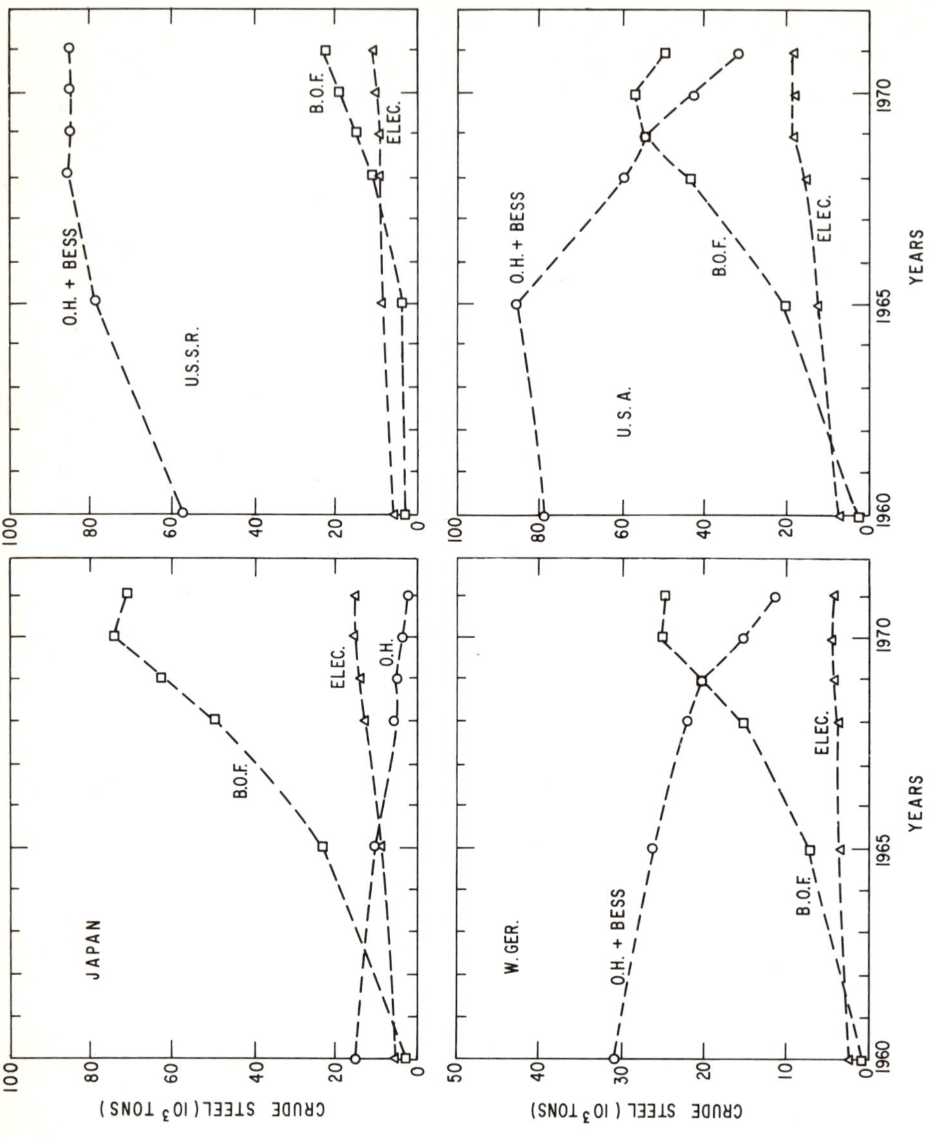

FIGURE 1: CRUDE STEEL PRODUCTION IN JAPAN, U.S.S.R., WEST GERMANY AND U.S.A. SINCE 1960

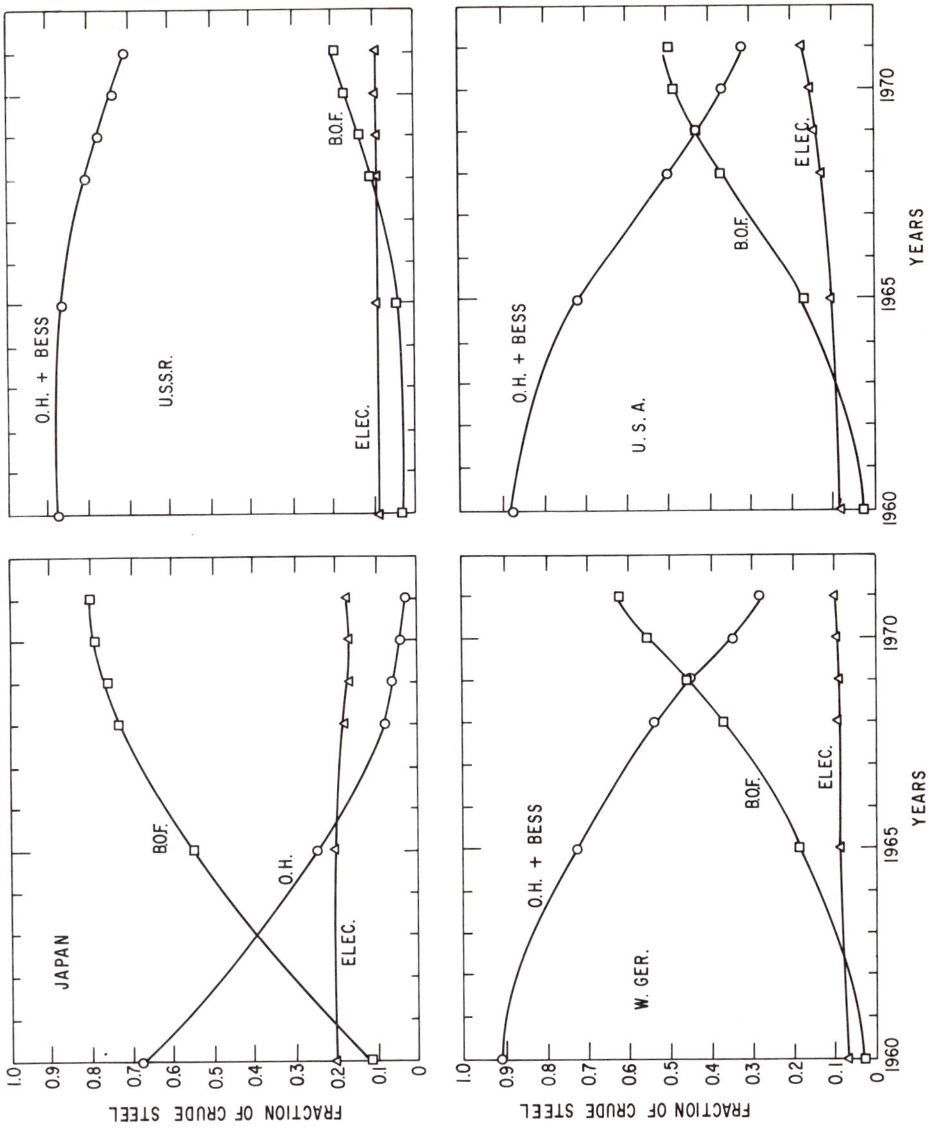

FIGURE 2: FRACTION OF CRUDE STEEL PRODUCTION MADE BY VARIOUS PROCESSES IN JAPAN, U.S.S.R., WEST GERMANY, AND U.S.A. SINCE 1960

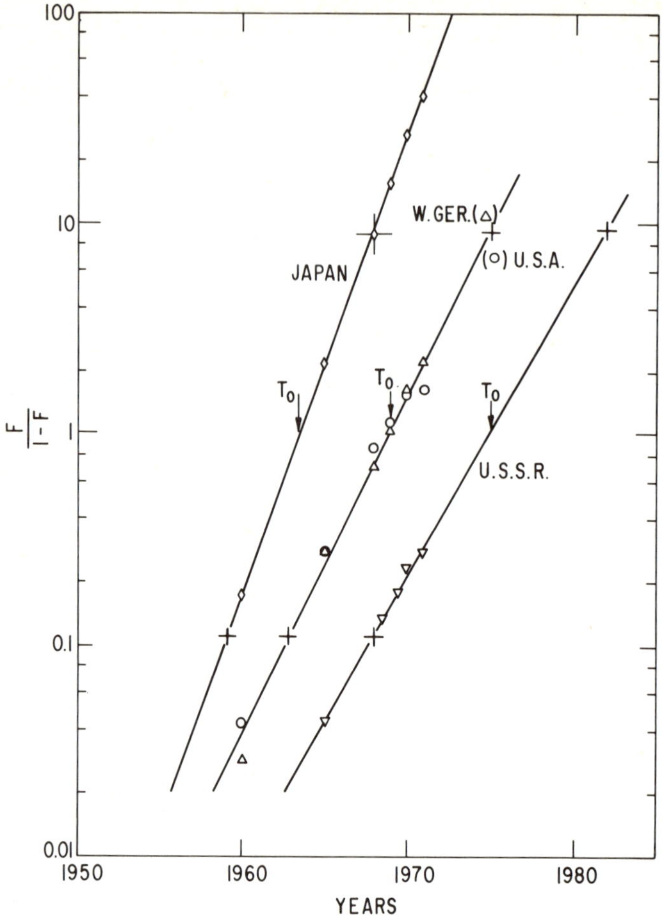

FIGURE 3: SUBSTITUTION PLOT OF BOF FOR OPEN HEARTH AND BESSEMER STEEL PRODUCTION IN JAPAN, U.S.S.R., WEST GERMANY AND U.S.A. SINCE 1960

these countries relative to Japan.

Knowing very little about the underlying facts in the recent advances in the U.S.S.R. steel industry, one can only speculate about the significant time delay in the U.S.S.R. substitution. Could it be that a lack of first-hand knowledge of the technical operating characteristics of BOF plants delayed the substitution for five to ten years? It was not due only to a different plant investment policy relative to plant life, since open hearth capacity grew during this period when Japan, U.S.A., and West Germany were only installing BOF capacity. Notice that the substitution slope is not much different (∿15%) than the U.S. and West Germany. Thus, the substitution, once undertaken, is proceeding almost as fast as in other advanced countries with large existing plant investments.

It is not the purpose of this example to dwell in detail on the underlying causes for the differences in technology acceptance and diffusion between these countries. The example has been developed to further indicate the power of this method in uncovering gross differences, pointing to interesting places to dig for the causes.

SUBSTITUTION OF DETERGENTS FOR SOAP

Another interesting example for which data is available is the substitution of detergents for soaps in the U.S. and Japan.[5,6] This case is of interest because of the radically different consumer-use histories in the two countries. The U.S. has had a stable market for washing products over at least the last three decades. Japan, on the other hand, produced commercially only a small amount of washing products until well after the end of the Second World War. Figure 4 displays the production of soaps and detergents in the U.S. and Japan from 1944 to 1968.

From about 1948 outward in the U.S., there was little doubt that detergents would dominate in washing products. The growth of synthetic fabrics, automatic washing machines and dishwashers added to other forces, driving the substitution during the 1950's. By 1960 detergents had substituted for nearly all soap uses except for the washing of skin. Bath and facial soaps almost completely account for the level of soap sales after 1960 in the U.S. Apparently, some additional technical advance is required for any real penetration of the facial soap market. As a consequence, facial soaps have been subtracted from both the U.S. and the Japanese data before constructing the substitution plot shown in Figure 5.

As can be seen from the substitution plot, Japanese detergent substitution lagged the U.S. by ten years. In this case, the lag was not because detergent technology was not available in Japan,

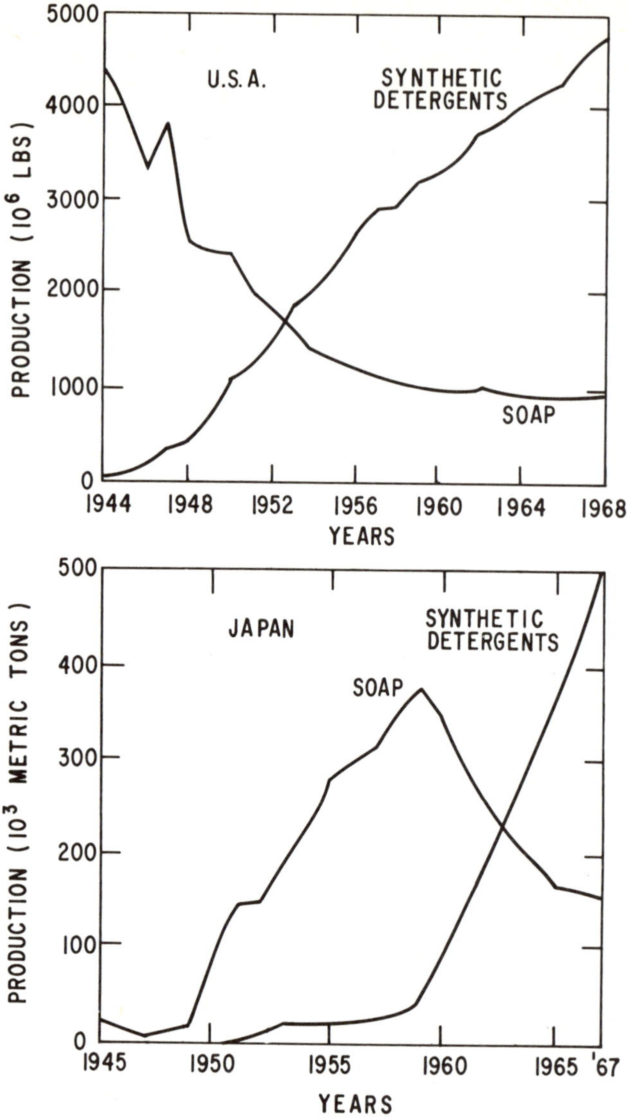

FIGURE 4: PRODUCTION OF MAJOR WASHING PRODUCTS IN U.S.A. AND JAPAN

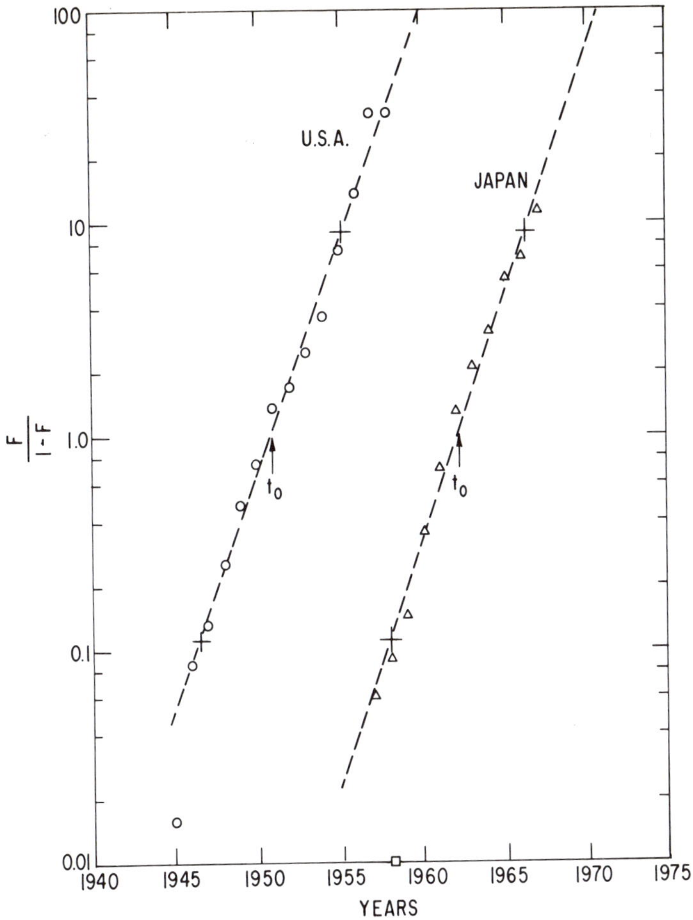

FIGURE 5: SUBSTITUTION PLOT OF SYNTHETIC DETERGENTS FOR SOAP IN U.S.A. AND JAPAN

since some detergents were produced in Japan prior to 1960. Most likely, the rapid growth was timed with respect to other factors such as the growth of synthetic fibers and washing machines. Once the substitution process began, the rate of substitution was almost identical to that of the U.S., probably limited principally by the ability to supply the market with the new products. This underlying reason is guessed to be dominant simply because of the widely different consumer purchase pattern and the similar producer structure in the two countries relative to washing products.

BASIC SOCIETAL SUBSTITUTION AND SOME CONSEQUENCES

In a recent paper, Montroll[7] displayed the substitution of non-agricultural workers for agricultural workers in the U.S. and Sweden, shown in Figure 6. Some comment was made about the deviation from a linear slope of the ln substitution ratio vs. time, occurring and continuing after the Second World War in both countries with an acceleration continuing up to the present. The authors, however, were considerate in pointing out that the slope was constant for a period of about 100 years and identical in both countries. This latter fact is indeed remarkable.

Two points may be worth emphasizing with respect to this type of substitution. First, substitutions requiring major changes in societal behavior, method of living, ethic, etc., tend to require more than a century to go from 10% to 90% complete. Second, substitutions of such a fundamental sort trigger other substitutions that have the same time dependence.

These considerations suggest that such fundamental substitutions should be studied country by country to try to determine the major rate-limiting factors and the effects of perturbations such as large investments of foreign capital and governmentally imposed social changes. Such a study should provide additional help to developing countries in understanding and planning for realistic rates of change.

An example of a companion substitution for the above case is the substitution of urban for rural living[8] accompanying the rise in non-agricultural employment. Such a substitution plot for the U.S. is shown in Figure 7. Two points are immediately noticeable. First of all, the slope of this substitution plot is nearly the same as the one of Figure 6. This is to be expected, since the shift to urban living depends upon a growing non-farm population. Second, the curves are shifted in time by approximately 40 years. That is, the midpoint in the farm, non-farm worker substitution plot is about 1880 in the U.S., whereas the midpoint in the rural-urban living plot is about 1920. Thus, the dominant behavior pattern seems to be: When a farmer quits farming, he works in town but lives on the farm. His sons and daughters, however, when they grow up, also move to town to live as well as to work.

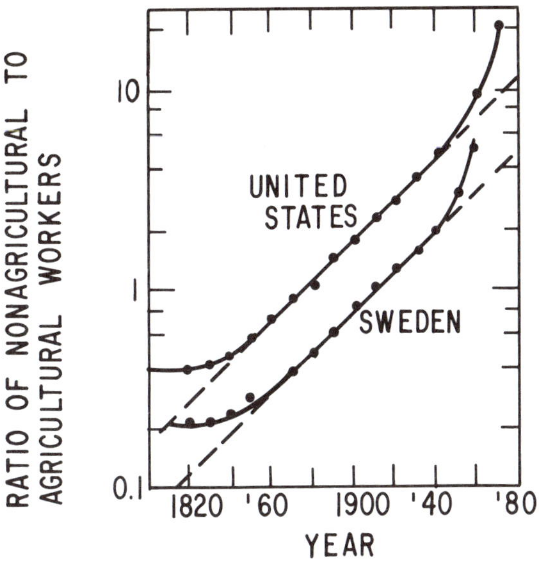

FIGURE 6: SUBSTITUTION PLOT OF NON-AGRICULTURAL FOR AGRICULTURAL WORKERS IN U.S.A. AND SWEDEN

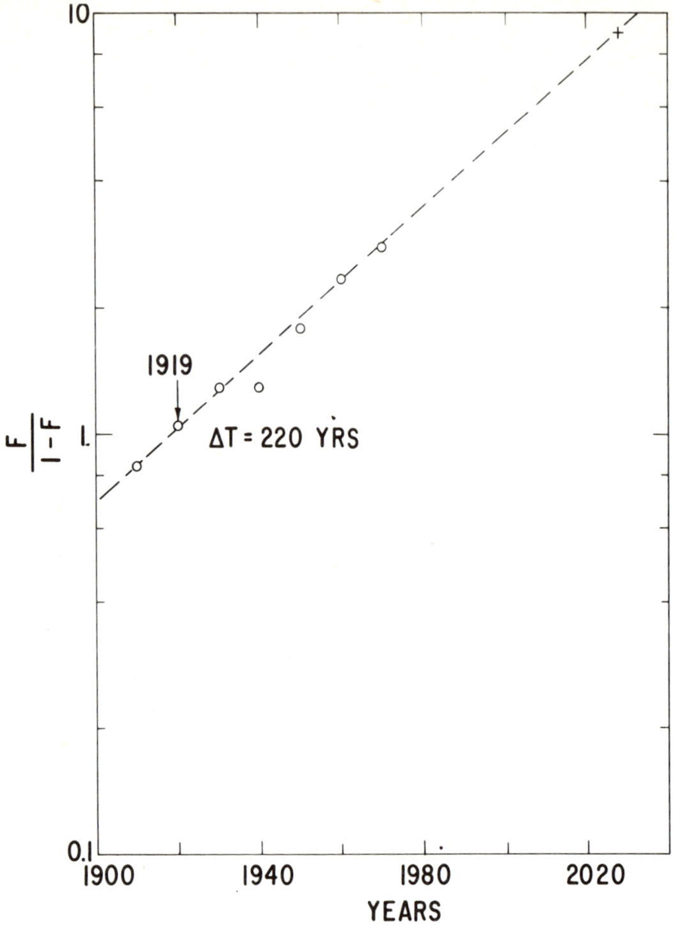

FIGURE 7: SUBSTITUTION PLOT OF URBAN FOR RURAL LIVING IN U.S.

This connection between the two substitution curves has only the charm of quaint simplicity to recommend it unless it is a true pattern. For if it is so, the significant deviation in the slope of the farm, non-farm worker curve will be reflected in an upsurge in the substitution of urban for rural living in the U.S. beginning around 1980, causing an acceleration in the demand for urban dwelling.

Of more significance is the consequence of a possible similar pattern in Sweden, since the worker substitution deviation occurred at the same time but at a lower point on the substitution plot. This would then bring about a much more significant perturbation of housing pattern changes in Sweden than the U.S., again starting about 1980.

INTERNATIONAL COMPETITION - AUTOMOBILES

Prior to 1960, foreign-car imports into the United States amounted to only a few percent of the U.S. car market. This situation was a result of two principal factors. First, the automobile had grown to be something of a symbol of affluence in a country where travel and mobility was much emphasized. As a consequence, the trend in consumer preference was toward the large and luxurious. Roads were widened to accommodate the trend and auto manufacturers encouraged it. In addition, the level of U.S. production in this cost-volume-sensitive business was an order of magnitude larger than other countries.

Foreign-car makers were faced with a different domestic market. At low-production volume in a less affluent society with high gasoline prices, little alternative existed to the small, low-power car.

What appeared to happen in the period between 1955 and 1965 was an approach to the same set of consumer conditions, at least in the U.S. and Europe. As Europe became more affluent, car production increased markedly. As superhighways and pocketbooks expanded, the average car size grew as well, approaching that of small U.S. cars.

At the same time, as measured by car purchase statistics, a significant segment of U.S. consumers' preferences started to shift away from increasing car size and cost, corresponding more to foreign-car offerings, which were now being made from a production base more equivalent to that of the U.S.

When sufficient inroads in the U.S. market were made to establish a feeling of reliability and a service availability, the subsequent substitution process was nearly inevitable. The results[8] are shown in Figure 8. Because of the trend by U.S. car manufacturers to import many of their smallest cars sold under U.S. labels, it is still more appropriate in the U.S. to view the substitution as an imported - vs. domestic-car substitution rather than a little-car vs. big-car substitution, which it will no doubt eventually become.

What makes this case interesting is not so much the foreign-competition aspects of the problem as such. Rather it is that the U.S. had, over the years, gotten into a kind of "lock-step" between producer and consumer in the automative field. At large volumes it was not economic to introduce small numbers of small cars and there was little incentive to do so. Thus, the popular, cost-effective (whatever effective means here), small car could not completely sustain itself. On the other hand, a foreign "invader" technology (or style) working from an advantageous cost-volume, offshore base could, and did, economically change automotive trends in the U.S. Simplistically stated, the only question now remaining is when the substitution should be properly called a small-car vs. big-car substitution rather than a foreign vs. domestic one.

CONSUMER APPLIANCES

Normally, discussions of technology diffusion by sophisticated technical people rarely includes much discussion of consumer appliances. A new appliance somehow does not have the aura of advanced technology to the scientist. After the satisfaction of food and health needs, however, the lightening of woman's burden must be considered as equally important to fulfilling man's desires -- at least these days it must.

In thinking of the growth in acceptance of a generically new item of consumer goods as a substitution process, the substitution quantity is the process performed rather than the purchase of the item performing it. Thus, for a clothes dryer, it is the process of machine vs. hand drying that comprises the substitution. As a consequence, it is the ownership of a clothes dryer that is important, not its purchase. The same argument can be made for most other appliances, such as air conditioners, dishwashers, vacuum cleaners, etc.

Figure 9 displays a number of such substitutions plots.[8] In each case, the quantity, F, is the fraction of U.S. households containing the named generic appliance. In contradistinction, if one were concerned with the substitution in purchase preference between two competing products performing the same function, then a substitution plot of sales ratios would be appropriate. For example, such a plot for manufactured vs. conventionally built houses [8] is shown in Figure 10. Manufactured houses here mean mobile homes in excess of 35 feet long, 8 feet wide and weighing

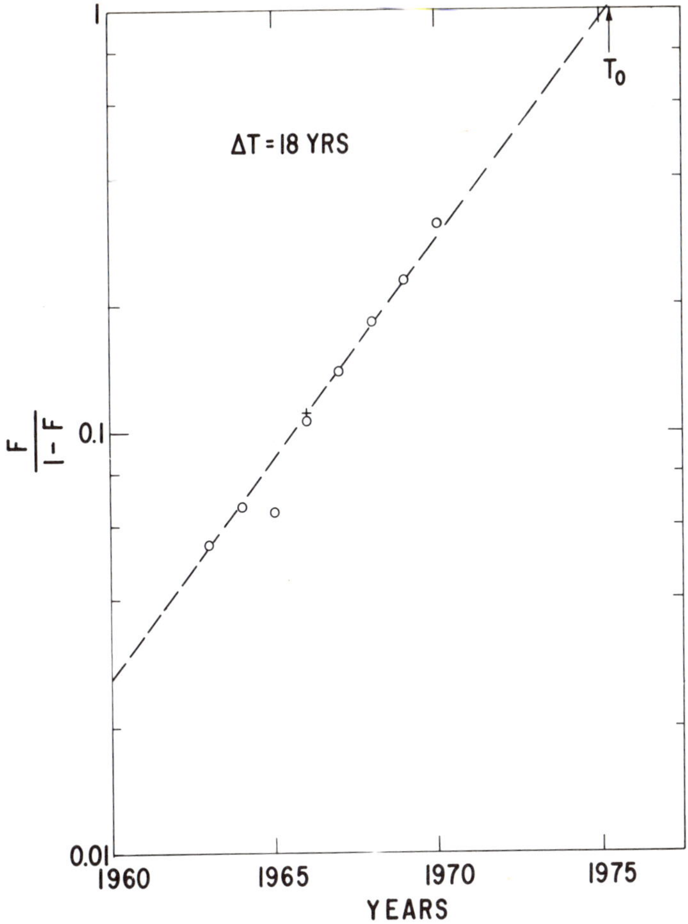

FIGURE 8: SUBSTITUTION PLOT OF FOREIGN FOR DOMESTIC CAR PURCHASE IN THE U.S.

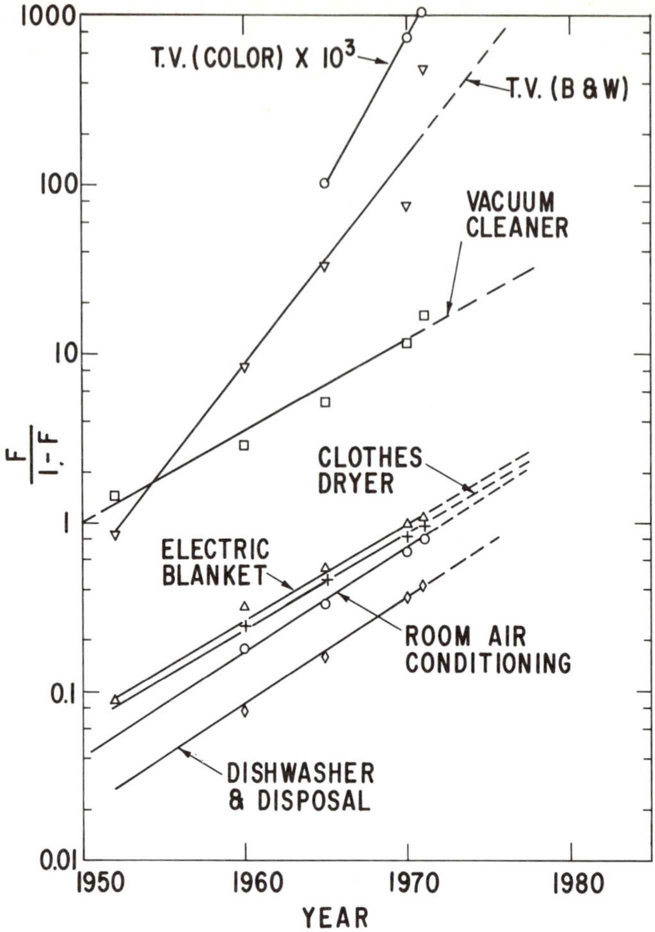

FIGURE 9: SUBSTITUTION PLOTS FOR "FRACTION OF HOUSEHOLDS THAT HAVE" VARIOUS CONSUMER APPLIANCES

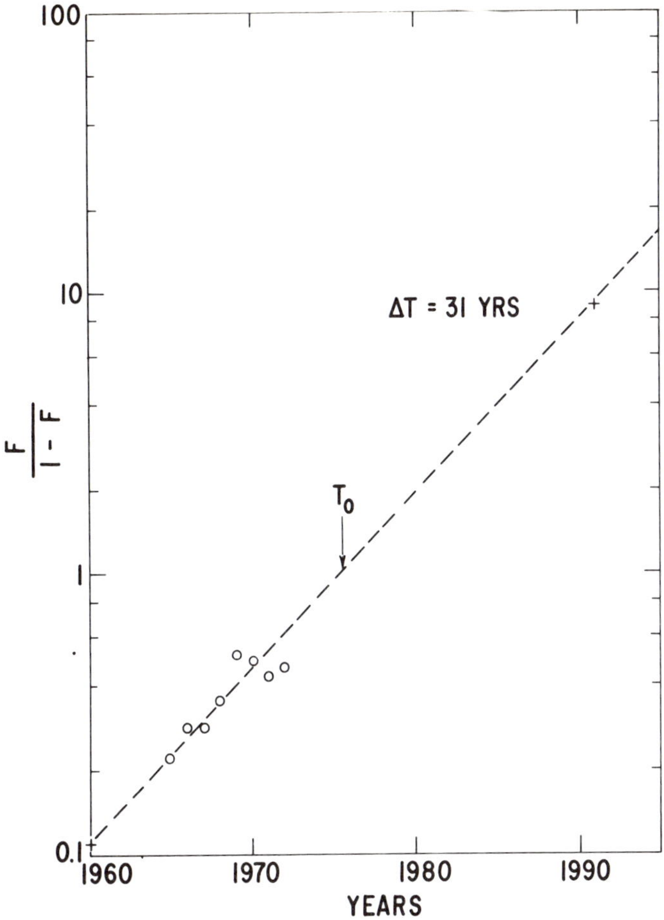

FIGURE 10: SUBSTITUTION PLOT FOR NEW SINGLE FAMILY DWELLINGS, MOBILE HOUSES VS. CONVENTIONAL CONSTRUCTION

more than 4500 pounds, the dominant manufactured house up to the present. Eventually, such a plot must include houses manufactured remotely but assembled on site.

The number of appliance examples of Figure 9 was chosen to bring out one interesting feature. Most appliances aimed at increasing comfort or reducing drudgery have a takeover time of approximately 30 years, from the vacuum cleaner that was in one half of the households in 1950 to the dishwasher that will not reach a similar position until approximately 1980. Notice in passing that dishwashers and disposals are accepted as an appliance at identical times and rates. Also notice, as with the TV example plotted, entertainment appliances do not follow this 30-year takeover rule.

This takeover time is nearly equivalent to the forty-year lag between the time the farm worker quits farming and his children move to town. Again, as a simple rule, already mature adults do not readily change their household habits with respect to appliance purchase and use, but their children do. Thus, it takes a generation of householders to accomplish a takeover for a generically new appliance function.

This observation applies to the U.S. only. Is this also true in Europe? Does the rapidly rising aspiration and achievement rate of the Japanese indicate a different behavior? These questions, obviously of interest to the consumer appliance industry, are only suggested here as ones for further study.

CORRELATION OF SUBSTITUTION CURVES - TV ENTERTAINMENT

The substitution of cable for air television reception in terms of "fraction of households having" is shown [8] in Figure 11. Notice that two takeover times have been shown. If all of the data shown is used, a takeover time of 24 years and a half substituted point of 1984 is predicted. If, because of the trend and certainty of substitution in the one-to-two percent range, only the data after 1964 is used, then a takeover time of 18 years and a half substitution point of 1980 is found. This more rapid substitution forecast is considered here to be more reliable, both for the above reason and for the reason that the takeover time of black and white TV ownership shows the same acceptance time. Color TV, on the other hand, is proceeding at the somewhat faster pace of an 11-year takeover. From these data it might be assumed that this 18-year-takeover forecast for cable TV is, if anything, conservative. Notice that the existence of a saturated black-and-white TV ownership should not be a drawback for the acceptance rate of cable TV, since by 1980, the time when one half of U.S. homes are predicted to have cable connections, more than 90% of U.S. households are predicted to have color sets.

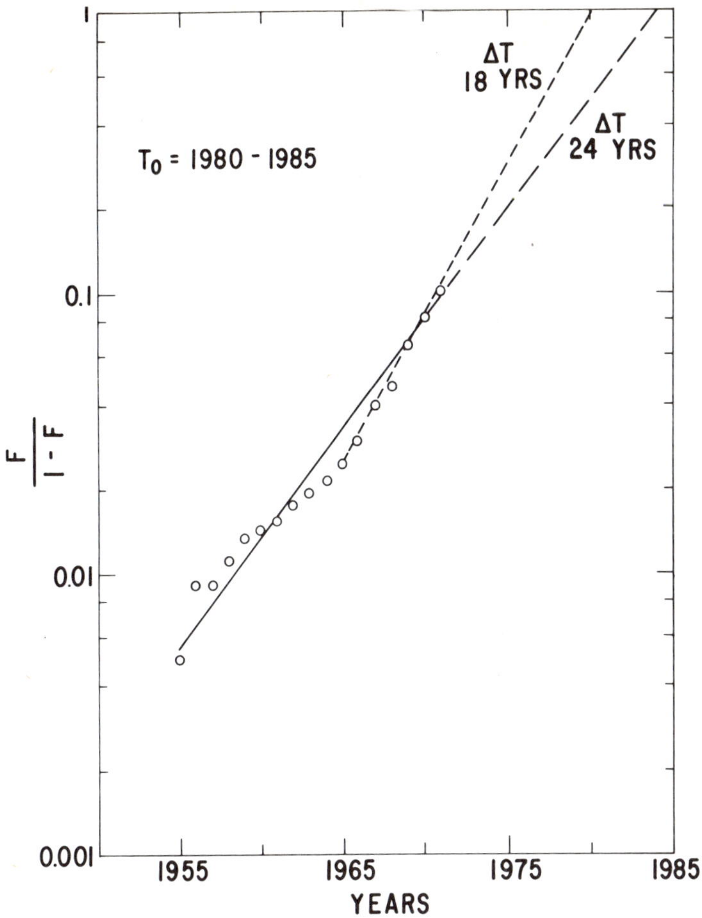

FIGURE 11: SUBSTITUTION PLOT FOR FRACTION OF HOUSEHOLDS HAVING A CABLE TV CONNECTION

This example was brought out to show that in the case of substitution as with other forecasting forms, a correlation of similar forecasts can be used as reinforcing information.

SUMMARY

A number of examples of technology, product or process advances, taken from widely published data have been presented here as simple substitutions to draw further attention to the simplicity and power of the substitution model in analyzing a wide variety of technical, social and product changes.

The comparison of substitution phenomena between countries and cultures was emphasized because it is believed that this approach can prove valuable in a macroanalysis of innovation and the diffusion of technology on a world scale.

REFERENCES

1. J.C. Fisher and R. H. Pry, *Technical Forecasting and Social Change*, Vol. 3, 75-88 (1971).

2. *Annual International Steel Statistics 1971*, Country Books, British Steel Corporation, Statistical Services, 12 Addiscombe Road, Broydon, Great Britain.

3. D. R. Dilley and D. L. McBride, *Iron and Steel Engineer*, October 1967.

4. W. Adams and J. B. Dirlam, *Iron and Steel Engineer*, July 1968.

5. *Chemical Week*, p. 69, September 26, 1969.

6. *Stanford Research Institute Chemical Economics Handbook*, Menlo Park, California, 1970.

7. R. Herman and E. W. Montroll, *Proceedings of the National Academy of Sciences*, U.S.A., 69, No. 10, pp. 2019-2023, October 1972.

8. *Statistical Abstracts of the U.S.*, U.S. Department of Commerce, 1972.

BASIC RESEARCH ON TECHNOLOGY TRANSFER

Albert H. Rubenstein

Northwestern University

In this brief paper, I would like to cover several points related to the field of basic research on technology transfer (TT). My remarks are addressed to a particular segment of the literature and ongoing activity related to TT and not the whole field, as represented by the many papers and discussions presented at this conference. In order to more clearly delineate the kinds of efforts being devoted to investigating and improving the TT process, I would like to distinguish three kinds of activities:

1. Research, of a descriptive, theory-building nature, whose main task is to explain and provide a basis for predicting the behavior of individuals and organizations involved in the TT process and the outcomes of that process.

2. Policy formulation, of a prescriptive nature, whose task is to set standards and criteria, to indicate what courses of action are desirable and "correct," and to provide guidelines for legislation and other kinds of social decision-making.

3. Design, the "nitty gritty" task of converting research findings and experiential generalizations into precise specifications for the design and operation of TT-related organizations and systems.

In my view, this distinction is vital to improving the state of the art of establishing and operating TT organizations and systems for these reasons:

a) There has been very little systematic, basic research on actual behavior and the consequences of that behavior for effective TT.

b) There is a large amount of literature on policy for this problem area, but very little of it is based on whatever theory and research results there are. In addition, much of the policy prescriptions are in direct conflict with each other (e.g., equally distinguished observers and practitioners directly conflict over such issues as: the kind of liaison people, procedures, and institutions that are needed; the kinds of training programs for TT people that are needed; the kinds of research programs and institutions that are needed to study and improve the TT process.)

c) Design of TT-related institutions is going on continuously, but much of it is unsystematic, cut and try "design" which involves little systematic analysis of the likely effects of different design configurations on the outcome of the TT itself and its potential side effects on the participating organizations and other sectors of the economy and the society.

So I will be discussing, primarily, the first kind of activity - research on the phenomena involved in technology transfer, which has the possibility of contributing significantly to improving the second two kinds of activities - policy formulation and design.

A number of issues arise for the investigator planning to undertake basic research in this field; I will discuss only a few of them here. Others have been treated in the more than 50 publications, theses, dissertations, and working papers related to technology transfer in our Program of Research on the Management of Research and Development at Northwestern University. The issues are these: 1) How deeply can and should a given investigator go into the phenomena underlying the TT process; 2) What theoretical or empirical points of view are relevant in developing and testing propositions and theories about TT; 3) What kinds of overall methodology are available and feasible; and 4) What specific kinds of field techniques are effective.

1. <u>Depth of Delving into TT</u> - There are a number of subissues here that involve questions of the "size" of the units to be used as a focus of the study and the depth of explanations to be sought for situations and events encountered in the TT process. Various investigators have chosen to focus on units such as "pieces of technology," "projects," "fields of technology," "the firm," "an industry," "a country," "a region," "a type of institution," etc. This is an important choice, because the level of explanation of events and situations can have a profound effect on comparability and generalizability of research findings. So far, the field suffers from a relatively low level of both.

The art of replication, as widely practiced in the natural sciences, is still at a fairly low level in the social sciences and in this particular field of research on TT. Some of the reasons are due to differences in methodology and data collection techniques, as discussed below; but others are related to a lack of a scientific tradition that values and rewards successful replication (as opposed to blind duplication) of research. The consequences of the wide diversity of units of analysis and depth of analysis is that we find non-comparability of many research results in the field - such as those having to do with: climate for innovation, kinds of institutional arrangements, the role of "entrepreneurs" and "product champions," the involvement of top management, the need for technically trained people in the receiving organization, the need for common languages, etc.

As an example of the concept of depth of analysis, consider the following levels of analysis of technology transfer between two groups within a given organization or between two organizations:

(Group or organization) A transferred technology X to
  (Group or organization) B
The following phases occurred during this transfer. . .
These specific people and their associated roles were involved
These specific events occurred, in which these people were
  involved (e.g., Mr. A communicated with Mr. B)
The participants had these individual motivations for participating in the TT process (Mr. A's real reasons for what he did were such and such)
The characteristics and skills of the involved individuals
  were of a certain specific nature
The macro-economics of the transfer were. . .
The micro-economic asspects were. . .
The general outcome for the firm was. . .
The specific outcomes for the individuals were. . .

As illustrations of the range of variables and propositions which can form the basis of research studies of the TT process, I present <u>Figure 1</u> and <u>Figure 2</u> - samples of propositions and variables forming the bases for some approximately 20 studies of the TT process over the past 8-10 years.

Obviously, one platitudinous criterion for depth of analysis is "sufficient to illuminate the phenomena being studied." Unfortunately, we are far from agreement in this newly emerging field of research on what is a "sufficient" level of analysis.

2. <u>Theoretical Bases for TT Research</u> - It is clear from the range of variables in the illustrative list in <u>Figure 2</u> that a great number of disciplines can contribute to our understanding

FIGURE 1: SOME ILLUSTRATIVE PROPOSITIONS ON TECHNOLOGY TRANSFER

1. Product and process innovations require different types of coupling mechanisms. Operating scientists and engineers are most important in the utilization of process type technologies, whereas top managements have to be sold on the idea of product innovations (Chakrabarti).

2. Formally structured methods (e.g., appointing liaison agents) will be less likely to result in high levels of innovation than informally evolved methods of directing the transfer of technology where application areas are similar (Bean).

3. An organizational innovation process may be under-structured so that progress is retarded by a lack of direction and it may be over-structured so that progress is retarded by over direction (H. C. Young).

4. The timing of the initial contact and subsequent maintenance efforts are equally important in successful coupling in a project of relatively short duration (e.g., under 8 months) (H. C. Young).

5. A multiple liaison mechanism (or coupling of several groups between the source and user of the innovation) operating at all phases of the technology transfer process and at all levels of both source and user organizations, is an effective means of insuring high utilization rates of new technology (J.E. Ettlie).

6. The more new technology is divergent from the technology the user has utilized in the past, the more need of a formal liaison/coupling arrangement between source and user (Jedlicka).

7. Pairs of groups that value collegial growth tend to have fewer problems, but this has to involve both groups (Douds).

8. Individuals and groups that preferred the "confrontation" method of joint decision making and conflict resolution perceived fewer communication problems than those who preferred the "forcing" or "smoothing over" methods (Barth).

9. Liaison agents whose organization affiliations and reporting relations permit the client a relatively high degree of control over the liaison agents' actions will be more effective than those over whom the clients have relatively little control (Bean).

10. Each transition point between phases of implementation (e.g., from shakedown of the technology to production) results in potential communication gaps (Ettlie).

11. The more aggressive the coupling group is in determining what the user wants, the more likely the coupling group will be effective (Jedlicka).

12. Where a single source of a new technology is coupled with several users of the innovation, it is more effective to locate major coupling functions in the source organization. On the other hand, where several sources of technology are coupled with a single user, it is more effective to locate major coupling functions in the user organization (E. C. Young).

13. Several forms of control are used that directly affect the management of a European laboratory by their American parent organization: An American is appointed as a laboratory managing director; American scientists are frequently sent to overseas labs to head specific departments or groups temporarily; and promising managers from foreign labs are sometimes brought to U. S. labs for a temporary training period (Forster).

14. The strongest single indicator of the level of communication effectiveness is the level of technical respect one person or group has for another (Douds).

15. When a client has high expected benefits from an innovation, he places an emphasis on the involvement and technical competence of the liaison agent (Davig).

16. If some members of <u>each</u> organization involved in a coupling arrangement are conversant in the primary languages (including technical) of the other organizations or groups, a more effective coupling arrangement will result. Common language increases the problem formulation capability of each organization in terms of the others and it lowers status barriers (perceived status differences) between organizations (E. C. Young).

FIGURE 2: <u>SOME ILLUSTRATIVE VARIABLES INVOLVED IN BASIC STUDIES OF TECHNOLOGY TRANSFER</u>

1. Timing of R&D - Marketing contact
2. Maintenance of R&D - Marketing contact
3. Confidence in the firm's product development process
4. Confidence in the respondent's dyadic counterpart
5. Confidence in product X which is under development by the dyad
6. The recognized need for information
7. The willingness to exchange information
8. Alternative information sources
9. Market conditions
10. Government regulations and influence
11. Economic environment
12. Top management commitment to product development
13. Top management objectives and attitudes
14. Cultural characteristics
15. Top management perceptions of constraints
16. Top management perceptions of consequences
17. Risk propensity of the idea evaluators
18. Exposure of the idea evaluators to outside contacts and sources of information
19. Firm's freedom to alter its product
20. Perceived risk of accepting the idea
21. Perceived availability of people to implement the idea
22. Perceived urgency of solution to the problem

23. Amount of search behavior associated with the problem to which the idea is a potential solution

24. Organizational commitment

25. Union and employee relations

26. Top management support

27. Interpersonal rewards

28. Task-environmental rewards

29. Achievement, affiliation, power, aggression

30. Task and interpersonal obstacles (stimuli) contributed by each group to intergroup system

of the TT process. Some of them address a very macro level (e.g., economic theory) and others a very micro level (e.g., personality or individual learning theory). Our own work, so far, has drawn upon these many distinct disciplines or subdisciplines for propositions, models, and measurement methods (not in any particular order except alphabetical):

Anthropology (enculturation, innovation, culture shock)
Bureaucratic Theory
Business and Public Administration
Economic Development
Economics of Innovation
Engineering and Physical Science (traditions, methods, values)
Entrepreneurship
History of Science and Technology
Individual Learning Theory
Information Science
Intergroup Relations
Organizational Climate
Personality Theory
Role Theory
Science Policy
Small Group Interaction
Sociology of Innovation
Subjective Probability
etc., etc.

It is no wonder that there is a babel of voices and viewpoints in the literature as well as an ever-increasing special language or sets of special languages and concepts. We are still in the "elephant-and-blind-men" stage of this field, and it is too soon to say which specialties will make the significant contributions to our understanding of this complex phenomenon.

3. Overall Methodology - By this I mean the general research strategy used by various investigators. Much of this depends on research traditions in the various fields mentioned above, as well as the style and preferences of individual researchers, the funding available for the research, the constraints on data collection, etc.

In general, there are a few major methodologies represented in the literature:

Abstract speculative theorizing with no recourse to empirical data
Manipulation of large amounts of data abstracted from the "real world" - e.g., data on exports, license fees, patents, etc.
Mathematical model building

Simulation in laboratories or by computer
Survey research by mass questionnaire or interview administration
Individual field studies (single cases)
Comparative field studies
Longitudinal behavioral studies (in "real time")
Field experiments

A major difference between these general strategies revolves around the extent to which they deal directly with the behavioral and attitudinal phenomena *in situ*, where the actual technology transfer is taking place vs. operating at one or more levels of abstraction from the actual phenomena as they are occurring.

Many factors determine the strategy used, and it is my feeling that many different approaches are needed and can be blended to improve the current scattered state of the art. One problem, of course, is the need for "critical level" efforts in research in this field, where budgetary, time, and manpower limitations do not completely dominate the choice of methodology.

4. *Field Techniques* - Where empirical research *is* undertaken, a wide range of field techniques or instruments is available. Among them are many kinds of questionnaires, interviews, observation schemes, self-recording, "remote-sensing", and other non-obtrusive techniques for data collection. The choice of techniques is a major issue, since on it rests the issues of measurability of phenomena and credibility of the results of that measurement. In my opinion, very inadequate attention is paid to this important scientific issue in this field and the large volume of mixed and non-credible results is a reflection of this neglect. As a minimum condition of our own research, members of our research program make every effort to use existing instruments that have already been tested in the field and for which some comparative results are available. A second methodological bias we have is toward extensive pilot testing of instruments before full-scale use. And a third is the use of multiple techniques or measurements to increase the reliability and credibility of our measurements.

Finally, *Figures 3* through *12* illustrate some of the conceptual models and study designs that have formed the basis for our research in this area.

In conclusion, I would like to comment (not unexpectedly, from my biased viewpoint) that much more basic research on TT is needed, but that even more urgently, we need more comparability and convergence in the research we are now doing and plan to do in the future.

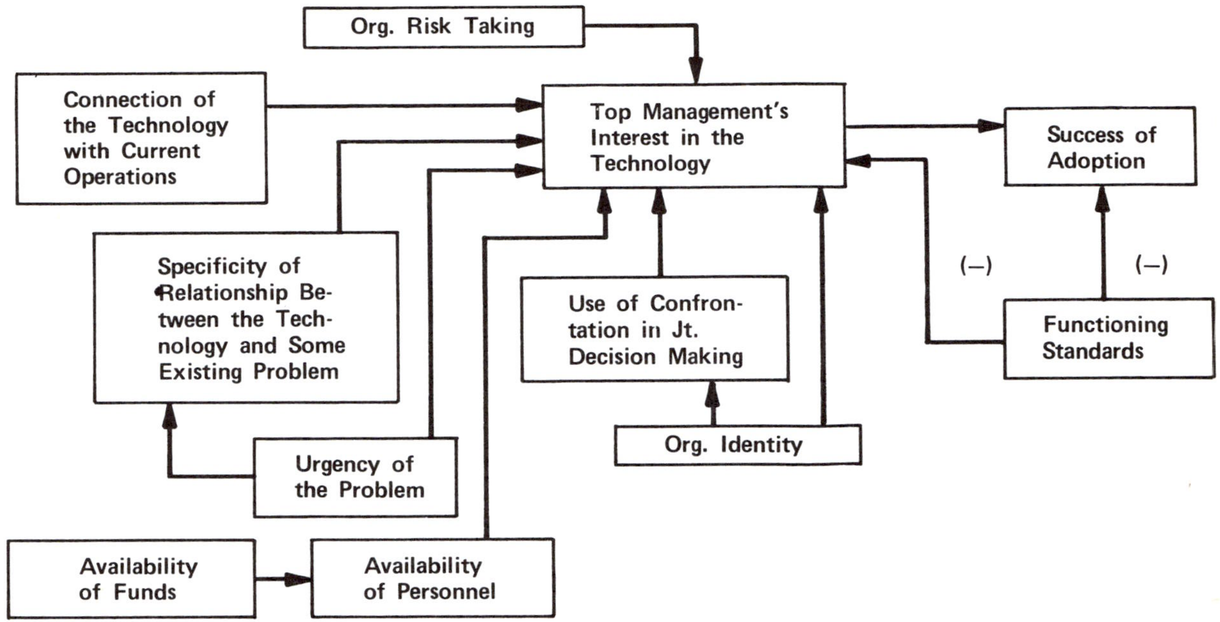

Fig. 3 Model for Product Cases

72/5 Chakrabarti

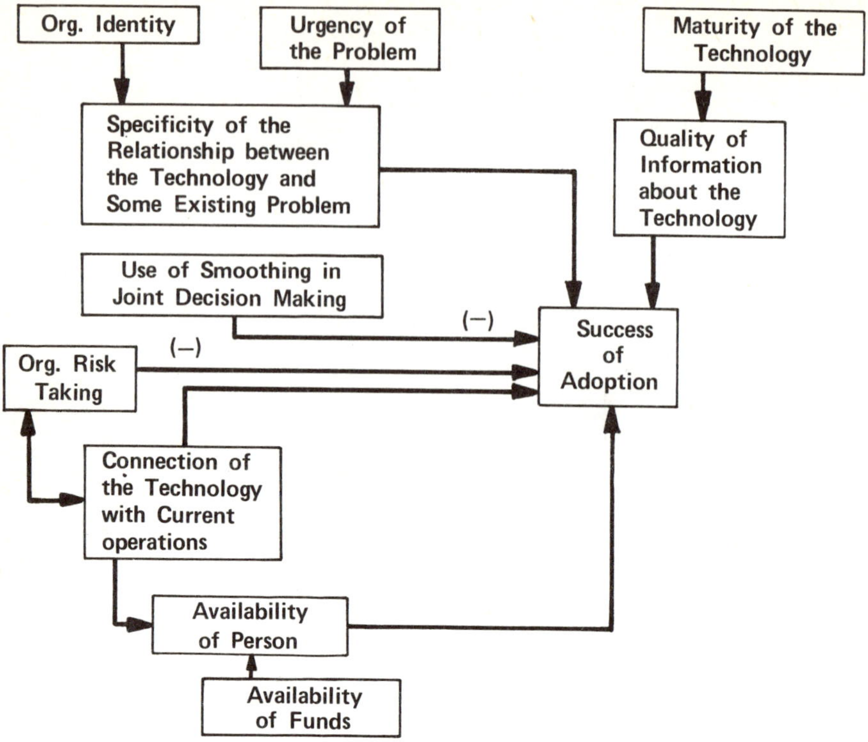

Figure 4 - Model for Process Cases

72/5 Chakrabarti

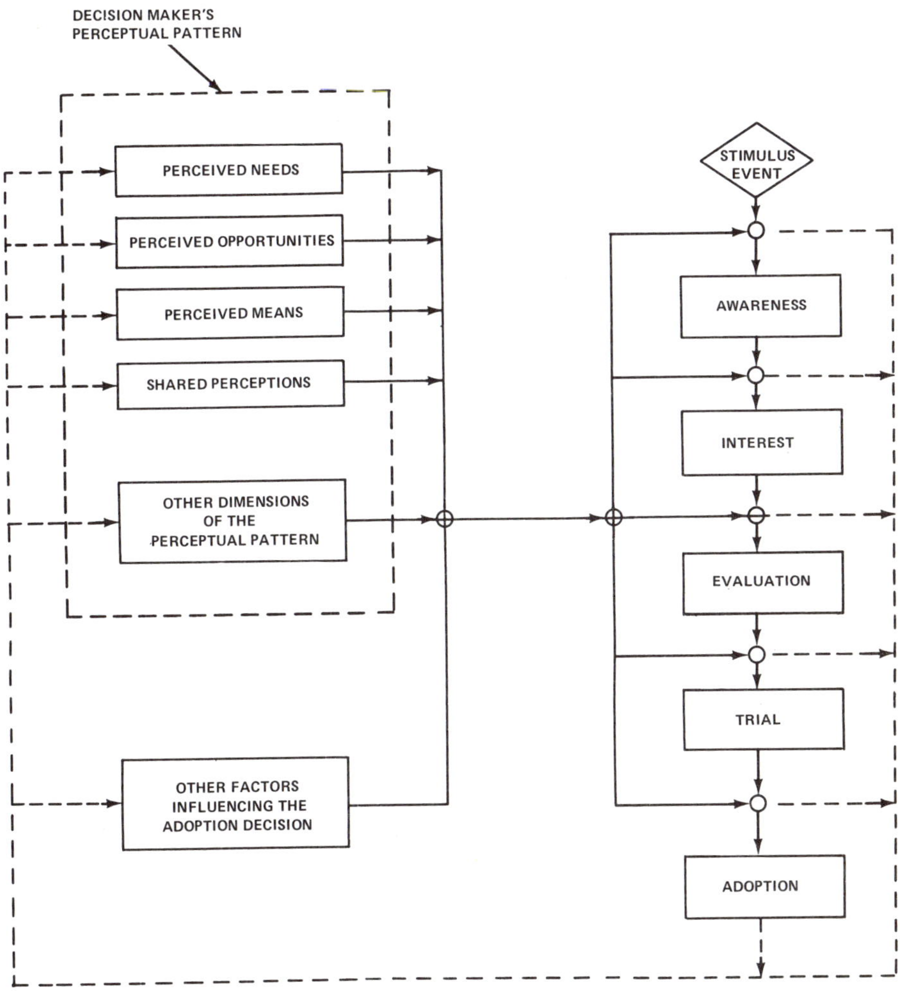

FIGURE 5

THE DECISION MAKER'S PERCEPTUAL PATTERN AS IT INFLUENCES THE DECISION TO ADOPT CHANGES IN PRODUCTION TECHNOLOGY

72/19 E.C. YOUNG

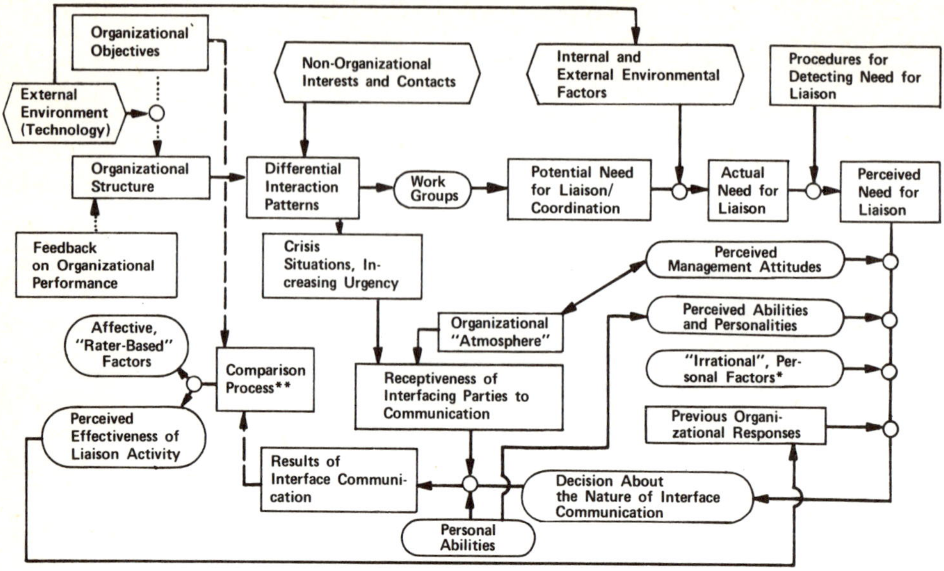

Fig. 6 - A BASIC INTERFACE MODEL

* The term "irrational" refers to organizational purposes and rationality rather than to individual rationality

**Done at both individual and organizational level.

——— Solid line implies a relationship of one variable influencing another

.... Dotted line implies a very weak relationship of one variable influencing another

— — — Dashed line implies a comparison process

Organizational Level

Individual Level

External Factors

67/33 Mills

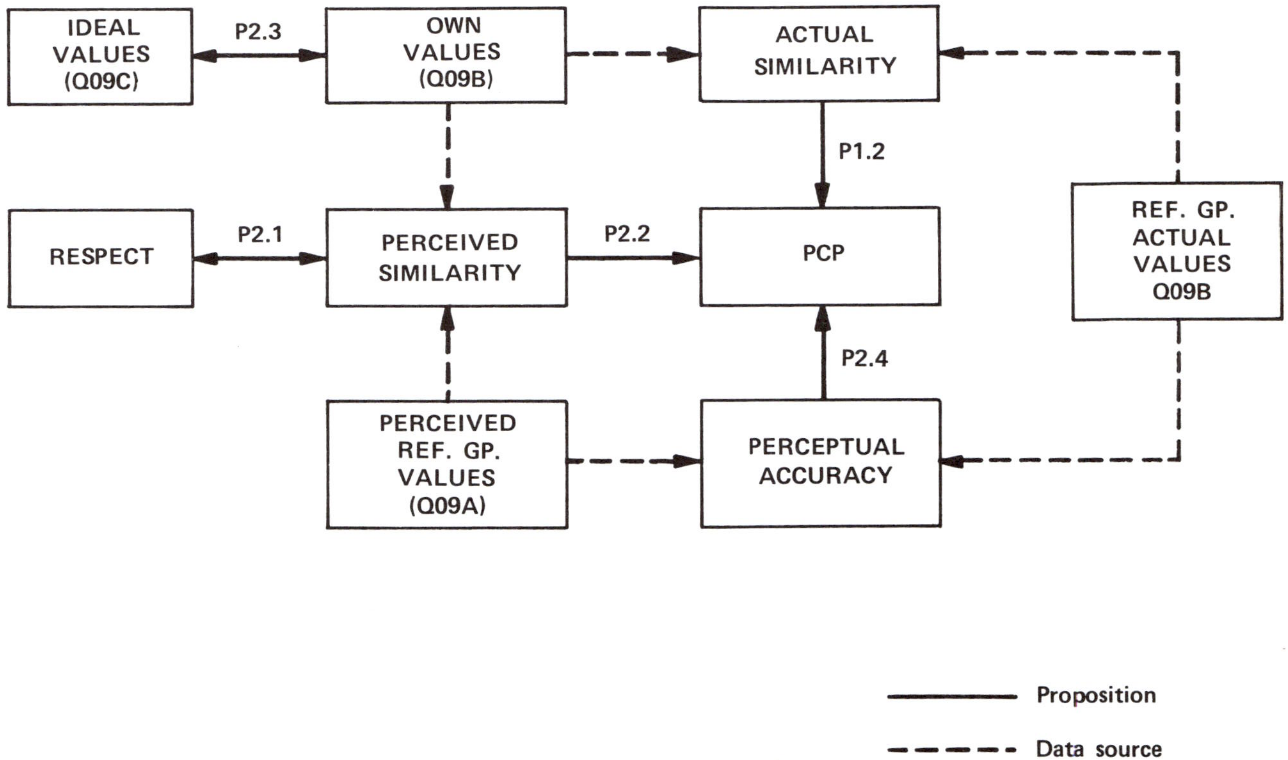

Fig. 7 - Relationship of "Values Perception" Propositions and Data Sources

70/32 Douds

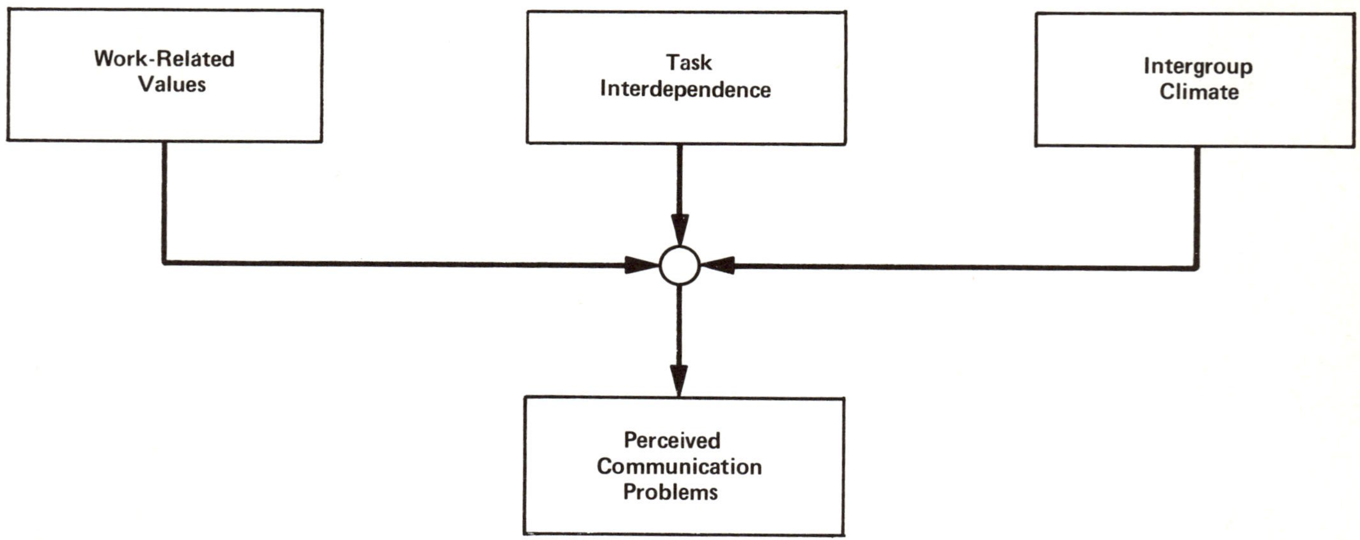

Fig. 8 - Major Variables of Barth's and Douds' Studies

70/32 Douds

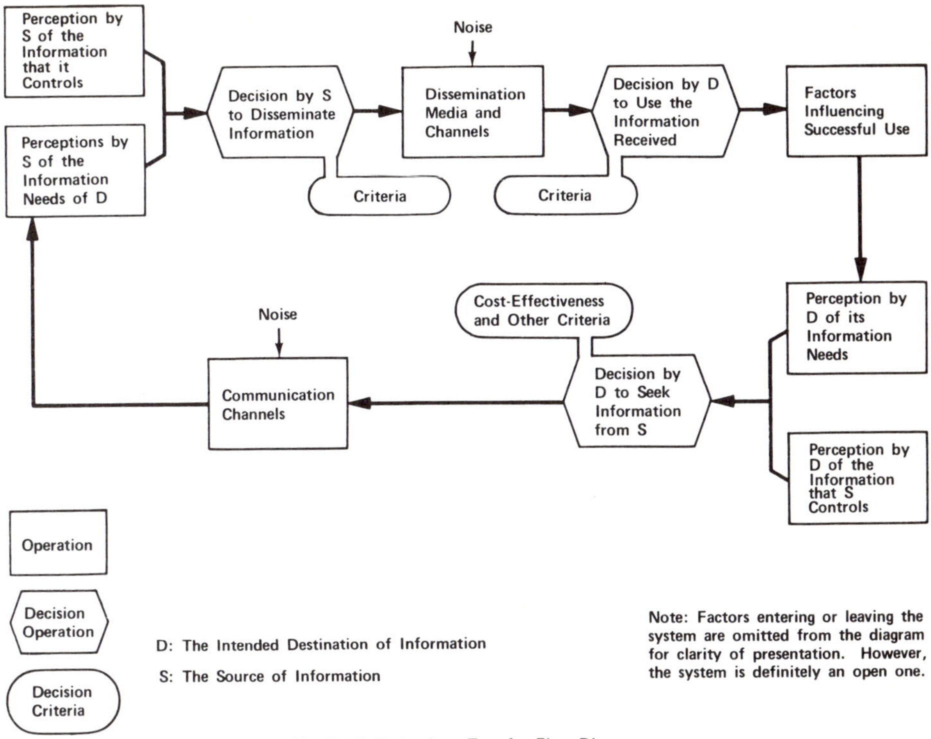

D: The Intended Destination of Information

S: The Source of Information

Note: Factors entering or leaving the system are omitted from the diagram for clarity of presentation. However, the system is definitely an open one.

Fig. 9 - A Technology Transfer Flow Diagram

69/8 Kegan

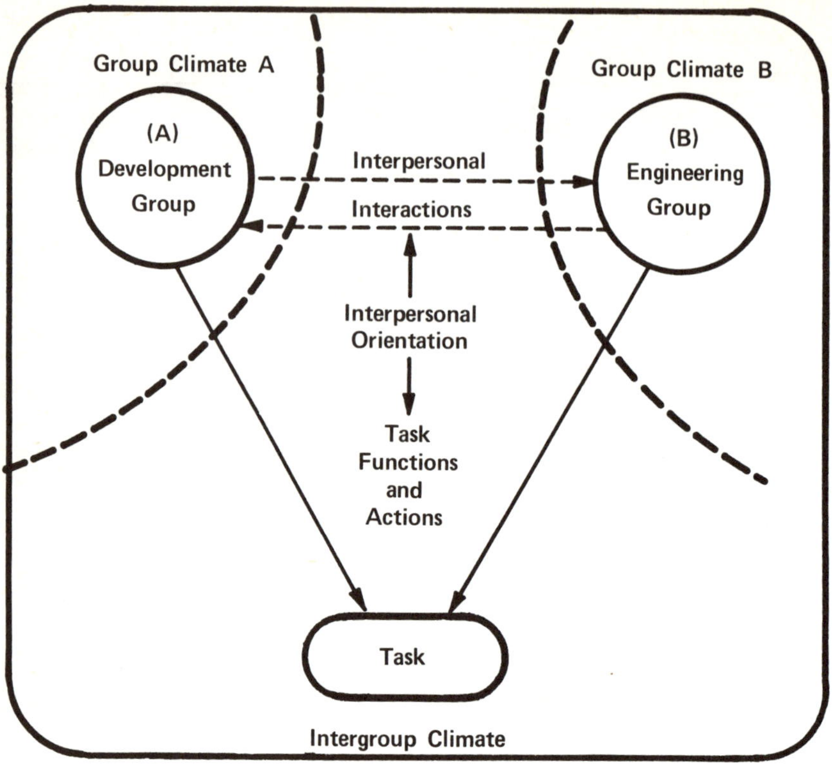

Fig. 10 : Intergroup Climate System of Two Task Interdependent R&D Groups

70/34 Barth

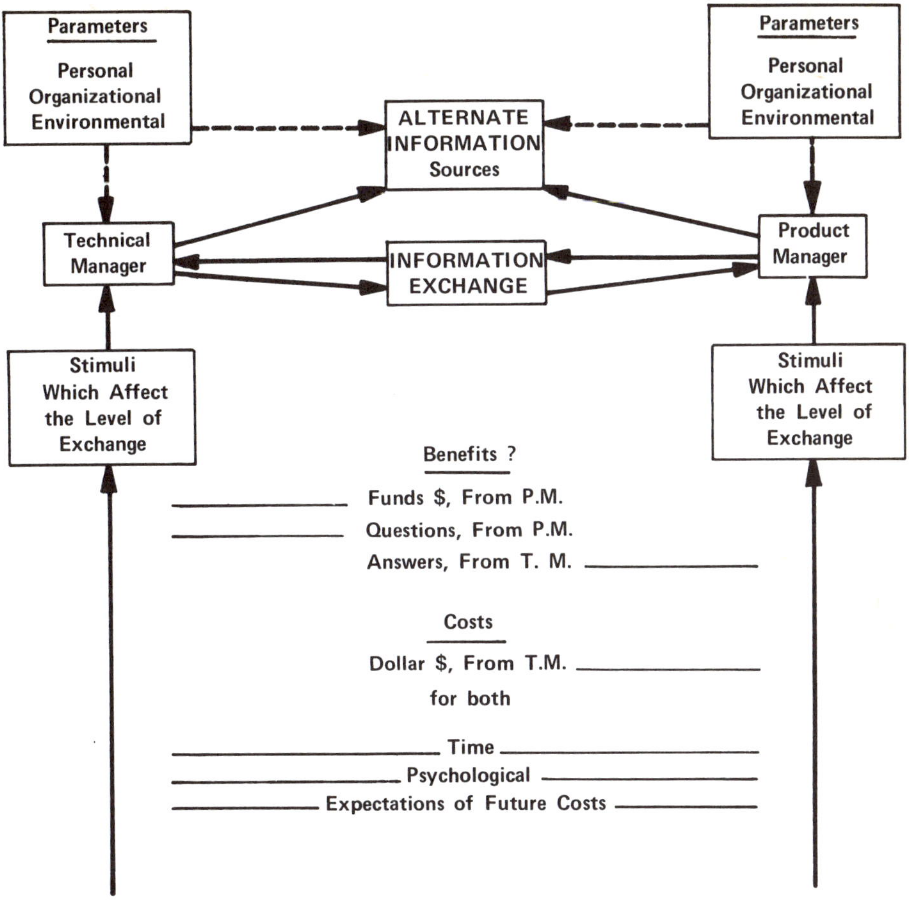

Key: Product Manager = P.M.
Technical Manager = T.M.

FIGURE 11

Typical Stimuli Which May Influence The Levels of Exchange

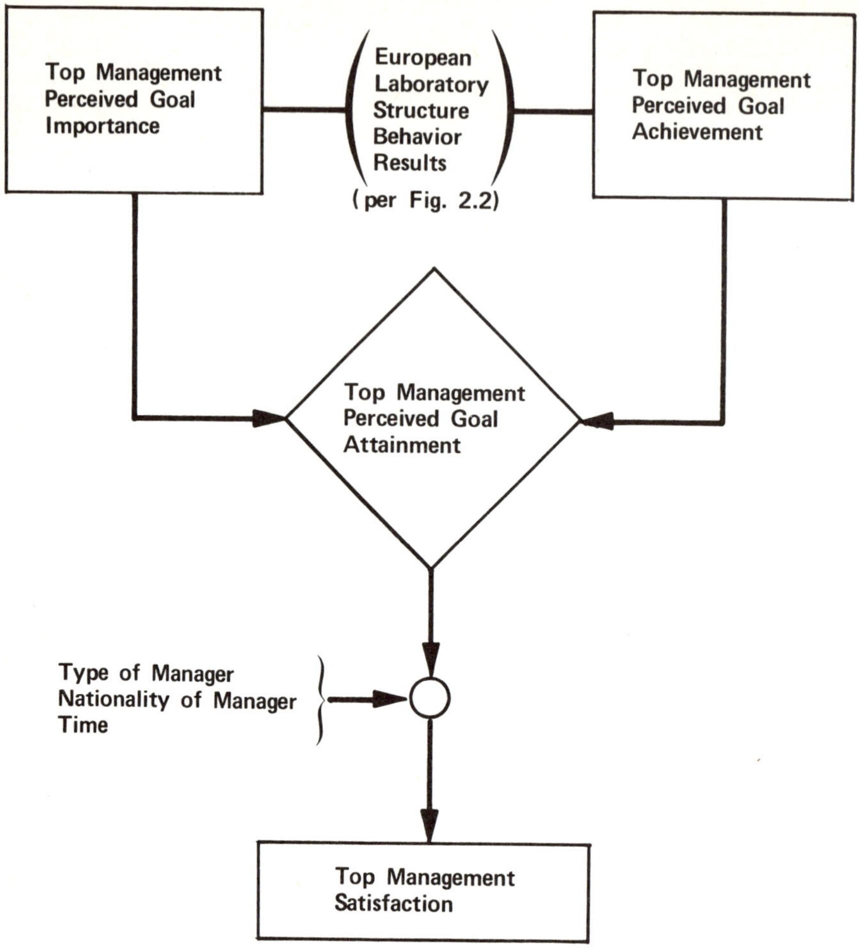

Fig. 12 - Proposition 1a

PART III: The Public Sector

Experiences in the public sector range from efforts to transfer commercial technology *into* the public sector, particularly at the state and local levels, to well organize and financed efforts to transfer technology from military and space activities to the civilian sectors of government and industry. Since all efforts involving the public sector must take their form from, or have a basis in, law; this section appropriately begins with Susan Doscher's review of current U.S. government initiatives in the development and stimulation of technology transfer mechanisms, with emphasis upon new and pending legislation in the U.S. Congress. John Mock discusses technology transfer problems at the local level with particular attention to the social and political barriers , while Karl Willenbrock points out the special problems of transfer from government laboratories to the private sector--a coupling receiving considerable emphasis in the U.S. in recent years. A series of case studies of innovation transfer in education and training are presented by Alicia Pagano, William Bierfelder and Ralph Dusek. In the context of transferring technology into the United States Courts, William Eldridge disucsses the innovations and barriers created by the history and culture of the adopting organization.

The unique problems of military to civilian sector technology transfer are discussed in papers by Samuel Bar-Zakay, Leroy Pagano, Marvin Cetron, and Harold Davidson. A common theme of organized effort emerges. Emphasing the importance of human channels, Ernest Short describes his experiences as a "Technology Transfer Agent" and John Wells describes the NASA experience--possibly the longest running effort of its type. Summerizing this section is a report on the "Government Policies and Technology Transfer Workshop" compiled by Paul Jervis and Susan Doscher.

GOVERNMENT'S ROLE IN THE TRANSFER OF TECHNOLOGY

Susan L. Doscher

Congressional Research Service*

INTRODUCTION

The process of technology transfer is a complex one. There are a myriad of factors which influence the conception, initiation and operation of the process. A wide range of commonly recognized elements and influences are implicit in a discussion of technology transfer; for example, there are different types of transfers; there are a variety of barriers and stimulants; there are operational and organizational influences; i.e., licensing practices and multinational corporations; and there are functional considerations, such as the capability of the recipient to adapt and apply the transfer of technology. Also important, of course, is the individual transfer agent, who couples the technology generator with the new user and who is generally regarded as the key to the transfer process.

In essence, the process of technology transfer aims at getting the knowledge from the generator into the hands of the user.

Even if all the practical and theoretical elements of this transfer process were understood, the success of such coupling would not necessarily be assured. In order for technology transfer programs to be successful, they must be conducted in a political environment which is conducive to the process. In other words, for good theory to be put into practice, there must be support by government policies which serve as a catalyst to stimulate, not hinder the transfer of technology.

The government's role vis-a-vis technology transfer is a vital one, and it is to that purpose that this paper is addressed.

Technology Transfer -- the Concept

In simplistic terms, technology transfer may be defined as the process by which a technology is applied to a purpose other than the one for which it was orginally intended. So stated, technology transfer appears to be a straight-forward, readily understandable process. A familiarity with past and present technology transfer efforts, and an examination of the pertinent literature, however, explodes this neat conception.

Technology transfer is a highly complex process involving a wide variety of interactive elements. Complexities inherent in the process are reflected by the many descriptions ascribed to it, and by the numerous and assorted factors which have been identified as either central or related to it.

This paper considers the role of government vis-a-vis technology transfer. Because of the intrinsic imprecision of the term, an understanding of its usage should first be established. Both "technology" and "technology transfer" are used in the broadest sense. "Technology" connotes both hard and soft technologies or techniques. As suggested by the General Accounting Office; "Technology or a technological resource may be defined as any hardware device; equipment or system; scientific knowledge, engineering, design, or process; special laboratory or test facility; or specially trained person."[1] Similarly, "technology transfer" encompasses a broad spectrum, including the myriad of theoretical and practical elements and stages which have been attached to it. To elucidate, some commonly identified transfer elements and several of the many concepts will be reviewed.

The concept of technology transfer varies in accordance with the terms by which it is defined. For example, technology transfer may be thought of in terms of mechanisms involved, i.e. information dissemination techniques, technical services, symposia, conferences, study institutes. It may also be considered in terms of the active stages of the process, i.e. technological generation, information dissemination, technological adaptation and application to new problem situations.

Technology transfer also varies in concept with the perspective of the viewer. For instance, some factors will be regarded as fundamental to the process by the technician or the developer of the transferrable technology; other factors may take precedence in the mind of the marketing agent; still others may be identified by the manager or entrepreneur involved in the transfer; and yet another set of considerations may be identified as significant by the ultimate user of the transferred technology.

Despite the confusion, fundamental definitions have been suggested which imply many aspects of the transfer process. Samuel Doctors offers the following, defining technology transfer as:

> ...the process whereby technical information originating in one institutional setting is adapted for use in another institutional setting. The transfer typically requires active participation by both the transferor and the transferee and implies more than a mere dissemination of technical information; it implies the adaptation of new technology through creative transformation and application to a different end use.2

In a recent report prepared by the U.S. General Accounting Office, the following definition of technology transfer was put forward:

> Technology transfer is defined as the secondary application of technology developed for a particular mission or purpose to fill a different need in another environment. 3

Passive Transfer Mechanisms: In the report of the General Accounting Office, a distinction is drawn between active and passive technology transfer techniques. Passive mechanisms involve collecting, screening, indexing, storing, and disseminating information in response to perceived or stated user demands.

Effectiveness of passive technology transfer techniques is largely dependent on the user's capability to articulate his needs, to recognize technologies which have potential applicability to these needs, and to be able to obtain further details of specific technology so that it may be adapted to the particular problem circumstance. Another determinant of success is the quality of the information selection, storage, retrieval and dissemination services.

Both theory and practice support the thesis that information services--collection, selection, dissemination -- are not sufficient in themselves to effect successful technology transfers. As stated by Dr. Peter Chenery:

> The distinction between technology transfer and information transfer should be emphasized. Whereas information transfer is an essential part of technology transfer, much information is transferred but never put to any application. A technology transfer occurs when machines, materials, or methods are actually used to perform a new task. 4

Active Transfer Mechanisms: Active technology transfer mechanisms usually involve the participation of transfer agents or imply the personal interaction between the developers and users of technology. Active programs often depend on the role of a third party or transfer agent who is able to assist the potential user in identifying requirements which could benefit from technical developments, who is able to help identify appropriate technologies, and who is able to promote communication between the user and the supplier, so that a particular technology may be adapted to the new problem situation.

The importance of active transfer mechanisms -- in particular, the employment of skilled individuals to form dynamic links between the developers and the users -- is stressed over and over again in the literature on technology transfer. In the words of Samuel Doctors; "Personal contact is significantly more important than the mere dissemination of literature." 5 Richard Lesher and George Howick attested to the importance of personal interaction in effecting technology transfers:

> It is increasingly apparent that a communications gulf exists, as a derivative of the technology gap, between the principal generators of new knowledge and large bodies of potential users. This is not a simple problem of language, but a complex problem involving attitudes, values, goals, work patterns, orientations, environments, and other variables. This results in a need for intermediaries or couplers who can operate effectively at the interface between knowledge and need, and who can communicate effectively with those at both ends of the pipeline. 6

"Push" and "Pull" Transfer Methods: While on the topic of active technology transfer mechanisms, it is of interest to note a distinction which characterizes two fundamental types of active methods. This characterization is informally, but aptly, referred to as "push" and "pull" techniques of technology transfer. "Push" techniques seem to predominate in experimental and operational transfer programs, but they have generally produced disappointing results. These efforts initially focus on the availability of specific technologies which are channeled to other groups and organizations in response to perceived needs or in anticipation that new markets could be created for these technologies. Because the perceived needs often do not exist, or because new markets do not emerge once these technologies become available, the "push" methods have had limited success in transferring technology. What is missing is the ready recepient to adapt and apply the technology to new problem areas.

"Pull" techniques, on the other hand, spring from user requirements. Problems or needs are identified and technologies having potential applicability to these needs, are then sought. These types of programs are more difficult to initiate, primarily because new technologies are not immediately thought of as offering viable solutions, or if they are, it is generally not known how to go about locating appropriate technologies. Although the initiation of "pull" programs is more difficult, the likelihood of success for the resulting transfer effort is greatly enhanced. These types of programs have not been tried to any significant degree in the United States to date.

There appears to be no set method to effect technology transfers; the process is too complicated to be subject to a patterned approach. In testimony before the Senate Select Committee on Small Business in 1967, Dr. Charles Kimball stated:

> Technology transfer of any significance will only occur when the right people, the right markets, and the right ideas coincide with usable technology at the right point in time. The technological content, per se, may be the least important element in the transfer process. 7

In an article on applied research, Harold Gershinowitz stated:

> The application of research is not a simple or straightforward process. It takes place at the interface between knowledge and action. At this interface there are barriers of language, of psychology, and of values. Fields of knowledge are fragmented in different ways than fields of action are. It is not generally realized that, even in those areas in which research and development are unambiguously technological in character, many of the problems involved in the application of research are not technological at all. 8

The method by which the technology is transferred and actually used to perform a new task, has taken the posture of an unanswered, perplexing question. As stated at the outset, technology transfer is a highly complex process involving a variety of elements. Although technology transfer must necessarily involve the availability of a technology to be transferred, often the available technology is overemphasized in its importance. As has often been noted, many transfer efforts seem to result from the situation of developed technologies in search of problems. In these situations, the technology is the focal point of the effort and consequently, other vital factors, i.e. the existence of a market or requirement; the availability of managerial or entrepreneurial skills, are often underemphasized or, at worst, overlooked. It is extremely important that the complexities of the process be underscored and kept in mind when considering the content of this paper.

The Role of the Federal Government

The Federal Government's interest in and responsibility for the dissemination of knowledge, including the transfer of technology, dates back to the time of the Constitutional Convention. At that time concern was expressed as to whether the new nation had sufficient financial resources and adequate scientific and technological capabilities to compete with the developed European nations. As a consequence, a provision was written into the Constitution authorizing the Congress to promote the progress of science and the useful arts; the provision became the foundation for the current copyright and patent system. 9

In recent years the federal role in knowledge diffusion and technology transfer has become more clearly defined, and its importance more pronounced. The accelerating pace of scientific and technological advance has demanded a comparable increase in the government's capabilities to effectively utilize the resulting resources. The fast pace at which our knowledge has been accumulating threatens to outstrip government's ability to keep up with it, and portends a situation in which resources will not be fully utilized, or will remain entirely untapped.

The rate of advance is not the only influence which has sharpened the federal role in technology transfer. More significant is the increasing level of federal investment in the nation's research and development activities. The vast majority of the United States' scientific and technological development derives from the expenditure of public funds. As sponsor of most of the research and development activities performed in this country, the government controls much of the technical information and data resulting from it.

During the past decade, federal expenditures for research and development have risen steadily. A downward trend since 1970 in federal support for research and development is now clearly established.

> The increase from $15.5 billion in 1971 to an estimated $16.8 billion in 1972 was expected to move the Federal R&D obligation total to a record high. The R&D total of $17.8 billion requested in the President's 1973 budget marks still another high. 10

At the present time, there is no central organization or office to formulate policy or to conduct programs for technology transfer within the Federal government structure, nor is there any agreement as to whether or not one should exist. Technology transfer programs are generally created and implemented by mission agencies having significant research and development support programs. Some examples of agency programs, both operational and experimental will be considered. It is not the intention of this paper to provide indepth analyses of the selected programs, but, instead, to highlight some salient characteristics, and to indicate the variety of approaches which have been taken at the federal level to implement technology transfer.

The first of the programs considered is the Agricultural Extension Service, administered by the Department of Agriculture, which constitutes perhaps the first federal program in technology transfer. Programs of mission agencies having major science/technology procurement expenditures, specifically the National Aeronautics and Space Administration, the Atomic Energy Commission, and the Department of Defense are subsequently described. Finally, the small technology transfer effort administered by the Small Business Administration, and the now defunct State Technical Services Program of the Department of Commerce, are discussed.

Department of Agriculture: The Agricultural Extension Service is one of the earliest programs specifically intended to transfer technology from the Government to the civilian sector of the economy. The extension service was established to operate in a rather unique environment which has been characterized as:

> . . . a large industry with tremendous potential for technological advance in which no single member or group of members possessed either the investment capital or the technological competence to develop or utilize the needed technologies. 11

The fundamental objective of the Extension Service is to assist individuals in identifying and solving farm, home and community problems by utilizing the research findings of the Department of Agriculture. The Extension Service is a cooperative venture of the Federal Extension Service and State Extension Services which are divisions of their land grant colleges. Extension personnel operate on three tiers: federal, state and county. The actual transfer of the technology takes place at the county level. County agents create awareness of the existence and advantages of new agricultural techniques and products and assist in probelm-solving upon request.

Unlike other federal transfer programs previously noted, the Agricultural Extension Service engages in vertical technology transfer; it is designed to serve the interests of only one industry, and has no intention or provision for horizontal transfer of technologies to other applications. The Program has enjoyed the benefit of considerable field experience in the transfer of technology, and because it has been operable for an extended period of time, it has, in effect, become a part of the farming community and farming industries in the U.S. 12 The interaction among the users, government, the state university systems and the private sector have fostered the effectiveness of this transfer program which has grown to include approximately 17,000 professional staff members.

National Aeronautics and Space Administration (NASA): NASA is the first major science procurement agency to establish an agency-wide program to promote the external transfer and application of its technological developments. 13 Written into NASA's enabling legislation was the requirement to provide the widest practicable dissemination of results of its research and development efforts. NASA's goal was to establish a program through which many of the technological advances made during the space program would find useful, commercially feasible, and -- where possible -- immediate adoption into the civilian industrial community. To meet this objective, the Office of Technology Utilization was established.

The NASA Technology Utilization Program consists of two major functional elements: an acquisition or input operation and a dissemination or output operation. The program consists of a number of other elements which serve in a supplemental role to the two primary functioning units. The Acquisitions Branch is responsible for new technology reporting -- that reporting by NASA employees, contractors, and subcontractors of innovations conceived while working under NASA sponsored research and development projects. The Dissemination Branch is charged with the responsibility of promoting the transfer of the technology primarily to the commercial industrial market.

To carry out the transfer function, the agency utilizes several different mechanisms. The agency maintains a current and accessible store of information on the results of aerospace-related R & D efforts conducted throughout the world. On the national level the agency functions through its field installations to assure availability of newly collected information to potential users both within and outside the aerospace community. NASA "Tech Briefs" are the most common form of announcement through which the agency provides a concise description of a technological innovation, supplying basic information on the fundamental concepts or principles. The agency makes additional information available about a particular innovation upon request from users.

Other types of publications made available by NASA are: "Technology Utilization Compilations", "Technology Utilization Reports", "Technology Surveys", and "Conference Proceedings". NASA disseminates information through six Regional Dissemination Centers which are based either in universities or not-for-profit research institutes. The Centers promote government-university-industry partnerships by serving fee-paying and retrieval services. In addition to providing information to industrial clients. The latter services are generally provided by Technology Utilization (T.U.) Officers, located at the Centers and in other NASA field offices (of which there are seven). T.U. Officers are theoretically able to identify, collect and evaluate the new technologies and are able to represent them to potential users.

Although the NASA Technology Utilization program provides visible testimony to the importance of technology transfer, many feel that the program has not been particularly successful. In an evaluation of the NASA transfer effort, Samuel Doctors offers some conclusions regarding possible causes of weakness in the effort. Doctors noted that the effort is fragmented and that it tends to isolate the Assistant Administrator for Technology Utilization from various segments of the program; thus his authority is weakened. Among the other conclusions which Doctors derived was that the program was founded primarily on a political base, and has continued to be employed as partial justification for the agency's R&D expenditures. Doctors argues that this has precluded experimentation and innovation within the program and has forced it into a library-like operation. He stated that the Regional Dissemination Centers have been forced to essentially provide library services; consequently, entrepreneurial activity and market assistance efforts have been restricted. Doctors also notes that the emphasis on commercial-industrial transfer has limited the program in that technical information of interest to other researchers outside this community receives little or no professional screening or selective dissemination. 14

Atomic Energy Commission (AEC): The AEC presently accounts for 8 per cent of the federal government's R&D expenditure. The Commission supports a variety of activities to disseminate unclassified information which derives from its R&D support program. Like NASA, the AEC's dissemination program was mandated by its enabling legislation.

The Division of Technical Information manages the Commission's publication activities and coordinates the Commission's distribution with the National Technical Information Service administered by the Department of Commerce. Among the variety of research reports,state-of-the-art surveys and R&D status reports, AEC publishes jointly with NASA the "Tech Briefs" which are distributed by both agencies.

The height of the AEC's active technology transfer program was reached in 1969, when it supported the operation of three Offices of Industrial Cooperation. The establishment of these Offices was designed as a pilot effort to encourage the application of AEC developed technologies to industrial needs. These offices acted as dynamic links between the industrial community and the Commission. The offices were directed to identify and announce technological innovations and to acquaint themselves with the needs of particular segments of industry. The funding level of the technology transfer program in 1969 was approximately $200,000. There has been no additional program money made available for this program since 1969.

The termination of funding for the AEC's transfer program may be viewed as an indication of the rather unimpressive results of the program. Although the past efforts of the agency have not produced the results anticipated, interest in promoting the circulation and alternative applications of AEC developed technology persists.

Plans for a new technology transfer effort are currently being explored within the AEC. As presently conceived, the program would have two primary aspects: information dissemination and laboratory cooperation. Its central objective would be to enhance the communication and cooperation between potential users and the AEC research laboratories.

Department of Defense: Although the Department of Defense (DOD) has long accounted for a major portion of Federal R&D expenditures -- estimated at 49 per cent of the Federal total in R&D funding for Fiscal Year 1973 -- the Department has initiated minimal efforts to transfer technology from the defense to the civilian sector. The Defense Documentation

Center provides technical information for DOD contractors, DOD itself, and other qualified users. The Center, however, performs basically a library function, and information services are performed only on request.

As stated in a Department of Commerce Report, DOD demonstrated little interest in promoting technology transfer activities. DOD's posture on technology transfer programs was characterized as follows:

> Any program of technical data dissemination is limited on the input side by this mission orientation, and on the output side by security requirements, overzealous concern for contractors' proprietary data rights, and a general lack of concern for dissemination of technical data. Of the small amount of technical information disseminated, none is professionally evaluated for its value to non-defense applications. 15

In December of 1972 a study was conducted by the General Accounting Office which examined potential means of transferring defense related research and development results to the public sector. The study was instigated by growing Congressional and Presidential interest in applying the federal government's scientific and technological resources to help solve public problems. The study examined various aspects of the technology transfer problem, focusing particularly on the constraints affecting DOD's potential transfer programs and recommending means by which such hindrances could be mitigated or circumvented, and suggesting positive program alternatives. One of the fundamental conclusions was that the Secretary of Defense should establish definite policy and procedures to encourage active transfer of DOD technology to civil agencies. 16 Perhaps this report will signal a change in DOD's interest and efforts in technology transfer.

<u>Small Business Administration</u>: Unlike NASA and the AEC, the Small Business Administration (SBA) does not support extensive research and development efforts and does not generate scientific and technical information. The Administration does however, have a technology utilization program which was mandated in the Small Business Act which authorized the Administrator to:

> ...assist small business concerns to obtain the benefits of research and development performed under Government contracts or at Government expense. 17

The purpose of this provision in the act was to compensate for the inability of small business to maintain in-house research and development capabilities which would enable them to adapt results of federal programs and apply them to their particular business needs. In response, the Small Business Administration initiated a Technology Utilization Program which provides support for ten technology utilization field officers.

The primary vehicle of the technology transfer program is a prepared technical publication which is disseminated through the field offices. The publication, which presents profitable new aerospace technology with potential industrial application, is distributed among small manufacturing firms. A reader service card is attached so that interested firms may acquire technical support packages for the abstracts initially provided. In addition, firms can request technical assistance from the SBA in addressing production and pollution problems. The technical abstracts and assistance are channeled from the Federal level through the ten field personnel.

In operating this program, the SBA works closely with several of the mission agencies providing the technical information which the SBA then distributes. The SBA works with NASA, the AEC, DOD and with the Department of Commerce National Technical Information Service.

The future of the SBA Technology Utilization program is uncertain at this time. Although the program has had some demonstrated success, the level of its operation has decreased in recent years. The program has Congressional support, but faces competition from other priority programs within the SBA. A primary difficulty is that the sharply limited size of the effort renders it almost ineffectual in the face of the enormous task which it is striving to perform.

Office of State Technical Services: The State Technical Services Program succeeded the Civilian Industrial Technology Program (CITP) which was originally proposed by Department of Commerce Assistant Secretary Hollomon in 1963. Objectives of the CITP were primarily: to foster innovation in "lagging" industries, including building and textiles; to examine information needs and the state of technological development in other industries; and to create an industry-university service to diffuse information and provide technical assistance. The CITP was never fully funded, although funds were appropriated to support, for a limited time, that part of the program involving the textiles industry. 18

The State Technical Services Act of 1965, also initiated by Holloman, called for the promotion of economic growth by accelerating dissemination and utilization of scientific and technological knowledge by industry. As a result of the Act, an Office of State Technical Services was established within the Department of Commerce, Technical Services operated a Special Merit Program, on a matching basis, which was used for special studies and experiments in technology transfer.

OSTS was funded to a total of $20 million during its existence, including close-out costs. This figure represents only Federal expenditures, and does not take into account matching State funds. Through the use of fairly small amounts of money, the OSTS increased the viability of technology transfer to industry for a significantly wide range of programs, particularly in the medium and small sized industries. OSTS identified the types of information and services needed by industry, and the most effective techniques for technology transfer.

The STS Program was ended by the Department of Commerce in 1970, after the Congress "zero-funded" the supplemental budgetary request which was required for continued operation. OSTS was formally abolished in April, 1970. The dissolution of the STS Program came about despite favorable evaluations of the Program which were conducted by Arthur D. Little, Inc. [19] and by the Public Evaluation Committee chaired by Carl Elliott. [20] A consensus emerged from these reports as to the desirability of certain objectives which the STS Program embraced. Some of the significant points have been summarized as follows:

> There is a proper governmental function to be performed in the transfer of the technology from publicly supported R&D activities and from other "generally available" (non-classified and non-proprietary) sources to the private sector.
>
> Federal, State and (more recently) local governments have had both active and passive roles to play in technology transfer. The private sector should be expected to contribute as well as benefit.
>
> Technology transfer actions return to their sponsors, and to society-at-large, benefits (in the form of increased productivity and tax revenues) in excess of program costs.
>
> Federal Government leadership, systems research and design, and "catalytic" sustained partial support are necessary for a viable nationwide effort.
>
> A vigorous Federal, State, local government (plus private sector participation) technology transfer program can be a significant, stimulating, and sustaining force in technological, economic and social enhancement. [21]

It is not possible to identify a single cause for the demise of the Program. Prior to the program's dissolution, it was contendeded that the activities of the OSTS should be conducted by other programs within the Department of Commerce, including the Economic Development Administration and the Office of Field Services. Another factor which has been identified is the fact that the STS Program may have suffered as a consequence of its being only one of a number of Federal conduits for technology transfer, and the program did not enjoy the backing of Federal programs of research relevant to specific industrial needs.

At the present time there is no STS organization within the Department of Commerce nor are there any funds directly appropriated for STS-type activities. Interest in the concept of a STS Program, however, is not completely extinct. Several bills were introduced to the House of Representatives during the 92nd Congress which had as their objective the reestablishment of the STS Program in the Department of Commerce. No action was taken on these bills, but they serve as an indication of continued interest in the support of programs designed to foster the dissemination of information and services from the Federal level to the States and localities and to industry.

The Role of States, Localities and the Private Sector

Interest in the initiation and implementation of technology transfer programs is not unique to the Federal government. New emphasis is being placed on the need for cooperative arrangements among the layers of the Federal government system, and universities and the private sector, to make possible the effective application of our scientific and technical resources to public problems. The requirement for greater participation on the part of the States and localities has become particularly pronounced in recent years.

In his address to the 1971 National Conference of the American Society for Public Administration, Congressman John W. Davis highlighted the confluence of three noteworthy trends in the U.S.:

> (i) a renewed interest in federalism, caused by the realization that all problems cannot be solved in Washington; (ii) a shifting of our national priorities from defense and space to domestic problems, which traditionally have been the responsibility of state and local governments; and (iii) a rapid increase in funds for applying science and technology to domestic problems, such as housing, transportation, health care, crime, and pollution. 22

Emphasis on the State (and implicitly the local) role was similarly placed by the National Governor's Council for Science and Technology in a letter to the then Presidential Science Advisor Dr. Edward David, Jr. The Council put forth the following thought:

> The utilization of technology to serve domestic public purposes is necessarily a political as well as a technological matter and requires the mobilization of all possible public and private resources. The need for state government to provide leadership, far beyond that of a canny purchasing agent is clear. 23

Other recent events and public statements have focused on the need to increase state and local participation in the application of science and technology. In his message to Congress on science and technology, the President stressed the requirement for each level of government to do what it is able to best in order to make the federal structure more responsive to public needs.

> The Federal Government . . . can usually do a good job of massing research and development resources. But State and local governments usually have a much better 'feel' for the specific public challenges to which those resources can be applied. If we are to use science and technology effectively in meeting these challenges, then State and local governments should have a central role in the application process. 24

This statement reflects the President's desire to give the states and localities more responsibility in coping with public problems, and it indicates the Administration's intention to forge new links between the various levels of government.

Last year a national conference was held to discuss intergovernmental policy and program alternatives to promote the application of science and technology to public problems. Among the participants were legislators and executives from Federal, State and local governments and representatives from universities and industry. The Conference resulted in a firm concensus on the need for active state and local involvement in applying science and technology to public problems. A resolve for action ensued and a seven-point program, mapping out a strategy to build necessary state and local capabilities, was devised. 25

In addition to the Conference, two recent publications report on studies of the application of science and technology to public problems. One, conducted by the Committee on Intergovernmental Science Relations for the Federal Council for Science and Technology, explored the interaction of Federal, State and local government research and development policies and programs. 26 The second study, conducted by the Council of State Governments, with financial support from the National Science Foundation, assessed the part that science and technology play in State government, and explored methods by which the interaction could be strengthened. 27

A general concensus emerging from the findings of the two study groups purported the real need for state and local involvement in the transfer of scientific and technical resources to the public sector, and demonstrated the present lack of state and local capabilities to perform this function. Recommendations endorsed the formation of intergovernmental programs to create and expand upon such transfer mechanisms.

Despite the general findings which purport a paucity of technology transfer capabilities at the state and local levels, there are a significant number of ongoing state and/or locally-based efforts which are working to channel technological resources into the public sector. Several programs have been selected for inclusion in this report to illustrate some of the approaches which have been taken.

Pennsylvania Technical Assistance Program: Among the most notable of State technology transfer activities is the Pennsylvania Technical Assistance Program (PENNTAP), which was created by the State in response to the Federal-State Technical Services Act of 1965. Its fundamental purpose is to foster the utilization of State scientific and technical resources by uniting technical interests, capabilities, and services, both private and public, existing within the State.

The operation of the program incorporates three fundamental principles derived from five years of study of the technology transfer process: (1) the use of information is not assured merely on the basis of its availability; (2) an interpreter is usually needed to translate the technical information for the potential user; and (3) the quantity of information transferred is directly proportional to the amount of person-to-person contact employed. 28 PENNTAP incorporates these principles in a three-pronged program which involves; the operation of a statewide information dissemination network; the administration of a staff of technical field specialists who carry out the transfer process on a person-to-person basis; and, the coordination of existing sources of expertise.

PENNTAP is administered through the Pennsylvania State University's Division of Continuing Education and receives financial support from the State Legislature through the Department of Commerce. 29

The Tacoma Washington Project: The Totem One Program which is the name of the technology transfer program in the City of Tacoma, Washington, is exemplary as an effective effort in the transfer of aerospace technologies to assist in the solution of urban problems. The city government of Tacoma began the program in early 1972 by requesting technical assistance from the aerospace industry -- Boeing Company specifically. During its year and several months of operation, the program has involved, in addition to the city government and Boeing, the University of Washington and federal and state agencies.

The technology transfer technique which has been utilized for this project is referred to as the "process approach." 30 This approach was formulated on the principle that a common base of understanding is required before a successful transfer of technology can take place. To bring technology into cities in a planned, systematic manner, an integration process must first occur to allow for the interaction of the users (city officials) and technologists to promote an understanding of the nature and scope of the problems and the applicable potential of available technologies.

Under the Totem One project, technology has been targeted at several specific problem areas for which the city government is responsible. Projects involving the introduction of technology are: (1) planning programs and base-line analyses of the city's transit system; (2) analysis of procedures and equipment required for fire fighting operations; (3) information gathering and processing and urban modeling techniques to improve city planning capabilities; and (4) the study of harbor service craft -- specifically identifying models best suited to the city's needs. 31

Public Technology Incorporated: Another noteworthy transfer effort is being conducted by a non-profit group - Public Technology Incorporated (PTI).

It was created by a joint action of public interest groups serving State and local governments. Its founding was in response to a perceived need that States and localities should have the capability to apply scientific and technical solutions to domestic problems. Initially Public Technology was the application program of the International City Management Association with strong support from the National League of Cities, and the U.S. Conference of Mayors. In December, 1971, after receiving a grant from the Ford Foundation to cover inital operational expenses, Public Technology was incorporated as a nonprofit tax-exempt organization by six public interest groups. The groups are: the Council of State Governments, the International City Management Association, the National Association of Counties, the National Governors' Conference, the National League of Cities, and the U.S. Conference of Mayors.

The purpose of the organization is to create and operate an effective mechanism that applies available technology to the solution of city, county and State problems. Public Technology Inc. identifies the city, county and State problems that can be solved by technology, develops realistic solutions which are acceptable to the particular governments and their citizens, and assures the widest possible dissemination and use of these solutions.

PTI carries out this mission by assuming the role of a "broker" or middleman. After identifying local government requirements, PTI brings the government representatives together with appropriate industry groups and helps them define specific problems. It then forms a user-design committee of department heads, budget officers and technical experts who work to develop suitable cost-performance specifications and prototype requirements. Simultaneously, a market profile is drawn up to be used as a base for assuring the marketability of the products and services which would be developed by the industry to solve the specified problem.

The funds for PTI's operations are provided by participating State and local governments, foundation grants and Federal subsidies. The problems with which the organization has worked to date have led PTI into seven hard and soft technology areas: fire protection systems, streets and highways, law enforcement, solid waste management, housing and construction, management science, and municipal information systems. Other areas will be added as participants present new technology-susceptible problems. In addition, PTI is building the capability to take on more complex long-term problems such as liquid and solid waste management, housing and health care. 32

<u>Los Angeles Technical Services Corporation</u>: The Los Angeles Technical Services Corporation is a non-profit, tax-exempt foundation, established in October, 1966, by cooperative effort of leaders in local government, the business-industrial complex, and the Greater Los Angeles Chamber of Commerce. The basic

mission of the Corporation is to assist the city in the solution of problems, providing more effective functioning by utilizing modern management methods and technology to strengthen city programs and simultaneously to reduce costs.

Financial backing for the organization has been provided by Foundation grants, Federal research grants, and modest local government allocations. In addition, potential market gains for local business and industry are implicit in the programs which are being initiated in collaboration with city departments.

During its five years of operation, the Corporation has made a significant number of accomplishments, ranging from the application of modern command and control techniques for emergency services (i.e. ambulance, fire, police services) to programs for the Youth Advisory Council and the development of the Los Angeles Technical Services Corporation, which have general applicability to cities throughout the Nation. Presently, the Corporation's activities are attracting widespread interest and attention from groups across the country. 33

Although these programs do not respresent a large sector of State, local and not-for-profit organizational activities, they serve as an indication of a widening interest in the viability of scientific and technological opportunities and solutions adaptable to the public sector. Scientific and technical developments which have spurred the creation of such transfer programs at the state and local levels, have also captured the attention of an increasing number of Congressmen. Current Congressional interest and actions of pertinence to technology transfer will next be considered.

Technology Transfer - Congressional Interest

Congressional interest in policy and programs which either explicitly or implicitly involve technology transfer, has become highly visible within the past couple of years. The rising investment in scientific and technical research and development has caused the Congress to focus on policy and programs which would allow for the most extensive and beneficial utilization of the resulting resources.

One indication of this burgeoning concern is last year's Congressional authorization of the new Experimental R & D Incentives Program which is being jointly administered by the National Science Foundation and the National Bureau of Standards. The fundamental objective of the program may be simply given as that of identifying means to promote rapid and effective transfer of technology to the public and private sectors. The Foundation's program is geared towards locating and testing viable Federal incentives which would encourage technology transfer in the academic community, state and local governments, not-for-profit and for-profit industry, and public service industries and institutions. The Bureau of Standards program is more specifically directed towards identifying incentives which would encourage technology transfer for commercial product development. In addition to the incentives programs, a related effort, the National R & D Assessment Program, administered by the National Science Foundation, was authorized to assess the value of investment in research and development to the national economy - specifically of how science and technology contribute to the achievement of national goals and objectives.

The importance of integrating science and technology into public policy and programs was given Congressional emphasis with the passage of the Technology Assessment Act of 1972. The Act established an Office of Technology Assessment within the legislative branch. The Office will be responsible only to the Congress and will serve to provide factual information on the broad range of impacts - physical, environmental, economic, social, and political - which could derive from the increasingly extensive and pervasive applications of technology. Assessments will be used to identify positive benefits and opportunities which might result from specified technological applications and to expose negative or harmful effects which could be anticipated.

Legislation which has been introduced in both the House and Senate during the first part of the 93rd Congress demonstrates a growing interest in technology transfer programs. Among the pertinent bills is the "National Science Policy Priorities Act of 1973" which was recently introduced to the Senate by Edward Kennedy. Kennedy's bill (S. 32) proposes a program of large scale efforts, designed to apply science and technology to civilian needs. Among other provisions, the bill proposes the establishment, within the National Science Foundation, of a Civil Science System Administration which would do research, design, development, and demonstration of systems in areas such as health care delivery, public safety, public sanitation, pollution control, and the like. Implicit in the terms of this bill is the need for effective programs to transfer existing technological resources into the public sector. No action has yet been taken on the bill, but it is anticipated that hearings will be held in the near future.

Another bill which does not specifically address the issue of technology transfer, but whose primary objective involves technology transfer mechanisms is the "Civilian Science and Technology Policy Act of 1973" (S. 1686) introduced to the Senate by Peter Dominick. The basic intent of the bill is to authorize the National Science Foundation to carry out a program facilitating the application of science and technology to civilian needs and assisting in the ordering of civilian research and development priorities. A major provision of the bill authorizes Federal support for State initiatives in establishing science/technology advisory mechanisms and utilization programs. No action has yet been taken on this bill, but hearings will be held on it in conjunction with those for S. 32.

Two bills have been introduced to the House during this session of Congress which constitute counterparts to Kennedy's S. 32. "The National Science Policy and Priorities Act of 1973" H.R. 32, was introduced by Representative Bell on January 3, 1973, and a similar bill, H.R. 2272, bearing the same title was introduced by Representative Davis of Georgia on January 18, 1973. Among the basic objectives of the two bills is the application of the country's scientific and technical talent and resources to national priority problems. A similar bill the Conversion Research and Education Act of 1973 (H.R. 615) was introduced by Representative Gubser on January 3, 1973. As in the case of S. 32, the technology transfer process is basic to the missions of the proposed legislation. Although no action has yet been taken on these bills, they have focused attention on the need for technology transfer policy and programs.

Another bill of interest is H.R. 1676, introduced by Representative Roush on January 9, 1973. The bill, entitled the "Federal Technology Transfer Act" calls for the establishment of an Office for Technology Transfer to be headed by a Director authorized to create and administer a program to facilitate technology transfer. As is the case of the bills previously cited, no action has been taken to date on the Technology Transfer Act. But, again, the bill serves as a focal point of interest.

The legislation and actions which have been mentioned testify to a widening pool of Congressional involvement in issues pertaining to the transfer of technology.

Some Policy Alternatives

A review of past and present programs and thinking in technology transfer leads to speculation about the future of the government-technology transfer relationship. The relationship is one which can be mutually supportive. The government can contribute significantly towards developing a better understanding of and methodology for technology transfer; similarly, technology transfer can enhance government's capabilities to perform its far-reaching responsibilities. Some suggestions as to how this cooperative interaction may be strengthened will be considered.

One policy option is to deemphasize the role of the Federal Government as the primary actor in the transfer process. It may prove to be advantageous to consider the role of the Federal Government as one of a number of forces interacting in a technology transfer network. As such, the Federal Government may assume the posture of a support function as opposed to that of a "prime mover" in transfer missions.

It appears that past Federal efforts have generally been thwarted by the basic inappropriateness of the national "high-level" approach to the solution of local "low-level" problems. In many cases, a broad, sweeping technique has been ill-suited to specific problem situations or has been ineffective as a stimulus to new development. It has been pointed out by Lesher and Howick that the pressing nature of urgent public problems tends to lead to proposals for sweeping but impractical solutions.

> Several times it has been proposed that a 'national system' [for technology transfer] be created. Far too little, it seems obvious, is known at this time to design a single national system, and it is unlikely that this would be the optimum solution in any case. 34

The high-level approach is not the only factor which may have impeded the success of various Federal transfer programs in the past. There are a number of generally accepted elements, which, according to many, should be factored into the design of Federal programs, but which have been overlooked or underemphasized in many Federal transfer efforts to date. Among those are three noted by Walter Hahn, in a report of the national meeting of the American Society for Public Administration, March 1972. Hahn stated that a concensus appears to be growing on three points pertinent to the transfer process: (1) broad-scale "pushing" technology developed for a specific purpose toward unspecified new uses is a very low pay-off activity; (2) storing (on paper or film) information about once-used or thought-of technology in clearinghouses in a necessary but insufficient element of technology transfer; and (3) technology transfer can occur once a need or opportunity has been recognized and some kind of "eyeball-to-eyeball" chain of events is launched to link the need with either the solution or the solver. 35

These findings, or accepted factors, pose problems for the Federal Government. Because the Federal Government is often far-removed from the source of the problem, it is difficult for it to ascertain specific requirements or to accomodate the diversity of needs which a particular program is intended to address. As a consequence, many Federal transfer programs have been formulated and operated on the basis of the "push" approach. Secondly, the Federal Government is in a position to provide extensive information services, but experience has demonstrated that information services alone cannot be expected to transfer very much technology. Thirdly, the need for face-to-face communication in the transfer process confronts the Federal Government with a formidable task, and one which it is not in the position to meet. It does not have the financial resources nor the manpower to distribute transfer agents throughout the country to work with all of the Federal Government's scientific and technical resources.

These thoughts argue for the distribution of technology transfer activities throughout the federal government system, and underscore the importance of State and local governments in the process. Support for this contention is found in a variety of reports and articles contemplating the implementation of technology transfer.

In an analysis of wide-ranging responsibilities of the Federal Government in technology transfer -- particularly in the application of defense and space technologies to public problems -- the General Accounting Office noted that most missions of the Federal civil agencies concern aggregations of nationwide local problems, with varying degrees of commonality and diversity. The report states that frequently the success and implementation of program responsibilities are contingent on acceptance and cooperation from State and local governments. The GAO stresses the importance of the State and local roles, which, it contends, should be coordinated by national leadership supplied at the Federal level.

> The primary role of the Federal civil agencies in technological leadership. . . involves identifying problems and potential solutions that have a nationwide commonality, adapting existing technology or sponsoring of appropriate research, demonstrating the feasibility of technological improvements, establishing national standards, and removing barriers to acceptance at State and local levels. 36

Characterization of the Federal role as one affording leadership and support, and fostering the utilization of transfer capabilities of other government and non-government organizations, is put forward in a position paper of the National Governors' Council on Science and Technology.

> . . .a majority of problems and opportunities to benefit from the introduction of technological innovations exist in individual companies or in State and local governments. A major reservoir of technical capabilities consists of faculties in public and private universities. A national program should assign appropriate tasks to each of these groups.
>
> An examination of experience under the State Technical Services Act of 1965 shows that field service programs were most effective in solving problems. Many more benefits could have been achieved had more people known about the program, and had there been greater abilities to modify developed technology for new applications. 37

Further stressing the need for lower-level government involvement in technology transfer, Gershinowitz offers the following thought:

> In many ways, the structure of a municipality or of a public authority seems to be more adaptable to experimentation in the application of research to social or environmental problems than are federal or even state governments. In the first place, there are some clear common interests and an obvious interaction among competing interests. ...Second, there is more immediate contact between those governing and those being governed. ...Third, within the city council (or equivalent body) and within the administrative structure of the mayor's office, there are frequent opportunities for those whose interests or responsibilities are affected by innovations to converse with each other. 38

The Council of State Governments, in its previously cited report on Public Technology, punctuated the role of the States, and implicitly of the localities, in the transfer of technology to serve public needs. It was noted that the Federal posture towards the nation's internal problems is distinct from that of the States in its willingness to support the work of scientists and technologists. Both are concerned with problem-solving, but the approaches taken at the two different governmental levels vary substantially. According to the report, differences in the respective roles of the Federal and State governments create problems. Such problems are fundamentally those that occur when suppliers and users have different concepts of what is specifically wanted and needed to accomplish an objective. 39

The thinking reflected by the selected excerpts suggests a direction for policy and program planning which promises a brighter future for technology transfer endeavors. The fundamental message is the decentralization of technology transfer responsiblities and activities and closer interaction among levels of government. The heavy concentration of technology transfer activities extant at the federal level should be diffused to other sectors of the federal government structure, and to universities and private industry.

Future policy planning and program formulation could foster the development of a technology transfer "network," in which all levels of government, the universities, and the private sector would each carry out essential functions. The network concept connotes a dynamic, interactive system in which technological resources may be channeled from the original sources to alternative uses, and in which feedback, or user participation in program decisions, may be effectively directed to the generators and suppliers of technology. Perhaps this idea smacks of utopianism, but pragmatic steps can be taken towards bringing the concept closer to reality.

Continuation and expansion of the Federal Government's information services is a desireable objective. Information dissemination programs should be implemented by the civil mission agencies as well as by the defense/aerospace agencies to enable the Federal Government to maintain a technological information base necessary to continued technology transfer activities.

Secondly, within the Federal structure, greater interagency coordination should be encouraged through the establishment of a coherent interagency technology transfer policy. At the present time, very few of the civilian agencies will explore existing scientific and technological resources before launching new research and development efforts suited to their specific needs. Improved coordination and interagency communication could preclude some duplication of effort, could expedite the pursuit of specific agency missions, and would allow for more widespread usage of available technologies.

Active roles, to be assumed by States and localities, and by the university and private sectors, should be central to new technology transfer policy and programs. It would seem advisable for States and localities to improve their mechanisms to locate scientific and technological resources applicable to their needs and to create capabilities to utilize these resources once identified. The effectiveness of State and local government activities can be strengthened by efforts generated from the respective governmental organizations, and by encouragement and incentives from the Federal Government. In promoting increased participation of States and localities, it is important that requirements and problems are understood at the program planning level. One means to secure a better understanding of need is through State and local government representation on Federal planning councils and advisory boards. Because, in many instances, States and localities are functioning close to the ultimate user-groups, or to the sources of problems to which technology is being directed, it is vital that they are active players in the technology transfer process.

Characteristics of an intergovernmental technology transfer network have been implied in previous discussion. The fundamental criteria for making the network operational, however, need to be summarized. The concept of a technology transfer network envisions a system in which the different sectors of government operate interdependently, and one in which the resources and capabilities of each governmental unit are utilized to compliment and reinforce those of the other units. In other words, no single governmental unit is expected to be responsible for the entire transfer process.

According to the network concept, the Federal Government, with its mission agencies, would be responsible for providing the scientific and technological resources. The Federal government procures research, development and experimentation projects from which the scientific and technological resources are generated and by which the utility of these resources is verified. The Federal Government also serves in an overall policy-making mode. The States and localities function as a dynamic interface between the suppliers of technology and the user groups. States and localities are responsible for recognizing and articulating needs and opportunities, and for linking them with potentially applicable technologies. The last major force in the network is comprised of user groups. The users provide the need for the transfer efforts, and the promise of market potential which stimulates innovative technological application and utilization.

Implicit in the network concept is the requirement for coherent policy and program guidelines to assure its efficient and effective operation. The role of policy formulation and program oversight and evaluation is one which would be most appropriate for the Congress to assume. The Congress would effectively be responsible for the design and implementation of technology transfer programs. To aid the Congress in carrying out its mission, ready access to the full range of theoretical and empirical data on the technology transfer process should be assured. This may be done through hearings, symposia and the like.

To conclude, the interest and importance attached to technology transfer are real, as are the vast stores of scientific and technological resources which await innovative application to new situations. As stated by Richard Lesher and George Howick, the effective utilization of our scientific and technical resources ". . . can increase the rate of economic growth, create new employment opportunities, help offset imbalances between regions and industries, aid the international competitive position of U.S. industry, enhance our national prestige, improve the quality of life, and assist significantly in filling unmet human and community needs." 40 Past efforts in technology transfer have generally produced unspectacular results. Despite the past, the future of technology transfer is far from bleak. There is no reason why technology transfer cannot be made to work -- intergovernmental and interorganizational cooperation in new transfer efforts may be one way of proving it.

ACKNOWLEDGEMENT

A special note of thanks is due to Walter Hahn, who along with other members of the Congressional Research Service, gave of their time and talent by critiquing the manuscript. These individuals have added to my knowledge and have given me the benefit of their expertise.

---

* The views expressed in this paper are solely those of the author and do not necessarily reflect those of the Congressional Research Service.

REFERENCES

[1] U.S. General Accounting Office, Means for Increasing the Use of Defense Technology for Urgent Public Problems (Washington, D.C.: U.S. Government Printing Office, December, 1972) p. 5.

[2] Samuel Doctors, The Role of Federal Agencies in Technology Transfer (Cambridge, Massachusetts: The M.I.T. Press, 1969), p. 3.

[3] U.S. General Accounting Office, Means for Increasing the Use of Defense Technology for Urgent Public Problems p. 5.

[4] Peter Chenery, "Technology Transfer and Utilization" A Background Paper prepared for a Workshop Session of the National Action Conference on Intergovernmental Science and Technology Policy, Harrisburg, Pa. June, 1972, p. 2. (Peter Chenery is the Director of the North Carolina Board of Science and Technology.)

[5] Samuel Doctors, The Role of Federal Agencies in Technology Transfer, p. 41.

[6] Richard Lesher and George Howick, Assessing Technology Transfer (Washington D.C.: U.S. Government Printing Office, 1966) p. 6 (Lesher is the former Assistant Administrator of NASA for Technology and George Howick is the former Director of the NASA Technology Utilization Division.)

[7] Charles Kimball, President of Midwest Research Institute, Statement for Hearings Before the Subcommittee on Science and Technology of the Select Committee on Small Business 90th Cong., 1st Sess. (Sept./Oct. 1967) p. 42.

[8] Harold Gershinowitz, "Applied Research for the Public Good -- A Suggestion" Science, April 28, 1972, p. 380. (Harold Gershinowitz is an affiliate in the faculty of the Rockefeller University, New York, N.Y.)

[9] Richard Lesher, Statement for Hearings on Technology Transfer before the Subcommittee on Science and Technology of the Select Committee on Small Business, 90th Cong., 1st Sess. Sept./Oct. 1967, p. 82.

[10] National Science Foundation, Federal Funds for Research Development and Other Scientific Activities, Fiscal Years 1971, 1972, and 1973, NSF 72-317 Vol. XXI, U.S. Government Printing Office, 1972. p. 1.

[11] George C. Nichols, et al., Government Enhancement of Civilian Technology A Report prepared for the Assistant Secretary for Science and Technology Department of Commerce, Washington, D.C., February, 1971, pp. 86-87.

[12] Samuel Doctors, The Role of Federal Agencies in Technology Transfer, p. 56.

[13] Ibid., p. 68.

[14] Ibid., pp. 161-163.

[15] Nichols, Government Enhancement of Civilian Technology, p. 85.

[16] General Accounting Office, Means for Increasing the Use of Defense Technology For Urgent Public Problems.

[17] Small Business Act, section 9 (b) (2); 15 USC 638 b.

[18] Dorothy Nelkin, The Politics of Housing Innovation, (Ithaca: Cornell University Press, 1971), pp. 2-3.

[19] "Program Evaluation of the Office of State Technical Services." Report by Arthur D. Little, Inc. to the Department of Commerce, Office of Program Planning. October 10, 1969.

[20] "Growth Through Technology." Final Report of the Public Evaluation Committee for the STS Act of 1965. Presented to the Secretary of Commerce, Nov. 14, 1968.

[21] Walter A. Hahn and Rosemary A. Chalk, "State Technical Services," a monograph, Washington D.C.: (Library of Congress, Science Policy Research Division of the Congressional Research Service, June, 1972), pp. 3-4.

[22] John E. Mock, "Where the Action Is", Science, Sept. 1, 1972, Vol. 177, No. 4051, p. 747.

[23] Letter from the National Governor's Council on Science and Technology to Dr. Edward David, Jr. March 22, 1972, p. 6.

[24] President's Message to Congress, "Science and Technology" March 16, 1972, Weekly Compilation of Presidential Documents, March 20, 1972.

[25] National Action Conference on Intergovernmental Science and Technology Policy, Harrisburg, Pennsylvania, June 21-23, 1972. A Summary Report and the Action Resolutions appear in the publication Action Now, (Pennsylvania: Pennsylvania Office of Science and Technology, 1972).

[26] Report of the Committee on Intergovernmental Science Relations to the Federal Council for Science and Technology, Public Technology - a Tool for Solving National Problems, (Washington, D.C.: U.S. Government Printing Office, May 1972).

[27] Council of State Governments, Power to the States - Mobilizing Public Technology, (Kentucky: Council of State Governments, May 1972.)

[28] The Pennsylvania Technical Assistance Program Annual Report 1972. (Pennsylvania: The Pennsylvania State University, 1972) p. 5.

[29] Report of the Honorable George P. Miller to the Committee on Science and Astronautics, Public Technology, U.S. House of Representatives, 92d Congress, 2d Sess. (Washington, D.C.: U.S. Government Printing Office, 1973). pp. 36-37.

[30] "Aerospace Technology Goes to Town", Industry Week, January 8, 1973, p. 55.

[31] Ibid., p. 57.

[32] George P. Miller, Public Technology, pp. 34-35.

[33] Ibid.

[34] Richard Lesher and George Howick, Assessing Technology Transfer, p. 7.

[35] Walter Hahn and Rosemary Chalk, "State Technical Services" pp. 11-12.

[36] General Accounting Office, Means for Increasing the Use of Defense Technology For Urgent Public Problems, p. 11.

[37] "Technology Applied to National Needs" Position Paper of the National Governors' Council on Science and Technology. October 22, 1971.

[38] Harold Gershinowitz, "Applied Research for the Public Good -- A Suggestion" pp. 384-385.

[39] The Council of State Governments, <u>Power to the States -- Mobilizing Public Technology</u>, (Background Report and Supporting Analyses) p. 84.

[40] Richard Lesher and George Howick, <u>Assessing Technology Transfer</u> p. 5.

BARRIERS AND STIMULANTS TO THE TRANSFER OF PUBLIC TECHNOLOGY

John Edwin Mock

Chairman, National Governors' Council on Science and Technology

INTRODUCTION

The mayor of one of America's largest cities recently observed that he would be delighted if the public technology in his city were brought up to the same level of sophistication as that of industrial technology in the 1920's. His view epitomizes the fact that public technology has simply failed to keep pace in the 20th century with the rapid progress of industrial technology.

Consider briefly the problems of any major city: (1) traffic jams, (2) poor garbage collection services, (3) dirty streets, (4) high crime rates, (5) inefficient fire control, (6) poor school systems, and (7) generally inefficient government. Is there any doubt but what improved public technology should be useful in solving such problems?

Consider that much of our public technology antedates World War II, e.g.;

(1) Cotton fire hose is still used although synthetic hose, introduced in the 1950's, is considerably lighter, and does not rot.

(2) The hand-held broom is still a basic tool of street sanitation -- long after the invention of the vacuum cleaner.

(3) Most traffic-control signals in use today are basically the fixed-time electromechanical devices of the 1930's.

These represent some of the more tangible -- readily visible-- problems in America today. Superimposed on top of them is an amorphous layer of large-scale, systems problems which are basically the responsibility of state government. Every governor and every state legislature are interested in finding viable, cost-effective solutions to:

(1) More efficient and effective health care delivery systems;

(2) A resolution of the "urban blight-rural flight" dilemma;

(3) More efficient and effective education systems;

(4) Promotion of economic development and an improved quality of life;

(5) More rational local, state, and regional transportation systems;

(6) Rational statewide telecommunications and energy policies;

(7) Improved environmental management systems; and

(8) Improved approaches to statewide land use planning.

An impartial observer might note that considerable technology already exists (or could readily be developed based on existing knowledge) to solve many of these problems. Why has it not been done? Why is the transfer of technology from the private into the public sector so difficult to accomplish?

The reason is that there are a host of barriers -- political, social, economic, and technical -- which impede the technology transfer process. Identification of the key barriers is sometimes a difficult process in itself; elimination or removal of the barriers is ofttimes an impossibility.

One might imagine that the transfer of technology <u>within</u> the public sector is easier than across the public-private sector interface. However, such does not appear to be the case, as the following two examples illustrate:

(1) The City of Scottsdale, Arizona set out several years ago to develop a new trash pickup system. With an investment of $200,000 they developed a system which reportedly saves the city an estimated $350,000 annually. Whereas the hand-loaded garbage truck previously took four men half a day to fill, the new system can be loaded in two hours by one man at the wheel of an air-conditioned cab listening to stereotape music.

(2) New York City's slippery water project was conducted jointly by RAND and New York City to help get water onto a fire as soon as possible. By using a high polymer chemical it is possible to significantly reduce the friction of water, making it easier to propel the water through a hose. Small amounts of the polymer can increase the flow rate as much as 80 to 100%! As a result of this development, New York City ordered polymer injection devices to be installed on 100 new pumpers, and Union Carbide is manufacturing the polymer for nationwide marketing under the trademark Rapid Water.

These two success stories occurred several years ago and have been well publicized. A fair question to ask is: How wide-spread has the new technology developed by Scottsdale and New York City been diffused to other cities in America? The answer is discouraging. It is reported that no other cities have accepted the innovations demonstrated by these pioneering efforts.

The transfer of public technology is certainly not an easy task -- sometimes, those of us who work in this area full-time wonder if it can really be done systematically, and if the results are commensurate with the efforts.

BARRIERS TO INNOVATION

To gain insight into the barriers which so effectively inhibit innovation and technology transfer in the public sector, let us examine the unique nature of many of the problems existing in the public sector.

(1) The problems are ill-defined with constantly changing boundary conditions. (For example, what is poverty? Welfare? Is poverty in 1960 the same as in 1970, or in 1980?)

(2) Even worse, the criteria against which we can measure a successful solution may change while we are still working on the problem. (For example, governmental criteria for a clean environment may change more rapidly than our technological efforts can keep pace with.)

(3) Responsibility and decision-making are decentralized and diffused. (For example, who is responsible for housing? The Federal government? `State or local government? Private enterprise? The individual citizen? Obviously all have some responsibility and some authority in the decision-making process. This is one reason such problems are so difficult to address, much less solve.)

(4) Potential solutions quite often may prove to be counterintuitive. (For example, Professor Jay Forrester of MIT theorizes that underemployed training programs and low-cost housing programs, although they seem promising, may actually hasten degeneration of our cities, while demolishing slum housing and replacing it with industry to create jobs -- a policy that is widely frowned upon as inhumane -- tend to set in motion a number of desirable, long-term trends.)

(5) The social, political, and economic aspects dominate in most of our critical problems; unfortunately our knowledge in these fields is not sufficiently advanced to give us neat, clean solutions.

(6) Solutions to many of the problems are exceedingly expensive -- approaching the trillion-dollar level. (For example, to solve the housing problem posed by some Federal officials would cost approximately one trillion dollars, without even considering the associated costs to improve services and community facilities.)

(7) There is a lack of consensus of goals and priorities. (For example, the disadvantaged groups in society tend to be much more interested in economic development than in pollution control, whereas the more affluent groups tend to favor a slow-down in our economic growth rate while increasing our efforts in environmental clean-up.)

(8) There is an unbelievable inertia to our complex social system. Edmund Burke once stated that it takes at least a generation to accomplish a major change in society. Perhaps with modern communications and "mass marketing" techniques this time lag can be reduced. The important point is to realize that beliefs, values, and attitudes change slowly -- and for some people, never. The public official can seldom be too far in "front" of the views held by his constituents -- else he will not be a public official for long.

(9) The dichotomous nature of most societal problems means that the decision-maker can never really win, e.g., economic development vs. environmental management; urban development vs. rural flight.

(10) There is an inherent dilemma in our political process with its emphasis on short-range planning. At the state and local level where many public officials are up for election every two or four years (and quite often with the restriction of not being able to immediately succeed themselves), and where the voters ask: "What have you done for me lately?", is it any wonder that we are a nation of short-term pragmatists rather than long-range planners and problem-solvers?

(11) An especially critical problem facing us is the highly fragmented nature of the public technology market. When we consider that there are over 80,000 political jurisdictions in America today, the majority of which have some purchasing authority, we can begin to visualize the marketing problem faced by the businessmen who wish to sell to this large, lucrative -- but highly fragmented -- market.

(12) There is a general failure to appreciate the magnitude of our domestic problems, and the cost, efforts, and sacrifices necessary to solve them. This point is best brought out in opinion polls wherein the majority of citizens favor cleaning up the environment, but yet are unwilling to pay more than a token amount to accomplish this herculean task.

(13) There is a general lack of communications and effective working relationships between those groups which are generating new science and technology and the potential users of it. (For example, working relationships between those Federal agencies and universities which generate new technology and the potential users at the State and local levels are sparse indeed.)

(14) We lack criteria to compare disparate problems and solutions. How does the public official decide whether to use his limited funds to fight drug abuse or improve mental health programs? Housing or mass transit? Welfare or pollution control?

(15) There is always a lack of sufficient funds -- a ubiquitous problem that will be with us always. Man's grasp will always exceed his reach.

(16) There is a lack of personnel at the state and local levels -- as to quantity, quality, and as to technical expertise needed to solve critical problems.

(17) There is an inherent conflict between strong Governors and strong Mayors. This tension is reflected down through the ranks causing needless problems in many cases where close cooperation is needed. It has its counterpart conflicts in many other countries.

(18) There is an almost complete lack of interest of news media in science and its application to the mundane problems of our real world. This lack of media coverage makes it difficult to build up a "science constituency" and to obtain taxpayer support for new projects and new programs.

(19) There is a lack of appreciation by top public officials and agency heads as to the value of science and technology in solving their problems. It must be admitted that from their point of view, science offers no panacea to near-term problems which are essentially economic and political in nature. They fail to appreciate the potential long-term solutions which science may offer.

(20) There is a lack of interest in state and local government by the scientific community. Quite often it is difficult for state or local agencies to obtain the advice of top scientists, even in their own state; those same scientists are ready at the drop of a hat to travel to Washington to confer with key Federal agencies. The glamour, the excitement, the funds are in Washington -- state and local government are but pale comparisons.

(21) There is a general lack of interest of state universities in state and local problems. Here again the Federal government, with a relatively small amount of research funds, is able to guide the research direction of state universities, even though the majority of the university support comes from state government. For example, what research scientist in past years could be enticed into studying creek pollution when a fat research contract from NASA to study life on Mars was in the offing?

(22) There is an absence of peer groups for scientists in state and local government to interact with.

(23) There is a lack of adequate, interactive n-dimensional communication (information) systems to tie together public officials and problem solvers at all levels of government.

(24) There are inherent tensions between Governors and legislators (Mayors and councilmen) which frequently interferes with developing optimal (or any) solution to critical problems.

(25) There is an almost complete absence of incentives in state and local government for innovation or creativity. There are no rewards in government for being innovative; on the other hand, the risks are exceedingly great. The public official knows that if he spends a million dollars to install a computerized purchasing system to save the state money, even if it is successful

no credit will ever fall his lot. On the other hand, if anyone can point out any flaw in the system -- no matter how minor -- the news media will tar him with a "million-dollar blunder" which the taxpayers will long remember. In state and local politics, the safe procedure is: "If you are not 100% sure the project will be successful, forget it!"

(26) There is a general lack of venture capital. In view of the fact that most public officials are not risk-takers, it is no wonder that sources of venture capital are not available for public technology. In fact the concept of venture capital is somewhat foreign to the thinking of the typical public official.

STIMULANTS TO INNOVATION

It is always easier to find fault and point out the weaknesses in a system than it is to make suggestions to improve the system. We have attempted to specify the nature of the obstacles to be overcome in innovating and transferring technology in the public sector. Let us now bite the bullet and attempt to be more specific as to potential solutions.

As we focus our attention on the field of public technology, where pre-World War II technology reigns supreme, we note that too many manufacturers have been spoiled by operating in the NASA-DOD market which is characterized by: (1) a highly aggregated and efficient market, (2) large scale funding -- where ideas and performance are often more important than cost, (3) sophisticated (from a scientific and technological point of view) customers, and (4) a customer who is willing to take risks and willing to accept a reasonable proportion of failures in order to achieve his final goal. These manufacturers are usually confused when confronted with the public technology market which is characterized by: (1) a highly disaggregated and inefficient market, (2) small scale funding -- where cost is often the most important factor, (3) a relatively unsophisticated customer who has been burned badly in the past by a few firms offering "Aerospace technology" as a panacea to any (or every) problem, and (4) a customer who cannot afford to take risks -- both because of his political vulnerability, and because of the scarcity of resources at this level of government.

Let us now examine more closely some of the barriers to the introduction of public technology at the local level and indicate possible methods of stimulating innovation and technology transfer:

(1) Lack of venture funds - most Federal agencies have available to them sources of venture capital (although of course it is not called that by Congress). But when DOD, HUD, DOT, or HEW needs funds for developing promising, new technology, they can be reasonably sure that the funds will be forthcoming. Not so at the state and local level where funding is much tighter and R&D is considered a luxury which cannot be afforded. If state and local agencies had a source of venture funds which could be used to promote innovation and new developments, we would undoubtedly see a great change in the rate of diffusion of modern public technology. The question of course is how can we provide the needed venture capital? One possible answer (which will undoubtedly be politically unpopular to Governors and Mayors--but of course we are not trying to win a popularity contest but trying to overcome obstacles wherever or whatever they may be) is to earmark a certain percent of revenue sharing funds to be used exclusively for developing new public technology. Because of the intense pressure placed on state and local government budgets in order to meet escalating health, education, and welfare costs, I see no other way to obtain such funds on a broad scale other than through revenue sharing or other federal funds.

(2) Lack of technically sophisticated people -- low salaries and slow promotions have taken their toll in state and local governments. This is not to imply in any way that the average employee in state and local government is not every bit the equal of his industrial counterpart. The point is that the more dynamic, risk-taking individual, finding his route upward blocked, moves out, leaving the less dynamic, risk-averters behind. On top of this, there are very few slots in state and local government for scientists and engineers. Thus we can assert that most state positions are strongly biased in favor of risk-averting, non-scientists. How can we rectify this situation? Unfortunately, there is no quick, pat solution. Intergovernmental personnel exchanges are just a drop in the bucket. I see no short term palliatives. In the long run, with federal encouragement -- such as NSF seed funding of state science agencies -- and federal requirements for improved environmental impact statements, land use plans, etc., we will gradually see more scientists and engineers brought into state and local government.

(3) Lack of a continuing dialogue between government and R&D people -- there is a need for new institutional relationships between industry, government, and universities if we are to improve the transfer of public technology. This is one area in which the states must play the lead role in bringing together the science resources already existing within their jurisdictions.

(4) Highly fragmented nature of the domestic market -- here again the states can do quite a bit to help market aggregation. For example, by establishing statewide building codes we can help builders standardize and expand their market area. The highly fragmented nature of the market does of course offer some advantages to the smaller, regional firms which can tailor their product to the local market. Many of these smaller businesses probably could not compete successfully if markets could be aggregated at the national level, and the large national or multi-national firms moved in. If we desire states to aggregate certain markets -- say, housing, sewage or water treatment plants-- a good approach would be for the Secretary of the appropriate federal agency, say, HUD, to invite each Governor and state legislature to send representatives to a national conference in Washington to discuss the problem and ways of solving it. If such conferences are well planned -- complete with suggested model legislation -- acceptance at the state level should be high.

(5) High inertia on the part of state and local governments-- most government employees and public officials are characterized by resistance to taking large risks because of the nature of the incentive/reward/punishment system in government, and the relatively short planning horizon of the elected officials at the state and local level. This again is a difficult area to deal with. Obviously we are not going to change risk-averters into risk-takers overnight. In this case, perhaps we can think more in terms of shifting the risk upwards; e.g., if the risk of a research project is absorbed primarily by federal funds, the state employee will not feel as vulnerable as if state funds are used exclusively and he will be held closely accountable for them by the state legislature each year. On the other hand, in the long run we simply have to build more incentive for innovation into our entire governmental structure. This is one area in which the federal government must take the lead and set an example for state and local governments to follow.

Some additional recommendations on promoting public technology are as follows:

(1) A public technology task force should be established at the federal or state level with the responsibility of: (a) finding ways to increase the productivity in the service sector of our economy which is abysmally low and a drag on the more productive sectors of our economy; and (b) developing incentives for improvements in our domestic technology.

(2) Establish a new concept of a "State" or "Regional" Observatory to study problems where they exist. This should be a joint government-industry-university undertaking, with government definitely in charge -- there is no sense encountering

the serious weaknesses of the urban observatories all over again. Each Observatory could consist of a small number of scientists, engineers, systems analysts, and economists who would respond to the guidance and requests of public officials in their state or region. The main objective would be to identify problems amenable to technological solutions, develop appropriate solutions, and translate them into action.

Conduct an in-depth study of a wide variety of problems having to do with the need for institutional change (e.g., to improve health care delivery systems and education systems), realignment of responsibilities commensurate with scope of problems, development of goal and priority consensus, etc. The point to be made here is that quite often technology is not the key to solving a problem; instead, new institutional arrangements or even a re-definition of the problem is in order.

In closing, I am reminded of a statement by the late Dean Acheson: "As a nation of short-term pragmatists accustomed to dealing with the future only when it becomes the present, we find it hard to regard the future trends as serious realities... to treat as real and urgent -- as demanding action today -- problems which appear critical at some future date."

In concurring with Mr. Acheson, I am nevertheless confident that scientists and engineers will make us -- as a nation -- much more "future-oriented" than we have ever been in the past, and we will solve our problems calling for innovation and technology transfer.

CHARACTERISTICS OF TECHNOLOGY TRANSFER (TT) FROM THE PUBLIC TO PRIVATE SECTOR

F. Karl Willenbrook

National Bureau of Standards

Introduction

To consider technology transfer a process, in an engineering or natural science sense, holds many dangers. The development of a check list of factors which identify all the relevant parameters needed for good technology transfer and to expect, upon completion of the check list, to accomplish a successful technology transfer, is not likely to succeed. Why? Because technology is too varied. It results in new devices and processes as different as a ball-point pen, integrated electronics circuits, or linear programming. Yet each of these represents an important type of technological development.

The context in which the technology is developed is also tremendously varied. Important technologies can come from independent inventors, small research and development (R&D) firms, universities, not-for-profit institutions, industrial laboratories, government-sponsored research centers, and government laboratories. Similarly, the context in which the technology is produced is highly varied; some industrial sectors have an aggregated market which is sensitive to, and poised to accept, new technology; in others, projects in which the new technology is successfully demonstrated to a great number of decision-makers may not result in the acceptance of a new approach; while in others there may be no systematic way to introduce new products, processes, or techniques; in still others, there may be governmental regulations which militate against the use of a new technology.

Knowledge about technology transfer is still fragmented despite the work of many talented investigators. Studies on technical information dissemination, tne role of technical information gatekeepers, the transfer techniques used in vertically-integrated industries, have cast some light on the process. However, the available means for taking technology from one context to another are not sufficiently understood. Therefore, predictions for success in this area are difficult to make.

Also of considerable significance is the attitudinal orientation of the transfer agent. Where does he start in the technology transfer process? Is it with the solution or is it with the problem? Starting with the solution has often led to fruitless meanderings and failures - yet this is where the conventional entrepreneur usually starts. Starting with the problem requires a clear identification of that problem - which is a tough task. However, if the identification is done with an adequate appreciation of the technological possibilities available in the possible solution, this approach can become a most effective way of transferring technology.

Framework for Discussion of Case Studies

This paper undertakes a study of technology transfer by examining a number of cases where technology was successfully transferred from the public sector to the commercial world. From a critical examination of these cases, we should be able to learn about transfer techniques and the important phases of the transfer process. Hopefully, our selection of cases will be such that the conclusions drawn from our study can be generalized and applied to other examples.

Technology transfer is taken to be the process of transferring knowledge out of the laboratory into the hands of the users. The cases examined emphasize a particular circumstance in which the developer of the knowledge has some awareness of the needs of the users. This circumstance is different from the usual one of a "solution looking for a problem." It is somewhat closer to the inverse. Success is more likely in this circumstance, but success is not guaranteed.

To structure the cases, a general set of steps is used.

The first step is the identification and articulation of a need. For example, we may want to measure the distance from the earth to the moon more accurately, or we may want to increase the durability of porcelain enamel coatings for building exteriors. The desire to improve a special technology of this type has a rather narrow focus; yet such requests occur frequently. They can come from industries, users, government agencies, or research organizations. They can also come from contacts made with peers at technical meetings. They usually come from people who have more than a layman's knowledge of the relevant science and technology or who have previously been involved with finding technological solutions to problems.

The second step is problem definition. From the general identification of a need, we start to collect more definite information. In the case of the earth-moon distance measurement, we may need to improve our knowledge of the velocity of light or of the index of refraction of air. This second step calls for interaction between those who have articulated the need and those who seek to solve the problem.

The third step is the selection of the method of approach. What techniques are available to work on the problem? Is data analysis sufficient or is laboratory work required? Can solutions be obtained from other fields of knowledge which have not as yet been used for the problem at hand? How much will the particular method cost? Considerable coordination is needed at this stage among the various groups involved: those concerned with the problem, those concerned with finding solutions, and those concerned with supplying resources.

The fourth step is the technical solution. At this point, formal feedback to the group articulating the need is usually started. This is done through progress reports and briefings. At this stage, there may be redirection of the R&D effort, either by those best informed about the need or by those most familiar with the technology to be employed in the solution. In addition, costs in time and resources must be estimated. Efficient management is critical at this stage.

At this technical solution stage, the advantages of developing the new technology must be identified. If the project is controlled too tightly, the solution is likely to be jeopardized and the "technological fallout" minimized. This could make the research program unprofitable since many significant technological advances come from technical serendipity. Conversely, if the project is not guided toward goals chosen from technological needs, the project seldom achieves the objectives and can become self-serving. Numerous techniques for management of technical problem-solving programs are documented in the literature and therefore will not be included here.

Following the identification stages is the fifth step, experimental trials or demonstration. When a technical solution is available, those who have had the most intimate involvement in the development of the solution usually feel very strongly disposed toward it. At this stage, other possible users must be made aware of the new solution and must be shown that this solution has advantages over the existing technology. They must develop confidence in the new solution. A demonstration which involves these users and disseminates the results of the demonstration can help to build that confidence in the possible users and bring about their acceptance of the new technology.

As well as a demonstration stage, incentives for accepting the technological innovation must also be identified which we will take as the sixth step. These incentives must be recognized by those who are to receive the new technology. Although this step can take place in some other sequence than the one just described, if incentives are lacking, it can become most painfully apparent at this point. Many projects, carried to the point of the technical success, are shelved even after a successful demonstration has been performed because of insufficient incentives for commercial exploitation. In this paper, special emphasis is given to examining those incentives which aided technology transfer in the cases to be studied.

The seventh step is a survey of the benefits of the new solution. Preferably this step should be performed by a group whose attitude toward the survey is unbiased and objective. This effort is of value not only to the users of the new technology but also to the managers responsible for funding technological research and development.

In summary, the seven steps which we have identified in the case studies of the transfer of technology from the public to the private sector are:

1. Identification and Articulation of Needs
2. Problem Definition
3. Method of Approach
4. Technical Solution
5. Experimental Trials and Demonstration
6. Incentives
7. Survey of Benefits

These steps are not to be considered completely separable from each other. However, each step has its own management problems, and neglect of a step can prevent the transfer process from being successful.

Series of Case Studies of Technology Transfer from the Public to Private Sector

The first study is the case of a technology transfer which has been in successful operation for an extended period of time and is still very active. The case will be analyzed according to the seven steps in the framework just described. The objective of the analysis is not to prove the validity of the framework but rather to identify characteristics of a successful example of technology transfer.

Dental Tools and Materials

The American Dental Association (ADA) and the National Bureau of Standards (NBS) sponsor a program which has produced significant useful inventions in dental materials and equipment and also provides for efficient transfer of the technology into dental practice. This cooperative program was begun 50 years ago and presently includes a staff of dental consultants and materials researchers working together in the NBS laboratories. Examples of technical innovations developed by this team include: the panoramic x-ray camera for dental examinations, the high speed turbine drill, spherical amalgam powders, and composite filling material.

A primary characteristic of this program is the close coordination between dentists and dental scientists who are intimately familiar with the broad spectrum of research capabilities of NBS and with the needs of practicing dentists. Several of the innovations developed stem from research specially directed to these needs.

Regarding the panoramic x-ray camera shown in Figure 1, this equipment was developed to improve the efficiency and to lower the cost of the standard dental x-ray examination. Figure 2 illustrates the panoramic x-ray. The Air Force was particularly interested in this development. By using the new equipment, the Air Force's dental clinics were able to reduce both the time and expense involved in the previous repetitive "bite" x-ray procedure. More recent users of the equipment include the Veterans Administration where estimated savings of $800,000 per year have been realized. The first commercial units available were made by dental equipment manufacturers just entering the x-ray field. These manufacturers thus were able to gain a significant market opportunity they had not had previously. When the commercial units appeared on the market, dental clinics rapidly assimilated the technology. However, the incentive to use the technology was not as great for the individual dentists due to the higher capital investment needed for the panoramic equipment.

The high-speed turbine handpiece invention shown in Figure 3 was in response to a very acute need in dental practice - pain! As well as reducing pain, the new handpiece also improved the accuracy and speed of the process. Previously handpiece drill sets were driven by long belts and, because of their inertia, speeds were limited to about 20,000 revolutions per minute (RPM). By placing the drive-mechanism in the contra-angle handpiece, much higher speeds became possible. Speeds of 40,000 RPM were achieved in 1954 using water turbines. Later, the incorporation of air turbines and air bearings moved the maximum RPM to the 400,000 range. Since these high RPM's permitted high tool surface contact speed, small diamond abrasive tools became effective. This innovation is considered by the ADA to be the most important improvement in dental health service in the last decades.

The early development of the water turbine was performed by the NBS-ADA team. However, Navy scientists continued the turbine design study and did outstanding work in developing an air-turbine drill.

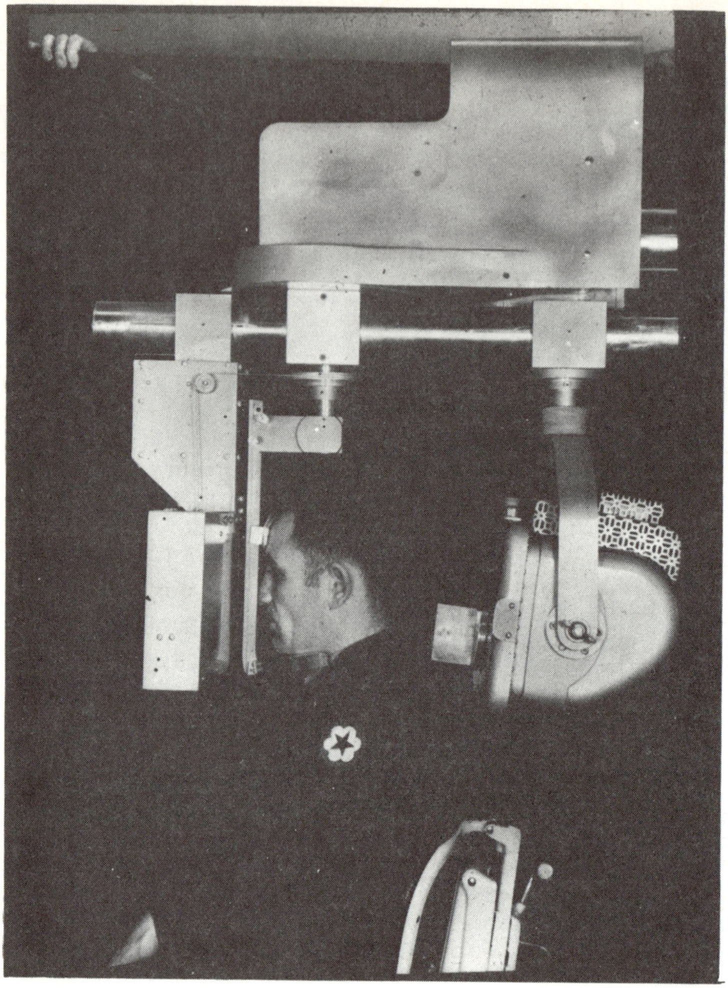

Figure 1

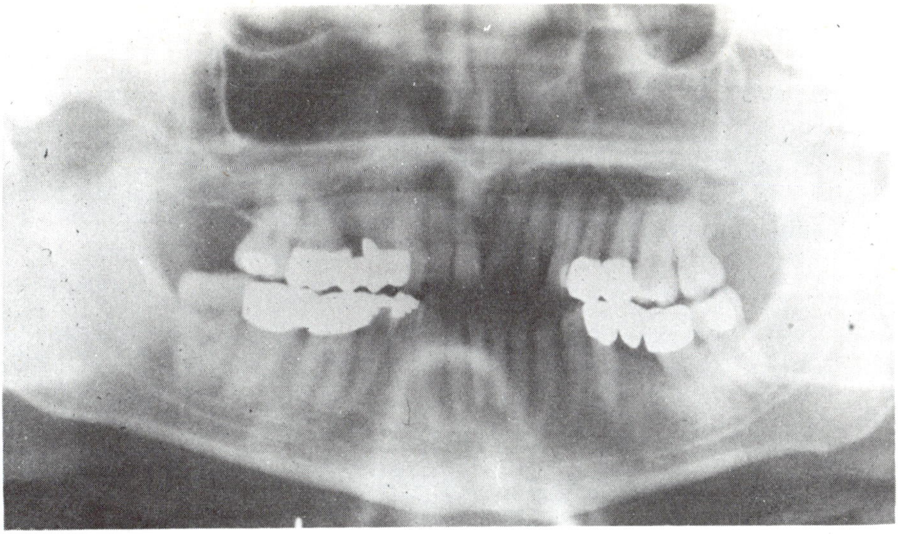

Figure 2

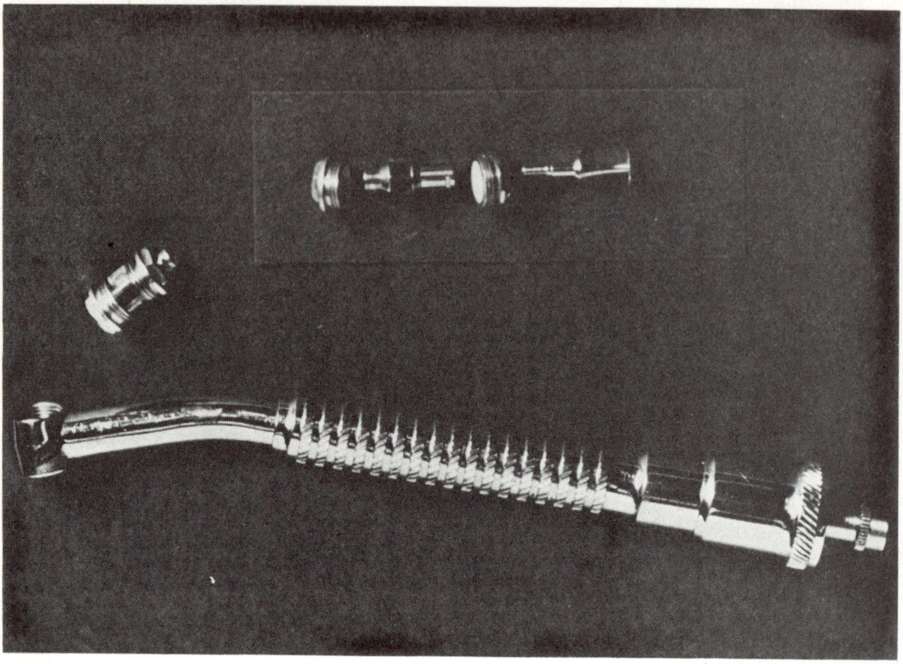

Figure 3

The Navy, as a user, was able to test the high-speed handpiece in large-scale programs. Today, this innovation is almost universally accepted in modern dental practice.

In another dental area, a study of the kinetics of the solidification of dental amalgams turned into an interesting innovation. Dental amalgams are made by mixing silver-tin powders with mercury. Formerly, these powders were machine-cut, but by using atomization, spherical silver-tin powders were produced which were superior to the machine-cut powders. The improvement in the amalgam with the incorporation of the spherical powder relate primarily to the lower forming pressures required for packing the amalgam into the cavity and the faster buildup strength for curing. The one- and two-hour compressive strengths of the amalgam are increased two to threefold by using the spherical powders.

The experimental work demonstrating the spherical amalgam powder was performed in clinics and the superiority of the new product established. Incorporation of the product into dental supply lines was made by a U. S. company, new in the field, and by several Japanese companies interested in expanding their share of the market. The hesitancy of established producers and suppliers to introduce an innovation in the field should be noted; it is a chronic problem in technology transfer. The incentive of a new or changing market is often a key factor in technology transfer.

This innovation was reported in the "Guide to Dental Materials and Devices," an annual publication of the ADA. This Guide has been an effective tool in the transfer of technology to application in the field. It is a well respected document written directly as a service to practitioners. To practicing dentists, it serves as a comprehensive technical and buying guide to the materials and the devices used in dentistry.

One of the most recent technological innovations to come from the NBS-ADA team is that of the composite filling material. Dental researchers have known for some time that the silicate filling materials used for the frontal areas have numerous drawbacks. The primary problem relates to the tendency of the silicate filling to dissolve and disintegrate in service. In addition, the strength of silicate is considerably lower than the amalgams and will not take as smooth a finish as is desirable. Silicates must also be kept wet or they will degrade rapidly.

These problems led to numerous attempts to use polymeric cements based primarily on poly-methyl methacrylate. These room-temperature curing resins had even more serious problems such as high contraction on curing leading to marginal gap, excessively high thermal expansion coefficient, bonding problems, and low strength. These problems were attacked by the NBS-ADA team. The team conducted research in organic materials to improve the resin and research in glass to perfect a glass filler for the resin. This composite material approach, using up to 80% glass filler, dramatically reduced the thermal expansion problem and the solidification shrinkage problem. In addition, the aesthetic characteristics of the filling materials were significantly improved through the use of the glass composite. Sealant addition and other resin improvements successfully solved the other problems.

The early experiments using the composite filling material were performed at NBS. In Figure 4, a comparison between the durability and appearance of the new composite filling (left) and the silicate filling (right) is shown. The advantages of the composite filling materials observed in these programs have led to very rapid acceptance by dentists in the U. S.

In summary, the NBS-ADA team for dental research and technology has developed a well-organized approach to technology transfer from the public to the private sector. Several of the salient points involved in their work are presented in Table 1.

Table 1

NBS Dental Technology Transfer

1. Producer and user community work in same laboratory with maximum communication about what needs to be done and what can be done.

2. Broad scientific and technical capabilities are available.

3. Approach determined by practicing dentists, dental research experts, and scientifically oriented chemists and engineers.

4. In-house demonstration projects performed.

5. ADA builds confidence and publicizes results.

6. An efficient information-dissemination process is present through the ADA Bulletin and Guide.

High Modulus Composite Materials

An interesting example of the transfer of technology which started as Government-funded research and then became a significant industrial technology is that of high-modulus composite materials. The central theme of this new technology is the development of a boron composite.

The work in this area can be traced to the early research done by the Federal government in their search for storable fuels of

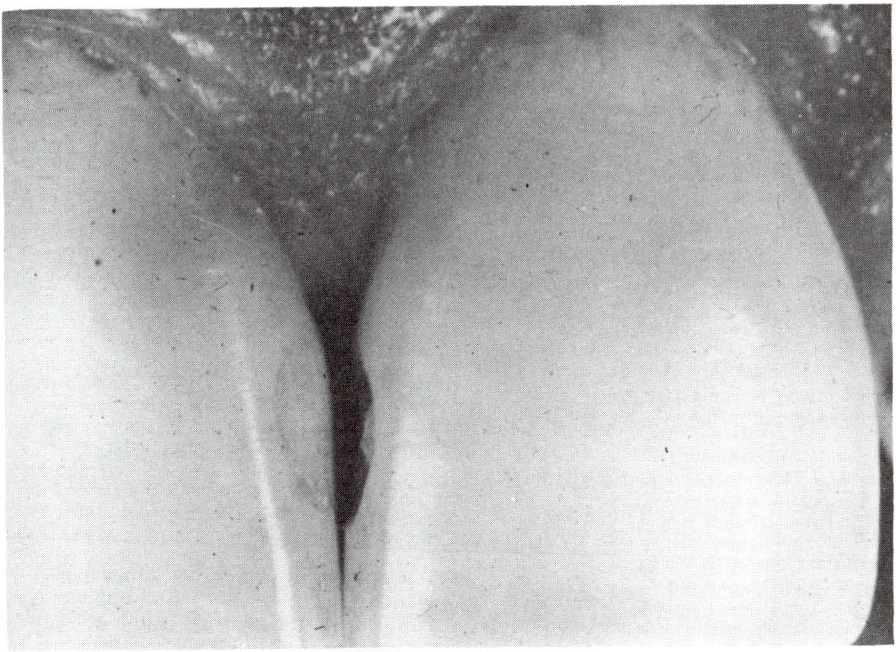

Figure 4

high energy density. The solid fuels which contain the element boron have a tendency to burn unevenly. The objective was to obtain an even, smooth burning form of the boron; the approach was to develop a fibrous, wick-like boron. One attempt produced a boron fiber synthesized by chemical vapor deposition on a hot tungsten filament from the trichloride of boron. Although the fiber produced was still difficult to ignite and did not burn evenly, a more significant observation was made. This boron fiber was extremely strong (10 times that of most high strength ceramics) and also had the high elastic modulus characteristics of a covalently bonded crystal. But it was not until 1963, that it was recognized that boron fibers might contribute to an even greater need, that is, the need for aerospace structural materials with improved specific strength and specific stiffness or modulus.

The problem was that the materials previously used in high-strength aerospace structures were all metals which, due to their atomic bonding, have a low intrinsic specific modulus. This metallic bonding not only limits the specific modulus but also limits the specific strength. The boron fiber, which has a covalent rather than metallic bonding, was found to have six times the specific modulus and five times the specific strength of high strength steel, aluminum, or titanium alloys. The advantage of the new composite materials with respect to the ratio of stiffness to density is illustrated in Figure 5. The approach to solving the need for improved structured materials was therefore determined to be the incorporation of the higher strength fiber in a matrix to form an aligned composite.

The U. S. Air Force sponsored research at the United Aircraft Corporation in Connecticut to perfect the boron manufacturing process and to work on other programs to develop boron epoxy composite fabrication procedures. In order to try out the composite system, The Air Force initiated a project to design, fabricate, and test F-111 aircraft tail sections made of boron epoxy. This program was most successful and permitted a weight savings of 30% in that structure. Although incentives for the use of the boron epoxy pertained mainly to weight savings, other advantages included improved maneuverability, range, speed, economy, and combat ability.

Confidence in the material was indicated by the use of that material in fifty F-4 aircraft tail sections. Figure 6 shows an F-4 aircraft tail section as it proved out in operation. The successful trials led to the incorporation of the material in virtually all the new high-performance aircraft such as F-14, F-15, and B-1. Further commercial application of the material has shown that weight savings also led to improved economy in performance and operating cost.

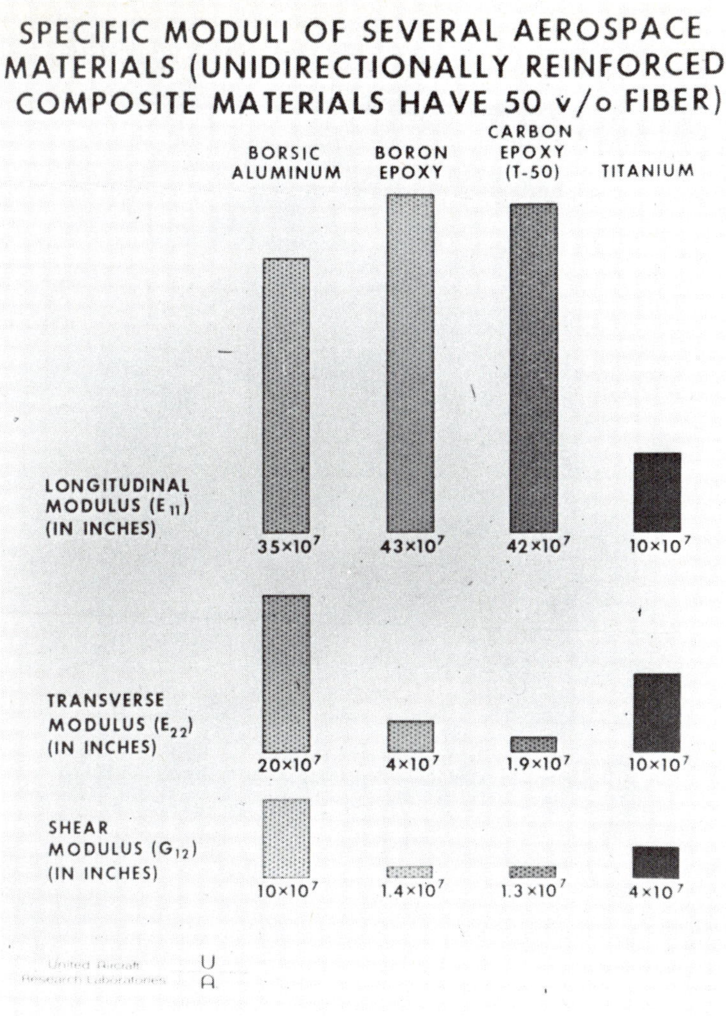
SPECIFIC MODULI OF SEVERAL AEROSPACE MATERIALS (UNIDIRECTIONALLY REINFORCED COMPOSITE MATERIALS HAVE 50 v/o FIBER)
BORSIC ALUMINUM
BORON EPOXY
CARBON EPOXY (T-50)
TITANIUM
LONGITUDINAL MODULUS (E_{11}) (IN INCHES)
35×10^7
43×10^7
42×10^7
10×10^7
TRANSVERSE MODULUS (E_{22}) (IN INCHES)
20×10^7
4×10^7
1.9×10^7
10×10^7
SHEAR MODULUS (G_{12}) (IN INCHES)
10×10^7
1.4×10^7
1.3×10^7
4×10^7
United Aircraft Research Laboratories
UA

Figure 5

BORON EPOXY F4 RUDDER
412
740
BORON
United Aircraft
Research Laboratories

Figure 6

The development of the boron epoxy high modulus composite has also spurred the development of graphite epoxy. The graphite fiber of which the composite is composed is processed from relatively cheap rayon or orlon material. This fiber has the advantage of being lower in cost than other aerospace composites and of having the potential of eventually being competitive with fiberglass. Graphite epoxy composites are receiving wide use in aerospace materials, sporting equipment, and other structures with high performance requirements.

This example can also be analyzed in terms of the steps previously discussed:

1. The U. S. Air Force recognized the need for high modulus composites early in their development and provided a strong driving force in their further development as structural materials for high performance aircraft.

2. The problems of fiber development, composite fabrication, development, design methodology, manufacturing technology and end-item test and evaluation were clearly defined by the U. S. Air Force and industry.

3. The method of approach assimilated talents from a wide spectrum of technology and involved substantial funding from the U. S. Government.

4. Problem solutions were accomplished both through U. S. Air Force sponsorship and aerospace companies investments.

5. Demonstration hardware was used repeatedly on Air Force end-items to investigate in-service durability, maintenance, and performance.

6. The incentive was originally only improved performance; however, it soon became evident that these composites were also cost effective. In addition, it was found that the use of the composites in weight savings also solved such in-service design problems as premature fatigue failure in structural components of helicopters and of F-111 fighter planes.

7. Cost-benefit analyses by government and industrial organizations were made in more detail than normal for high performance military aircraft. The additional information in these studies assisted the commercial aircraft industry and the National Aeronautics and Space Administration in their application of these new composites.

Semiconductor Technology: Measurement of Resistivity

The third example is a classical NBS-type of development. It is essentially the development of a more precise test-method for determining impurity concentration in semiconductors. The increased precision had important commercial implications.

The need for improved resistivity measurement techniques was well-recognized in the semiconductor electronics industry. This industry has usually demonstrated a willingness (or eagerness) to develop and exploit new technologies as soon - or even before - their validity is well-established.

Silicon, and to a lesser extent, germanium, is used in the fabrication of most current semiconductor devices. Resistivity is the crucial material property for device purposes, for device design, and for control of the fabrication process. When silicon and germanium are bought and sold, the most important single property specified and measured is the resistivity.

The lack of accurate, reproducible resistivity measurement techniques severely hampered the industrial operations. Measurement disagreements between vendors and customers were so commonplace that files of adjustment factors were maintained by the major suppliers in order to compensate for the large discrepancies between the manufacturers and their various customers. Measurement disagreements within the same organizations were so frequent that there were deleterious effects on the process control for device fabrication. Therefore, in 1961, the industry, through the American Society for the Testing of Materials (ASTM), a leading standards-writing organization, asked NBS for assistance in improving the precision of the measurement.

Two techniques were in use at the time, the two-probe method and the four-probe method. Measurement of resistivity by the two-probe method requires a special specimen while the four-probe method does not. Despite the preference for the four-probe technique, it was generally believed in the industry that the two-probe method was the more precise.

NBS was able to improve the precision of the four-point probe method so that it was ten times better than before and, as a consequence, the more expensive and cumbersome two-point method became unnecessary. A diagram illustrating the experimental technique used with the four-point probe is given in Figure 7.

Partly due to the ASTM Committee, a consequence of the close coupling between NBS and industry has been the dissemination of industry of the information and techniques before a cost-benefit analysis was performed. However, an analysis (made by NBS talking

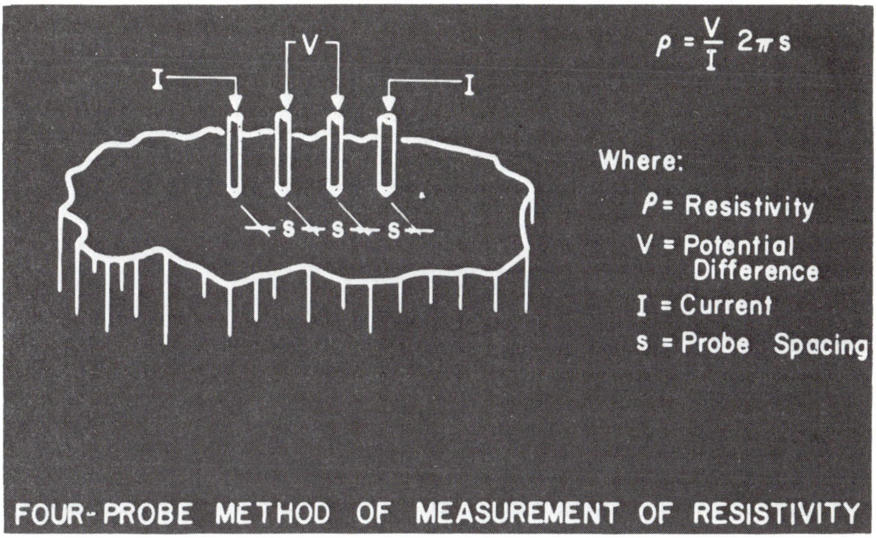

$\rho = \frac{V}{I} 2\pi s$
I
V
I
s
s
s
Where:
ρ = Resistivity
V = Potential Difference
I = Current
s = Probe Spacing
FOUR-PROBE METHOD OF MEASUREMENT OF RESISTIVITY

Figure 7

with industry representatives) suggested that a benefit-cost ratio of over 100:1 existed and that savings over a ten-year period could amount to $30 million from reduction in material rejection and waste alone.

Low Voltage Switching Circuits

A fourth example relates to a National Aeronautics and Space Administration (NASA) spin-off which is presently under development and appears to be an excellent example of identifying the problem first and then seeking the solution. In this case, a third party, ABT Associates of Cambridge, Massachusetts, played the marriage-broker role. This organization brought together a technology and an industrial need which resulted in a commercially successful venture.

The technological innovation is a new, low-voltage switching device that could well reduce the cost of house, office, and industrial electrical wiring systems.

At present, electrical circuitry in buildings is composed of a power distribution and also a switching system. The latter system conventianally operates at the same electrical potential as the former and therefore requires the same costly safety measures. In most building wiring, both the power distribution and the switching circuits are installed inside a building's walls and thus a high labor cost is involved.

In this innovative wiring system, the switching circuitry is of low voltage and so can be mounted entirely on the outside of the wall. This exterior system consists of a wall switch, a power unit, and a .005 thick conducting tape. The tape, which connects the switch to the power unit, is backed with adhesive. Thus the circuit can be pasted on the wall. The tape is so thin that you can paint over it. Due to the low voltages employed, there is no electrical shock hazard.

The labor savings aspect of this innovation is evident. In actual building projects, up to $30 per fixture on the labor costs can be saved in new construction. In building renovation, the savings are even greater since the low voltage technique eliminates the need to chop into the walls for switching circuits.

The New York State Urban Development Corporation (UDC) is an agency currently designing or building on the order of 50,000 housing units annually. They are interested in innovative low cost techniques for their new construction and were particularly receptive to the concept of the NASA electrical low voltage switch wiring. After studying numerous alternatives, UDC adapted the NASA innovation for new installation in its new construction. The purchasing and installation of the innovative wiring system is about 60 per cent of the cost of a standard system. The State of New York will thus save hundreds of thousands, if not millions, of dollars. The market projections of the company that is now producing and selling the new system doubled in its first six months (10,000 units) of operation. Projections for the current year, the first full year of operation, exceed 200,000 units.

In addition to the direct installation cost benefit, a major cost/efficiency benefit will be the lessening of "trade-phasing" during construction projects. In current practice, there is a rigid time-dependency among carpenters, masons, plumbers, electricians, and other trades. Each trade must wait for the others to finish certain tasks before it can begin work. The construction process can be speeded up significantly, if a major in-wall process is eliminated by the use of a surface-mounted electrical switching system.

For this technology transfer, the users had a clear economic incentive for incorporating the innovation. They also had capabilities for performing an experimental trial. Thus both the technological capabilities and the economic incentives could be established in a well-controlled project.

Conclusions

A number of conclusions can be derived from the examples we have examined:

A. The transfer of technology from idea formulation to research to development to utilization can be aided by a management coordination and anticipation of future requirements.

B. A focusing of the method of approach of a research program in a recognized need by the ultimate user of the technology can aid in accomplishing the transfer.

C. Careful definition of the problems can avoid wasted and unfruitful efforts in all phases of the invention and innovation process.

D. Review of the research and development effort can lead to changes of emphasis unanticipated in the early stages of the effort. Such analysis can help to avoid accepting the first workable solution rather than seeking the best one.

E. Careful planning of the experimental trials or demonstrations of a development should include considerations of what methods will encourage confidence, cultivate interest in key users of the technology, and clarify incentives for the application of the technology. Mechanisms must be found to deliver the new technology to users; standards-writing organizations, government procurement, and peer group acceptance have all been found to be effective delivery mechanisms.

F. Cost/benefit analysis is an effective management tool in government research and development; it also can help overcome the inertia of the status quo.

G. The development of technologies in the public sector which are useful in industry can be best done in an environment where various disciplines involved interact with freedom at both working and management levels.

H. Technology transfer can be successfully effected across the public/private border; however, sufficient attention must be given to the constraints inherent in each of the sectors.

Recommendations

The rather general nature of these conclusions points strongly to the need for greater study and investigation of the process of technology transfer. It is for this reason that the new experimental technological incentives programs at the National Bureau of Standards (NBS) and the National Science Foundation (NSF) are of such importance. Hopefully, these programs will be able to put technology transfer in the USA on a more solid foundation of knowledge from which practitioners of technology transfer would be able to carry through transfers with a degree of assurance comparable to that of the scientists and engineers carrying through the development of a new technology.

The need for such new knowledge was specifically recognized by the U. S. Government in the President's message on science and technology in March 1972. It was at that time that the two new experimental programs were announced which assigned NBS and NSF the responsibility of finding out how the Federal Government could stimulate technological invention and innovation in the USA. The fact that these programs were labeled experimental was an explicit recognition by the government of the lack of available knowledge on how best to increase technological invention and innovation. The fact that two quite diverse Federal agencies were given essentially parallel responsibility for these programs is another indication by the Federal Government of the need to examine many alternatives. NSF is an agency which has primarily supported basic research in universities but is now developing applied research programs and is interacting increasingly with industry while NBS is a broadly-based laboratory with primarily in-house programs which performs research for other Federal agencies and gives technical assistance to industry.

To gain an insight into the experience of other countries in this area, NBS undertook a preliminary study of technological incentive programs in Japan, Canada, the United Kingdom, France, West Germany. Thirty-eight different programs were identified. It was clear from this study that much experience had already been gained but that many of the programs were only applicable within the context of a particular country.

The NBS program was started with the clear recognition that technology transfer is strongly influenced by Federal Government policies in such areas as taxes, patents, procurement practices, regulatory practices, and trade restrictions. It was also apparent that the activities of a number of mission-oriented Federal agencies (in addition to space, atomic energy, and defense) are directly concerned with improving and advancing specific technology. Some examples are:

1. The Department of Transportation which is involved in automobile and bus safety,

2. The Department of Justic which is involved with improved equipment and procedures for law enforcement officials,

3. The Department of Housing and Urban Development which is involved with industrialization of housing construction, and

4. The Environmental Protection Agency which is involved with environmental control technology.

The operational question which the NBS program addresses is: How can the Federal Government modify its present practices so that on-going programs and procedures can more effectively stimulate, and advance technological developments in the private sector? The answer to this question is to be investigated through analyses and experimental projects. Three initial areas from which the experimental projects will be chosen have been selected. They are: (1) Federal Government procurement, (2) Federal Government regulations, and (3) Federal Government assistance to the small research and development firm and independent inventor.

In the area of Government procurement, the Federal Government spent 25 per cent of its budget ($58 billion in the 1972 fiscal year) for goods and services purchased from the private sector. In view of this large expenditure, it is prudent to examine present procurement procedures to determine their impact on technological development. Some preliminary studies were made which revealed many inhibitors. For example, the present procurement practice is to base decisions on minimum cost, competitive bidding, and design specifications. This approach frequently results in the freezing of existing technology or inhibiting new technology. To encourage new technology, procedures which reward both the innovator and the purchaser of the innovation are needed. One technique could be to pay higher prices for advanced technology products which have, for example, lower noise output or lower energy consumption. Another technique could be to write procurement specifications based on performance rather than on design specifications. Hopefully, in this way, the development of more technologically advanced goods and services would be encouraged by effective use of governmental purchasing power.

In the area of governmental regulations, the regulatory power of the Federal government has had a strong effect on technological developments. For example, the Consumer Product Safety Commission and the Environmental Protection Agency have recently issued regulations which have effectively required the development of new technologies. Other Federal Government regulations have distorted or inhibited the introduction of new technology. For example, building codes and standards have often made it difficult to incorporate significant innovations in building construction. Therefore, the goal of this program is to determine how best the Federal regulatory power can be used to stimulate new technology. One technique could be for a combined industry-government group to identify areas in which there is inadequate technical knowledge and to sponsor appropriate research. For example, in the flammable fabric area, the Federal Government and the various textile and clothing manufacturers might team together to learn how to decrease the flammability of clothing and textiles. Another approach is to locate technological-inhibiting regulations and to try to rewrite them to permit new technological solutions. In many cases, the substitution of a performance-based specification for a design-based specification would suffice.

The third area for the NBS experimental program relates to the small research and development (R&D) firm and the independent inventor. There has been considerable speculation on the importance of the contribution to technological advancement of the small R&D firms and the independent inventors. Some indicators suggest that small R&D organizations and independent inventors play a crucial role in improving technology. If such is the case, what experiments should be carried out to improve the innovative capability of these groups? One approach could be to involve the Small Business Administration (SBA), a Federal agency which supports small businesses. (Small businesses are defined as those having fewer than 500 employees.) A major effort of this Agency is to assist small businesses by supplying capital to them in the form of bank-insured loans. The SBA presently bases its decisions for making the loans primarily on the financial soundness of the firmand on the capability of its management. No special attention is given to technologically innovative businesses, and so they frequently find it difficult to gain needed financial support. If special attention and services such as surveys of markets and studies of technological feasibility were given to small businesses concerned with developing new equipment or services which relate to environmental protection or energy conservation, it might be possible to foster the development of the small businesses. Through the experience-based knowledge gained by such an experiment, SBA could become a more effective supporter of small businesses involved with the development of new technology.

These three areas represent the initial thrust of the ETIP program at NBS. A number of other areas are under consideration for future investigation. They include the use of industrial associations to attack technical problems in industries characterized by relatively static technology, the role of anti-trust considerations in technological innovation, and a more systematic methodology for making government-held technology of greater benefit to the public.

An increased understanding of how to facilitate the transfer of technology within the private sector and also from the public to the private sector can be of major benefit on both a national and international basis since many problems facing individual nations can be solved with better technology. Therefore, any improvement in techniques for advancing technological developments are in the best interest of everyone.

The author acknowledges the significant contribution made by Dr. Kenneth Kreider and Dr. Ian Bartky of the Institute for Applied Technology in the National Bureau of Standards.

TECHNOLOGY TRANSFER: FROM THEORETICAL PSYCHOLOGY TO MANAGERIAL TECHNOLOGY

Alicia L. Pagano, Ed.D.

Medgar Evers College

Introduction

This paper presents a transfer of technology from basic research to applied situations. It discusses management of groups by identifying micro-units of verbal interaction, categorizing them under various styles of management, and assessing the effects of the various styles of management on the socio-emotional climate and the behavior of members of the group. The events of the technology transfer will be described in a time sequence from the development of psychological theory to application. A specific research in an academic setting will be described, and finally a suggestion will be made for the transfer of the techniques of this quantitative technology from the academic setting to other operational environments.

Transfer of Scientific Knowledge into Technology

The role of the leader and the effects of styles of management in group interaction have always been a fundamental concern of social psychologists. One view of leadership may be stated in terms of the social climate it creates and of the behavior of the groups within various social climates. By social climate is meant the prevailing atmosphere within which the group functions. This atmosphere is, in turn, responsible for the generalized attitudes which the group members have about themselves, each other, their leader, and the group's goals and purposes.

In a study of social interaction in 1937, H. H. Anderson theorized that there are two techniques of intra-group behavior that have been experimentally demonstrated to be psychologically different.[1]

These two techniques of behavior are (1) domination and (2) socially integrative behavior. Domination is characterized by "a rigidity or inflexibility of purpose...stifles differences and attempts to make others behave according to one's own standards or purposes" while integrative behavior designates "behavior that is flexible...expenditure of energy in common purposes...makes the most of individual differences." Anderson found that the domination and socially integrative behavior of preschool children could be observed in their interplay. In later studies with kindergarten and elementary school children in the classroom he found that the main direction of influence in classroom interaction is from the teacher to the pupil. He demonstrated that it is possible to obtain reliable patterns of teacher and pupil behavior through categorization of their overt behavior . The findings also demonstrated that the children's behaviors in the classroom were consistent with the style of behavior of the teacher.

The development of a consistent theory of interaction analysis was advanced when Kurt Lewin working with Ronald Lippitt and Ralph White designed an experimental research to answer questions concerning social climate and styles of leadership.[2] From their study, they concluded that leadership style was a primary factor in producing climatological differences and that group membership was of secondary importance. Their findings suggested that different leadership styles produced different social climates and resulted in different group and individual behaviors. They also found that verbal categories differentiated behavior techniques of the leaders more adequately than other social categories.

John Withall, in 1949, tested the hypothesis that an objective, valid, and reliable index of socio-emotional climate in a classroom could be obtained by means of a categorization of teacher statements. He developed seven mutually-exclusive categories that encompassed all types of statements teachers used in the classroom.[3] His seven categories were analyzed on several continua from problem-centeredness to person-centeredness, from leader-centeredness to teacher centeredness, and from objectivity to subjectivity. Withall found that socio-emotional climate in the classroom could be assessed and described from the analysis of teacher verbal behavior. He also found that his social-climate index could be used by trained observers with an adequate measure of agreement in the categorization of teacher statements and that the index gave a consistent pattern of verbal behavior for a given teacher from day to day.

Ned Flanders developed a more comprehensive system for categorizing verbal behavior in the classroom that described dominative, initiative behavior as "direct" and acceptant, responding behavior as "indirect." He concluded that direct teaching sets a socio-emotional climate that results in student hostility, apathy, and negative attitudes while indirect teaching results in decreased interpersonal anxiety, integration, and emotional readjustment. In 1957, Flanders began placing interaction analysis information on a matrix whereby he began to see the interrelationships and the interdependence between successively coded statements and how they could explain a teacher's influence pattern in the classroom.[4] The purpose was to seek a scientific theory of instruction that could assist in teacher education programs and in the improvement of instructional communication whereby one could specify conditions in which learning is maximized.

The above mentioned researchers created their own terminology to describe essentially the same patterns of behavior. Each of the studies described behaviors that were shown to be different in quality and that were observable for categorization during group interaction. The different behavior patterns were found to produce different social climates and different responses from the groups. The time sequence for a transfer process from basic research by social psychologists to an applied technology for utilization in the academic environment slightly exceeded thirty years.

A Specific Utilization of the Technology

In 1972, the Pagano Interaction Analysis Categories were developed for the purpose of describing and analyzing teacher influence in the highly specizlized academic area of music education. The system shown in Figure 1 was based upon the Flanders' system and includes categories for utilization in music classes.[5]

I. TEACHER BEHAVIOR
  Response Influence
    1. Accepts feeling
    2. Praises or encourages
    3. Accepts or uses ideas of students
    4. Asks questions
    4a. Questioning performance
  Initiation Influence
    5. Lecturing
    5a. Lecturing performance
    6. Giving direction
    7. Criticizing or justifying authority
II. STUDENT BEHAVIOR
    8. Student talk-response
    8a. Student performance-response
    9. Student talk-initiation
    9a. Student performance-initiation
III. SILENCE OR CONFUSION
    10a. Silence while listening
    10. Silence or confusion

Figure 1. Pagano Interaction Analysis Categories

This research was a transfer and adaptation of existing interaction analysis technologies to new situations. Prior to the creation of this system there was no adequate system to analyze the total classroom interaction in an environment where many of the activities varied from the usual verbal interaction. Studies of music classes had been made by Nolin[6] and Whitehill[7] but they both stated that one of the basic problems involved the extensive amount of time in which behaviors were expressed musically. The Pagano category system was designed to include all verbal and all musical behaviors within the context of the system.

The research sought to categorize and describe the classroom interaction of selected music teachers in two grades: the first and the sixth, and to analyze and interpret the variation of the classroom interaction patterns of the music teachers in the same grades and between the music teachers in grade one and the same music teachers in grade six.

The sample included ten music teachers and their first grade and sixth grade music classes. The general music of classes of these ten teachers were audio-taped a minimum of five different class sessions. The information was categorized and tabulated on 15 x 15 interaction analysis matrices. The statistical treatment of the data included the Kendall coefficient of concordance and the Wilcoxon matched-pairs signed-ranks.

The results indicated that although the teachers in the study had flexible patterns in their use of the fifteen behaviors in the matrix, they tended to be direct using initiation in their influence patterns. Musical behaviors were used extensively, but teacher talk dominated classroom behaviors in both grades. The musical behavior of student initiation occupied less than one per cent of total classroom behaviors in both grades.

The teachers in the study were consistent in their patterns of interaction in the same grades and there was a significant relationship in the patterns of interaction among teachers in the same grades. There were no significant differences between grade one and grade six in influence patterns. The teachers who were direct tended to remain direct in both grades; the teachers who were indirect tended to remain indirect in both grades.

Conclusions were made that the teachers in the study were flexible and direct in interaction patterns in grades one and in grades six. They were consistent in their patterns among classes in the same grade levels and consistent in their flexibility and influence patterns between grade levels. There were differences in their uses of the fifteen categories between grade levels.

The description and analysis of the micro-units of interaction and of the present behavior patterns of teachers in this specialized field provide the basis for prescriptive suggestions to improve the management of music classrooms toward specified goals such as the expansion of student initiation.

Suggestions for Transfer to Other Operational Environments

The applied technology of interaction analysis is a direct product of the theories of social psychologists. The development of the interaction analysis system which permits a scientific description and analysis of micro-units of behavior has been a valuable technical exploitation of a scientific discovery because it provides a tool to assess the effect of specific leadership behaviors in interaction environments and to prescribe more effective management behaviors for directing and controlling the environment toward stated goals.

The quantitative techniques of interaction analysis could be transferred directly to group situations other than the management of academic environments. Industry, government, and other organizations could use interaction analysis for achieving more effective means of developing leadership skills that will motivate employees or group members toward the stated goals of the organization. With a minimum of modification, managers in other operational environments could use interaction analysis to describe verbal behavior as it occurs and could analyze the effect of the varied interaction patterns upon the socio-emotional climate and upon the behavior of the members of the group. This direct transfer of technology could be useful in the refinement of managerial skills in many operational environments.

REFERENCES

1. Anderson, Harold H., "An Experimental Study of Dominative and Integrative Behavior in Children of Pre-School Age," Journal of Social Psychology, VIII, (1937) 335-345.

2. Lewin, Kurt and others, "Patterns of Aggressive Behavior in Experimentally Created 'Social Climates'." Journal of Social Psychology, X (1939), 271-299.

3. Withall, John, "The Development of A Technique for the Measurement of Social-Emotional Climate in the Classroom," Journal of Experimental Education, XVII (1949) 349.

4. Flanders, Ned A., Analyzing Teaching Behavior (Reading, Mass.: Addison-Wesley Publishing Company, 1970), pp. 33-37.

5. Pagano, Alicia L., "A Study of the Classroom-Interaction Patterns of Selected Music Teachers in First-Grade and Sixth-Grade General Music Classes" (unpublished doctoral dissertation, The American University, 1972), pp. 5, 48.

6. Nolin, Wallace, "Patterns of Teacher-Student Interaction in Selected Junior High School General Music Classes" (unpublished Doctoral dissertation, The Ohio State University, 1969), pp. 8-10.

7. Whitehill, Charles, "The Application of Flanders' System of Classroom Interaction Analysis of General Classroom Music Teaching" (unpublished Doctoral disseration, West Virginia University, 1970), pp. 6, 7.

EMPIRICAL STUDY ON BARRIERS TO TECHNOLOGICAL TRANSFER IN THE FIELD OF EDUCATIONAL TECHNOLOGY

Wilhelm H. Bierfelder

University of Stuttgart

PRELIMINARY REMARKS

In starting an empirical investigation, a very important strategic decision has to be made. Out of a set of hypotheses a subset has to be chosen whose key character can be supported. Different sources offer conjectures which are used to explain resistance to change. To examine them all would not only be an endless, but a dubious task.

In looking for a criterion for decision, discriminations must be made between hypotheses of high and low potential increase of knowledge.

The most suitable criteria can be offered under a highly developed theory of innovation. But supposing that such a theory already exists, the task of empirical theory testing remains a challenge even using very sensitive instruments of investigation. The arrangements leading to fruitful mutual control between theoretical concept and empirical test--or more generally between speculation and experience, so successful in physics and other fields of science--soon reached a "sound barrier" in the social sciences. The horizon of expectation was therefore reduced to theories of medium range (Merton).

Within the field of innovation research, such theories of limited range are not yet visible. To overcome this dilemma an empirical approach had to concentrate on a single, or a subset hypothesis, which can be favored because of its supposed key characteristic in respect to further findings of new knowledge (heuristic potential).

The subset is chosen from the set of conjectures about stimulants to innovative processes in socio-technical systems. It is related to the informative, communicative or cybernetic abilities of such a system. Such abilities seem to play a key role because in innovative systems, or subsystems, the necessary planning output rests "especially on gaining and conversion of relevant information."[1] The limited research capacity asked for priorities. It seems to be reasonable to allot it to an area as mentioned above.

SETTING UP HYPOTHESIS

Barriers to innovation irrespective to technological transfer can be expected in and between all phases of innovative processes.

A model of such a process as an abstract scheme will include the following phases or modules in logical order:[2]

MODULE	ACTIVITIES
1	analytic activities in an existing system
2	stimulants to change arising out of detected defectiveness transformed into challenges to the system
3	goal setting for change or adaptation
4	activity patterns for reaching the preferred goals
5	introducing changes (implementation) in several stages
6	realisation under every days condition

MODEL OF THE INNOVATION PROCESS

The better that barriers to innovation can be explored and predicted, the greater are the chances of finding ways for successful innovation and strategies for elimination of such barriers. They can partly arise out of the new role which expectation plays in a revised system, or they can be deeply rooted in the history of the old system. A theory of transferring planned organizational change within a complex socio-technical system depends on support from within the organization--to overcome barriers and resistance to change successfully. The modular structures of innovative processes imply reflection of module 5 (implementation) as far as the problems of application are concerned. Prediction of barriers increases the possibility of supporting measures covering the complete modular structure being formed.

Of what kind are the barriers which do not allow implementation of change? What are the specific attributes? Which abilities must a management system possess in respect to these barriers, to make an innovator successful? The last question completes the circuit again. Following this circuit of questions, the first task is to describe the phenomena in which barriers to innovation can be observed.

According to the current state[3] of innovation research, an essential condition of successful innovative processes within socio-technical organizations seems to be the acceptance of informal groupings of innovators. The core of such groups can consist of a single member of the organization, for a stipulated time. The dominant effect on innovative behavior of an organization neither depends on formal expert competence nor on formal authority of the line managers. However it can be supposed that the contribution of an information agent (Shapero) is of utmost importance. Because he is able to simplify his knowledge, increasing numbers of organizational members request his services.

The single agent can be replaced by a group, which, as the single agent, gains informal appreciation within the organization. Attempts toward professionalizing the members of the informal innovative group show negative results.[4] As soon as the informal status becomes explicit, the agent or group is brought under the authority of the management system. According to the observation of Argyris the innovation impulses soon come to an end too choked by the formal network of authority.

The benefits an innovative group can contribute depend, as Hinterhuber suggests, on "an efficient network of information and communication." This network assists in linking the informal innovative system with the formal management--and decision system. Through this new relations are created which tie an "information agent" (Shapero) or a "change promoter" (Witte) to another member of the organization. The extent to which the members of the organization can play the part of "power promoter" (Witte) affect chances for an increase of the innovative potential. New opportunities for realization of innovative processes are gained. At the same time implementation becomes easier.

If there is a narrower interrelationship between the innovative and cybernetic attributes of a system than can be supposed, barriers to innovation can especially be detected where information flow is hindered, dammed up or completely blocked. Communication then faces the same difficulties and cybernetic abilities cannot be developed.

This type of barrier within an organization is located in specific interpersonal relationships and in interorganizational relationships. Or, in more general terms, relationships between subsystems, as well as the relationships between subsystem and systems, and relations between systems and the environment.

Out of the profusion of possible barriers, a few have to be chosen. Their influence on the innovative behavior on organizations will be the subject of empirical investigations, now in preparation or already started.

INTERPERSONAL BARRIERS

Within interpersonal relationships, the information flow can be disturbed or can totally break down, when the communicator and receiver have no common language or knowledge. The barriers to mutual understanding can be specified as different expert language, such as the language of technician and administrator, or a different stock of knowledge, such as different know-how. Insofar as such know-how barriers do exist, any attempt of interactive relationship is questionable.

To achieve proper communication, means which propose to eliminate such barriers have to be looked for. One possible way to achieve this aim is to motivate members of the organization to learn a common language. But nobody in any larger organization possesses the motivative power asked for to reach this aim. The observations which Pirker makes in the German machine tool industry lead to the assumption that only the active members of an innovative group are willing to make sacrifices towards this goal.

This reference demonstrates the important contribution of an information agent to eliminate such know-how barriers as far as he is accepted as translator and interpreter within an organization. So far as barriers of this kind are eliminated, only in the way mentioned above can the organization accept the bottleneck arising out of the information proceeding capacity of such an agent.

INTERORGANIZATIONAL BARRIERS

The boundaries between organizational units can become preferred places for another kind of barrier. Such units can be labelled as departments, sections or subsystems within a broader system. Using departments as possible organizational elements between which interorganizational boundaries can be erected, Argyris[5] defined barriers of this origin as "interdepartmental barriers." They can owe their origin to different reasons as follows:

--to preserve the boundary by exercising boundary control
--as a defence against organizational members increasing their sphere of influence (power)
---preference for competition with other departments (Argyris)
---dynamic of gaining and losing in receiving scarce resources (Argyris)

Barriers of this type can best be eliminated by organizational structures promoting coordination and cooperative leadership.

INTENTIONAL BARRIERS

Intentionality as an intended directiveness is a necessary presupposition for communication between individuals. Whenever communicator and communicant have different intentions, successful communication cannot be expected.

It is the view of T. M. Newcomb[6] that communication between individuals or organizational units has also the function of co-orientation. Whenever somebody needs orientation and depends on the support of others, he has to accept the orientation of the medium. The source of information must be credible otherwise it would be impossible to depend on others in this respect. But human beings tend toward a one-sided selection of informational sources in accordance with intentions. Attitudes and beliefs are preferred, those which do not offer co-orientation are avoided. This is the way intentional barriers between members of organizations come into being. Those barriers can be eliminated to some extent by increasing the credibility of information resources. As it is known by the theory of cognitive dissonance (Festinger)[7]--a theory very often empirically tested---a behavior has to be expected which prefers a one-sided selection of information resources. A sufficient elimination of intentional barriers therefore can only be reached on the basis of common beliefs and attitudes. Very often this basis is determined by ideologies. A strategy between production of credibility and ideology is supposed to be an explicit conflict concerning values and goals of an organization.

EFFORTS TOWARDS ELIMINATION

The efforts which have to be undertaken to eliminate the barriers mentioned are quite different in extent and intensity. Because organizations seldom raise efforts of this kind for more than a short period, phenomena of resignation and fatigue have to be taken into account. Under the aspect of dynamic aspects of system behavior it can be assumed that the three barriers presented here have quite different levels (threshold).

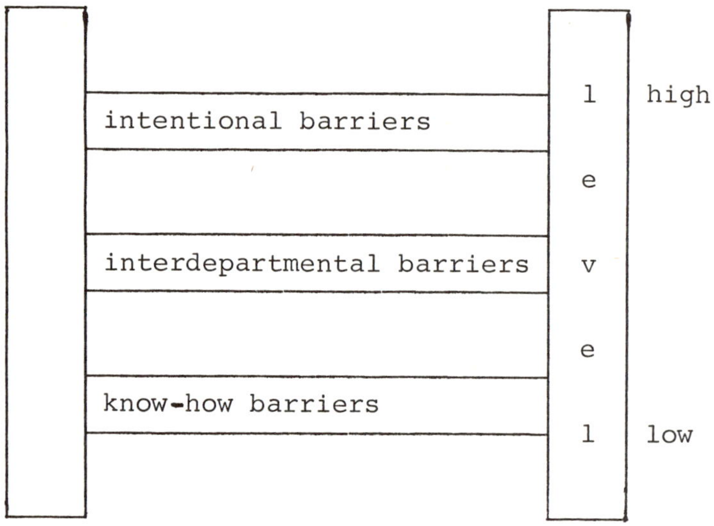

LEVEL OF RESISTANCE TOWARDS COMMUNICATION

The highest level (threshold) is marked by the intentional barriers. This is followed by the interdepartmental barrier, and the level of know-how barrier is the lowest. When each barrier makes a contribution to resistance to change, then the contribution of intentional barriers is the highest and that of know-how barrier is the lowest. Herewith the central hypothesis which has to be empirically tested is formulated.

If it is possible to identify not only those barriers but also their different levels (thresholds), then it should be possible to explain relations between ability to innovate and technological transfer and attributes of organizational structure.

If all this is done, innovative behavior cannot be observed directly. Communicative barriers must be detected by means of theoretical constructs sometimes called "intervening variables." Present research work concentrates on formulating such a construct and developing multivariate analytical tools for mass data handling. Only the descriptive part of the research work has already been fulfilled.

ABILITIES OF MANAGEMENT SYSTEM TO REALIZE PLANNED ORGANIZATIONAL CHANGE

The empirical investigations took place on four different organizational settings, such as industry, civil service, management training centers and universities. All are users of educational technologies. Most of them are short of skilled manpower in the field of education, training, or management development. Introducing self-instructional multi-media systems as subject of study and comparison, they too, seldom allow innovative action by incremental change only.

Comparable to the induction of computer-based information systems, the strategy of muddling through is not very helpful. Organizations preferring such deep-rooted changes (basic innovations) can up to now only choose between improvisation and the strategy of planned organizational change. Provided that several conditions are given, the strategy of planned organizational change is applicable to complex socio-technical systems. The following table allows one to link strategies to their presuppositions so far as the management system is concerned.

The cybernetic capacity is one of the four necessary capabilities and includes gaining and processing information, as well as planning and control. They have to be developed to such a stage that ill-structured situations can be mastered.

A further capability is consensus finding within the management system. Diverging interests and values render this task more difficult. The other two tasks include the exercising of power and gaining sufficient support by those members of the organization confronted by change. All the input data of this study were gained within the management systems of the organization already mentioned. The persons interviewed coult not be asked if they possess such capabilities or not. The following hypothesis was gained from sources such as direct experience or on the results of the number of analytical studies of organizations. It can be expected that the capabilities mentioned (see table) can be found, to different degrees of intensity and more or less exploited, in different organizations.

Organizations demanding high capabilities in the areas mentioned are expected to realize planned organizational change better than those demanding less. Following this second hypothesis and applying it to the four organizational settings under study, it seems more probable that business enterprises innovate more successfully in the field of educational technology than public authorities. It is supposed that universities will not very successfully execute planned organizational change in this area.

The low rate of success in respect to universities can have still another reason. Universities are objects of reform and their reformers are recruited out of those organizations such as parliament and government which have never demonstrated planned organizational change in matters where they were personally responsible. A reform without the consent of those directly affected is something imposed but without much chance of being the concern of all those who have to contribute to it.

REFERENCES

[1]Schmitt-Grohé, Jochen, *Produktinnovation*, Wiesbaden, 1972, p. 32.

[2]Compare a model limited to product innovation development by Jochen Schmitt-Grohé, *ibid.*, p. 51 (Figure 3).

[3]Argyris, Chris, "Management Information Systems: The Challenge to Rationality and Emotionality," *Management Science*, Vol. 17, 1971, p. B-275-292.

Becker, Selwyn W. and Whisler, Thomas L., "The Innovative Organization: A Selective View of Current Theory and Research," *The Journal of Business* (Chicago), Vol. 40, No. 4, 1967, p. 462-469.

Hinterhuber, Hans H., "Technische Innovation und Unternehmensführung," *Unternehmensführung und Organisation*, Bericht von einer wissenschaftlichen Tagung in Innsbruck vom 23-27 Mai 1972. (ed) Kirsch, Werner, Wiesbaden, 1973.

Kirsch, Werner, *Unternehmensführung und Organisation*, Bericht von einer wissenschaftlichen Tagung in Innsbruck vom 23-27 Mai 1972 (ed.) Kirsch, Werner, Wiesbaden, 1973.

Mensch, Gerhard, "Zur Dynamik des technischen Fortschritts," *Zfb* 41 (1971), p. 295-314.

Steiner, G. A. (ed.), *The Creative Organization*, Chicago-London, 1965, p. 4.

Witte, Eberhard, "Innovationsfähige Organisation," *Zfb* 42 (1973), p. 17-24 and *Die Organisation komplexen Entscheidungsverläufe*, Mannheim, 1968.

Woot, Ph. de, *La Fonction d'Enterprise: Formes Nouvelles et Progrés Economique*, Löwen, 1968, p. 102.

[4]Argyris, *op.cit.*, p. B-278.

[5]Argyris, *op.cit.*, p. B-279.

[6]Newcomb, T.M., "An Approach to the Study of Communicative Acts," *Psychological Review*, 60, 1953, p. 393.

[7]Festinger, L., *A Theory of Cognitive Dissonance*, Evanston, 1957.

NECESSARY ABILITIES OF A MANAGEMENT SYSTEM IN CASE OF (1-3)

Kind of ability	(1) Stabilising the steady state	(2) Incremental change	(3) Planned organizational change
1 cybernetic ability	Ø	-	+ (?)
2 exercising power	Ø	Ø	+
3 securing support by members of organization	Ø	Ø	+
4 to built up consens	Ø	-	+ (?)

EXPLANATION:

1 + methods of planning, heuristic procedures
2 + immaterial stimulants
3 + managing conflict, anticipating resistance to change
4 + increasing possibilities for consens finding by mean of planning
Ø average
\- not necessary

TECHNOLOGY TRANSFER IN HUMAN RESOURCES RESEARCH

E. Ralph Dusek

U.S. Army Research Institute for the Behavioral and Social Sciences

In our increasingly complex and specialized technologically oriented society we are faced with a continuous challenge of our ability to select, train and utilize men and women effectively and efficiently in productive work. The armies which these advanced industrial societies support are faced with a similar challenge in using the manpower they have. Thus the society, and the army as an institution within the society, have and are continually seeking systematic and more effective techniques for the utilization of the available manpower. In part this is done through research carried out in relevant disciplines and particularly in psychology, and to a lesser extent in sociology and anthropology. However, these developments involving increasing application of research in the above disciplines to problems of personnel management, training and utilization are relatively recent. Moreover, such research was greatly stimulated during World War II and has received increasing support in the U.S. Army since that critical period.

To highlight some of the more significant developments in military support of human resources research as well as to show its relationship to research in universities and industry, I propose to follow the history of psychology in the U.S. Army with occasional references to significant and related anthropological and sociological research. Uhlaner has published a very illuminating history of the research psychologist in the U.S. Army which adds much detail to what I am presenting here (10). Prior to World War I there was no interest by the military in psychological research. However, psychology as it applied to industry and business, had received vigorous support by the turn of the century by a number of psychologists, among them the famous Hugo Munsterberg of Harvard. During this same period many of the famous pioneers in experimental psychology in the United States were conducting research in some of our most prestigious and famous universities. Thus when the United States entered World War I in 1917 American psychologists, led by some of the most distinguished psychologists of that day, mobilized to offer their services to the defense effort. Drawing on their knowledge of psychology they contributed to perfecting apparatus for selecting and training naval gun pointers, studying and developing techniques for predicting successful fighting aviators, and drawing on their knowledge of group testing, they developed the Army Alpha (for classification of literates) and the Army Beta (for illiterates) tests. In addition they launched efforts to introduce systematic methods of judging and rating qualifications of officer candidates and to train soldiers who showed some ability to learn. While these accomplishments were impressive, following demobilization the Army reverted to its pre-World War I procedures using an apprentice system of selection and assignment and the research psychologists returned to their positions in the universities and industry.

In the interim between the two World Wars psychologists continued to improve their techniques for testing in the general intellectual and aptitude areas, for vocational guidance for training and education and for understanding the effects of environmental stress on performance. In this period the rapid development of aircraft for commercial use led to studies of the effects of high attitudes and associated hypoxia on the physiological responses and psychological behavior of the individual.

In 1939 World War II was raging in Europe and Asia and the United States was awakening to the threat. As Uhlaner has stated "By 1939, the American Psychological Association had 618 members and 1909 associate members. They would be needed for the new crisis facing the United States." In 1939 a whole series of actions affecting the role of psychology in the Army occurred (Note 1). By August of 1939 the Army General Classification Test (AGCT) was ready for standardization. A second form of the test was ready in April 1941. Meanwhile non-language tests for illiterates, mechanical and clerical tests, and trade tests had been prepared. Thus the classification for enlisted men until the end of the war depended primarily on civilian experience. At the same time efforts to develop selection techniques for officers were pursued. Tests for determining ability to meet academic requirements for officer training were utilized; however, although combat leadership aspects of officer candidates were recognized, preliminary studies did not result in satisfactory prediction.

In 1940 the first selective service inductees were administered the AGCT and the mass testing of men in World War II was initiated. In addition achievement and aptitude tests were constructed for use in the Army Air Corps program for selecting and training of aircraft pilots. Later the Aviation Cadet Qualifying examinations were developed. Finally, as Uhlaner reports, in November 1942 one hundred and thirty Army Specialist Corps psychologists were commissioned in the Army of the United States.

Up to now the discussion has concentrated on testing, selection, classification and training, areas in which psychologists had already established a pool of knowledge through research conducted in their civilian jobs. But World War II greatly expanded the areas into which psychologists and other social scientists applied their talents. In 1942 six clinical psychologists were commissioned as first lieutenants in the Sanitary Corps of the Army Medical Department and assigned to general hospitals for duty with neuropsychiatric services. This was the beginning of the use of clinical psychologists in the U.S. Army and it continues to this day. Since I do not intend to discuss this further, I should point out that subsequent to World War II there has been a tremendous expansion of the use of clinical psychologists in the United States as diagnosticians and therapists and many of the leaders in the field were associated with the Army during World War II. Since World War II the Veterans Administration has also employed many clinical psychologists.

One pioneering effort of the U.S. Army in human resources research can be traced to the establishment of the U.S. Army Quartermaster Climatic Research Laboratory at Lawrence, Massachusetts in 1943. It was a pioneering effort to conduct research on environmental protection of the soldier. The problems encountered by the Army in the Aleutian Islands were an immediate stimulus for this research. As our Army began operating in the extreme cold it became imperative that we develop better clothing, sleeping bags, footgear, tents and other personal equipment. Although these initial efforts were primarily physical and life science oriented, subsequent to World War II psychology and anthropology became important contributors to the research effort. At the end of World War II in 1946 this Quartermaster Laboratory conducted an anthropometric survey of approximately 85,000 U.S. Army white males. (8) This classic study was designed to provide information that would lead to a better integration of the sizes of clothing and equipment with the dimensions of the men using the items. At the same time the Quartermaster anthropologists were conducting their studies, other laboratories were conducting anthropometric studies on aviators and other special populations. Since World War II this type research has continued and has included cooperative efforts among different countries. A recent reference work, jointly sponsored by the Army-Navy-Air Force, Human Engineering Guide to Equipment Design (12) cites much of the anthropometric research pertinent to the design of equipment. Much of this research was supported by the military service. Yet today we find the civilian sectors of our industrial societies making increasing use of such research. The design of chairs, automobiles, aircraft, missiles, electronic equipment and many other types of equipment has been influenced by results from anthropometric studies. Research in dynamic anthropometry has influenced design of aircraft controls, and more recently, design of work space and physical work load associated with industrial jobs. Moreover, as such data continues to be collected within a country longitudinal comparisons can be made across generations. Thus we have reports of increasing stature of some populations, etc. Finally, cross cultural or cross country comparisons yield anthropometric data important for proper design of equipment to be used by the consumer or user population.

During World War II the Army conducted research on the effects of heat on soldiers in support of units trained for desert warfare. Much of this research was summarized in a book by Adolph, Physiology of Man in the Desert, which is a classic in the area. (1) As a result of this and subsequent Army research there is now a much better understanding of comfort as influenced by heat on physiological and psychological responses and limits. Such military supported research has had important implications for the air conditioning industry, for those who must labor under hot conditions, and for those who venture into extreme environments such as space.

The rapid expansion of the Army Air Corps during World War II and the rapid advances in aircraft technology were the impetus for greatly expanded research efforts on human physiological and psychological capacities and capabilities. In the 1930's Ross McFarland had studied the effects of high altitudes and hypoxia on human physiological and psychological responses in the Andes mountains of South America. (6) By the early 1940's altitude chambers had been constructed by the military to systematically study the effects of variables such as hypoxia and barometric pressure on human responses. In addition the effects of sudden decompressions could be studied. A great many human studies involving the above variables were conducted in the 1940's; however, with the advent of pressurized cabins and the development of high altitude aircraft military interest in such research was reduced drastically. Tune has commented that the period 1950 to 1965 did not result in the same advances in our understanding of the psychological effects of hypoxia as had the two preceding decades. (9) However, in the 1960's, as a result of the Indian and Chinese clashes in the Himalayas, there was a renewed interest in such research by the U.S. Army. (4) One must conclude that the military services have supported some of the most important research on the effects of hypoxia on physiological and psychological behavior. Moreover, because of this research it was possible for our space flight program to advance rapidly.

I have already referred to the psychologists contribution to training research during World War II. When the Air Force was created as a service separate from the Army a large part of the Army's capabilities in human resources research including training, went to the Air Force. Consequently the Army was faced with the problem of integrating, strengthening, and expanding the existing Army program of research in the behavioral and social sciences. The Army's solution for research on training was HumRRO, created by George Washington University in 1951 at the specific request of the Department of the Army. Among the many significant contributions of HumRRO since that time, one should include the programs listed in Note 2. Although the psychology of learning has been an important area for laboratory research by psychologists for at least fifty years, generalizing the laboratory results to specific applied requirements for training is difficult. As a result HumRRO psychologists have conducted an enormous amount of empirical research on training which has important implications for training of civilians as well as soldiers. In addition HumRRO psychologists have advanced training capabilities through research on the latest media and techniques including computer aided instruction. It is clear that this Army sponsored training research has been and continues to be a leading source of new ideas and techniques contributing to our knowledge in training technology.

Another area of human resources research which was born during World War II and grew tremendously thereafter is human factors engineering or engineering psychology. It was stimulated in particular by the complicated weapons and the water, air and ground vehicles which have emerged in ever increasing numbers and sophistication since 1940. The field itself began to be identified in the mid and late 1940's and early 1950's. Kraft (5) has listed some of the significant documents which appeared during these formative years (Note 3). The data found in these early documents came from many sources, but the sensory and perceptual research in audition and vision done mainly in university laboratories was a particularly important source. However, research done in laboratories formed by the Army, Navy or Air Force became an increasingly important source in the 1950's. Moreover, the military services became a primary source of funds for human factors engineering research in industry and universities after World War II. This area has a vitality to this day as is evident by the continuing publication of important documents (Note 4). However, one important trend in human engineering in the United States begins to emerge when one compares the two editions of the Human Engineering Guide to Equipment Design. The scope of research and development in environmental support and control, closely associated with medical and physiological research, was too great to be included in the revised edition and the readers are referred to excellent sources in this area. Thus specialization is creating new research areas--in this case closely related to the space flight efforts supported by NASA. It should be noted that the Army had lost much of its human engineering capability when the Air Force was formed. By 1951 the Army founded the Human Engineering Laboratory at Aberdeen Proving Ground which continues to conduct research in support of the Army Material Command.

Kraft has compiled a brief history of human factors research in industry and business in the formative and early years. (5) As a result of the increased numbers of professionals in this research area the Human Factors Society was formed in 1957. By 1971 it had about 1500 members. During this same period the Ergonomics Society was being organized in Great Britain and Europe. In recent years members of both societies have increased their emphasis in applying the research methods and techniques of human engineering to products for the civilian marketplace.

In 1957 Melton presented a review of military psychology in the United States. (7) His review across all the services discusses all of the major service laboratories devoted to human resources research. It is significant that he estimates 729 psychologists were working for the agencies of the military departments. This is from a total of about 15,000 members listed in the 1957 American Psychological Association (APA) Directory and represents an enormous growth from the total 1939 membership of 618 APA members. To further emphasize the growth of the profession the APA now has well over 30,000 members. Melton's review was one of the first to recognize a trend in human factors research which has received important emphasis in the last decade. He points out that "The needs of the military establishment for military psychology may be conveniently structured under three headings: (a) improvement of the military personnel management system, (b) contributions to the development and effective operational use of new weapons and weapons systems, and (c) the interactions of the personnel management system and the weapon systems." The critical word is systems and the significance is profound for no longer are we talking of personnel selection, classification, training and human factors engineering as independent areas of research. Instead we recognize that the primary variables in each of the areas interact and if our research is to have significant impact on real problems the interactions must be studied and evaluated. The effects of the systems concepts on the traditional approaches of psychologists to these areas were clearly described by Uhlaner in his presidential address to the Division of Military Psychology of the APA in 1970. "For the applied setting, the research psychologist must move from preoccupation with covariance of abstract measures, which often yield little in the way of practical knowledge, to a means of bringing into the laboratory a better and more meaningful representation of the jobs and systems to be studied. The psychometric heritage of measurement techniques has provided such concepts as construct validity and predictive validation. The last 50 years have seen the accrual of a rich methodology of empirical methods and, more recently, a sophisticated base of computer techniques which together make simulation a more mature technology. With the systems measurement bed and systems psychology principles, men, training and experience, and management styles can be varied in relation to specified outcomes, thus applying the more rigorous methodology of the study of jobs and human performance in systems settings."(11) Research sponsored by the military services was instrumental in the development of the systems approach as is made clear in books which have been published on the subject (Note 5); however, the concepts involved are rapidly influencing research in industry as well as in our major universities.

The development of the computer has greatly facilitated research in many areas of psychology. A short listing of some of the areas includes simulation studies, modeling of man-equipment systems, computer assisted training, computer assisted counseling, modeling of thought processes, analysis of multivariate experiments, and studies of human decision-making. In a very real sense computer technology is a driving force behind a rapidly increasing research capability in human factors engineering research. Chapanis has presented some interesting analyses of problems associated with research on the interactions of man, machines and models in some of the areas above. (2,3)

Finally, I want to discuss the growth of social psychology and sociology and the study of attitudes in the Armed Forces. Prior to World War II Gallup and Roper and others had begun the study of attitudes of the American people with their public opinion polls. During World War II the Army formed the Research Branch of the Information and Education Division of the War Department. A primary purpose of this organization was to provide the Army command quickly and accurately with facts about the attitudes of soldiers with the object of using such information in policy formulation. This applied social engineering effort conducted a great many studies, but most importantly, out of this effort came the classic studies in social psychology in World War II (Note 6). Thus as World War I had seen the Army advance the psychology of mass aptitude testing, World War II saw the Army advance the psychology of mass attitude testing. Many of the famous names in American social psychology and sociology were associated with these projects and the four volumes became standard references in many of the social science departments of our universities. The methodology and techniques used were particularly useful for studies of attitudes. The group of social scientists at the Research Branch also conducted the classic studies on attitudes which resulted in the point system for determining the order of individual soldiers being discharged from the demobilizing Army. Inasmuch as this was a sensitive issue at the time and the demobilization was accomplished without serious incidents, the social scientists can be credited with an outstanding demonstration of the utility of their research. After World War II the Army lost this capability for social science research until it again built up a capability through the Special Operations Research Office (SORO) of the American University in 1956 which was responsible to the Chief of Psychological Warfare. SORO was later renamed the Center for Research in Social Systems (CRESS). The new organization, however, was oriented more toward studying foreign armies and countries and less toward attitude survey techniques. It does seem clear from the rapid expansion of attitude and opinion surveying in the United States in the last 20 years that the Army research of World War II did have a stimulating affect in the attitude measurement area as did the mass testing program of World War I, on later use by industry of aptitude tests for personnel selection, classification and counseling.

In 1946 the Army initiated the study of soldier's attitudes and preferences for food at the Food and Container Institute for the Armed Forces. The famous American psychologist, L.L. Thurstone, and his students were important contributors to the development of a food acceptance program based on research involving psychological and psychophysical scaling techniques. While wine tasters and gourmets have a long history, the application of scientific psychological measurement to support food technology and research techniques and to evaluate end products, including both rations and menus, was new. This work has continued to this day and the psychological research has expanded to include research on the chemical senses and flavor. Today many of our major food companies have acceptance testing departments and some are staffed with scientists formerly associated with the Army research. There is a healthy and continuing exchange of information between Army, industrial and university related psychologists, chemists and food technologists interested in this particular area of research.

In mentioning World War II research I should point out that cultural anthropologists made significant contributions to the defense effort. Ruth Benedict's <u>Chrysanthemum and The Sword</u>, was a scholar's contribution to our understanding of the Japanese and their culture and had an impact on the American Forces occupying that country. In some sense the establishment of SORO in 1956 was an effort to increase this type of social science research in the Army. There have been other Army supported studies in social science, but I have tried to highlight those which I feel have had the greatest impact. Today the Army is continuing support in this area particularly as it has bearing on social areas such as race relations and equal opportunity, drug abuse prevention, career motivation, social change, and quality of Army life. Once again some of these pioneering research efforts by the Army should provide the stimulus for efforts in other institutions in our society. There already exists an Inter-University Seminar composed of scholars in universities, the military services, and other institutions who vigorously support social research on our military institutions.

In summarizing I would point out that in the area of human resources research there has been continual technology transfer between our civilian and military institutions since at least World War I; however, the direction of transfer has varied depending on time and the specialized area involved. Since World War II the military services have been a source of continual stimulation in nearly all areas of human resources research. Where the investment in laboratory facilities is extremely large and specialized, military laboratories have been in the forefront of the research effort. Since the investments in and sophistication of military equipment has continued to increase, requirements for human factors research have increased also. As a result industry has shown great growth in human resources research. We are now seeing the effects of human factors research show up in many of the products designed for the civilian consumer. It would appear that human resources research in the area of human factors engineering and protection against environmental extremes is most readily transferable between countries, research in social psychology, cultural anthropology and attitude measurement being more country specific. Personnel selection, management and training research seems to have an intermediate amount of transferability between countries.

In my discussion I have limited myself primarily to research associated with the U.S. Army; however, the Air Force and Navy each have had and continue to have outstanding human resources research laboratories and programs. I have not discussed them since I am less familiar with them, but there have been many contributions to the human resources area from those services.

In closing I would point out that human resources research is frequently driven by technology. Unfortunately and too frequently the need for the research has become so critical that a short term approach is the only one acceptable. It is most important that those in decision-making positions recognize the need for support of systematic research on potential problems some years in the future as well.

REFERENCES

1. Adolph, E.F. Physiology of Man in the Desert. New York: Interscience Publishers, Inc., 1947.

2. Chapanis, A. Men, machines, and models. American Psychologist, 1961, 16, 113-131.

3. Chapanis, A. On the allocation of functions between men and machines. Occupational Psychology, 1965, 39, 1-11.

4. Dusek, E.R., & Hansen, J.E. Biomedical study of military performance at high terrestrial elevation. Military Medicine, 1969, 134, 1497-1507.

5. Kraft, J.A. A 1961 compilation and brief history of human factors research in business and industry. Human Factors, 1961, 3, 253-283.

6. McFarland, R.A. The psychological effects of oxygen deprivation (anoxemia) on human behavior, Arch. Psychol. N.Y., No. 145, 1932.

7. Melton, A.W. Military psychology in the United States of America. American Psychologist, 1957, 12, 740-746.

8. Newman, R.W., & White, R.M. Reference Anthropometry of Army Men. Report No. 180, Environmental Protection Section, Research and Development Branch, Military Planning Division, Office of the Quartermaster General, September 1951.

9. Tune, G.S. Psychological effects of hypoxia: Review of certain literature from the period 1950 to 1963. Perceptual and Motor Skills, 1964, 19, 551-562.

10. Uhlaner, J.E. The Research Psychologist in the Army--1917 to 1967. Technical Research Report 1155, U.S. Army Behavioral Science Research Laboratory, April 1968.

11. Uhlaner, J.E. Human performance effectiveness and the systems measurement bed. J. of Applied Psychology, 1972, 56, 202-210.

12. VanCott, H.P., & Kinkade, R.G. (Eds.) Human Engineering Guide to Equipment Design (Revised Edition), Washington, D.C.: Superintendent of Documents, U.S. Government Printing Office, 1972.

NOTES

1. Events important to human resources research which occurred in 1939 (Uhlaner):

   March-April 1939. Personnel Testing Section established in the Office of the War Plans Officer, The Adjutant General's Office, War Department. Reserve Officers with psychological training were assigned to the unit.

   September 1939. Emergency Committee in Psychology established by the Division of Anthropology and Psychology of the National Research Council (NRC).

   October 1939. Committee on Selection and Training of Aircraft Pilots established by NRC at the request of the Civil Aeronautics Society.

   May 1940. Committee on Classification of Military Personnel, Advisory to The Adjutant General, established by NRC, at the request of TAG.

2. Examples of Army training affected by HUMRRO research:

   --current course of instruction in individual rifle marksmanship

   --current course of instruction in infantry squad training

   --programmed instruction booklets for training in aerial observation

   --techniques for teaching land navigation

   --training programs and miniature devices for armor

   --techniques for training low aptitude level individuals

3. Significant documents which appeared during the formative years of human factors research:

   Human Factors in Undersea Warfare; Human Factors in Air Transport Design; Tufts Handbook of Human Engineering Data; Applied Experimental Psychology--Human Factors in Engineering Design; Human Engineering Guide for Design Engineers; Visual Presentation of Information.

4. Significant documents which appeared during the late 1950's the 1960's, and early 1970's in human factors engineering:

   Research Techniques in Human Engineering; Human Factors Engineering--Concepts and Theory; Human Engineering Guide to Equipment Design; Human Engineering Guide to Equipment Design (Revised Edition).

5. Recent books on systems concepts and human factors research:

   Psychological Principles in System Development
   Systems Experiments

6. Studies in Social Psychology in World War II:

   Volume I, The American Soldier: Adjustment During Army Life;

   Volume II, The American Soldier: Combat and Its Aftermath;

   Volume III, Experiments in Mass Communication;

   Volume IV, Measurement and Prediction

BARRIERS AND INCENTIVES TO TECHNOLOGY TRANSFER INTO THE UNITED STATES COURTS

William B. Eldridge

Federal Judicial Center

In all likelihood there is no area in the American social fabric that has greater incentive to seek technology transfer than the area of law, and particularly the area of court administration. The incentives are not peculiar to courts; indeed, they reflect the same problems that have led many other groups to explore the possibility of importing into their fields the solutions adopted by others. Moreover, the importation of technological innovation is probably the most promising and the least threatening of the courses open to us. Technology offers an opportunity to continue to perform the same functions we have been performing, despite increasing quantities of tasks and increasing complexities of situation. There are other responses besides technological innovation, but many of them run far greater risks of substantial distortion of function.

The courts of the United States, at federal and state levels, are faced with tremendously increased caseloads. These caseloads already outstrip the capability of court processes as now constituted. It is a virtual truism to point out the three possible alternatives:

1. We can increase the resources for handling the caseload,

2. We can reduce the number of cases permitted to enter the system, and

3. We can do something different to the caseload once it enters the system.

The first alternative offers possibilities of increasing the personnel of the judicial system at the supporting level, which most certainly will be done, but ultimately it would involve substantial increases in the number of judges in the court systems. Very serious reservations are held on this point by many of the most eminent representatives of the legal community. Henry Friendly, Chief Judge of the United States Courts of Appeals for the Second Circuit, which sits in New York City, has expressed these concerns, as regards appellate courts particularly, in his recent book on federal jurisdiction. Justice Felix Frankfurter raised the point at least as early as 1928. Tripling or quadrupling the size of courts would, it is argued, reduce the honorific quality of the position, making it more difficult to draw to the bench the very best lawyers. Further, increasing the size of the court would seriously interfere with those aspects of court functioning that have a collegial quality. _En banc_ proceedings by appellate courts or sentencing councils by trial courts become nearly impossible with 30, 40 and 50 judges. The authority symbol of the judge in the society depends to some extent upon having relatively few judges. There is no suggestion that we have reached a danger point in court systems generally, but that we must be aware of the significance of continued expansion of judicial resources as a response to increased caseloads. This avenue presents a substantial risk of altering the role of court and judge. Such alteration may be required from time to time, but when it is, it should be approached directly with the objective clearly in mind. Role change should not be an inadvertent by-product of response to other problems.

The second alternative, the reduction of intake, involves even more clearly a distinct alteration in the court role. Reducing the intake for the courts means either: (1) redefining certain disputes as no longer cognizable by courts, or (2) presenting enough obstacles to involving court procedure that courts will be less often resorted to. Either approach means that the part courts play in conflict resolution would be changed, possibly drastically so. Again, there are good arguments in favor of radical reduction in court intake. There is a general consensus that Americans are the most litigious people in the world, calling upon our courts to resolve everything from differences between neighbors over who can trim a tree limb to differences between branches of government over who controls national policy. If reduction is to be undertaken for any of the fundamental policy reasons advanced, then the reduction effort should be straightforward and aimed at reducing what is unnecessary or undesirable for courts to do. That court burdens might be lessened should be a happy side effect.

Thus, the third alternative altering the processing of cases once they have entered the system offers the most promising, most practical, least disruptive alternative. By no means do I suggest that all the problems can be met by resort to improved processing, only that it is the first approach to be employed and that it should be fully explored and exploited so that other alternatives will be turned to only as absolutely needed and that other alternatives will be aimed directly at the problems to which they are most responsive.

New ways of handling our caseloads are not limited to the introduction of new technology into our systems. There is considerable opportunity and considerable incentive to find economy and efficiency in better organization of what we are presently doing. Some of the basic work of management science in reallocation of responsibilities so that the right man is doing the right job is a much needed improvement in the judicial system support services -- but this is not the kind of thing I would see as technology. But many such innovations can be realistically considered only when we have a much stronger launching-pad than we presently have. The launching-pad will depend very heavily on the translation of technology from other fields into the legal area.

Simulation modeling of court processes offers a significant opportunity for technological advances in other fields to broaden the options of the judicial field for experimentation. The increasing complexity of our social structures means that judgments in what were once individual cases now often have immense potential impacts. As a result, every suggestion for a change in the <u>way</u> in which judgements are reached becomes much more heavily freighted with impact considerations. We cannot devise laboratories analogous to those of the physical sciences, therefore we must turn to simulation if we are going to do any experimentation before implementation. There is nearly no experience with simulation and modeling in operation of legal institutions. Therefore, if we are going to have any capability in the foreseeable future, we will have to get it through transfer and adaptation. It seems not unreasonable to look for benefits through transfer. Lawsuits are adversary procedures involving constant reevaluation of position by each adversary followed by decisions to advance, retreat, surrender, increase pressure on the other side, etc. Court rules have some effect as parties weigh their decisions. Are these decisions enough akin to those involved in defense strategy and international diplomacy to permit transfer of the immense sophistication developed in those fields to the courts. If so, years of development may be saved.

In order to utilize the power that might be realized from the transfer of established technology into the courts, we must have far better information systems than we now have. While a number of significant advances have beem made in information systems for legal institutions in recent years, they can only be characterized as surface scratches compared to what others have done and to what we must do. The most sophisticated models would be of very limited use to us at present because the information bases and the dissemination systems do not provide adequate grist for the simulated mill. Experience in other fields should provide great assistance, not only in defining what it is we need to know, but also how to acquire it. Forecasting provides another good example. The present lag between recognizing the need for additional resources in the federal judicial system and acquiring those resources approaches six years. By the time the resources have been acquired, the need has taken on different dimensions. A portion of that six year lag could be shortened by truncating the resource acquisition process, but there is no realistic promise of substantial relief there. The real hope lies in anticipating the need well in advance of its actual occurrence so that the resource acquisition process could be triggered far sooner than at present. Again there is every reason to believe that the technology of forecasting that has been developed by everybody from economists to agronomists could be transferred to the court systems. The incentive here is one of immense pressure. Legal institutions simply cannot continue to respond only after the fact. Here, too, information development is as much a part of the court needs as models that express the relationship among information bits.

Every researcher smiles behind his hand when the subject of his research begins to talk about how distinctive he is -- especially do we exchange knowing looks when the subject begins to assert that he is unique or bordering upon it. So, I am acutely aware of the risk involved in tentatively and diffidently suggesting that there are certain barriers and obstacles to innovation, technological or otherwise, in the legal field that are probably distinctive, and possibly, just possibly, unique. I only seek to raise the question, not to speak definitively on it. For, whether our field is distinctive or unique or not, it is best to know what we think we are and why we think it.

I have alluded to a need to find ways of acting that are not simply reactive to crises. I believe that need is understood by the legal community generally and judges in particular, but agreeing that we need a change and getting it are two different things. Our reactive style is not simply the result of passivity or unaggressive leadership; the reactive style is a part of the very core of legal institutions and of courts in particular. Indeed, it is much more than style, it is a part of the role of the law as a stabilizing institution in society. Courts act upon controversies before them, and, in theory at least, only upon those controversies. Consequently, if an appellate court, through improved information systems, was alerted that a great mass of litigation was building around some important issue, and if the court knew that an early decision by the court would save immense amounts of lawyer and judge time, still the court can only wait for the ripe issue to wend its way to the appellate court.

The inherent conservatism of courts toward changes can be illustrated by the polygraph. Without any comment on whether the results of lie-detecting machines should or should not be received into courts as a basis of decision-making, we can all note that the machines are used in a wide range of human institutions and have been for many years. There is still no indication that courts are likely to make any change. It is not that the machine is demonstrably less reliable in identifying liars than are contradictory witnesses. It is rather that we are accustomed to receiving contradictory testimony and reaching a subjective judgement about who is telling the truth. With the machine, our subjective judgement making apparatus is thrown out of whack -- how to resolve the conflict between a machine and a man. We don't know, so we reject the machine; we don't disbelieve it as we do a contradictory witness, we refuse to hear it. The courts always operate in binary form. Its decisions are yes-no whether on truth telling or who wins a lawsuit. We have no capability for assessing risk of error, nor any capability for acting on probabilities.

The approach of the law is craft oriented rather than system oriented. We are concerned about our production, of course, but a higher value is placed on an individualized response than on a quantity of responses. This is not merely the quality vs. quantity tension that arises in all forms of human activity. There is a high value placed upon the mode in which service is rendered that is to some extent independent of the quality of the service. This value finds its expression in such terms as "the lawyer-client relationship" which implies that there is a substantial value in the individually crafted service for the client even if the tangible product of that service for the client were identical to another produced more quickly and more cheaply by routinized systematized methods. Thus, a will or trust prepared from clean sheets of paper following a series of interviews between lawyer and client is superior to an identical will or trust prepared from information furnished by a client on questionnaire forms and prepared with reference to hundreds of similar production efforts arising out of the same facts. The notion of the law as one of the ancient learned professions is threatened by the assertion that the services can be routinized. Even where systematic procedures for dealing with repetitive problems have been instituted, there is a strong tendency to avoid recognition of them or even to deny their existence. Thus, we have a significant barrier of accepting an honest description of what we do which must be overcome before we can apply technology for the improvement of what we do.

If these observations be true about the activities and value structure of legal practice, they are quintessentially true of judicial operations. While precedent is an extremely important element of the judicial process, individualized interpretation and application are absolutely indispensable to the discharge of judicial responsibility. We believe that this value is held by the general society as well as the legal and judicial communities. Thus, to feed into a computer a set of facts in a controversy and to find a prior controversy with an identical set of facts and then to render the resolution of the prior case in the instant case is not a satisfactory or satisfying methodology. A series of justifications is instantly available; conditions may have changed, perception of the relative importance and interaction of the bare facts may be different. Indeed, the effort in legal analysis is much more heavy on demonstrating the distinctiveness of each new situation than in finding similarities -- at least there is little effort to find systematic similarities pervading large portions of legal activity. The result is that the perception of legal work is anecdotal. One never hears lawyers or judges talk about the reiterative character of their work. Or, more accurately, when that perception is forced by the facts, the response is that the task partakes but little of the real character of judging or lawyering. By way of illustration, in the federal courts there has been during the past ten or twelve years a tremendous growth in the filing of petitions by prisoners seeking release on grounds of some flaw in the procedures that put them in prison. These cases do present repetitive and often frivolous assertions that are generally handled in short order by reference to past experience. But, we do not hear a call for the development of standardized and routinized processes for handling them. A frequent answer, rather, is that these cases do not belong in the courts; their very character says that they belong somewhere else, such as in an administrative agency, for which routinized procedures are appropriate.

The result of this perception of the dynamics of the legal process is an almost completely anecdotal view of court operations. It is difficult to demonstrate the value of technological application to non-recurring situations.

Finally, the greatest barrier of all is the absence of clearly articulated goals. The goal of the legal community is to achieve justice. On that point we have not a dissident voice, but when we try to move beyond that consensus, we make no progress at all. There are no agreed upon elements of justice. Consequently, we have no measures of predicting the impact of technological innovation on our work product to guide us in selecting the best approaches. Nor, once having embarked upon change, technological or otherwise, do we have measures of impact. That is, we cannot measure it in terms of the critical goal that "justice shall be done." We are, of course, able to measure impact in certain terms of efficiency. We can demonstrate that the video-taping of testimony by expert witnesses in advance of trial saves a measurable quantity of time lost when the expert witness cannot appear at the scheduled time due to emergency brain surgery, but we cannot say that the end result of such cases was more or less in accord with the justice goal. Until we can, we cannot assess the impact of the technological innovation.

There is enough intuitive feel for the "justice" of court proceedings to permit certain innovations, such as the video-taping, without demonstrating its ultimate effect in quantifiable terms. But even these advances become inhibited by the fact that all technological applications are restricted to the things measurable by efficiency impact. The feeling wells up that we are placing all our emphasis on efficiency to the neglect of quality. Since we are unable to move effectively on the quality front, and since we are sensitive to the alleged overemphasis on quantity and efficiency, we arrive at a stalemate. Part of our work is mystical and unknowable; obviously we don't tamper with that. To work with the more concrete and more knowable is to slight the importance of the mystical; obviously we don't want that.

The barrier is a tension between goals and the operating rule is that neither shall relax until the other has given way.

TECHNOLOGY TRANSFER FROM THE DEFENCE TO THE CIVILIAN SECTOR IN ISRAEL

Samuel N. Bar-Zakay

Manager, Center for Technological and Interdisciplinary Forecasting (CTIF), Tel-Aviv University, Israel

"Knowledge is a process of piling up facts; wisdom lies in their simplification."

Martin H. Fischer,1879

INTRODUCTION

For quite a few years it has been the feeling of policy makers at various government ministries in Israel, that the civilian sector could benefit from defence technology transfer. Several committees discussed the subject but operational conclusions were not attained. Several months ago the Center for Technological and Interdisciplinary Forecasting (CTIF), whose charter will be described briefly below, was asked to conduct a research program on the topic, with the general goal of analyzing alternative policy recommendations to enhance the latter process within Israel.*

---

* Staff members of CTIF working in the program include: Samuel Bernholtz, Yair Eylam, Eitan Harduf, Irit Portugali, David Rabinovitz and Shoshana Stiftel.

The research program is still far from completion. However, the development of the program, its methodology and preliminary results will be described here with the objective of stimulating new ideas (and different mistakes) in the general field of technology transfer research.

ABOUT THE CENTER

On the initiative of Tel-Aviv University, the Ministry of Defence, the Ministry of Commerce and Industry, the National Council for Research and Development at the Prime Minister's Office, and the Ministry of Development, the Center for Technological and Interdisciplinary Forecasting at Tel-Aviv University was established in January 1971. The work of the Center is guided by a "Users' Committee" on which all these bodies are represented.

The objective of the Center is to assist government policy makers in their functions by conducting technological and interdisciplinary forecasting studies on issues of national interest and by translating the findings into short term decisions.

The permanent staff of the Center is currently composed of the following disciplines: Chemistry, Economics, Aeronautical Engineering, Physics, Mechanical Engineering, Electronics, Mathematics, Operations Research, Information Science, Sociology, Chemical Engineering, Statistics and Business Administration.

INCEPTION OF THE PROGRAM

The "Users' Committee" appointed a "Steering Committee" for the research program. It was decided that the major responsibility of the "Steering Committee", which convenes every three months, would be to choose objectives of the research program from a list of alternatives that will be presented by the CTIF. It was also decided that it would be the CTIF's responsibility to choose the research tools required to achieve the program's objectives.

During the first month of the program the CTIF prepared a table of Alternative Research Objectives with their corresponding Alternative Research Methods. Two examples from the long and detailed table will be given here.

Research Objectives ranged from (1) Increasing the understanding of the technology transfer (TT) process to (19) Influencing the TT process immediately. Correspondingly, it was suggested by the CTIF that objective (1) would require a comparative study of TT in other countries while objective (19) would require that the CTIF play a role of a catalyst in the TT process.

The final choice was the following:

1. To recommend government policies which will improve the process in the future by analyzing past cases of transfer.

2. To indicate to the civilian sector in Israel the benefits of technology transfer by inducing three successful transfers.

Two research teams were created within the program. The first, dealing with objective (1) includes an economist, mathematician/statistician and a sociologist. The second dealing with objective (2) includes a mechanical engineer, electronics engineer and a chemical engineer. More than ten consultants are also involved in the program. These interdisciplinary teams meet at least once a week for thorough discussions. The project leader is the author of this paper.

FIRST OBJECTIVE - OVERVIEW

The research team searched the literature for policies to enhance any type of technology transfer - in any country. It is amazing how poor the crop was. (Appendix One lists a few of the items reviewed). It is beyond the scope of this paper to analyze the reasons for the gap. One reason seems to be obvious:

Since the technology transfer process is characterized by a huge number of variables (technological, sociological, economic, etc.), it is extremely difficult to analyze it, let alone to arrive at operational conclusions, or policies on how to effect it.

The CTIF is currently attempting to overcome the above difficulty in the following manner:

ASSUME:

1. that a finite number of variables in the technology transfer process could be identified.

2. that data on these variables in a representative sample or total population of transfer cases could be gathered.
3. that a success measure (output measure) for technology transfer cases could be developed.

THEN:

1. By some method of statistical multivariate analysis, the data for the (sample) population could be analyzed.
2. By some kind of regression analysis of the success measure over the various variables, one could find the relationship between success and the variables.
3. Based on these variables, the success for new technology transfers could be forecasted.

Indeed, this in itself would seem like a worthy piece of research. However, we must remember that the CTIF's interests center around policy recommendations which poses an important constraint:

The recommendations should be of a general guideline type (and should not deal with details), especially since these are expected to be government policies.*

For this reason the CTIF decided to look into the possibility of performing factor analysis [1] . The main result of factor analysis is the delineation of distinct clusters of interrelated data. This technique gives a secondary advantage (sometimes primary to the research): data reduction without loss of information. One can perform regression on the factors instead of on the original variables.

---

* WEBSTER: A policy is a definite course or method of action selected from among alternatives and in light of given conditions to guide and determine present and future decisions.

The basic idea is therefore the following:

A. To collect data on relevant variables, to perform factor analysis with subsequent regression analysis of the success measure on the factors and to identify a few meaningful factors.*

B. To try to <u>translate the factors into policy decisions</u>. If this could be done then a real <u>raw policy</u> would emerge:**
   1. It will be of a general type (built of details but more encompassing).
   2. It would enable decision makers to interpret the policy in their <u>own environment</u>.

For the above to be achieved several conditions must hold:
   1. The factors translated to policies must be meaningful.***
   2. One must assume that the exogenous conditions for TT will not change in the future (certain types of analysis may eliminate the need for this assumption).
   3. The analysts must take into consideration the common difficulties with factor analysis (analysis over time, analysis with additional variables, ratio of number of variables to sample size, etc.)

FIRST OBJECTIVE - WORK CARRIED OUT

In the previous section we assumed that certain data would be available for the analysis. It will be described here how the data was gathered.

---

* The question of how many factors and why is beyond the scope of this paper.

** By "raw policy" I mean policy based on analytical tools before the addition of human judgement to policy formulation.

***In the sense that (1) they account for a significant amount of variation among the variables; (2) they are highly correlated with success; and (3) they can be interpreted.

1. The finite number of variables involved in the TT process was based on a Technology Transfer Model (2) and on other studies listed in Appendix 1. Each variable shown to affect the TT process was translated into a question for either the donor or recipient. (Two questionnaires were prepared).
2. The CTIF team interviewed various defense donors and collected data on past transfers. Most of the answers were ranked -1, 0, +1 (or otherwise)*, though there were a few continuous ones.
3. Since a success measure is still in the process of development, the analysis to date refers to success as positive or negative (yes or no).

Like all analysts the CTIF is uneasy in presenting unfinished analysis, especially since even data collection has not yet been completed. However, in order to demonstrate the analytic process intermediate results will be given here.

Table 1 presents factor matrix (after rotation.)[1] It can be seen that the input consists of data on 15 variables and five factors were obtained. Computer program B M D X 72 for factor analysis was used. The sample size in this example is 35 technology transfer cases of which 30 were successful. We asked the program to perform the analysis on the correlation matrix. The success variable was excluded.

---

* CTIF is aware of the difficulty of performing factor analysis with ranked variables rather than continuous ones.

TABLE No. 1*

VARIABLES	FACTORS 1	2	3	4	5
1	-.40	.12	-.06	.63	.23
2	-.88	-.12	.02	-.05	-.01
3	.88	.11	.19	-.27	-.05
4	-.52	.10	-.31	.03	.04
5	-.05	.17	.07	-.08	-.52
6	.09	.89	.07	-.09	-.24
7	-.05	.90	-.21	-.01	-.12
8	-.12	.05	-.78	-.14	-.05
9	-.06	-.03	.77	-.17	-.43
10	.21	.06	.17	.32	.13
11	.55	.04	-.12	.27	-.14
12	-.25	-.02	-.01	-.07	.58
13	.28	.50	.49	-.01	.34
14	.38	.49	.20	-.54	.22
15	.11	-.19	.02	-.69	-.10

For the purpose of the examples in this paper variable 14 in factor 2 and variable 13 in factor 3 will not be discussed. However, their meaning size and sign do not contradict our conclusions and recommendations.

Next we performed Stepwise Discriminant Analysis (BMD07M) on the factor scores into two groups: success and failure. This was chosen instead of linear regression since our dependent variable "success" is for the time being dichotomic.

* The numbers are rounded to two significant digits. They represent the correlation between the factors and the corresponding variables.

The factors entered in the order 2, 3, 4. After the third step the program classified the cases as follows:

TABLE No. 2

Program Decision

Groups	"Failure"	"Success"	Total
Failure	4	1	5
Success	5	25	30

We see that the classification was right in 29 out of 35 (83%) cases. This means that the factors are meaningful.

In order to understand factor 2 and 3, the most important ones, as our two examples, we go back to Table 1.

FACTOR 2 - Discussion

We see that variables 6, 7 and 13 have high loadings.
The corresponding questions were:

(6) For how many years did the donor enterprise work in the general field of this technology?

(7) For how many years did the donor enterprise use the technology before it was transferred?

(13) How was the technology explained to the recipient? The possible answers are:

1. Verbal description
2. Written description
3. Demonstration
4. Specifications
5. Loan of donor's personnel to work at recipient enterprise and vice versa.

Since the loadings of the above variables are high and positive we understand this factor to mean that enterprises which have used a technology for a long time before transferring it, use more advanced means in the transfer process. We call the factor "the experience factor". From the Discriminant Analysis (and from the correlation matrix) we saw that high factor scores are associated with success.

One may have intuitively thought that better explanation of a technology to a recipient will increase the probability of success. Yet, one could not forecast that the "experience factor" would be the most important one for discriminating between success and failure out of the 15 variables which were considered relevant for technology transfer.

FACTOR 3 - Discussion

We see that the variables 8 and 9 have high loadings. The corresponding questions were:

(8) Did the technology generally fit the needs of the recipient before the transfer?
Yes = 1 No = -1

(9) Did the donor invest in the adaptation required for the technology transfer process?
donor = 1 neither = 0 recipient = -1
(Note: the ranking does not imply "good" or "bad" and multiplication by -1 is possible).

In this factor one loading, of variable 8, is negative. Namely if the technology did not generally fit then the donor invested in the adaptation. Indeed, from the original data we know that in most cases the technology did not generally fit, and the donor invested in the adaptation.

From the Discriminant Analysis we found that high factor 3 scores are associated with success. To increase factor scores in cases which do not fit, the donor should invest in the adaptation. We call this factor "the economic factor".

RAW POLICIES

Keeping in mind that we are only concerned with examples the two factors will be translated into policies in this section.

Factor 2 implies the following recommendations:

> For successful technology transfer* defence organisations should prepare technology specifications and accept recipient personnel for work, as well as allow defence workers to work at recipient organisations.

* In other words "in order to increase the probability of a successful technology transfer".

This recommendation may be "revolutionary" because it implies the need to open gates of classified organizations to civilian workers, and to the possibility of defence losing personnel after loaning it to recipient organizations for transfer projects.

One may ask why we have not dealt with the finding that the longer the experience with the technology the higher the success. The answer may be simple: We cannot recommend that organizations should become experienced in some field. However, what should be considered is to attempt transfers only in fields in which some experience has been accumulated at donor's organization. But then again the novelty of the technology may be lost.*

In order to derive a policy recommendation from factor 3 some preliminary remarks are required. The 35 technology transfer cases were of two types which we designate:

A) "Proper" transfer
B) "Order" transfer

In the first type, defence transferred some technology without any accompanying order of goods, the production of which would necessitate the use of the technology. In the second type an order was placed for goods which would require the use of a technology for their production. We know that successful technology transfer is crucial in both types. Our finding that the donor invests the money when there is need of adaptation is not at all obvious. It shows that Israeli enterprises are reluctant to make investments in technology transfers even when the transfer is accompanied by an order. This may be due to the small size of the civilian Israeli enterprises, or to the fact that they are "spoilt" in view of the general economic environment in Israel.

It is also possible that the Israeli manufacturers' final price of goods to defence, including cost of adaptation, is still lower than purchase abroad. Furthermore, it is possible that the Israeli price is higher than foreign but that the delivery time is shorter. These points have not yet been studied by CTIF, and in any case are beyond the scope of this paper.

---

* See last paragraph in this section on following page.

We found previously that in order to increase factor 3 scores, in cases which do not fit, the donor should invest in the adaptation. From the original data we know that the frequency of donor investment in adaptation in cases of the "proper" type is significantly lower than in the "order" type. Therefore we could have forecast failure for most of the "proper" cases. Indeed, this is the case.

Factor 3 therefore implies the following recommendation:

For successful "proper" technology transfer, in cases which do not fit, defence organizations should invest in adaptation.

After the technology transfer takes place defence may or may not decide to charge the civilian recipient for the adaptation costs.

The discussion in this chapter illustrates one important point: the idea of translating factors to policy recommendations was not meant to be an automatic process. Human judgement will always be necessary but with the help of factor analysis intuitive ideas are complemented by logical analysis and vice versa.

SECOND OBJECTIVE - OVERVIEW

As was mentioned earlier the objective here is to induce three successful transfers. It is clear that one will not be able to generalize on the basis of the three. However, some useful lessons may be learned and most important, these transfers, if successful, will alert the civilian sector to the existence of potential technologies.

From a methodological point of view it would have been desirable to carry out this stage after the completion of the first one. For reasons of time saving this is not being done, and the two objectives are aimed for simultaneously.

The three engineers in charge of the three transfers may be termed "matchmakers" or "experimental entrepreneurs".

Their responsibilities include:

1. Search for potential technologies at donor's organizations.
2. Analysis of which technologies found in (1) should be transferred.
3. Search for recipients for the technologies chosen in (2).

In spite of the fact that this type of activity is not common for the CTIF, the three "matchmakers" seem to enjoy the challenge. It should be emphasized that they are not expected to use any "useful connections" which may be available to them as staff members of CTIF.

Each one of the three engineers keeps a "diary" in which all his activities and findings are listed for future evaluation.

SECOND OBJECTIVE - WORK CARRIED OUT

A questionnaire was developed for "potential technologies" and various donors were interviewed. A list of more than 60 "potential technologies" was compiled in a relatively short time. The search is continuing to date. In the meantime authorization was sought to present the findings to civilian recipients: (1) security authorization (2) commercial authorization.

It may be mentioned that security problems could be overcome since CTIF deals with technologies and not with their usage in some defence context. The technologies range from corrosion resistant paints to a quick push type lock.*

Once proper authorizations were received, the list of technologies was sent to 500 enterprises employing 30 workers or more. The answers will point out potential recipients and will also be analyzed statistically. (Similar statistical studies of respondees were carried out by NASA).

The team responsible for the transfers does not wait for the responses to the list, and are using consultants, experts in the various industrial branches. The consultants offer advice concerning: (1) The value of each technology and (2) who in the industry may be interested in what.

At this stage of the project the matchmakers cannot report of any "matches".

Yet many lessons have already been learned, some of which are in accordance with the findings of the factor analysis. For example, the three engineers found that a great number of the potential "proper" technologies for transfer were in the minds of defence personnel. Their explanation was difficult, and was carried out through laborious verbal communication.

---

* The lock, (called Poperlock) disengages simply by turning.

EPILOGUE

Even if all the problems raised in this paper were to be solved, others are still waiting for answers. For example: How to make the civilian sector, which is decentralized, adopt technology transfer policies? Would threat/opportunity analysis based on sound forecasting help in this mission? What are the characteristics of an Israeli entrepreneur? Could he be identified from among defence researchers? Should he be given incentives to transfer defence technology to the civilian sector?

Above it all hangs the most difficult question: What are the cause-effect relationships in technology transfer phenomena.

With a lot of hard work (and a little bit of luck) the result of the CTIF program may be of value to both Government and private enterprises. The Israeli Government would be given a set of alternative policies to enhance the technology transfer process, as well as a list of potential technologies. Private enterprises will be able to answer some 15 questions concerning technology transfer projects which they are about to launch, and the CTIF will forecast their probability of success.

APPENDIX

1. Layton, Christopher, *Ten Innovations*, George Allen & Unwin Ltd., London, 1972.
2. Gruber, William H., Marquis, Donald G., *Factors in the Transfer of Technology*, The M.I.T. Press, Cambridge, Mass. 1969.
3. Hall, G.R. and Johnson, R.E., "Transfers of United States Aerospace Technology to Japan", *The Technology Factor in International Trade (Report of national bureau of economic research)*, Columbia University Press, 1970.
4. Rosenbloom, Richard S., *Technology Transfer- Process and Policy*, National Planning Association, Special Report No. 62, July 1965.
5. *Policy Planning for Technology Transfer- A Report of the Subcommittee on Science and Technology*, 90th Congress, 1st Session, April 6, 1967, Washington: U.S. Government Printing Office.
6. *Space Programs Benefits*, Hearings before the committee on Aeronautical and Space Sciences, April 6, 1970, 91st Congress, 2nd Session, Washington: U.S. Government Printing Office.
7. OECD, *The Conditions for Success in Technological Innovation*, OECD, Paris, 1971.
8. NASA, *Transforming and using Space-Research Knowledge*, NASA-UCLA Symposium and Workshop, Los Angeles, June 2,1964.
9. Leonard, Robert C., *Evaluation of the Effectiveness of a Conference on Problems of Technological Transfer and Economic Growth*, College of Engineering, University of Arizona, Tucson, Arizona, 1968.
10. Sovel, M.T.,*A Descriptive History of the office of Technology Utilization of the National Aeronautics and Space Administration*, University of Denver, May 1969.
11. *The Environment and the Action in Technology Transfer 1970-1980*, Report of a Conference sponsored by Denver Research Institute, Denver Colorado, Aspen, Colorado, September 26-28, 1969.
12. Lesher, Richard L. and Howick, George J., *Assessing Technology Transfer*, National Aeronautics and Space Administration, NASA SP 5067, Office of Technology Utilization, 1966.
13. Bar-Zakay, Samuel N., "Policymaking and Technology Transfer: The need for National Thinking Laboratories," The Rand Corporation, P-4497, December 1970. *Policy Sciences*, 2 No. 3, September 1971.
14. Bar-Zakay, Samuel N., "Technology Transfer Model," The Rand Corporation P-4509, November 1970. *Technological Forecasting and Social Change*, 2, Nos. 3/4, 1971.

15. *Program Evaluation of the office of State Technical Services*, Report of Department of Commerce, the office of Program Planning, Arthur D. Little, Inc., Cambridge Mass. October, 1969.
16. Melman, Seymour, Five-column series entitled, *Conversion of Industry from a Military to Civilian Economy*, Praeger Publishers, 1970.
17. Denver Research Institute, *Project for the Analysis of Technology Transfer (Patt)*, Six Reports prepared by Denver Res. Inst. for Nasa, 1968.
18. *Selected Bibliography on Technology Transfer*, Denver Research Inst. University of Denver, 1970.

REFERENCES

[1]For a basic description of factor analysis see Rummel, R.J., "Understanding Factor Analysis," *Journal of Conflict Resolution*, Volume XI, No. 4, 1968. For more details see Harman, H. H., *Modern Factor Analysis*, University Press, 1967.

[2]Bar-Zakay, S. N., "Technology Transfer Model," The Rand Corporation, P-4509, November 1970, *Technological Forecasting and Social Change*, Nos. 3/4, 1971.

THE TRANSFER OF SYSTEMS TECHNOLOGY FROM MILITARY TO INDUSTRIAL APPLICATIONS

LeRoy E. Pagano, Ph. D.

Newark College of Engineering

INTRODUCTION

This paper briefly presents the research of Henry Ursic and LeRoy Pagano in transferring Systems Technology from military security operations to organizational security applications.[1] From a philosophic overview, the research plan proceeded from Baconian definition through Hegelian dialectic to Newtonian synthesis.

THE DEFINITION

Security is a primary personal and an ultimate societal concern. Thus, the United States Constitution in its Fourth Amendment of the Bill of Rights guarantees the individual's rights to security. Social psychologists rate security as a fundamental human requirement. In Maslow's Hierarchy of Needs, safety and security are ranked as a basic physiological need preceding and preempting higher social and psychic needs. Despite the high priority and concern for security, one searches in vain for no comprehensive definition can be found. Even the fundamental law of the United States Constitution merely sets forth violations and prohibitions, but it does not present an inclusive or complete definition of the term nor a description of the condition.

A contemporary view of security holds that its concerns cover the entire spectrum of conditions and issues that involve individual, property, and group integrity ranging from international affairs to personal safety. The social responsibility for this state of integrity which we term security, appears to have three levels:

1. international affairs,
2. law enforcement, and
3. organizational security.

These levels are most directly and simply represented by:

1. the military,
2. the public police, and
3. private security guards.

While all three groups are concerned with similar security concepts, issues, problems, functions, and management; it is clear that soldiering, policing, and guarding have different parameters and constraints.

THE DIALETIC

Since the advent of the technological age, security has become increasingly a technological matter, and all three security levels have experienced phenomenal growth in scope, content, and severity. From the point of view of technological development and innovation, it is immediately recognized that the area of international security commands and controls the vast amount of work done in research and development and has been the determinant for the heavy emphasis on the physical sciences and technological development. Apparently, no ready transfer of this technology occurs across the boundaries of the three security levels. It could be argued that much technology that is developed for military security is inappropriate for public police and private guard use and that the general public dislikes linking these various security levels and their applications. Examinations by Kahn and others of this argument have revealed that the "unthinkable" technological solutions can only be accepted for use across national boundaries and that extensive public funding cannot be obtained or justified for personal and property security innovations for police and guard programs in the intranational sphere.[2]

Why should national governments be so predisposed to invention and innovation in both the product and process areas for international security, and at the same time be so resistant to change (especially change involving the implementation of conceptual innovations) in the operations and management of the public police and private guards? This attitude is very clearly demonstrated in the histories and commentaries on security technology. For example, compare the glowing descriptions of new military weaponry (such as the atomic bomb) and military management innovations (such as PERT) with the critical descriptions of the introduction and use of police hardware innovations (such as electronic surveillance equipment) and the municipal and corporate management conceptual innovations (such as quantitative productivity measures). The last group of innovations which includes public administration processes are subject to classification under the label of "reforms" which often connote a moralistic pejorative overtone. Process or conceptual innovation in the areas of law enforcement and organizational security are almost universally described in normative terms that appear to imply that an improvement in human nature, or at least of human values, must occur before new methodologies can gain public acceptance and be implemented. At the same time, no such justification appears necessary for product or process innovation for military security.

This widespread social attitude poses an almost insurmountable barrier in the form of mental blocks or taboos to the process of technology transfer across the levels of security management. This cultural gap introduces a serious technological lag in the areas of law enforcement and private security. It has become necessary for researchers to avoid this barrier because of the urgent need for new technology in the form of product and process innovation in public police and private guard operations.

Today the public views crime as the most serious of all domestic problems. According to the FBI, crime in the United States is increasing at a rate of roughly ten percent annually.[3] Although the accuracy of crime statistics is challenged by critics who cite improved methods of detection, more extensive and complete reporting and record-keeping, racism, and political "empire building" of police authorities as sources of exaggeration, security concerns increase. What is unquestionable is that Americans at all socio-economic levels seem to be driven to the brink of hysteria with their fear of crime and that technological innovations in both product and process form are an obvious need to improve police and guard operations.

THE SYNTHESIS

Aware of this need and of the structural and institutional bias against conceptual innovation, Ursic and Pagano considered what might be done to expedite technology transfer between the three levels of security. They examined the innovations in the most technologically advanced level--military security--to determine the most logical candidate for transfer.

Of the many managerial innovations in military security operations since World War II, the most applicable technology appeared to be the discovery, development, and diffusion of a sophisticated body of conceptual knowledge that has come to be called Systems Technology.[4] Military organizations have evolved a comprehensive and inclusive multidisciplinary conceptual framework for the applications of Systems Technology to the planning, organizing, direction, and control of military security.

Just as Military Science developed two main divisions comprising Strategy and Tactics, Systems Technology has developed two main branches consisting of Systems Analysis and Operations Research. These two branches of Systems Technology are analogous in scope and outlook to the two Military Science divisions. For military security programs, this technology has developed a remarkable precision and definition since the early 1960's.

Since that time, Systems Technology has begun to diffuse throughout government and industry; but because of the barriers to technology transfer in the security field, very little serious work has been done in public law enforcement and none in the realm of private security. For this reason, and responding to the exigency, the researchers chose the area of industrial security to attempt a conceptual technology transfer. Their work adapts Systems Technology developed for military security systems to the management of organizational security systems.

In endeavoring to transfer Systems Technology to private security management and operations, emphasis was placed upon viewing the security program as a total (considering the ogranizational entity as a whole) and open (to the larger society) system. This approach made it possible to define organizational security in an all-inclusive manner. Organizational security is a dynamic and open psycho-social and economic-technical system designed to create a security awareness that fosters mutually acceptable patterns of attitudes, behavior, and relationships within an organization's environments. The systemic objective of security programs was set as a service function whose purpose is to develop a social consciousness for the protection of life and property within the organization as a collective responsibility complementing law enforcement.

General Systems Theory, Systems Analysis, and Operations Research techniques were modified and adapted to the functions of private security. The researchers did not attempt to achieve the ultimate formulation of the extremely complex set of variables, which encompasses the human, technical, and economic organizational elements. Instead, they utilized the transferred technology to delineate the psycho-social and economic-technical subsystems and thus present a pragmatic connection between knowledge from the social and physical sciences and the functional demands of daily security operations. The researchers sought to equip the professional practitioners in the organizational security field with a systemic framework and operational techniques transferred from military Systems Technology.

REFERENCES

1 Ursic, Henry and LeRoy Pagano, *Security Management Systems*, Springfield, Illinois: Charles C. Thomas, 1973.

2 Kahn, Herman and Anthony J. Weiner, *The Year 2000, A Framework For Speculation on The Next Thirty-Three Years*, New York: Macmillan, 1967.

3 Gray III, Patrick L., *Crime in the United States*, Washington, D.C.: U.S. Government Printing Office, 1972.

4 Kahn H., and others, *Techniques of Systems Analysis*, Santa Monica, California: The Rand Corporation, 1957.

A METHODOLOGY USED TO TRANSFER FUTURE TECHNOLOGY FROM THE MILITARY TO THE CIVILIAN ENVIRONMENT

Marvin J. Cetron

Forecasting International, Limited

Harold F. Davidson

U. S. Department of the Army

Military research and development in the United States bears a unique responsibility. While its original functions may have been strictly military, history has broadened these functions inexorably. Results of military research have crept into widespread use in civilian life, commerce, medicine, and technology. And though many of these transfers of knowledge originally came as pure windfalls to civilian life, the steady flow of military research and development has now become a major tributary to the mainstream of the American economy. Furthermore, the interdependence of civilian and military research, while politically delicate and socially risky, represents a logical extension of the fundamental purpose of the American military forces: to enhance the security of the American people, their culture and their property.

There are many forces, other than those of military aggression by foreign enemies, which can threaten the security of a nation. They include natural catastrophes, famine and disease, environmental depletion and contamination, economic depression, cultural dissolution, and social malaise. Research designed to bolster the national security or well-being in one of these areas generally has widespread applicability in a number of others. Since the "acceleration" of life and its incumbent problems has made lead time an increasingly critical factor in research, and since research funding is becoming severely constrained in some sectors, it is to the advantage of both civilian and military institutions to make research efforts as efficient as possible in producing results which will be useful to the nation as a whole.

Unfortunately, the efficiency of Army's research and development in achieving both its and societal goods cannot be assessed simply by measuring the amount of effort being expended in different projects having the same general goals; it is also necessary to consider the effects of a project *beyond* its intended purpose. For example, the value of research conducted for the purpose of eliminating venereal disease among servicemen cannot be measured by its effect on servicemen alone. Furthermore, the effects of a project may be indirect as well as direct. The development of impact-absorbing devices for aircraft landing gear, for example, has made possible the production of vastly improved bumpers for surface vehicles.

While the transfer of benefit between organizations is often inhibited by secrecy (whether it be corporate or governmental) it can be argued that for most research efforts this inhibition is only temporary. Within a particular sphere of research, however, the transferable benefit spreads quickly throughout the sphere. In this respect the costs and benefits of R&D closely follow the model of the general economy: just as input-output analysis has shown that a significant change in the purchasing patterns of one sector causes a slight shifting in the purchasing patterns of all other sectors, two similar techniques -- cross impact analysis and cross support analysis -- have shown that a significant change in the technology of one sector causes a shifting in the technology of many, if not all, of the other sectors which are exposed to it. A good example is the development of electronic alternatives to first class mail, such as remote facsimile services. By cutting into the Postal Service's most profitable market, these technologies force changes in the price structure for postage. This in turn forces changes in the market structure for magazines, which are heavily dependent on postal rates. Changes in the magazine market effect changes in national advertising patterns, with heavier reliance on television. The increased demand for commercial TV time raises advertising rates, thereby increasing the demand for reliable information about advertising psychology. The labor markets for psychologists, actors, and film makers are significantly restructured -- as are the markets for technologies employed in these professions. These changes in turn create similar, albeit smaller, changes in other industries. Ultimately, hundreds of sectors can be affected by a single technological development.

For planners in government and industry, the question is not only whether a new development will have impacts in many areas other than the one for which it is planned, but what those impacts will be -- and how they can be measured. In order to improve the coherence and efficiency of a research program -- i.e., to restructure the program so that its total impact on relevant goals is maximized -- it is necessary to apply quantified values to both the direct *and indirect* impacts of its component projects.

Several years ago the U. S. Army undertook a search for a method that would enable it to measure the value of various Army R&D projects not only in terms of their success in achieving immediate goals, but also in terms of their contributions to other Army goals and to the goals of American society in general. It was clear that at least some of the Army's vast R&D effort was mired in inefficiency with respect to long-range objectives, even where its achievement of particular shorter time frame project goals was proliferating. In 1972 the Army Research Office retained Forecasting International, Ltd., an Arlington, Virginia firm which specializes in technology assessment and forecasting, to develop a technique for assessing the Army research effort.

F. I. based its approach to the problem on two premises which underlie most of its work: first, that government and business can no longer afford to wait for technological developments and changes in their operating environment to occur before reacting to them; and second, that by evaluating their relationship to the total environment instead of becoming preoccupied with immediate goals, government and business leaders will find that they can act -- and not just react -- in determining the country's future. Accordingly, F. I. brought two complementary efforts to bear on the Army problem. One part of the study was a forecasting effort -- an attempt to identify future trends over a wide range of technological, social and economic phenomena, as well as to assign probable dates of occurrence to hundreds of specific events which are likely to occur by the year 2000.

A second part of the study was the construction of an analytic method by which the Army could assess its own R&D programs by measuring the extent to which each project (or area of research, etc.) contributes both to the achievement of other Army objectives thans its own, and to the achievement of societal goals independent of those set by the Army. For this methodology, F. I. made one of the most ambitious uses of cross support analysis ever attempted. (The technique had been used previously as a project selection and resource allocation technique, but never before for such a complex and multi-faceted subject as the Army's research and development program.) Since the F. I. study entailed not only constructing but also demonstrating the new methodology, the Army was presented with some clearly quantified perspectives -- described below -- on the efficacy of its R&D effort and on the relationship between its own objectives and those of the nation.

According to forecaster Christine Ralph, cross support analysis can be simply defined as "a method for determining the degree to which one variable of a set contributes to another variable of the same set. Examples of this can be the degree to which the achievement of one goal supports the achievement of another, or the support one technology program has for another." The analysis thus takes the form of a matrix in which the values entered in the individual cells represent subjective judgements of the support one variable gives to another (e.g., the degree to which achievement of one objective within a given future time frame will contribute to the achievement of another objective within the same time frame.) Judgments, which are made by groups of experts in the field, are converted to numerical values on an adjectival ratio scale. The degree of support, of course, can be either positive or negative. By adding the contributions a single variable makes to all other variables, a "weighting" factor is derived which is a more accurate measure of value than the assumed value with which the analysis began.

Because of the complexity of the Army assessment problem the weighting of various programs does not in itself provide the information required for resource allocation and project selection decisions. The weighted variables merely become inputs to a new matrix, where they are lined up against a new set of variables in a second round of cross-support analysis (this is a departure from Ralph's description of the process as involving a single set of variables). In this way the method eventually takes into account, as quantified inputs, all of the many factors considered necessary to make optimally efficient and socially responsible planning decisions.

In the F. I. study, six categories of inputs were processed. These included:

(1) Army Strategic Operations Plan Objectives (examples: "surface mobility," "air defense," etc.)

(2) Army Technology Areas (examples: "electronics", "biological and medical sciences", etc.)

(3) Technology Forecasts (examples: "degradable plastic by 1981";"protein from petroleum by 1985", etc.)

(4) Societal Trends (examples: "trend of R&D funds as a % of GNP"; "trend of new housing starts", etc.)

(5) Areas of National Concern (examples: "national defense";"urban development")

(6) Areas of National Satisfaction, or Societal Indicators (examples: "equality of opportunity"; "quality of the cultural environment", etc.)

Although it is beyond the scope of this brief summary to define these inputs precisely, two general observations are in order. First, the different inputs are characterized as much by their sources as by their content. These sources include the Army, the Census Bureau, the National Planning Association, and a number of independent research organizations. Second, the content of the inputs touches on virtually every dimension of American life. No system of values -- no matter how remote from Army policy -- is excluded from consideration by the structure of the method. Of course, no claim can be made that any single variable receives consideration in fair proportion to its importance to the nation. Ultimately, the results will reflect the biases of the method's users -- those who make the judgments on which the weightings are based at the cell level. On the other hand, no planning technique has yet been developed that does not rely on the use of subjective (albeit expert) judgment as an essential input. The trick, of course, is to structure this input in such a way as to minimize human tendencies to overlook or distort important considerations.

The six types of input to the F. I. study were processed through a total of seven cross-support analyses. In the first matrix, Army Objectives were weighed in terms of their support for each other. In the second, Army technology areas were also weighed in terms of their support for each other. Then, in the third matrix, the weighted technology areas were weighed in terms of their support for the weighted objectives. The same general procedure was followed in matrices 4, 5, and 6, in the weighting of areas of national concern and areas of national satisfaction. The process is structured much like an elimination tournament (although sets of values are subsumed rather than eliminated), culminating in a "final" matrix in which Army objectives are weighted according to their support for various areas of national satisfaction (matrix 7). The flow of analysis is summarized, in somewhat oversimplified form, by the following diagram:

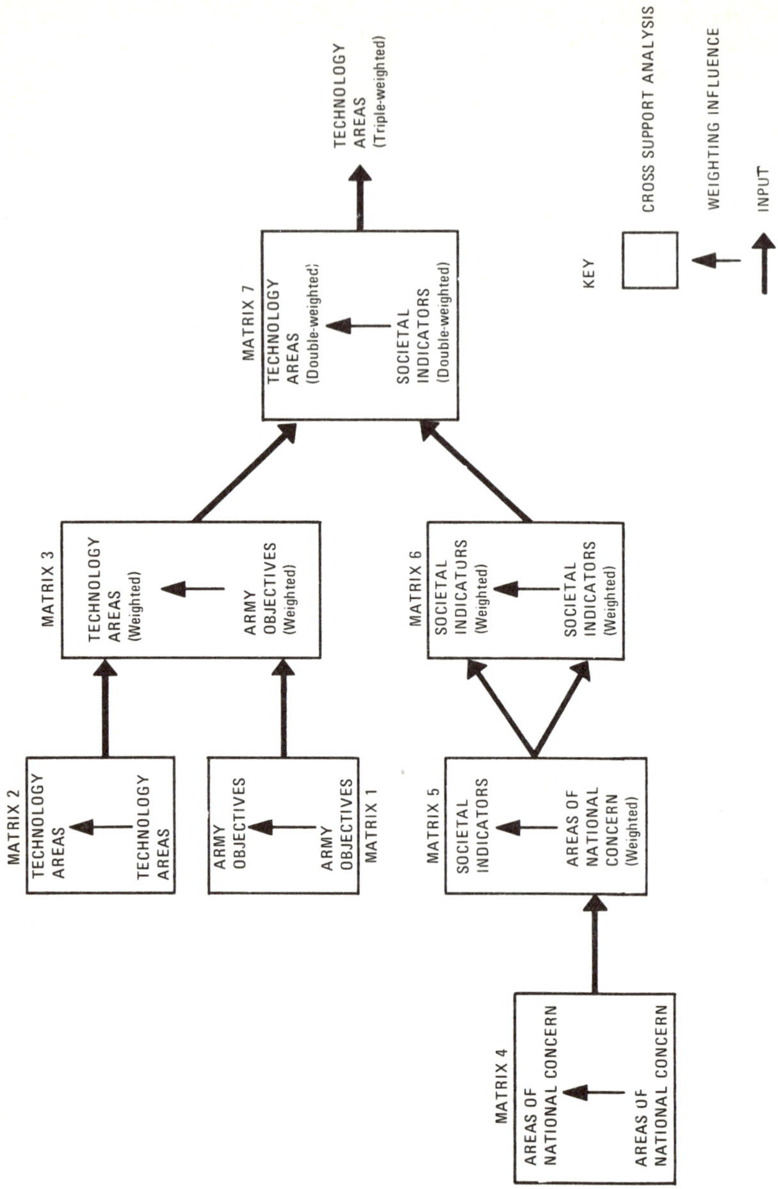
MATRIX 1
ARMY OBJECTIVES
ARMY OBJECTIVES
MATRIX 2
TECHNOLOGY AREAS
TECHNOLOGY AREAS
MATRIX 3
TECHNOLOGY AREAS (Weighted)
ARMY OBJECTIVES (Weighted)
MATRIX 4
AREAS OF NATIONAL CONCERN
AREAS OF NATIONAL CONCERN
MATRIX 5
SOCIETAL INDICATORS
AREAS OF NATIONAL CONCERN (Weighted)
MATRIX 6
SOCIETAL INDICATURS (Weighted)
SOCIETAL INDICATORS (Weighted)
MATRIX 7
TECHNOLOGY AREAS (Double-weighted)
SOCIETAL INDICATORS (Double-weighted)
TECHNOLOGY AREAS (Triple-weighted)
KEY
CROSS SUPPORT ANALYSIS
WEIGHTING INFLUENCE
INPUT

The net effect of this progressive refinement is to give the Army Research Office a clearer basis for evaluating various R&D programs in light of their contributions to the larger objectives of both the Army and the nation. The difference between the values attributed to various R&D efforts which have been weighted by the full range of inputs, and those which have been weighted only according to their contribution to Army R&D objectives, is graphically illustrated by the contrast between the two sets of results obtained in the third and seventh matrices respectively.

Matrix #3 weights Army technology areas in terms of their relevance to Army R&D Objectives only. When the contribution of each area to all Objectives is added up, the 10 most heavily weighted areas are as follows (weights are normalized so that the total weight of the 20 Objectives is 100):

Rank	Technology Area	Normalized Weight
1	Air Mobility	10.85
2	Environmental	9.03
3	Materials, Mechanics, and Structures	8.67
4	Munitions and Weapons	7.66
5	Electronics and Energy Conversion	7.25
6	Biological and Medical Sciences	6.07
7	Combat Support	5.82
8	Surface Mobility	5.48
9	Nuclear	5.11
10	Missile	5.02

Matrix #7, the final analysis in which both Army and "societal" variables have been considered, produces a somewhat different list of priorities:

Rank	Technology Area	Normalized Weight
1	Materials, Mechanics and Structures	12.08
2	Air Mobility	10.02
3	Surface Mobility	7.66
4	Environmental	6.91
5	Combat Support	6.57
6	Human Factors	6.06
7	Biological and Medical Science	5.99
8	Electronics and Energy Conversion	5.93
9	Chemical Systems	5.72
10	Detection and Location	4.86

There are few surprises in the shifts which take place. Areas which are weighted less heavily are as follows:

Technology Area	Change in status (normalized final weight)	Change in rank
Missile	-4.40	10th to 19th
Munitions and Weapons	-4.01	4th to 14th
Nuclear	-2.21	9th to 16th

Areas which are weighted *more* heavily in the final analysis are as follows:

Technology Area	Change in status (normalized final weight)	Change in rank
Materials, Mechanics and Structures	plus 3.21	3rd to 1st
Chemical Systems	plus 2.63	14th to 9th
Communications	plus 2.49	19th to 11th

These shifts show a general de-emphasizing of strictly military technology areas, with a corresponding increase in the weighting of those areas for which the benefits of research and development are more readily transferable to a wide range of fields. Materials, mechanics, and structures research, of course, finds an endless list of uses in the construction (both public and private, military and civilian) of housing, roads, bridges, vehicles, and the like. Chemical systems and communications developments are likely to have significant ramifications in every major industry, as well as in every branch of the armed forces.

One somewhat enigmatic result is the loss of weight, from 9.03 (2nd) to 6.91 (4th), of the Environmental Research area. A partial explanation may be that of the two items which were elevated above it -- Materials, Mechanics, and Structures, and Surface Mobility -- simply have many more uses for new technological developments in the judgement of those who conducted the study. The implication is not that societal considerations make environmental research less important but that environmental progress depends more heavily on economic priorities and public attitudes than on technological solutions; in contrast an area like Surface Mobility depends for its progress on heavy investments in the development of new hardware.

A second interpretation is that environmental research may have indirect impacts -- on Army objectives -- which have been underestimated in the public sector (i.e., the shift of the Environmental area should not be surprising). Thus, meteorological and air chemistry studies, weather satellite, technology and associated techniques of surveillance, detection, and monitoring all have direct relevance to Army Objectives -- and all are richly enhanced by environmental R&D.

The results of the Army/F. I. study (of which only a portion are reported here) still leave questions as to the meaning of the quantified values obtained. As with any system by which human judgments are transformed into hard numbers, there are unanswered questions of reliability and reproducibility. In a field as overwhelmingly complex as that of the Army's R&D program, these questions are difficult to answer definitively. But a close reading of the Forecasting International methodology strongly suggests that the Army now has the means at least to (1) assist in preventing the kind of neglect of societal considerations many Americans have long feared, and (2) begin to achieve a degree of coherence and efficiency, through comprehensive planning of its R&D programs, that has never before been possible.

TRANSFERRING EXISTING TECHNOLOGY INTO THE PUBLIC SECTOR: A CASE STUDY

Ernest H. Short

National Center for State Courts

Introduction

The Systems and Technology Division of the National Center for State Courts was set up to help the state courts in the United States apply modern technology and management systems to their operations. Based on standard definitions of technology transfer, the Systems and Technology division may therefore be viewed as an agent of technology transfer into and throughout a component of the public sector, the courts. It is hoped that documentation of our experiences, including observations on the environment in which our work takes place, the problems which have been and are being encountered, and the program which we have launched to overcome these problems, will be useful to groups or individuals attempting to understand or actually effect the transfer of technology in general and the transfer of technology into the public sector in particular.

For the purposes of this paper, the following defintions advanced by Gruber and Marquis[1] will apply: "technology" is defined as the means or capacity to perform a particular activity; the "transfer of technology" will be taken to mean the utilization of an existing technique in an instance where it has not previously been used - either the acceptance by a user of a practice common elsewhere ("adoption") or an application of a given technique in a new way ("innovation"); the "diffusion of technology" is defined as the spread of adoptions of a particular practice; "invention" consists of bringing technical elements together in a new way such that a new technology results; and "use" connotes utilization to satisfy a demand or need for a product or service. Thus, innovation occurs each time a technology is first employed for a particular kind of use, and diffusion occurs when this new way of using a technology spreads to additional users. Of course, innovation must precede diffusion in the transfer of technology; a given technique must be used <u>once</u> in a particular way before it can achieve widespread use in satisfying a general demand.

In transferring existing technology into the courts, the National Center for State Courts is necessarily concerned with both innovations and the subsequent adoption and diffusion of innovations. But in the courts, as in general, "social innovation in the public sector must depend upon private as well as public resources."[2] As the U. S. Department of Commerce Panel on Invention and Innovation has pointed out: "There is a mutual interdependence between social and private innovation."[3]

To effect innovations in the technology-deficient courts, appropriate technologies, existing in the private sector, must be identified, adapted where necessary, and put to use in the courts. The interactive character of the transfer process is even more pronounced in situations where the flow from technology to use is followed by feedback from users in the public sector to technology vendors (as practical experience exposes the inadequacies of the technology in use). This feedback, hopefully, results in a second flow from technology to use (constituting a new and more useful innovation). This interaction is illustrated below:

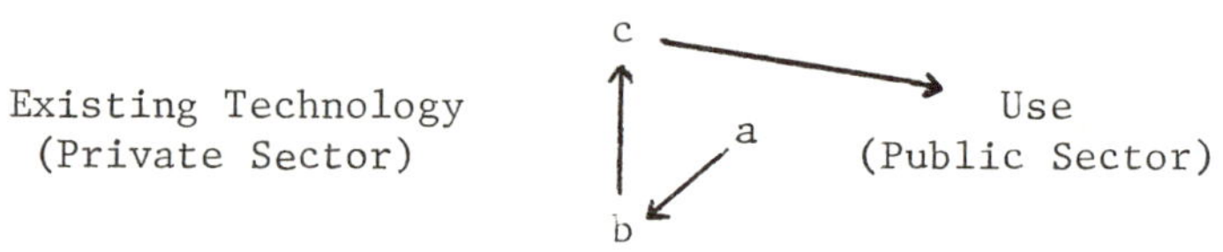

a. Initial Innovation
b. Demand for Modification of Technology in Use
c. More Useful Innovation Developed from Existing Technology.

The transfer of technology from the private to the public sector involves diffusion as well as innovation. This diffusion occurs when technology in private sector use is transferred in its existing from to the public sector for a similar use. In addition, diffusion from court to court within the public sector may be viewed as the final step in the transfer of technology from the private to the public sector.

Both the innovation and the diffusion processes described above can be represented by an expanded version of the Communications Model of Technology Transfer presented in the Introduction to <u>Factors in the Transfer of Technology</u>:[4]

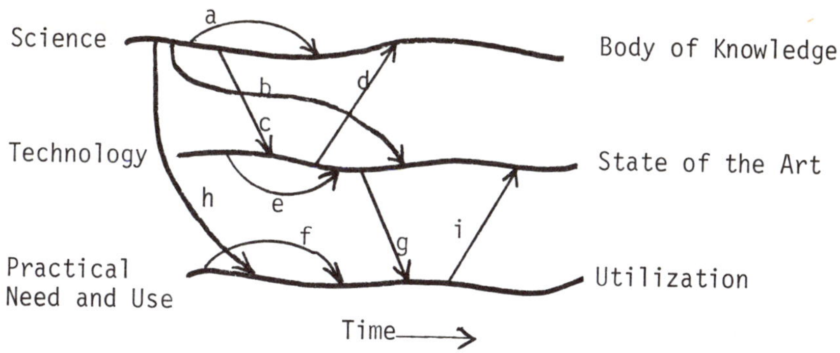

(a) Science to science; (b) Science to technology (slow); (c) Science to technology (fast gap filling); (d) Technology to science (e.g., instruments); (e) Technology to technology; (f) Use to use (diffusion) - within the context of the courts, this would include private to public and public to public diffusion; (g) Technology to use - within the context of this paper, this represents the flow of technology from the private to the public sector; (h) Science to use*; (i) Feedback from users to technology vendors.

Within the broad conceptual framework of technology transfer discussed above, this paper will analyze the environment in which technology transfer in the courts takes place and will outline the approach adopted by the Systems and Technology Division of the State Courts Center to overcome the obstacles to the transfer process. In so doing, the empirical observations will be matched to several models of technology transfer found in the literature. Special attention will be paid to Gruber and Marquis' model of the utilization of existing technology. According to Gruber and Marquis, the "process of technical advance vis-a-vis the utilization of the inventory of available technical knowledge is a function of (1) demand recognition (2) technical feasibility recognition and (3) the ability/willingness to achieve utilization provided by recognition and subsequent development of a solution."[5] Here, the first factor posited in the transfer of technology is the recognition of demands that are not being satisfied by the technology currently in use. Technical feasibility recognition, the second factor posited, is defined as the discovery that available technical information (with perceived modifications) would provide a new technical capability. According to this model,

---

* The representation of this flow has been modified from that presented in Factors in the Transfer of Technology.

"innovation is most likely to be initiated when recognition of a demand and a feasible technical solution for the specified demand are fused and a satisfactory technical response to a user requirement becomes available."[6] The third factor, ability/willingness to take action, is often discussed in the literature as "resistance to innovation."

Gruber and Marquis delineate six key determinants of their three factors of technology transfer. These six determinants, whose values vary according to the level of organization involved in the technology transfer, are: 1) training and experience; 2) individual personality characteristics; 3) communication patterns; 4) organizational effects; 5) mission orientation; and 6) motivation.[7] Shapero, in an article discussing innovations and their adoption, outlines a set of five key variables which, in spite of some overlap with Gruber and Marquis' determinants, provide a slightly different perspective on the transfer process. Shapero's five variables are: 1) innovation characteristics; 2) adopter characteristics; 3) sources of information; 4) information channels; and 5) contextual variables.[8] Other models and theories in the literature will be invoked when they are helpful in organizing and understanding the empirical observations advanced in this paper.

II. Environment

Perhaps the most salient feature about the state court systems is their diversity. Attitudes toward innovation vary widely across and within court systems. Motivational, experiential, organizational, informational, and contextual determinants of technology transfer carry different relative weightings according to the individual state or jurisdiction involved. Even more fundamentally, different court systems have different needs. In addition, local statutory constraints in some cases constitute stubborn obstacles to change in traditional court procedures. The result is that court jurisdictions differ considerably from each other in their propensities to innovate and/or adopt the innovations of others. Actually, this variance is not surprising; it has frequently been noted in the literature that rates of technology transfer vary from region to region, and individual to individual.[9]

An important factor underlying the variation in rates of innovation and adoption is the wide range of administrative structures found in the state courts. They fall on a continum between centralized and decentralized organizations. At one extreme are "unified" court systems in which all courts in a state are under one central administration. At the other extreme are court systems in which each court is a separate administrative entity, with the judge of the court as the primary administrative authority. State court systems may also be characterized by the number of "tiers" in their organizational hierarchy. There may be as few as two tiers:

1) a single-level trial court of general jurisdiction; 2) a court of last resort (usually called the Supreme Court). On the other hand, there may be as many as six tiers: 1) "inferior" trial courts handling less serious civil or criminal matters; 2) trial courts of general jurisdiction having authority to try all civil and criminal cases; 3) trial courts of intermediate jurisdiction usually organized on a county basis and possibly overlapping with "inferior" and/or general jurisdiction; 4) specialized trial courts in areas of large population; 5) intermediate appellate courts between the trial and supreme courts or as a division of the trial court; and 6) a court of last resort (usually called the Supreme Court) which reviews the actions of the lower judicial tribunals of the state and may also exercise superintending powers over all the courts of the state.

Thus, technology transfer in the courts takes place in a heterogeneous environment. With this in mind, the courts environment will now be analyzed in terms of the key determinants identified by Gruber and Marquis and also in terms of the variables specified by Shapero. The environment in the relevant portions of the private sector, as well as the environment in the courts, will be discussed since much of the relevant technology must be transferred out of the private sector before it can be introduced into and diffused throughout the courts.

As Burns has observed: "The mechanism of technological transfer is one of agents, not agencies; of the movement of people between establishments, rather than the routing of information through (formal) communications systems."[10] This is certainly true in the courts, where innovations are adopted only if court administrators, judges, and lawyers are favorably disposed toward them. Unfortunately, these people are trained primarily in the law and often do not understand technology. Many judges, lawyers, and administrators are not accustomed to turning to technology for solutions to their problems. Hence, while many of the personnel connected with the courts recognize the need for improvement of court operations (i.e., demand recognition is present), they often do not have the training and experience necessary to recognize or develop potential technological solutions to court problems.

For example, court personnel may recognize the need and demand for improved record-keeping, but they often do not recognize the potential solution offered by computers and even more seldom are they capable of adapting computer technology for their use. Those jurisdictions which have pioneered in the implementation of computer information systems have either had individuals with prior training in the development of data processing systems on their court staff or have brought in consultants with such training. This experience supports Gruber and Marquis' assertion that "it is more efficient to bring into an organization someone experienced in a given technology than to try to develop expert personnel in an organization."[11]

In any case, the training and experience situation in regard to successful technology transfer is improving to the extent that judicial and administrative training programs are beginning to include information and courses on technology.

Just as training and experience of judges, lawyers, and administrators limits their ability/willingness to recognize the potential of technology, the training and experience of technicians in the private sector limits their ability to recognize court situations in which technology could effectively be employed. The technical personnel usually recognize the capabilities of their technologies, but they may not understand court operations well enough to adapt existing technology so that it is suitable for court use. This has been a continuing problem in the development of suitable audio recording equipment for court reporting purposes. Until the equipment manufacturers are given definite standards and capability requirements, it does not seem likely that appropriate equipment will be marketed; the manufacturers do not seem to understand courtroom operations well enough to develop by themselves an adequate set of standards.

Shapero's adopter characteristics variable is very similar in scope to the training and experience determinant as applied to judges, lawyers, and administrators. Shapero states that younger individuals are more likely to adopt innovations than others in a potential adopter population.[12] This has not been convincingly demonstrated in the courts. Age seems to have little or no correlation to individuals judges', administrators', and lawyers willingness to innovate. For instance, the age distribution of the Center's Videotape Advisory Committee, composed of leading innovators in the application of video technology in the courts, seem to be representative of the age distribution of judges, lawyers, and administrators in general. However, Shapero's assertion that new <u>organizations</u> are more likely to adopt innovations than others in a potential adopter population has been borne out in the courts.[13] The relatively new court system of Alaska was the first to adopt electronic recording as a means of producing an official record of court proceedings. The Oregon District Courts, recently made courts of record,* are also planning to use electronic recording as a method of court reporting and, in addition, are planning to experiment with

---

* A "court of record", as that term is used in this paper, is a court wherein it is required by statute that all criminal proceedings <u>must</u> be recorded and that civil proceedings <u>may</u> be recorded.

a new method of court reporting that combines "voice writing"** with an audio record.

Individual personality characteristics also play an important role in technology transfer in the courts. Some innovations, such as entirely pre-taped civil trials, have been the direct result of the forceful leadership, imagination, and aggressiveness of one individual, in this case, a judge. On the other hand, much resistance to innovation has stemmed from the static conceptions of courtroom dignity held by some judges. For instance, a significant amount of opposition to audio recording seems to be based on judges' desires for personal court reporters, which lend added prestige to the position of the judge. On the part of the private sector, personality characteristics enter to the extent that technology vendors do not always exhibit imagination in perceiving potential markets for their products. This, of course, is also due to experiential, organizational, and informational factors.

Whether it is labelled as the communication pattern or broken down into sources of information and information channels, the quality of communications in the courts heavily influences the rate of technology transfer. Starting with the demand recognition phase, the communication pattern may be characterized as indirect, informal, and multi-channelled. Demand for change often originates with the "consumer"; the public outcry about the inefficiency of the courts has combined with the complaints of judges and lawyers about the deteriorating quality of justice to create a clearcut demand for change. But the demand is not formally registered or measured and it is therefore difficult to assess how much the public is willing to pay for improving court operations. In addition, communication between technology vendors and the courts is often nonexistent; thus, the private firms are not always aware of the courts' problems and the courts are often not aware of the existence of technologies which might help solve their problems.

Several of Shapero's generalizations about sources of information apply to the courts' environment. Shapero states, "Late adopters are more likely than early adopters to depend on informal sources

---

** Voice-writing is a method of reporting wherein the reporter records proceedings by dictating in court a taped audio recording from which either he or a typist can later transcribe. Microphones are used to pick up the actual trial proceedings on one or more tracks of audio tape, while the reporter simultaneously dictates the proceedings on a separate channel of the tape, employing stylized diction to give transcriber instructions, speaker identification, and punctuation.

of information and to be influenced by the number of previous adopters."[14] Courts almost always want to know if a particular innovation has been tried elsewhere; the more widely accepted an innovation has been, the more likely a particular court will choose to adopt it. Shapero goes on to say that the "social and professional status of an individual who is the source of information on an innovation is associated positively with its adoption."[15] Courts are impressed not only with the number of previous adopters, but also with the prestige and status of the adopters. Schon's observation about technical administrators applies equally well to judges and court administrators: "Often the technical administrator must base his judgements on the soundness of the man at least as much as on the promise of the development itself."[16] Within a given court system, the prestige or status of an internal source of information is partially a function of the court's organizational structure. In a highly centralized system, court administrators, presiding judges, and appellate judges often have much more authority, both formally and informally, than their counterparts in a decentralized system.

Information about citizens' attitudes and about existing technologies reaches judges and court administrators through many and various information channels. Citizens' attitudes may be inferred from informal communications, courtroom behavior, written complaints, or merely from intuition. Information about possible applications of technology comes through the mass media, through groups such as the American Bar Association and its committees, from documented studies on projects, and importantly, from lawyers, administrators, and other judges. Technology vendors also rely heavily on informal communications and the mass media for information about court problems.

Many of Shapero's generalizations about information channels are as applicable to the courts as his generalizations about sources of information.[17] He states that "hierarchical networks of communication, in which innovation information travels from top to bottom, impeded the diffusion of an innovation or slow the rate of diffusion." This aptly describes the situation in many multi-tiered courts. Shaper also asserts: "Informal networks of communication are associated with the adoption of innovation when empirical evidence regarding the innovation is contradictory, ambiguous, or lacking." This has certainly been the case in regard to the adoption of audio recording as a method of court reporting. And finally, Shapero states: "When the adoption of an innovation requires changes in traditional values, a person-to-person channel of communication between the supplier and the potential adopter is required to effect adoption." Since most innovations in the courts do require changes in traditional values, this statement is particularly relevant. The person-to-person channel of communication suggested here has in fact been successfully utilized by the Systems and Technology Division as a means of expediting the transfer of technology in the courts. This will be discussed in detail in the section entitled "Approach."

The organizational effects in both the public and the private sector provide one of the strongest sources of resistance to innovation in the courts. The courts often refuse to introduce an innovation unless its efficiency has been thoroughly demonstrated; the technology vendors refuse to develop and market a technology without reasonable guarantees that the technology will sell. This conflict has severely hindered the development of technologies, such as audio recording equipment for court reporting, that require special adaptation for the court environment. The risk-averse private firms are not willing to invest in adapting a product for such a limited market without assurance that the courts will purchase their product.

Gruber and Marquis' first and third hypotheses about organizational resistance to innovation seem particularly applicable to the courts. The first states that "once an activity has become established with a given pattern, the response of the power structure in an organization is to reject innovations that would alter the existing activities."[18] This kind of institutional inertia is often found in the courts, in the name of tradition and dignity. Tradition is a powerful and even valid force inasmuch as the law is necessarily a slowly evolving body of rules and procedures based on precedent. In order to be effective as a guideline for behavior, the law must be relatively stable. Schon's observations thus ring especially true when applied to the courts:

> Resistance to change is not only normal but in some ways even desirable. An organization totally devoid of resistance to change would fly apart at the seams. It _must_ be ambivalent about radical technical innovation. It _must_ both seek it out and resist it. Because of commitments to existing technology and to forms of social organization associated with it, management _must_ act against the eager acceptance of new technical ideas, even good ones. Otherwise, the technical organization would be perpetually and fruitlessly shifting its gears.[19]

Gruber and Marquis' third hypothesis holds "that organizations acting in a given way have a vested interest . . . in maintaining the status quo."[20] This can readily be seen in the courts. Some judges and clerks resist computer information systems since they would result in closer regulation of judge and clerk time. Court reporters strongly resist the introduction of audio recording systems since they perceive the systems as threats to their jobs. Lawyers resist any methods of court reporting, such as audio or audio-video recording, which might restrict their mobility in the courtroom. As Schon points out, this kind of opposition to change "springs from the normal human instinct to protect oneself and one's way of life." Schon goes on to say that "[novelty ... challenges accepted ways of doing things and long-established skills. It may throw a company (or a court), including top management, into areas where it feels inept and uninformed . . . . changes in technology tend to carry with them major changes in social organization, threatening established hierarchies, undermining the security of positions based on old products (here, the old mode of court operation)."[21]

The courts furnish a good example of what Schon characterizes as "a ground swell of covert resistance to change (coming) into conflict with official enthusiasm for it . . . . Underground resistance paves the way for disguised defenses against change. In screening new ideas, many large corporations (and courts) employ formal committees which are, in effect, buffers against new ideas."[22] The numerous committees set up across the country in investigate audio recording as a method of court reporting have come very close to forming such a buffer.

Shapero's innovation characteristics variable covers some of the same ground discussed in regard to organizational effects. His statements about innovation "compatibility" correspond closely to Gruber and Marquis' first and third hypotheses about organizational effects and thus describe the court environment accurately:

> The less an innovation requires changes in pre-existing socio-cultural values and behavioral patterns, the more likely is its rapid diffusion. The less an innovation requires change in existing facilities, equipment, and procedures, the more likely is its rapid diffusion.[23]

In addition, the innovation characteristics variable deals with an area, namely the efficiency of the innovation, not covered by the human factor determinants advanced by Gruber and Marquis. Shapero asserts:

> The technical efficiency of an innovation, and/or its superiority to existing technology, facilitates its diffusion. The more apparent the technical efficiency of an innovation, the more rapid is its adoption.[24]

In many court jurisdictions, technologies such as the computer-aided transcription of stenotype notes, selected computer information systems, and audio recording systems have not as yet demonstrated sufficient technical efficiency to overcome the human factor obstacles to innovation.

Few court systems have personnel specifically assigned to the task of seeking out and implementing innovations which could help satisfy demand for improved court operations. Court administrators perhaps have the largest degree of mission orientation among court personnel in that they are generally responsible for discovering demand situations to which the courts are best able to respond and for developing within the courts the commitment for translating demand recognition into organizational action.[25] However, only the most sophisticated court systems have adequate resources to devote to the task of linking available technical resources, financial considerations, and perceived capabilities to meet perceived court needs.

Consequently, the courts must rely on outside sources to supply them with mission orientation. These outside sources are part of what Shapero calls contextual variables. In the courts, the federal government is the primary supplier of advice and financial assistance for research, development, and implementation of innovations. The Law Enforcement Assistance Administration of the Department of Justice furnishes technical assistance to the courts; the National Institute of Law Enforcement and Criminal Justice, a division of LEAA, sponsors research and development on court programs; and the State Planning Agencies, set up to administer LEAA block grants to the states, often have courts specialists who are charged with the responsibility of helping courts develop programs to improve their operations. Shapero's general statement applies well to federal support for innovations in the courts: "Government financial assistance to potential adopters of innovation is conducive to the diffusion of innovations."[26] As will become apparent from the "Approach" section, federal government assistance in the form of project grants has played a leading role in assisting technology transfer in the courts.

At this time, the motivation determinant outlined by Gruber and Marquis often seems to slow technology transfer out of the private sector but expedite transfer into the courts. The courts often do not consiitute a large enough market, and consequently a large enough potential reward, to motivate private firms to take the risks involved in investment for adapting existing technology to court use. As a result, the courts have been more successful in obtaining technologies, such as computer information systems, which rely on equipment which is already marketed in suitable form than they have been in obtaining technologies, such as audio recording systems, which require modifications of existing equipments.

In contrast, the motivation for courts to implement innovations is relatively strong. With the congestion and delay, and the attendant air of crisis, currently existing in the courts, successful innovation produces highly visible results. In a sense, courts are also in competition with each other to produce the best product, i.e., the highest quality of justice possible. As alluded to earlier, if some courts adopt innovations that improve their operations, then other courts will be motivated to follow suit. But this kind of motivation is mitigated by the vast differences which exist between court systems. Often a court in one state is not convinced of the desirability of an innovation merely because courts in other states have found it to be useful. Frequently, the only way to convince a court system, and its budgetary sources, is to demonstrate the innovation in that particular court's environment. The variation in constitutional and statutory constraints on particular court systems further limits ability/willingness to adopt innovations demonstrated in other court systems.

III. Approach

To help introduce modern technology and management systems into the courts, the Systems and Technology division of the State Courts Center is pursuing a three-part program consisting of: 1) research and demonstration projects on promising technologies; 2) direct, short-term technical assistance in implementing technology; 3) dissemination of information on the use of technology to improve court operations. This program has been tailored to the technology transfer environment discussed in the preceding section of this paper. The research and demonstration projects investigate and develop innovations that individual court systems are usually unable to pursue because of financial and/or personnel limitations. As a part of the experimentation, these projects usually include on-site applications of the technology under examination, thus carrying out actual innovations and setting the stage for their diffusion. Short-term technical assistance is used mainly as a means of helping court systems evaluate and implement innovations already at a fairly advanced stage of development. Information dissemination is necessary in order to make courts aware of existing technologies in all stages of development.

Demonstration projects are extremely powerful tools in facilitating the translation of technology in the private sector into use in the courts. A demonstration project, especially one operating on a grant basis, allows courts to experiment with an innovation and evaluate its effects before adopting it. Institutional conservatism can be overcome when an innovation demonstrates dramatic potential for improvement of court operations. At the same time, the technology vendor is given the financial support that allows him to undertake any necessary adaptations of the technology without risk.

Consistent with the literature on technology transfer, pilot operations which provide a model of product innovation have been effective in introducing and diffusing innovations in the courts.[27] After introducing a technology in one court, via on-site demonstrations, this experience is documented and serves as an "in" to other court systems, expediting the diffusion of innovation from court to court. Thus the demonstration projects encourage transfer of technology to a first use in the courts and then on to repeated use in the courts, all in one process.

Systems and Technology demonstration projects bridge the communications and training gap between private firms and the courts. Field demonstrations are brought to those courts which are amenable to experimentations. Schon's concept of a "product champion" is followed to the extent that the projects are placed with judges and administrators who could be, or are, effective advocates of the new technology.[28] The project staff helps the technology vendors to adapt existing technology to court use and helps the courts to understand and properly utilize the new technology. This often involves the development of procedures for utilizing technology in accordance with constitutional and statutory constraints.

The Systems and Technology Division is currently engaged in two major demonstration projects:

1) the "Video Support in Criminal Courts Project" - is designed a) to explore those applications of video recording most likely to be effective in cutting congestion and improving the quality of justice in criminal courts, and b) to clarify the legal and procedural issues surrounding the use of video technology in criminal courts; and

2) the "Court Reporter Training Rroject" is testing a program for training court reporters in a new "voice-writing" technique of reporting.

Both projects involve transfer of existing technology to use in the public sector and are funded by the National Institute of Law Enforcement and Criminal Justice of LEAA.

The experiences on the Video Support and Reporter Training projects confirm the observations made about the court environment earlier. It has been found that informal, person-to-person communications are the most effective way of securing locations for field demonstrations. Although motivation for change seems to be rather high throughout the state courts, it is the bold, progressive judges and administrators who take the lead in accepting experimentation; the more cautious ones wait to see the outcome of the initial experimentation before allowing demonstration projects in their courts. In any case, the fact that the courts do not have to fully pay for the experimentation out of their own budgets (since the federal government supplies the bulk of the funds on a grant basis) precludes a barrier to innovation that might otherwise be insurmountable.

The Systems and Technology Division's third major project at this time is the "Argersinger Project," a research project analyzing alterntive methods by which states can provide counsel to indigents in misdemeanor cases. This project, also funded by the National Institute of Law Enforcement and Criminal Justice of LEAA, is designed to help states respond to <u>Argersinger v. Hamlin</u>, 92 S. Ct. 2006

(1972), in which the U. S. Supreme Court extended the right to court-appointed counsel to defendants in any criminal case in which conviction might result in incarceration. The study will include a survey of nine states which were purported to provide court-appointed counsel to nearly all misdemeanants before the announcement of the Argersinger decision. The approaches taken in each of these states will be described, as will several additional models to be recommended for consideration by other states that need to expand their indigent defense services.

In contrast to the Video and Reporter Training projects, which are concerned primarily with innovation and only secondarily with diffusion, the Argersinger Project is concerned primarily with encouraging the critical analysis and diffusion of the techniques in current use and only secondarily with developing new techniques for supplying counsel to indigents. The Argersinger Project will hopefully speed diffusion by improving the flow of information among court systems and by outlining alternatives in such a way as to simplify each court system's selection process.

The State Courts Center hopes to receive funding for many additional research and demonstration projects. The Systems and Technology Division has already applied for several projects which involve further experimentation with technologies such as video-tape, computer-aided transcription of stenotype notes, and computer information systems. These projects would involve adaptation of existing technology for use in the courts and actual implementation of technology on a demonstration basis. The adoption and diffusion of innovations should be speeded if these projects are completed successfully; they have been designed to build on previous work, thus eliminating wasteful duplication. The Center is also hoping to undertake several projects to encourage the diffusion of innovations already well explored.

The Systems and Technology Division's second major activity is short-term technical assistance, either to execute the final stages of an innovation or to assist a court in adopting an innovation already implemented elsewhere. The Center staff provides the technical expertise without which the court would be unable to recognize and effect an appropriate technological solution to the particular problem at hand. When a court asks for assistance, the State Courts Center utilizes its extensive contacts throughout the country to gather the latest information available and to employ expert consultants when necessary. Technical assistance is usually met with less resistance than larger technology transfer efforts; such assistance usually comes in response to a request by the court, which implies that demand recognition has already taken place.

To date, the Systems and Technology Division has completed work on two important technical assistance efforts: 1) the development of a data processing plan for the 21st and 22nd Circuit Courts of Missouri and the publication of a Center report detailing the plan;[29] 2) an analysis of the transcript needs of the District Courts of Oregon and the formulation of recommendations concerning the most efficient means for these courts to meet the new statutory requirements that they become courts of record.[30] In the former effort, a project team headed by Systems and Technology staff assisted in adapting and implementing a technology already in use in several other court jurisdictions. In the latter effort, a similarly constituted project team set up rules and procedures for properly utiliznng a technology (multi-track audio recording) in use in other courts. At the same time, procedures for experimentally introducing a new technology, multi-track voice writing systems, were formulated. In addition to these efforts, the Center has responded to numerous requests for guidance by giving advice and information about the technologies of interest.

The Systems and Technology Division's third sphere of activity is the dissemination of information about existing technologies which are of potential use to the courts. This function is of vital importance in aiding the flow of information among court systems. Since one of the largest obstacles to the adoption and diffusion of innovations in the courts has been the lack of knowledge about existing technologies, improvements in the communications pattern should have a large impact on the rate of technology transfer into and among the courts. Although it is difficult at this time to assess the full impact of the Center's information dissemination program, it seems apparent that, if nothing else, substantial interest in hitherto unexplored technologies has been generated in judges and administrators throughout the country.

Information is dispensed through several channels, including formal reports, informal person-to-person phone communications, and written correspondence. The formal reports serve to document the results of research and demonstration projects or technical assistance efforts. By sending these reports to members of the Center's Board of Directors, Advisory Council (composed of representatives from twenty of the major judicial and judicially related institutions), Council of State Court Representatives (one officially selected representative from each state), State Chief Justices, and State Court Administrators, it is hoped that all state court systems will at least be well informed about some of the latest technologies. Informal person-to-person communications and written correspondence add an essential personal element to the information dissemination process, creating a more favorable atmosphere for information acceptance and assimilation. The personal contacts also allow individual questions and problems to be addressed, which facilitates the adoption of complex innovations.

IV. Summary

The Systems and Technology Division of the National Center for State Courts is an active agent of technology transfer into and among the state courts in the United States. To effect and diffuse innovations in the technology-deficient courts, appropriate technologies now in existence must be identified, adapted where necessary, and put to use. This transfer process must take place in a heterogeneous environment characterized in general by limitations in mission orientation and training and experience of relevant personnel; resistance to change stemming from vested interests, risk-aversion, and individual personality characteristics; and relative lack of effective communication between courts and technology vendors. To help introduce modern technology and management operations into this environment, the Systems and Technology Division is pursuing a three-part program consisting of 1) research and demonstration projects; 2) technical assistance efforts; and 3) information dissemination. The overriding characteristic of this approach, and the key to any success which the Systems and Technology Division may have achieved, is the active, person-to-person channel of communication which has been employed to break down resistance to change.

REFERENCES

1. William H. Gruber and Donald G. Marquis, "Research on the Human Factor in the Transfer of Technology," in Factors in the Transfer of Technology, eds. William H. Gruber and Donald G. Marquis (Cambridge, Mass.: The M. I. T. Press, 1969), pp. 255-56.

2. U. S. Department of Commerce Panel on Invention and Innovation, Technological Innovation: Its Environment and Management (Washington, D. C.: Government Printing Office, 1967), p. 12.

3. Ibid.

4. William H. Gruber and Donald G. Marquis, "Introduction," in Factors in the Transfer of Technology, p. 5.

5. Gruber and Marquis, "Research on the Human Factor," pp. 261-64.

6. Ibid., p. 262

7. Ibid., p. 266

8. A. Shapero, "Diffusion of Innovations Resulting from Research: Implications for Research Program Management," in Research Program Effectiveness; Proceedings of the Conference Sponsored by the Office of Naval Research, Washington, D. C., July 27-29, 1965, ed. M. C. Yovits, et al (New York: Gordon & Breach, Science Publishers, Inc.), pp. 378-80.

9. See, for instance, U. S. Department of Commerce Panel, op. cit., p. 13; George M. Beal and Everett M. Rogers, "Informational Sources in the Adoption Process of New Fabrics," Journal of Home Economics, Vol. 49, No. 8, pp. 630-34; and Gruber and Marquis, "Research on the Human Factor," pp. 260-61.

10. Quoted in Gruber and Marquis, op. cit., p. 266.

11. Ibid., p. 267.

12. Shapero, op . cit., p. 378.

13. Ibid., p. 379.

14. Ibid.

15. Ibid.

16. Donald A. Schon, "Champions for Radical New Inventions," Harvard Business Review, March/April 1963, p. 81.

17. See Shapero, *op. cit.*, p. 380 for all the cited generalizations about information channels.

18. Gruber and Marquis, *op. cit.*, p. 270.

19. Schon, *op. cit.*, p. 82.

20. Gruber and Marquis, *op. cit.*, p. 271.

21. Schon, *op. cit.*, pp. 82-83.

22. *Ibid.*, p. 83.

23. Shapero, *op. cit.*, p. 378.

24. *Ibid.*

25. See Gruber and Marquis, *loc. cit.* for this characterization of mission orientation.

26. Shapero, *op. cit.*, p. 380

27. See Schon, *op. cit.*, p. 85 and also Bryce Ryan, "A Study in Technological Diffusion," *Rural Sociology* 273 (1948) at 284.

28. Schon, *op. cit.*

29. *Courts' Data Processing Action Plan for the 21st and 22nd Circuits of Missouri*, National Center for State Courts, Report Number: NCSC 0001, October 13, 1972.

30. *Selection of a Court Recording Method for the District Courts of Oregon*, National Center for State Courts, Publication Number: NCSC R0003, May, 1973.

CONTRIBUTIONS TO TECHNOLOGY AND THEIR TRANSFER: THE NASA EXPERIENCE

John G. Welles

University of Denver Research Institute

INTRODUCTION

The Issue

For over a decade, controversy has surrounded the question of the extent to which "down-to-earth" or "secondary" benefits will result from the billions of dollars spent by the United States on developing technology for the civilian aerospace program. Optimists have promised untold riches for commercial manufacturers, medical practitioners and housewives in spinoffs from aerospace research and development (R&D). Pessimists say these promises are grossly exaggerated, pointing out factual errors in certain claims and maintaining that little of secondary value will come out of technology developed to meet aerospace missions. Such technology, they maintain, is too exotic and specialized to find general application in nonaerospace sectors.

This issue arose during the early 1960's shortly after President Kennedy announced plans to land a man on the moon "in this decade." Similar questions have been raised about other large-scale, mission-oriented Federal programs with high technology content such as those of the Atomic Energy Commission and the Department of Defense.

The civilian space program in the United States is administered by the National Aeronautics and Space Administration (NASA), which was created by Congress in 1958. For the last 13 years, the agency's large-scale R&D efforts have been the subject of considerable examination regarding secondary benefits. It has the longest history among Federal agencies of overt attempts to stimulate secondary applications of technology generated to meet primary missions. It should be possible, then, to draw inferences from its experiences for other government programs.

The discussion focuses on three questions. First, to what extent has the performance of NASA's primary mission produced identifiable contributions to different fields of technology? Second, what is the nature of these contributions and are they finding secondary applications outside aerospace? Third, what efforts have been made to stimulate the flow of these contributions to secondary applications?

Specifically excluded from this discussion is consideration of planned or mission-oriented benefits resulting from such segments of the space program as the communications satellite system, the weather satellite system, earth resources satellite (ERTS), and the future educational satellites. Rather, consideration is limited to secondary, "spin-off" or "fall-out," benefits such as products or technological information developed for space which have found application in nonaerospace markets.

Definitions

Discussions of the secondary benefit issue have been severely handicapped by lack of uniformity in definitions of key words. For the purposes of this paper only, the following definitions are used:

Technology - the systematic application of knowledge to meet the needs and desires of man, including its embodiment in machines, tools, products, processes, and "software" (i.e., the organization and management of ways of doing things) and including the possession of such knowledge by skilled labor, professionals, managers, and libraries.

Technology diffusion - the process by which science and technology flows through human activities and eventually becomes applied. This concept of the overall flow process includes the narrower concept of technology transfer.

Technology transfer - the process by which science and technology developed by one group or organization for a specific purpose becomes adopted or adapted and applied by another group or organization often, but not necessarily, for different purposes. For technology to be transferred, in this usage, it must be actually applied by a secondary user. The secondary application may take place as a result of traditional diffusion mechanisms (e.g., education, conferences, professional journals, customer-supplier relationships) or formal technology transfer programs (e.g., the NASA Technology Utilization Program).

The Concept of Technological Contributions

Also handicapping discussions of the secondary benefit issue has been insufficient information concerning the technological contributions made by the primary government-sponsored program. What contributions to technology are available for transfer in the first place, assuming secondary uses exist?

Some people assume that such contributions are limited to specific physical products, such as solar batteries, a heat resistant paint, or a rocket engine. But technology as defined above is not so limited. The definition also includes the initial concept of how knowledge of a basic phenomenon can be applied practically, the end product, the manufacturing machinery, materials, and processes used to produce the product and, in turn, the background machinery, materials, and processes which made possible the end product production, and so on.

Technological developments usually move through different stages before application takes place. Most developments, major and minor, build upon countless precursor developments, combining contributions from many sources by adoption, adaptation, or improvement. Tracing these flows is complex, and assigning credit for specific technological developments is difficult. When technology transcends the sector in which it was initially applied and flows into another sector for secondary usage, tracing flows becomes even more difficult, especially when the contribution is incremental and not major in magnitude.

The research upon which this paper is based has been concerned with identifying and tracing flows of technology originating in NASA programs to secondary applications (1,3,4,5,6), and with identifying ways commercial enterprises acquire technology (2). One of these (5) departed from the usual approach of tracing the technology forward from its point of origin and started instead with identifying major technological developments in selected fields and traced their origins backwards to determine the influence of NASA on the advancement of these developments. This proved to be a fruitful line of inquiry in providing a relatively broad view of what kinds of technology were being generated by the space program. The findings of this study are described next.

SELECTED NASA CONTRIBUTIONS TO TECHNOLOGY

Technological Fields Selected for Study

Twelve fields of technology were selected for study. Initial objectives were to identify major recent developments in each of the 12 fields and, second, to identify NASA contributions to each of the major developments. The 12 fields were:

Cryogenics - Encompasses the production, maintenance, and application of very low temperatures. The major part of the field is concerned with the liquefacation of gases and the storage, transport, and use of these liquified gases.

Electrochemical Energy Conversion and Storage - Encompasses the storage and direct conversion from chemical to electrical energy through electrochemical processes, including devices such as fuel cells, primary batteries and secondary (rechargeable) batteries.

High-Temperature Ceramics - Includes ceramic materials able to resist the hostile environment of high temperature and the corrosive effects of highly reactive systems. Included were those materials capable of withstanding temperatures greater than 2,750° F. and the processes used to form these materials.

High-Temperature Metals - Includes metals designed for use above 1,200° F., broadly classified as superalloys (cobalt or nickel-base, depending upon the predominating alloying element) and refractory metal alloys (chromium, columbium, molybdenum, tantalum and tungsten), as well as the technology of forming and fabricating these materials (casting, forging, sintering, etc.).

Integrated Circuits - Based upon fabrication techniques where the silicon semiconductor material itself is the substrate. The active and passive circuit elements are formed by either chemical or physical junction forming processes. The technology of the field includes the design and production of these monolithic devices.

Internal Gas Dynamics - Limited to the technology of gas turbine engines and the use of such engines in nonaircraft applications. Gas turbines use hot gases directly from a burning fuel to turn the turbine, as opposed to steam turbines which use burning fuel to generate steam which, in turn, drives the turbine.

Materials Machining and Forming - Entails the technology of shaping, forming, and finishing materials, and encompasses a broad complex of methods, processes, and equipment. Including materials cutting and forming machinery, cutting tools, coolants, and lubricants.

Materials Joining Processes - Includes such major joining processes as gas, arc, and resistance welding; electron beam welding; brazing and soldering; diffusion bonding; and adhesive bonding. Materials, as well as processes, are included.

Microwave Systems - Embraces very short radiowaves with a frequency range that falls between the UHF (ultrahigh frequency) used for television and IF (infrared) light bands. Microwaves have characteristics similar to light in that they travel in straight lines, can be reflected, and can be focused in a beam. This field includes the technology of the systems for producing, transmitting and receiving microwaves.

Nondestructive Testing - Covers techniques for the analysis of material properties without impairing the properties (NDT) and the concomitant step of evaluating the material thus analyzed, which is termed nondestructive evaluation (NDE).

Simulation - Involves the development and use of mathematical and physical models of various complex systems for purposes of prediction, analysis, decision-making, and training. Many of the models use computers (digital, analog or hybrid) to simulate the interactions of physical, managerial, political and social systems.

Telemetry - Deals with the presentation of measured data at a location remote from the source of the data and involves the generation of a signal which measures the pertinent physical variables, transmission of the information, and conversion of the data into a form appropriate for display, recording or further processing.

Types of NASA Contributions

Within the 12 fields, 263 major developments occurring within the past decade were identified in interviews with 161 recognized technical leaders in the fields. Based on further analysis and interviews, the developments were narrowed to 109 for additional study, to which 366 NASA contributions were identified. High-temperature metals led with 61 NASA contributions, and telemetry had the fewest with 14 identified NASA contributions.

Contributions by NASA to these technological developments did not fit neatly into one or two categories. Instead, the complexities of technological developments were reflected in the varied nature of the contributions. Thus, ten categories or types of contributions were identified and used for analysis purposes. Most individual advances embodied two or more of these types of contribution.

Table 1 lists the types of contribution and indicates which were dominant in each of the 12 technology fields. As shown in the Table, the first two listed types were the most prevalent: developing new scientific and technological knowledge and developing new technology.

The NASA contributions were also grouped according to the level of technological change represented: class I - step function change or significant forward advance; class II - incremental or systematic advance; and class III - consolidation of knowledge. Class I contributions were found to occur relatively infrequently, while class II were the most common.

Table 2 shows, however, that there was considerable variation from field to field among the selected technologies in class of contributions. For example, NASA was forced to pioneer in meeting requirements for deep space communications, and this is reflected in a high proportion of class I or step change contributions to the field of telemetry. In already well developed fields such as machining and joining, NASA's contributions tended to cluster in class II or incremental advances.

Significance of Contributions

Subjective assessments were obtained during the interviews as to the significance of the NASA contributions. Significance was defined as the effect the contribution had on the occurrence of the development or advancement. Did the contribution have any effect on whether or when the advancement occurred? Five different effects were used to characterize significance. They are portrayed in Table 3 for each of the 12 selected fields of technology.

Of the 109 developments studied, five were attributed solely to NASA's efforts including: high-frequency power transistors (in the field of microwave systems); computer enhanced radiography (nondestructive testing); and simulated lunar landings (simulation). Table 3 shows that four fields contained relatively high percentages of advancements that probably would not have occurred without NASA's contributions: cryogenics, integrated circuits, simulation, and telemetry. The dominant category of significance, however, was the third: advancements that occurred earlier than they would have without NASA contributions.

When analyzed as to which stage of developmental activity (i.e., concept, exploratory research, advanced development,

Type of Contribution	Cryogenics	Energy conversion	Ceramics	Metals	Integrated circuits	Gas dynamics	Machining	Materials Joining	Microwave systems	NDT	Simulation	Telemetry	12 Fields Averaged
Developing new scientific and technological knowledge	X	X	X	X	X	X	X	X	X	X		X	X
Developing new technology (products, processes, methods, etc.)	X	X	X	X	X	X	X	X	X	X	X	X	X
Demonstrating the application of new technology					X				X		X	X	
Augmenting existing technology	X	X	X	X	X	X	X						X
Applying existing technology in a new context		X	X		X	X			X		X	X	X
Stimulating industry	X	X			X	X					X	X	
Evaluating and disseminating scientific and technical information	X			X		X	X	X		X	X	X	X
Providing technical assistance	X				X	X	X					X	
Identifying problem areas	X				X	X	X		X			X	
Creating and supporting new markets	X				X	X	X		X			X	

TABLE 1 DOMINANT TYPES OF NASA CONTRIBUTIONS TO MAJOR DEVELOPMENTS IN SELECTED FIELDS OF TECHNOLOGY

Field of Technology	Class of Technology, Percent			
	Technological Change Class I (step function)	Technological Change Class II (incremental advance)	Technological Change Class III (consolidation)	Nontechnical Contribution
Cryogenics	6.3	56.2	25.0	12.5
Energy conversion	3.8	73.2	19.2	3.8
Ceramics	25.0	54.2	20.8	--
Metals	6.6	68.8	24.6	--
Integrated circuits	6.7	73.3	13.3	6.7
Gas dynamics	4.0	60.0	24.0	12.0
Machining	--	59.1	27.3	13.6
Materials joining	--	76.8	20.9	2.3
Microwaves	--	76.5	17.6	5.9
NDT	9.7	54.8	29.0	6.5
Simulation	--	62.5	37.5	--
Telemetry	21.4	57.2	21.4	--
Twelve fields averaged	6.3	64.8	23.2	5.2

TABLE 2 CLASS OF TECHNOLOGY OF NASA CONTRIBUTIONS TO MAJOR DEVELOPMENTS IN SELECTED FIELDS OF TECHNOLOGY

Field of Technology	Percent Distribution				
	Entire development attributed to NASA	Advancement would not have occurred without NASA contributions	Advancement occurred earlier than would have due to NASA contributions	Advancement occurred due to parallel contributions of NASA and others	NASA contributions had minor effect on advancement
Cryogenics	--	22.9	77.1	--	--
Energy conversion	--	7.4	33.3	37.1	22.2
High-temp. ceramics	--	--	95.8	4.2	--
High-temp. metals	--	8.2	88.5	3.3	--
Integrated circuits	--	20.0	72.5	7.5	--
Gas dynamics	--	--	92.0	8.0	--
Machining and forming	--	9.1	86.4	4.5	--
Materials joining	--	9.3	90.7	--	--
Microwave systems	5.9	5.9	70.5	11.8	5.9
NDT	9.7	--	80.6	3.2	6.5
Simulation	4.2	29.2	54.2	12.5	--
Telemetry	--	35.7	57.1	7.2	--
Twelve fields averaged	1.4	12.0	77.6	7.4	2.5

TABLE 3 SIGNIFICANCE OF NASA CONTRIBUTIONS TO THE ADVANCEMENT OF MAJOR DEVELOPMENTS IN SELECTED FIELDS OF TECHNOLOGY

aerospace or military application, and commercial application) the NASA contributions were in at the time of the study, more than one-half were finding aerospace or military applications and about one-fourth were finding commercial applications. By field, commercial applications ranged from a low of 9.7 percent of the contributions in nondestructive testing to a high of 53.3 percent in integrated circuits.

Measuring Impacts of Contributions

Objective, quantitative measurement of the economic benefits of the NASA contributions encountered insurmountable obstacles within the study scope (5, pp. 25-27). However, four types of impacts were subjectively assessed based on a simple linear scale of high, moderate, and low. These impacts were:

Technological impact. The effect a contribution has in changing the matrix of products, processes, techniques, methods, and materials that make up the development.

Scientific impact. The effect a contribution has in changing what is known or how well the basic phenomena related to a technological development are understood.

Economic impact. The effect a contribution has in changing the economics of the development or the economics of the system in which the development is applied, including the availability and cost of the technology that makes up the development.

Social impact. The effect a contribution has in changing the immediate social environment in which the development exists. (For the purposes of this study, only first-order or primary social impacts were included.)

Figure 1 presents the results of these subjective assessments for the four types of impact, averaged for the twelve selected fields of technology. It is interesting to note that the impacts of NASA's contributions were judged to be greatest in the technological category, followed in order by economic, scientific, and social impacts.

When judging how many of the potential impacts had been realized at the time of the study (1972), however, Figure 1 indicates that almost all of the scientific impacts had already been realized, over three-quarters of the technological impacts had occurred, while less than half and less than one-third, respectively, of the potential economic and social impacts were estimated to have occurred. These findings of time-lag differences experienced in the flow of technology indicate the limitations inherent in focusing on economic impact to the exclusion of other types of impacts.

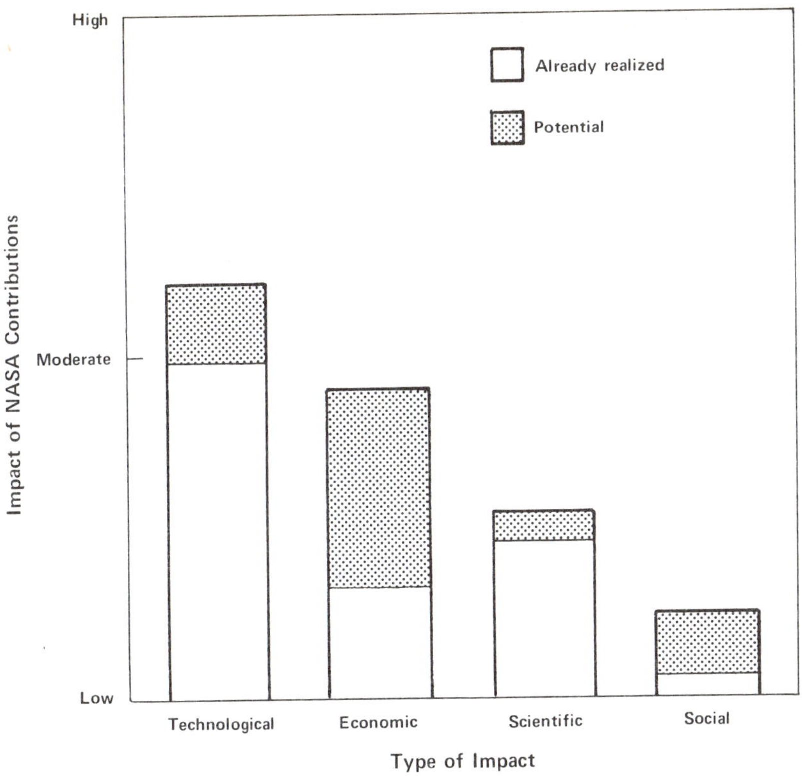

FIGURE 1. IMPACTS OF NASA CONTRIBUTIONS ON MAJOR DEVELOPMENTS IN 12 SELECTED FIELDS OF TECHNOLOGY

Information Flow

Almost no mention of secondary applications was made in the preceeding discussion of NASA contributions to the twelve selected fields of technology. This was by design as the purpose of the discussion thus far has been primarily to point out that the space program _has_ made contributions to selected fields of technology and to describe the _nature_ of these contributions. Once these aspects of the issue have been clarified, we then can look at the subject of secondary applications.

Even though secondary applications received limited attention within the scope of the contributions study (5), sufficient information was collected to conclude, as noted previously, that one-fourth of the NASA contributions identified were being applied commercially. Additionally, it was estimated that the amount of economic impact experienced up to 1972 was less than one-third of the potential total economic impact.

In examining secondary applications, it is important to know how information is being transmitted from the space program. The major communication channels used for disseminating information about NASA contributions to the 12 fields of technology are shown in Table 4. A different pattern of channels emerged from the findings for each field. Fields such as joining and machining made heavy use of NASA channels (described later), while other fields, such as gas dynamics and high-temperature metals, relied more on formal non-NASA channels such as conferences, professional journal papers, and professional committee activities.

	Cryogenics	Energy conversion	Ceramics	Metals	Integrated circuits	Gas dynamics	Machining	Materials joining	Microwaves	NDT	Simulation	Telemetry	12 Fields
Formal NASA report channels													
TR			X						X				
TN		X	X	X		X		X	X		X	X	X
TM	X	X		X		X	X	X	X	X	X	X	X
CR	X	X	X	X	X	X		X		X	X	X	X
SP	X	X		X		X				X	X		X
NASA TUD Program													
Tech Briefs	X	X	X	X			X	X		X		X	X
Tech Surveys											X		
Other formal NASA channels													
Request for proposal (RFP)			X	X									
NASA-sponsored conferences		X		X	X	X			X				X
Patents		X		X			X		X		X	X	X
In-house documents	X			X	X	X				X		X	X
Public relations talks		X											
Congressional testimony		X											
Awards				X									
Formal Non-NASA channels													
Conferences and meetings		X	X	X	X	X			X			X	X
Courses taught by NASA				X		X			X				
Other government reports						X							
Texts and handbooks				X		X							
Professional journals		X	X	X	X	X			X		X	X	X
Trade magazines	X	X	X	X	X	X			X	X		X	X
Committee activities				X		X						X	
Professional awards						X						X	
Informal NASA channels													
Contractor relationships	X	X	X	X	X	X	X	X	X			X	X
Technical assistance	X		X	X			X						X
Visitors and telephone calls	X		X	X		X	X					X	X
Informal non-NASA channels													
Movement of NASA personnel									X				
Personal contact		X	X	X	X	X			X		X	X	X

TABLE 4 MAJOR CHANNELS USED FOR DISSEMINATING INFORMATION ABOUT NASA CONTRIBUTIONS IN SELECTED FIELDS OF TECHNOLOGY

Most frequently used categories of channels for disseminating information about NASA contributions in all 12 fields were:

	Percent of Contributions Disseminated Through the Channel
NASA contractor reports	62.8
Technical interchange with contractors	35.8
Tech Briefs	29.2
Conference and meeting presentations	28.1
Professional journals	24.3
Personal contact during meetings, conferences, and other organized groups	24.0
NASA Special Publications	23.0

(Percentages add to more than 100 since multiple dissemination channels were used in most instances.)

The findings indicated that a direct communication channel seldom links the originator of a NASA contribution with a secondary user. Rather, such information usually is diffused through a complex network of channels involving multiple receivers, redisseminators, adapters, and adopters.

When the secondary user is a commercial enterprise serving both aerospace and nonaerospace markets, the communication linkages are often more direct and simple than when the secondary user is not a member of the aerospace community. However, in certain fields of technology such as cryogenics, high-temperature metals, integrated circuits, and telemetry, close professional and commercial ties exist between the aerospace and nonaerospace communities. Those ties facilitate the flow of information about NASA technology. In other fields such as machining, which is broadly applied by many industries, few ties can be found between aerospace and secondary user industries.

NASA TECHNOLOGY DISSEMINATION PROGRAMS

Background

Scientific information developed in NASA-sponsored research has flowed to a large extent via the highly developed, general purpose communication networks which have evolved over many years to serve the scientific community. Dissemination of this information, however, has been greatly aided by the specialized channels that comprise the information system set up by NASA and described next. This system differs little from those established by most other large-scale science and technology-oriented Federal agencies in the United States, except perhaps in its degree of automation and its scale.

Similar comments can be made about technological information developed by NASA-sponsored R&D. Here, however, NASA has supplemented conventional Federal practice by adding additional specialized channels to the dissemination process, attempting to stimulate the flow of NASA technology to secondary users (i.e., technology transfer) through its Technology Utilization Program. This is discussed later in the section.

The Information System

NASA has developed a highly automated system for storing, retrieving, and disseminating scientific and technical information (STINFO) about aerospace activities. The STINFO system was designed to serve the government-industry-university complex which comprises the aerospace industry as its primary audience. The importance of this system in disseminating information was shown previously in Table 4.

Major components of the system include: the NASA Scientific and Technical Information Facility in College Park, Maryland which stores, microfilms, abstracts, and computerizes hundreds of aerospace documents daily; *Scientific and Technical Aerospace Reports* (*STAR*) and *International Aerospace Abstracts* (*IAA*); *Selected Current Aerospace Notices* (*SCAN*) which notifies registered individuals of new publications in technical areas in which they have expressed interest; NASA/RECON, a nationwide computerized network that provides remote console access to the NASA STINFO system; and the technical report program of NASA-sponsored in-house and contractor investigations, translations, and other technical documentation.

The Technology Utilization Program

The Space Act of 1958 called for the ". . . widest practicable and appropriate dissemination of information concerning its [NASA] activities and the results thereof." In conjunction with several other passages in the Act, this has been interpreted by NASA as a mandate to encourage the use of its contributions to technology by both the primary community and by secondary users. Therefore, in the early 1960's it established a Technology Utilization

Program to aid public and private organizations, which are not in the aerospace community, gain access to NASA technology.

The TU program has developed a number of major activities. One involves an extensive publication activity including:

Tech Briefs - one or two page announcements of innovations, concepts, devices, and techniques which receive wide dissemination. Every NASA contract contains a "new technology clause" requiring contractors to report innovations in writing to NASA; these clauses, together with a similar in-house reporting requirement, form the reservoir of technical information for the Tech Brief reporting system.

Compilations of Innovations - dealing with a field of technology such as welding practices and transducers.

Technology Utilization Reports - describing innovations of special significance or complexity such as tungsten powder metallurgy.

Technology Surveys - consolidating NASA contributions to fields of technology such as inorganic coatings, solid lubricants, and advanced valve technology.

Conference Proceedings - of special meetings sponsored by NASA for particular industries and groups such as the petroleum industry.

A second major activity of the TU program is its sponsorship of six Regional Dissemination Centers, geographically dispersed around the United States. Five are affiliated with universities, and all provide local subscribers with tailored information services: current awareness searches, retrospective literature searches, standard interest profiles, and special publications. With access to the NASA computer tapes plus a number of other computer-based information sources, the data base of these centers approaches 1,500,000 documents.

A third activity is sponsoring the Computer Software Management and Information Center (COSMIC) at the University of Georgia which collects, evaluates, and distributes tapes, card decks, program listings, and machine run instructions. These are sold at handling costs to potential users, eliminating costly duplication in software development.

Other activities include sponsoring multidisciplinary Application Teams, technology applications projects, interagency technology agreements, research on the technology transfer process, and Technology Utilization Offices at 10 NASA field installations throughout the United States.

OBSERVATIONS

To the extent that successes in NASA's primary missions are indicative of technology diffusion, the flow of space technology to primary users can be judged to be at a high level.

The "adequacy" of flow to secondary users, however, is less obvious. It is true that hundreds of items of spin-off technology from space programs to non-space use have been identified (e.g., in 1, 3, 4, 6, 7, 8, 9, 10). They range from such well publicized examples as flame retardent paints and materials to more obscure contributions such as improvements in welding and joining, semiconductor manufacture, and controlling nitrous oxide production by stationary power plants. But when these are arrayed, what do they add up to? It is obvious to some of us who have studied the specific occurrence of transfer that the use of design and other types of information far exceeds, in frequency and importance, tangible and discrete transfers (products, materials, processes). Most frequently, design information is applied during the problem solving activities of the user. Its significance lies in its contribution to technical solutions. To date, efforts made to quantify such impacts have not been particularly successful.

Monitoring the Tech Brief program has produced some insight into the scale of nonaerospace problem-solving activity in which NASA technology has had relevance. Most Tech Briefs are backed up by Technical Support Packages (TSP's), available on request, that provide supporting information about the innovation reported in the Tech Briefs. Since 1963, about 6,000 Tech Briefs have been published and during the calendar year 1972, more than 100,000 requests were made for these TSP's through a variety of channels. By means of questionnaire surveys, supplemented by telephone interviews with selected TSP users, and personal interviews with experts in specific technological fields, substantial flows of technology generated under NASA sponsorship to secondary applications have been traced in such fields as plastics, lubrication, contamination control, fire safety, cryogenics, and nondestructive testing (3).

The following tabulation indicates the extensive nature of NASA documentation produced by the STINFO system and the Technology Utilization Program in three of these fields (7):

Type of Document	Cryogenics	Nondestructive Testing	Fire Safety
Contractor Reports	531	186	15
Tech Briefs	192	121	33
Technical Memorandums	123	54	27
Technical Notes	117	10	-
Other	175	54	14
Total	1,138	425	89

Criticism of the NASA TU program has often centered on the "push" nature of its activities. Critics have stated that Federal activities should, instead, concentrate on legal, regulatory, tax, financial, and institutional (including patents) incentives to boost the "pull" of space technology by potential nonaerospace users. This indeed, is one goal of two new Federal programs: the Experimental R&D Incentives Program at the National Science Foundation and the Experimental Technology Improvement Program at the National Bureau of Standards. The experience of both programs to date, however, indicates that they involve difficult and complex assignments. In other words, approaching the secondary user problem from the "pull" side may be equally as arduous as the "push" side.

Nevertheless, studies to date of the NASA experience indicate that, given multi-billion dollar annual budgets on aerospace activities ($3.2 billion in fiscal year 1973):

1. Significant contributions are being made by NASA to a wide variety of technologies.

2. A sizable portion (one-fourth in the case of the findings in the study reported above)of these contributions are finding secondary applications.

3. This technology transfer is sizable enough to be worth reasonable expenditures to enhance it (the NASA TU budget has totaled about $4 million annually in recent years) so that the taxpayers and secondary users alike will receive more of the potential benefits of already-developed technology.

4. Understandings of what contributions NASA is making to technology and how secondary users normally acquire such technology are important inputs to designing effective programs to stimulate technology transfer.

Additionally, inferences can be drawn from the NASA experience to other large, mission-oriented, government-funded R&D programs. When a mission requires that the government engage in extensive R&D to advance a field or fields of technology beyond the state-of-the-art practiced in industry, significant contributions to developments in such fields are likely to occur. The mix of technologies to which pursuit of the mission contributes will be determined by the nature of the mission itself. The space mission, for example, will have a quite different profile of technological contributions than a transportation or environmental mission.

Sizable time-lags, however, should be expected in most instances between the time these contributions are generated until they find secondary applications in other sectors, since the other sectors are usually not immediately prepared to absorb and apply these contributions in the form of economic goods and services. Considerably shorter time-lags can be expected between the generation of the contributions and the time they require to have impacts on the scientific and technological communities in other sectors. These latter impacts do not necessarily imply embodiment, for example, of the contribution in commercial products.

FOOTNOTES

1. Welles, John G., _et al._ _The Commercial Application of Missile/Space Technology_. Denver, Colorado: University of Denver Research Institute, 1963. [N64-24335; NTIS]

2. Gilmore, John S., _et al._ _The Channels of Technology Acquisition in Commercial Firms, and the NASA Dissemination Program_. Washington: National Aeronautics and Space Administration, 1967. [N67-31477; NTIS]

3. Kottenstette, James P., _et al._ _Applications of Aerospace Technology in Industry; A Technology Transfer Profile: Plastics_. Denver, Colorado: University of Denver Research Institute, July 1971. [N72-12515; NTIS]. In addition, _Lubrication_, July 1971. [N72-12417; NTIS] _Contamination Control_, July 1971. [N72-12418; NTIS]; _Fire Safety_, July 1971. [N72-12966; NTIS]; _Cryogenics_, December 1971. [N72-15624; NTIS]; _Nondestructive Testing_, April 1972. [N72-25482; NTIS]

4. Milliken, J. Gordon, and Edward J. Morrison. _Aerospace Management Techniques: Commercial and Governmental Applications_. Denver, Colorado: University of Denver Research Institute, November 1971. [N72-14971; NTIS]

5. Robbins, Martin D., John Kelley, and Linda Elliott. _NASA Contributions to the Advancement of Selected Fields of Technology_. Denver, Colorado: University of Denver Research Institute, January 1972. [DRI]

6. Kottenstette, James P. and James E. Freeman. _Project for the Analysis of Technology Transfer; Final Report_. Denver, Colorado: University of Denver Research Institute, July 1972. [DRI]

7. Committee on Science and Astronautics, U.S. House of Representatives. _For the Benefit of All Mankind, a Survey of the Practical Returns from Space Investment_. (92nd Congress, 1st Session) Washington: U.S. Government Printing Office, 1971.

8. National Aeronautics and Space Administration. _Useful New Technology_. Washington: U.S. Government Printing Office, 1971.

9. Alexander, T. "The Unexpected Payoff", Fortune. Vol. 80, No. 1, July 1969.

10. Ordway, F.I. III, C.C. Adams, and M.R. Sharpe. Dividends from Space. New York: T.Y. Crowell Co., 1971.

PART IV: International Issues in Technology Transfer

The first paper in this section presents an overview of the relationship between technology and foreign affairs. Eugene Sholnikoff suggests that effective technology transfer will become a requisite for effective international political decision making. Several papers then look at issues and efforts in particular countries. Dieter Schumacher describes a new program in the Federal Republic of Germany designed (based upon the U.S. experience) to bring technology to the citizen. Jacques Raffin describes the government-industry licensing mechanism as a low cost means of "border-to-border" transfer among developed countries and Win Straube discusses strategies for maximizing profits through licensing efforts. Within a single industry the transfer of technology is often made easier by the commonality of users in various countries; Marvin Cetron and James Dudar discuss the experiences of the aircraft industry.

The successful transfer efforts of Japan are described by Keichi Oshima. Papers by Samuel Bar-Zakay, Leslie Cook, and D. Babatunde Thomas provide a fitting conclusion to this section with discussions of the barriers and problems of technology transfer to developing nations--a problem for the 1970's and 80's.

TECHNOLOGY AND INTERNATIONAL RELATIONS: A VIEW BACK AND AHEAD

Eugene B. Skolnikoff*

Massachusetts Institute of Technology

INTRODUCTION

It is clear that advances in science and technology, particularly technology, have wrought major changes in society. The effects have in many cases been direct, as in the case of the development of major new artifacts such as the automobile, the telephone, or nuclear weapons. In addition, and often of greater importance, major social effects have been found in second-order consequences traceable over the years to new technology: urbanization, mass culture, pollution, industrialization, and many more. International affairs have been part of this process of social change induced by development and application of technology. The postwar development of bipolarity, the expansion of capabilities for communications across and within borders, the emergence of new subjects for international politics such as space and atomic energy, the basic change in the physical meaning of military power, the alteration of the significance of geography in conflict, and others all stem in large part from technology.

---

* Eugene B. Skolnikoff is professor of political science at the Massachusetts Institute of Technology, Cambridge, Massachusetts. The material for much of this paper is drawn from The International Imperatives of Technology, published in 1971 by the Institute of International Studies, University of California, Berkeley. An earlier version of this paper appeared in International Organization.

It is now accepted by most governments that the ability to develop, adapt, and exploit technological knowledge is a critical element in achieving a nation's objectives, whether those objectives be economic development, international power and prestige, or economic independence. Thus, national policy for technology, and issues associated with the movement of technological information across borders, have become important concerns for all countries, whatever their stage of development.

With all the developments triggered by technology, and with the high level of governmental interest, I think it can nevertheless be argued that the changes in international politics have not been as fundamental as is usually assumed. The subject matter of international affairs has been greatly expanded; the parameters of traditional issues have been altered; distinctions between domestic and international affairs have been blurred; and the interdependence of national economies has become steadily more striking. Yet the underlying assumptions which govern international relations, in particular the predominant view of the nation-state system as the organizing concept of the international system, has not been seriously modified. The United Nations has if anything encouraged the proliferation of sovereign states, since the fundamental unit of the UN and the source of its power and resources continues to be individual nations. In effect, the UN as it is presently functioning is a deterrent either to integration of independent states or to the development of countervailing international power. Even the development of atomic weapons, which might have been expected (indeed, was expected by many) to cause a fundamental change in the international system by changing drastically the meaning of war as a viable policy alternative, has over time become an umbrella under which more traditional or conventional warfare flourishes. The emergence of bipolarity, which was greatly enhanced by the development and cost of nuclear arsenals, slowly recedes as a dominating fact of international life as other power centers develop and as the inutility of nuclear weapons encourages the development of multiple, smaller power centers. Even technology itself, in the words used earlier, is now seen as critical to economic development, international power and prestige, and economic independence <u>of a nation</u>.

Against this background one must view with a skeptical eye the possibility of dramatic change that could flow from future technological developments and applications. In fact, I believe that over the next decade advances in science and technology are unlikely to cause anything but marginal, though important, changes in international affairs, barring, and it is a major caveat of course, any catastrophic events in military or political areas. Looking farther ahead, and policy over the 1970's must obviously be concerned with expected developments beyond that time, it is possible to see continued intensification of some trends whose cumulative impact may be very much more significant. The normative question of whether there should be more fundamental change in the 1970's, especially in the light of technologically related trends going beyond the next decade, is a more debatable proposition.

This conclusion that change will be moderate is in contrast to some of the apocalyptic literature that has appeared recently which argues that we are on the brink of disaster. Such a picture is debatable at best, though it must be noted that we lack adequate information about what is actually happening to our global environment. In any case, uncertain and unprovable apocalyptic predictions do not lead to useful analysis or prescriptions for change. Such predictions cannot be completely ruled out, however, for they are not amenable to proof one way or the other. The weight of the evidence does not yet support them, but the possibility of serious physical danger to our global environment cannot be ignored in foreign policy planning.

For purposes of the following discussion, using the 1970's as the primary time frame, we must inevitably be much more concerned with technology than with science. It is in any case the applications of technology that have the dominant impact on international relations. There are occasional exceptions (for example, the implications of geophysics for an agreement on a comprehensive nuclear test ban) but in general, and certainly on a short time scale, technology is the major interest.

Technology as used here includes not only the actual techniques and artifacts developed but also the side effects which accompany the application of technology. As we look ahead it is the side effects which grow out of intensive use of technology by burgeoning population and economically growing states that are likely to pose the most serious issues for international affairs.[1]

GENERAL AREAS OF INTERNATIONAL POLITICAL IMPACT

In order to identify future trends we will do well to start with a brief review of the major effects of technology on international relations in the recent past.

One such effect is the dramatic growth of the interdependence and interpenetration among societies. The appetite of technological societies for resources which are widely and unevenly scattered throughout the globe contributes. So, too, do the "revolutions" in communications and transportation that have led to the international movement of people, information, and ideas on an unprecedented scale. The growth of industry, especially high-technology industry, has expanded the reliance on markets and capital beyond national borders. The development of global technologies requires multinational cooperation and creates cross-national dependencies.

Another major effect of technology has been the blurring of the distinction between domestic and foreign affairs. Interdependence is a significant causative factor in this phenomenon as is the ease with which domestic developments are widely publicized and related to foreign policies. Further, the expansion in the subject matter of international relationships has brought more elements of society and government into direct roles in foreign affairs, to the extent that a Foreign Office's primary role in foreign policy-making among Government Agencies has been seriously eroded. Television has had a substantial impact in making foreign policy a subject of intense and immediate debate, and thus has made it easier to relate foreign and domestic issues within a nation.

Strategic/military affairs is another area obviously strongly affected by technology. Changes in the destructiveness, speed of delivery, and cost of weapons systems have altered the conditions of warfare, the meaning and usefulness of military power, and the time scale of the decisionmaking process. As noted earlier, the effects of these military developments have not wrought the fundamental changes in international relations as might have once been expected, but there obviously have been important effects, especially in the relations between the superpowers.

Technology has had major effects as well in the creation of new patterns of relationships among nations, in particular, between the developed and developing worlds. The gap between North and South, as measured by ability to accept and apply modern technology, is presently large and is growing larger. Even among developed nations the somewhat discredited notion of a "technological gap" implies concern for the growth of a new kind of dependence relationship, a dependence based on effective monopolies in high-technology industrial fields (such as computers, electronics, or nuclear applications). Technological monopolies, essentially knowledge-based, can give the nation with the advantage economic, political and even cultural influence. Such dependence relations may be more subtle and difficult to overturn than have been past patterns of domination through direct economic or military power. The significance of the movement of technology and technological information among nations clearly has direct relevance to the character of these dependence relations and to the ability of nations to maintain advantages or monopolies.

In institutional terms technology has led to the rise of new elites involved in the decisionmaking process and has contributed to the proliferation of international organizations built around new technologies or newly concerned with the application of technology. With regard to foreign policy machinery itself the advance and use of technology has contributed to marked alteration in the number of actors involved in the making of policy, in the subject and character of the issues, in the information that must be processed, in the role of former major actors (such as ambassadors), in the time allowed for decisionmaking, and in the pressures and interests that must be considered.

There are other ways by which the impact of science and technology on the international scene can be evaluated and additional specifics that could be enumerated. For purposes of this essay, however, this categorization is adequate. A more interesting question now is how any of these trends from the past are likely to change in the future and whether new trends of substantial importance will emerge.

Almost certainly, the interdependence and interpenetration fostered by technology will continue to be one of the results of advancing technology. In fact, the degree of interdependence is likely to be intensified by several kinds of technology-related developments. One is the likelihood that multinational corporatior which are in many respects the prime carriers of technology across borders, will continue to expand in size and scope. In effect, this will result in substantially more economic integration, and will force the development of international political cooperation to meet the relatively independent power of multinational firms. It will also likely lead to more extensive international economic, and particularly monetary, cooperation among governments as a means of regulating the firms.

Another cause of intensified interdependence will be the development of global technologies that automatically create dependence relations: communications, weather forecasting and modification, geologic surveillance, direct broadcasting, and others. In the long run the most significant may be the development of low-cost communication that encourages extensive international data transmission and the growth of electronic technologies for the handling and use of large amounts of data. It is difficult to specify in any detail the effects of this technology, but the growth of "data dependency" relations may prove of crucial importance in tying economies and societies together. In particular, these global technologies will require dissemination of sophisticated technological knowledge if the nations that take part are to have reasonably equal opportunities to participate in decisionmaking responsibilities.

Other trends, such as the further blurring of distinctions between domestic and foreign affairs and the continued widening of the technological gaps between the developed and developing countries, are also likely to be intensified. In addition, notwithstanding a growing popular reaction against technology, one can expect a continued enhancement of the role of intellectual elites as essential and powerful elements in the governmental decisionmaking process.

Some trends, specifically the impact of scientific or technological developments related to weapons, are likely to be substantially unchanged in the next decade. More will be said about this later.

Finally, we can look toward some new trends and some trends that may not be new but that are so changed in degree as to be different in kind. One of the latter is the requirement for international political machinery to assume much greater functional responsibilities (i.e., regulation, control, allocation and management of technology or its costs and benefits) than is the case today. This conclusion receives more detailed discussion in the next section.

Another likely development flowing from technology will be the creation of new patterns of conflict among nations, patterns that grow out of conflicts of interest generated by technology. The unequal use of unevenly distributed natural resources, the increasing gaps in technology utilization (or modernization), the problems of adequate transfer and adaptation of technology, the reaction to imported technology seen as cultural domination and economic imperialism, the divergent objectives over such issues as pollution, the criteria by which global technologies should be designed and utilized, and the allocation of resources found in an international environment, will all be sources of dispute and controversy in the future. The lineup will be largely the technologically advanced countries facing those aspiring to reach comparable levels.

Examples of these conflicts of interest are cropping up frequently even now: disputes in the World Bank over the weight to be given to the environmental effects of large development projects such as dams; the prolonged negotiation over the permanent charter for Intelsat because of questions of control of system design and allocation of benefits; and coordinated action by oil-producing nations to improve their bargaining position against oil-consuming nations. These kinds of conflicts of interest almost certainly will be more prevalent and more serious in the future.

The next decade is also likely to see the further substantial enhancement of the role of television, and thus public involvement, in the making of foreign policy. Increased ease and reduced cost of providing instant and vivid exposure to worldwide events may make the role of public opinion a much more decisive element in foreign affairs than even in the past. Television may even make the moral values of a society more relevant elements in the policy process.

We will also move farther down the road of building technologically vulnerable societies in which certain kinds of intervention, willful or inadvertent, can cause the collapse of systems affecting many people at once. The northeast power blackout, or urban chaos resulting from strikes by workers who provide critical services, are indicative of the sensitivity of society to small-scale disruption. The increasing technological interdependence of modern societies employing ever more complex technologies within growing concentrations of population is likely to exacerbate this situation and heighten its political importance. It must be noted, however, that technology can also be designed to reduce vulnerability through redundancy or system independence. A substantial cost may be involved to achieve reduced vulnerability, however.

Finally, I would single out as a new factor of considerable importance in the 1970's a growing reaction in technologically advanced societies against technology itself. It is difficult to predict the effects of this attitude; conceivably, it could work to slow technologically related change and thereby delay the meaning of some of the trends mentioned here. In time its effects could materially erode the strength of the underlying technological capabilities of advanced countries and in particular it could affect the ability of those countries to use technology to meet some of the longer term needs created or exacerbated by technology (e.g., substitution of nuclear power for conventional power sources to meet problems of atmospheric pollution). More likely, this new anti-technological bias will assert itself, for the 1970's at least, as a major unsettling societal force leading to difficulty in asserting leadership, to sharper questioning of values and of the political process, and to lack of clear national objectives. The effect could be to contribute to the inward-turning already evident in the United States and to the development of a desire to use technology to achieve autarky rather than interdependence. A trend toward autarky is not likely to be successful (though technology itself would make a considerable degree of autarchy a feasible goal) because present technological developments and desires to use technology are leading toward greater interdependence. But the tension between the two pressures could produce major internal contradictions and conflict over goals and use of resources.

SPECIFIC IMPORTANT CATEGORIES OF IMPACT

Armaments

During the 1970's it is unlikely that there will be developments in science and technology that will seriously affect the power balance between the superpowers. Such developments cannot be completely ruled out, of course, but the more probable course of events is a continuation of the present standoff between the U.S. and the U.S.S.R., though with substantial developments in weapons systems and substantial additional expenditure of resources.

The technologically related area with the greatest potential for destabilizing the present balance lies in the development of means to neutralize offensive missile forces, either after they are fired (ABM) or through a capability for location and destruction of fixed and mobile missile forces (such as Polaris). It is hard to imagine at this time the development of an ABM capability that would give sufficient assurance of protection against a missile strike as to imply the capability to neutralize another nation's missile force. Such a development by one, or even both, superpowers would certainly change the strategic relationship but it is most unlikely well into the future. The construction of an ABM system using present technology could also be destabilizing, of course, but the net result would almost certainly be reciprocal construction of ABM systems and more intercontinental missiles, thus resulting only in restabilizing the strategic balance at a higher level of expenditure.

A capability to locate and destroy missile forces, particularly undersea forces, is another matter and one which is receiving much attention. Clearly, the capability to destroy undersea forces, which are the most secure component of the deterrent, before they can be fired, could have a destabilizing effect on the superpowers' relationship. But, if such a capability emerges as a serious possibility, the probable result would be the development of new countermeasures and additional offensive weapons. It would be more useful, in fact, to consider any new technology for surveillance of undersea forces as a candidate for arms control rather than for armaments.

Other developments of more exotic weapons (weather control, genetic weapons, etc.) are beyond the 1970's if indeed they ever emerge as usable weapons systems. Surprise is always a possibility, but it is difficult to imagine any category of development in the near future that, at worst, would have more than a minor and transient effect on the basic power relationship between the superpowers.

Other political developments related to strategic armaments are likely. Specifically, the rise of the People's Republic of China (Communist China) and conceivably Western Europe to "superpower status," as measured by strategic weapons systems, are possibilities to be contended with. But the technologies involved are unlikely to be different in kind from what the United States and the Soviet Union now have or are building.

Weapons developments which would affect the power position of other countries in major ways can be ruled out with less certainty. Leaving proliferation questions aside (the problem of controlling proliferation is likely to increase along with the increase in nuclear power installations), more likely developments will be in limited war weapons. In that area, developments pioneered by the United States in Vietnam are likely to be copied by others when costs permit or when the big powers pay the bill. In addition, use of new antiaircraft weapons may for a time change the significance of air power in small wars, but countervailing tactics and weapons will be likely to redress the balance before long.

Thus, the 1970's are not likely to see armaments developments that appreciably alter the situation of today. There may well be increased expenditure to pay for new developments. There is likely to be one and perhaps two new superpowers. And arms trade among nations is likely to increase. There may also be some transient effects due to new technologies with significance for particular limited situations, but there will be nothing in the armaments themselves that would constitute a quantum change.

Nation-State

One of the major effects of science and technology on states will clearly be the need for increased reference to the international community in a growing number of politically and economically important fields, as detailed below with respect to international machinery. In effect, this need to place decisions in an international environment will constitute an erosion of traditional concepts of national sovereignty, an erosion already long in process but likely to be much exaggerated in the next decade.

The growing realization of this erosion by governments is in the first instance likely to make them more jealous of the prerogatives they retain so that international agreement is likely to be more rather than less difficult, especially over issues with major economic impact.

The more intense interdependence between nations for resources, markets, audiences, capital, and information, and the continuing process of coalescence of domestic and foreign affairs are also leading toward a reduction of the significance of the government of the state per se as an independent actor on the international scene. The state will remain the unit of negotiation and the point of political and diplomatic contact, but its freedom and independence of action in most areas will be gradually diminished in effect, even if there is no outward recognition of the changed situation.

The costs and technological base required to realize the benefits of some technologies will also create pressures both for political developments designed to pool resources on an international scale and for protection of technological advantage where it seems to exist. In some cases the technological sharing may contribute to political integration; in most instances it will more likely result only in increased cooperation among sovereign states--a "pooling of sovereignties" though also increased interdependence. But, the pressure for protectionism with regard to technology is likely to be the more prevalent phenomenon in the next few years, with potentially serious barriers to the flow of information.

The rise of competing centers of power in the multinational corporations, many based on leadership in sophisticated advanced technology, and concern over their role in technology exportation, are likely to lead governments to common cause in attempting to control international business, and to nationally-based attempts to impede their free transfer of information. Such attempts are likely to be disruptive, but futile unless they succeed in seriously cutting down the multinational firms themselves.

Thus, in several different ways, the 1970's will be a period in which the ability to control events and the independence of action of national governments will become increasingly compromised in technologically related areas, though without any formal alteration in structure of the nation-state. In fact, the existing structural forms are likely to be intensified as involvement in the substance of issues is diffused among additional actors on the scene. No specific type of technology is predominant in this development; all, and particularly the side effects of more intensive use of technology, are implicated.

Less Developed Countries

The problems of development cannot appropriately be dealt with here, but two points relevant to technology and its implications are worth making. The first point is that the role of technology in development is still little understood. There is much said and written about the importance of transfer of technology from the developed and developing countries and of the necessity for sound scientific and technological infrastructure, but a good theoretical underpinning to guide policy and investment of resources is lacking. Countries want modern technology as symbols of modernization, and they need new technology such as modified seed strains or communications systems to achieve their development objectives and to participate in global technology. But the scale of resources required, their mix, and the institutional needs to make them effective, remain matters of art.

A second point has to do with so-called technological gaps. Whatever the true state of affairs in the politicized gap between the United States and Western Europe, it is as nothing compared to the technological gap between North and South. And, as noted, most commentators believe that this gap is likely to widen rather than narrow.

In time, the problems created by disparities between the rich and poor nations are likely to become even more serious an international issue, not only in its economic terms where it is already with us, but in a new dimension: are advanced technological nations <u>precluding</u> the possibility that others can <u>ever</u> achieve comparable income levels? The issue here is whether the appetite of highly industrialized societies for raw materials, and the environmental contamination the use of the resources causes, will in the long run set a ceiling on the movement of other societies in the same direction. For example, will necessary natural resources for industrialization be depleted; will there have to be a total global limit on CO_2 emission or heat dissipated to the atmosphere, with the developed nations already using most of the quota? There can be no definitive answers to these and similar questions today, nor are these problems whose dimensions will be fully clear in the 1970's. But it behooves us to begin to understand that situation, for its political implications could be profound: at the very least, an irresistible demand to share more equitably the wealth of advanced societies, backed up, perhaps, by a threat on the part of raw material-providing countries to withhold the resources advanced societies must have.

International Machinery

The clearest and most certain international impact of advancements in application and use of technology will be with respect to the demands placed on international organizations, including nongovernmental organizations as well as the UN, regional and other multilateral bodies. In effect, there will be a rapidly-growing requirement to deal with the many functional implications of technology on an international rather than a national basis. New technologies or the side effects of increased application of technology will "force" a degree of international management and control that goes well beyond what is already in evidence. Nations will find it necessary to seek international agreement and to accept internationally-determined constraints on their actions in a variety of fields of growing political and economic significance to an extent that raises important questions about the capability of international machinery and governmental practices as they exist today to meet the new requirements.

A brief look at a few global issues related to technology will illustrate the point.

If we start with the environment, the current issue that has so prominently captured public attention, many different aspects are relevant.[2] Atmospheric effects caused by the activities of an increasingly technological society are suddenly being recognized. These effects, most often pollution caused by industrial effluents, are largely of regional concern at the moment, but there are also global implications that are as yet little understood. Accumulation of CO_2 or particulate matter could have in time profound effects on the heat balance of the planet and thus alter climate in unknown ways; so, too, could exhaust from high-altitude aircraft or continued transfer of land from agricultural to urban use.[3]

Similarly, pollution in the rivers and oceans from agricultural use of pesticides and insecticides, or from disposal of atomic waste, or from oil spills could seriously interfere with natural processes essential to life and with the ability to claim living resources from the seas.

These and other pollution problems will undoubtedly require international action. As a minimum, global monitoring will be necessary in order to obtain the necessary information for analysis, and to warn of impending problems. Regulations will have to be instituted, at first on a regional basis, but ultimately on a global basis to control those "domestic" activities that cause problems across borders or pose dangers more widely. These regulatory and monitoring functions will have to be internationally based to be acceptable, and to insure equity of application and effect. In turn, such regulatory activity will require the development of institutional capabilities for monitoring compliance, inspection, enforcement, and settlement of disputes.

The UN Conference on Problems of the Human Environment in 1972 in Stockholm was seized with exactly these issues of developing international machinery for regulation and control.

Inadvertent modification of the environment is not the only concern. Large-scale experiments such as Project Westford[4] or underground nuclear explosions in the Aleutians will eventually have to come under some kind of international regulation because of the danger they may pose to all nations. The U.S. may well come to regret its present assumption of unilateral authority with respect to carrying out such large-scale actions. On a longer time scale (but not much longer, for preparation is required during the 70's), new technology of planned weather and climate modification will require international means for regulation, allocation of costs and benefits, and perhaps management of the technology. Much of the required research and development is already being conducted by nations under the auspices of the World Meteorological Organization and the International Council of Scientific Unions as international experiments. Both of these organizations will certainly be involved in the resulting application and control of the technology once it is practical.

In the oceans, it is already clear that not only with regard to the resources of the seabed but also with respect to the living resources of the sea, internationally based controls and management are and will be required. The world is far from innocent with regard to international machinery dealing with ocean matters, but the requirements to be placed on that machinery will grow inexorably. Deep drilling technology is rapidly advancing as a direct result of substantial economic incentives, and fishing technology is equally, though not as quickly, moving ahead. Both already require means to control access, allocate resources and benefits, apply and monitor regulations, adjudicate disputes, and so forth. The need for such functions to be performed by international machinery will only be further emphasized each day that passes. The UN Conference on the Law of the Sea scheduled to begin in 1974 will be faced with that range of issues.

In outer space, we can expect continuing development in the applications of technology, particularly with reference to communications (including direct broadcast, though it can be argued that that will be a much more limited technology than is generally assumed), meteorology, resource mapping, agriculture, navigation, and data transmission.[5] Each of these will have important implications for international machinery. For example, the technology emerging for improved weather forecasting will require international management to insure the equitable distribution of benefits and to reduce dependence on the continued good will of any single state. The U.S. may provide most of the original technology, but as in the case of communications satellites today, as soon as the technology becomes important to other countries, continued domination of the system by the U.S. becomes politically unacceptable as a permanent arrangement.

The pattern will be the same for the other space applications because of the nature of the space environment, the potential economic value of some of the information generated, the need to assure equity and impartiality of operation, the need to allocate radio frequencies and orbital slots, and the cost of the technology. We already have, of course, one major new international institution--The International Telecommunications Satellite Consortium (Intelsat)--in this area.

The depletion of natural resources is becoming once again a subject of growing concern. Oil shortages have already become a prominent issue in international affairs, and others may soon arise. The issues here are not only the obvious questions of assuring adequate supplies of needed resources but also those of equitable distribution of benefits from resources and of wise management of a necessarily limited supply. The problem is that an increasingly technological society has been, for example, demanding metals at a rate of growth of 6 percent per year. The entire metal production prior to World War II is about equal to what has since been consumed.

Whether or not we have immediate shortages of resources for which there are inadequate substitutes (itself a controversial issue), it is clear that the world is sitting on a finite resource supply that must eventually be rationed and "managed." Moreover, it may well be that our present use of resources is highly inefficient from the perspective of future needs. The uneven distribution of resources over the globe (the heart of the present oil problem, since in fact there is no short-term shortage of known reserves) inevitably raises questions of control and equity. At some time, and perhaps the process should start within this decade, international regulation and allocation of natural resources on a major scale will have to begin.

The interaction of population growth and the need for food and other resources is another area that will press on the international community, notwithstanding the temporary respite of the "green revolution." The danger of widespread famine in 1973 is, in fact, very real. Estimates vary, but the UN Secretary General's report of world population predicts close to a 50 percent growth in population by 1985, and the Food and Agriculture Organization estimates that to meet population growth (assuming no change in per capita income) food production must increase between 3.2 and 3.8 percent per year against an average 1962 figure of 2.8 percent.[6] Coupled with some growth in income, the necessary increase in food supplies is even higher.

There is little point in attempting to spell this situation out in detail here. It is clear that even if population growth created no other problems, the magnitude of the task of supplying necessary food supplies will be enormous and will increasingly be of concern to the international community. Whether international organizations will or should be charged with greater responsibility for the production and distribution of food supplies is a matter of judgment.

But there are other complications as well. Increased production of food implies substantial increase in the use of fertilizers and pesticides, both of which can have major detrimental ecological effects. If DDT were to be prohibited, without an equivalent substitute at the same price, the net effect would be a reduction in the production and availability of food supplies. No qualifying substitute is yet available. Furthermore the new seed strains that are the basis of the green revolution are highly vulnerable to disease, so that sudden catastrophe could strike if much of the world's population were to become dependent on these varieties.

Population growth itself will of course lead to more pressure on natural resources, to increased production of energy, transportation of fuel, and industrialization, and thereby to more deleterious environmental effects. It will thus add to the need for international management, regulation, and allocation of global resources and of national activities that have international environmental and other effects. On a longer time scale population growth itself will eventually be a candidate for international regulation. It should be noted that in these areas the conflicts of interest between the developed and the less-developed world are likely to become a major source of tension.

The major conclusion to be drawn from these brief illustrations, each of which could be elaborated in great detail, is that the locus of decisionmaking in many technologically related areas "<u>must</u>" move (and is moving) from the national to the international sphere. The imperative results from the international nature and effects of forthcoming technology-related developments and from the apparent inevitability of these developments. They are inevitable because of the dispersed decisionmaking characteristic of technology and also because so much is already in the process of development or application. Most of the technologically related developments cannot be turned off even if there were the political will to do so.

The need for international machinery is not a new phenomenon. The world is well accustomed to international management, or regulation and even monitoring of some sort in many areas--telecommunications, postal service, weights and measures standardization, health, communications, and others. What will be different in the future is the degree of international interaction that will be required for many more subjects, the extent of the regulation needed, the fact that substantially more of what have been considered matters of purely national decision will have to come under some kind of international jurisdiction, with corresponding limitations on national sovereignty, and the related fact that the political and economic importance of subjects requiring international decision will sharply increase.

A major question for the 1970's, therefore, is whether existing international institutions, and attitudes of governments toward them, are adequate to meet the responsibilities they will have to assume during the decade or beyond. It is critical to remember that even if the major institutional impact will be after 1980, any institution building necessary must be going on well in advance if there is to be any hope for success.

SOME POLICY ISSUES

International Institutions

The first and the clearest question is whether existing international machinery is capable of meeting the functions it will have to perform by the end of the decade and beyond. We can think of these functions broadly as 1) provision of service, 2) norm creation and allocation, 3) rule observance and settlement of disputes, and 4) operation. In all of them there will be substantial increased requirements, with the politically most difficult lying among the latter three.

In fact, an examination of present machinery shows that all of these functions are performed at least to a limited extent today. A flight of imagination is not necessary to visualize these functions being performed by an international system greatly expanded in scale and importance in the future. Nation-states today do reach agreement on international regulations and controls (telecommunications, air traffic, health, fisheries) and by and large they adhere to those agreements. There are examples of international inspection (fisheries, atomic energy, health) that work reasonably well. International operation of large-scale technology has begun (communications satellites, atmospheric research programs).

But, there is cause for grave concern. The institutional mechanisms that exist, whether the United Nations and its specialized agencies, regional organizations, or small narrow-focus international machinery, are geared to a scale of activity and degree of responsibility far lower than will be necessary. As these issues grow in political and economic saliency there can be little confidence that the existing machinery is adequate to cope with the changed demands. The United Nations system is too often characterized by highly variable competence, jurisdictional conflict, weak and diffuse authority structures, burgeoning bureaucracy, lack of interest and inadequate coordination machinery, among others. This is not to say that nothing works, for in fact several of the United Nations agencies perform quite well; the reasons for good performance on the part of some are capable of analysis that can be quite instructive. Effective performance usually correlates with small organizations, technical nature of a mission, relatively high influence of knowledgeable member states, degree of agreement on scope and objectives, relative reliance on national experts vs. international bureaucracy and other such characteristics.

Regional organizations or other non-UN bodies often work more effectively than universal-membership organizations by focusing on a limited range of issues or by operating only among a limited group of like-minded countries. Among these as well, wide variation in performance and competence are found, and they of course present different kinds of problems and opportunities for a nation's foreign policies from those of the universal membership organizations.

Shortcomings of institutional machinery cannot be blamed simply on the institutions. The attitudes of member states are critical and are equally a basic cause of the existing malaise. Nations relinquish apparent attributes of sovereignty only reluctantly, especially to organizations in which their control is seriously diluted. The participation of many states in international organizations and particularly in the United Nations is too often characterized by little knowledge and much politics. The quality of national missions is often weak. The domestic coordination of policy is often poorer even than the coordination between international bodies.

In sum, today's international machinery and national attitudes toward that machinery are inadequate to meet the responsibilities that will have to be performed, somehow, by international machinery in the future. The situation is made considerably more serious when we realize that future responsibilities will be of much greater political significance for they will involve allocation of important resources, management of large systems, inspection and control that reaches into domestic activities, and international settlement of disputes that have substantial economic and political impact.

We are dealing here with a major issue of foreign policy that is only beginning to receive the attention it deserves. Realistically, no nation is likely to surrender willingly important prerogatives to institutions over which it has little influence and which do not have the competence to carry out their responsibilities. Nor should they. Some nations may have little choice, but the United States has. It must take the lead in planning for the development of international institutions that are capable of assuming responsibility and of adequately representing the interests of all nations. Is that a realistic prospect?

Here we have a major job of institution building for the 1970's. We have the time, but not much. "Revolutionary" changes in the international system are not necessary now; however, if there is inadequate institution building, revolutionary changes are likely later when global dangers and inequities release irresistible political forces.

It is not a foregone conclusion that the United Nations ought to be the prime candidate for reshaping the machinery necessary to meet the new functional needs. Perhaps more "Intelsats" are needed; perhaps more regional bodies; perhaps additional bilateral arrangements. Some proliferation of international organizations may be necessary and desirable, especially when existing machinery is weak and inflexible. Distasteful as proliferation may be, it could quite easily be preferable to granting increased responsibility to machinery unable to fulfill that responsibility.

But, such proliferation is likely to mean domination of new organizations and hence technology by a small number of technologically-advanced nations, on the argument that other nations do not have the knowledge, or interest, to participate effectively. Increasingly, we may be faced with a choice between, in the abstract, technological domination and "participation" on the international scene. The transfer of technology becomes a central route to avoiding that kind of choice which has obvious and serious political dangers.

Control of Technology

The question of the control of technology, or its direction into "more useful" channels, necessarily arises. In particular, one can ask whether some kind of internationalization of science and technology is possible that will help both to control and to direct technology more wisely.

Science and technology have been characterized in this century not by their internationalism, but by their use for the achievement of national objectives. This national orientation will undoubtedly continue as a major focus of technological development. However, even on the national scene there is a growing attempt to anticipate the side effects of technology, and to include those anticipated effects as part of the original decisionmaking for new technology. This movement is in its infancy, and has a long way to go before there are appreciable discernible results.

There are also good arguments, however, for attempting to put some portion of the scientific and technological enterprise under international auspices. There have been proposals, for example, for an international science foundation. One argument for it is that it is the only way to get adequate research and development on subjects set by international needs rather than by national objectives. That argument would be particularly relevant to less developed countries (for example, to meet the need for an economical DDT substitute, or to develop health systems based on $100 per capita incomes), but also would be relevant to the development of global technologies that can serve to ameliorate the undesirable effects of technology.

Another argument for internationally sponsored research and development is that such programs can help to keep "dangerous" subjects in the open, international domain, thereby making their ultimate appropriation for purely national purposes less likely and the prospects for their international control correspondingly more likely. The development of genetic technology, with potential dangerous as well as beneficial applications, is a case in point.

Lastly, internationally sponsored research and development may also be of great importance in the transfer of technological capability so as to develop the "international" expertise likely to be essential for the operation of the international machinery necessary to cope with the effects and regulation of technology. Looking ahead, one can see the importance of developing some international institutions as the arbiters and controllers of technology that are, and are seen to be, impartial. Expertise will be required, and international research and development may be one way to develop such capability.

CONCLUSION

The thrust of this essay has been that, barring major surprise or catastrophe, advances in science and technology and their application in the 1970's are likely to cause substantial but not revolutionary changes on the international scene. However, the potential for cumulative, more drastic change in the subsequent decade is very real. In particular, substantial international institution building and changes in national attitudes will be essential during the 1970's and must be understood and set in motion without delay.

The role of the transfer of technology and technological information is seen as a critical component in dealing with the problem of building international institutions that allow effective representation from the technological have-nots, and that avoid excessive domination by a few technologically-advanced nations.

REFERENCES

1 For extensive detail, see among others, Gordon J. D. MacDonald, "The Modification of Planet Earth by Man," Technology Review, Oct.-Nov., 1969, Vol. 72, No. 1, pp. 27-35; T. F. Malone, "Current Developments in the Atmospheric Sciences and Some of Their Implications for Foreign Policy," Papers presented at National Academy of Sciences, June 16-17, 1968, Washington, D.C.; and Man's Impact on the Global Environment: Assessment and Recommendations for Action, Report of the Study of Critical Environmental Problems (SCEP) sponsored by the Massachusetts Institute of Technology, The MIT Press, Cambridge, Massachusetts, 1970.

2 SCEP Report, *op. cit.*

3 The deployment of a belt of copper "needles" in the upper atmosphere for radio communication experiments.

4 For details, see among others, National Academy of Sciences/National Research Council, "Useful Application of Earth-Oriented Satellites," Washington, D.C., 1969; and "Selected Space Goals and Objectives and Their Relation to National Goals," Battelle Memorial Institute Report No. BM1-NLVP-TR-69-2 to NASA, July 15, 1969.

5 G. Child, "Famine or Sanity," New Scientist, October 23, 1969, Vol. 44, No. 672, pp. 178-181; World Population Situation, Note by the Secretary-General, UN Document C/CN.9/231, p. 117.

INTERNATIONAL COMPARATIVE TECHNOLOGY TRANSFER

Dieter Schumacher

Systemplan e. V., Institut fur Umweltforschung und Entwicklungsplanung
Heidelberg, Federal Republic of Germany

INTRODUCTION

Once again the international technology community has convened to find out how to make better use of technologies. The models, charts and case studies presented are concerned primarily with two kinds of TT processes, namely;

- the transfer from advanced technology sectors into public and private sectors, and

- the more classical innovation process leading from a given R&D base into the market place via product development.

When an international organization like NATO proceeds to convene some 100 technologists and technology managers from a number of countries, there is an implicit hope that such cross-national and cross-cultural effort will have a benefit for participants, for participant countries and for "the men on the street" in these countries. This paper is concerned with identifying and assessing the boundary conditions and suitable approaches for such exchange. The scope of the paper is limited to those areas of TT where public administrations play an important direct or indirect role in the TT processes.

Setting the Scene

It is not surprising that the bulk of TT efforts, as far as public administration is involved, has been performed so far in the United States, i.e., in a country in which the science policy of the 1960's, especially in the space and defense sectors, had led to a considerable amount of technology which was generated under rather unconventional conditions:

- It was designed to serve primarily apparent national needs rather than social and individual needs in the first place.
- It was therefore not oriented towards a multiclient market but to a consumer of a monopolistic nature.
- It was therefore not held to meet cost effectiveness requirement well accepted in the *private* market area.

The transfer of such technology into "real life market" and consumer structures could at no time be considered an easy job. In fact, any attempt to achieve a fair amount of return on investment would involve rather optimistic feelings about the feasibilities. The actual history of TT has shown that the hopes were exaggerated in quite a few areas. One can safely forget about the few instances where TT into private applications was possible *without* additional reengineering cost. For a given TT process, a comparative cost/benefit analysis would therefore have to look at the cost share of advanced technology development plus reengineering cost vs. development cost if there hadn't been anything to transfer at all, i.e., for conventional R&D support.

Transfer efforts like those of the NASA Technology Utilization Teams and of other agencies have no doubt, amongst others, a function of alibi, supposed to demonstrate to whom it may concern that this huge national expenditure has a vital impact on the people also in areas beyond the actual agency missions.

At the same time, however, they enable us to identify and to strengthen institutional mechanisms and opportunities by means of which a better use of technology can be made *regardless* of the source from which the technology originates. Thus, there is the hope that, by establishing new technology development mechanisms in the *presence* of a technology surplus, we may succeed in putting the same mechanisms to work under conditions where there is a *'normal'* technology supply only. The latter case corresponds, of-course, to the above-mentioned 'classical' innovation process. This aspect and hope is probably the most useful medium range objective of the whole TT campaign. It is a *forward-looking strategy* relevant to future technology development.

A Few Potential Biases

It is good to put these remarks at the beginning since

present case studies and organizational schemes in TT seem to suggest that they are narrowing the scope of TT down to the ex post revitalization of the technology surplus only. In relation to this 'backward' orientation of TT we can identify several biases which we should not impose on the future oriented part of TT efforts.

First, the need to have to get rid of *existing* affluent technologies from the past tends to undermine efforts to *avoid* such surplus in the future (ex ante), e.g., by better matching R&D and needs in the first place (which would of course be much more economic than doing all kinds of adaption work).

Secondly, dealing primarily with *advanced* technologies tends to reduce the attention for the development and transfer of *non-sophisticated* technologies. A good typical example was given where it was found to be difficult to transfer the zero-gravity-sewage-disposal system of NASA to U.S. highway relaxation facilities. A straightforward approach to the needs on highway parking lots might indeed be less expensive and more effective, if a skilled team would attack the problem right away.

Thirdly, the 'backward' oriented TT campaign is placing too much emphasis on pushing products into the market, the criteria of sales and productivity given almost exclusive priority in evaluating the success of the process. There is very little talk about quality of life impact, employment issues involved or consumer aspects on an individual level. TT people often forget about these aspects since the technology to be transferred is already on their desk, and the agency sponsoring the TT efforts leaves them little freedom to look for technologies somewhere else. In this respect, *TT can be conservative where it should be innovative*.

The 2nd Generation of TT

TT efforts which are not primarily concerned with reducing the technology surplus initiated and compiled in the 1960's, have to avoid the above-mentioned biases and should meet the following objectives:

- The process should not be just a one way street towards the market, but there should be a moderating feed-back loop on the amount of and the way in which new technology is generated.

- It should be dealing at least to a considerable extent with non-sophisticated technologies since in quite a few cases they are closer to the problem solution than one would assume (and possibly cheaper, too).

- TT needs to be placed into a social and societal context. The main guidelines and strategic concepts should be designed according to consumer needs where consumer stands for an 'optimum mix' of individual, group and public interests.

There is not much room for illusions about the feasibility of such demands. These objectives are, however, identical to what is being expected from *science policies in the seventies;* so there is a close relationship between the quality of the science policy of a country and the nature of the TT processes going on or being desirable: A well-balanced science policy (balanced with respect to sources and sinks) does not need a high TT effort of the 1st generation since there is no burdening surplus technology. It can rather concentrate on favourably advancing the 2nd generation type TT which puts prime emphasis on the consumer/market side and on providing the TT community with objectives (to reduce their random walk). This does naturally not mean that all R&D would be utilized. One would expect, however, that there is less misfit between sources and sinks of new and old technologies. One would hope to minimize the disequilibria which we generated during the 1960's.

A Remark on Methodology

The previous remarks call for a few words on methodology. The problem of affluent technologies has so far been rationalized largely by using *statistical indicators* like macro-economic input-output-relations, balance of trade data, figures on patents and licences flow etc. These categories of thinking are usually taken to present a case and to put pressure on governments to do something about all kinds of gaps seen by the technology push community.

On the micro-level, i.e., the level of individual companies, technologies or products, the tools of classical *business administration* and conventional accounting systems are being employed to describe transfer needs and opportunities.

In addition to these two approaches, there are quite a few model builders active in the field who provide us with a variety of models for the innovation process, its success and failures. Since models on such complex systems are compelled to involve a rather high level of aggregation, they are always likely to lack operationality.

While each one of these three approaches has a justification and merit for special applications, they are not able to cover the whole story for TT processes of the 2nd generation. More specifically, they cover only insufficiently the following two aspects:

- The consumer aspects, for which tools like social indicators, consumer satisfaction indices, etc. should be employed.
- The institutional mechanisms which vary from country to country and for which a general formalized description scheme is unfortunately not readily available.

The technology community is so far still lacking a comprehensive package of all methodologies named above: It would certainly put more weight to their public relation efforts.

As a kind of intermediate methodology, the case study approach is enjoying considerable recognition amongst professional conference speakers. The pragmatists among ourselves are fascinated by things which *have* been done, and they imply in presenting their case that we can learn a lot on how future things *could* be done. This hope is quite valid within a given industrial or policy sector *and* within a domestic institutional environment. It takes, however, more sophisticated and didactical skills to get one's ideas and experiences across to people operating in a different domestic and sectoral environment.

This leads us to the question of what we can learn from each other when convening for an international exchange of experience. Those presenting case studies frequently talk essentially to their own domestic colleagues (who presumably know the problems anyway). They leave it thus to the (international) listener to draw the conclusion he needs for his work. This procedure is fine for those who well know their own country and that of the case study. It leaves, on the other hand, quite a few participants in a somewhat isolated and merely passive situation.

Thus, the case study approach is not a bad thing, but we should put more pressure on case historians to include into their presentations an evaluation for the benefit of those who are not too familiar with the micro-world of the speaker. While this applies certainly to people from other countries, there is reason to believe that it would also apply to the people in one's own country who do not belong, however, to the well-established technology community.

A Framework for International Comparative TT

As a tool to achieve stronger communication links and learning effects across frontiers, a framework for analysis is suggested in the following table. In the vertical columns, we list the five major functions to be performed consecutively in a TT process, namely;

- Problem identification and specification
- Technology specification and search
- Match evaluation including technology assessment and feasibility analysis
- Product design and development
- Marketing

Any case study can readily be structured into these categories. Let us suppose now that four different public administrations, for example:

- The State of Georgia, U.S.A.
- The United States
- A European Country like the Federal Republic of Germany (FRG)
- The Bruxelles Common Market

have identified one problem such as the need for improving the fireman breathing apparatus: A problem of common concern, i.e., in principle a 'common problem market', but involving quite different ways and means to solve it within the jurisdictions, respectively.

The table shows for each TT function a check list of aspects and categories that needs to be evaluated in each of the four case studies. A *joint* effort of different experts to fill out this synopsis would clearly have the following advantages:

- to confront one's own experience with that gained in other environments, and thus

- to assess critically one's own operations and simultaneously

- to stimulate the concepts of the other colleagues.

This procedure would seem to at least improve to some extent the international coherence of cooperation which is inherent to the problem to be solved and to the technological know-how on problem solving anyway. The international technology community should make an appreciable effort to counteract renationalization of problems of common concern, leading to a rather artificial and suboptimal fragmentation of markets and skills.

The Design of an International Comparative TT Workshop

As a consequence, future international exchange meetings of TT experts should follow the following scheme:

Table 1

Framework (Check List) for International Technology Transfer

Field of Operation → / TT Process Phase ↓	The State of Georgia, USA	The United States	The Federal Republic of Germany	The Common Market
Problem Identification and Specification	Improvement of Fireman Breathing Aparatus			
Technology Search	Domestic/Abroad. Space/Defense Sector. Governmental R&D Facilities/ Agencies. Academic R&D. Industrial R&D. Etc.			
Match Evaluation incl. Tech.Ass. and Feasibility	Domestic/Nondomestic Evaluation. Patents/Licenses. Consumer Requirements. Jurisdiction, Regulations, Standards. Implementation Barriers. Other Constraints. Etc.			
Product Development/ Reengineering	Domestic/Nondomestic Exploitation. Allocation of Public Support. Venture Capital/Risk. Technical Expertise and Skills. Manufacturing Cost. Etc.			
Marketing	Domestic/Nondomestic Marketing and Sales. Market Structure/Size/ Aggregation. User Groups. Public Sector Purchases. Etc.			

1. Pre-Conference Planning

- Select problem areas to be looked at (such as public safety, medical technologies, etc.)
- Identify problems of common concern in participant countries.
- Develop framework and check list for synopsis and comparative analysis.
- Invite preparation of case study inputs in terms of the framework.

2. Conference

- Form working groups for each problem.
- Present in a synopsis various domestic solutions and/or
- Tackle the problem in *one* domestic environment by inviting contributions and comments from foreign environments.

3. Post-Conference Evaluation

- Evaluate the results with respect to domestic strategies, domestic science policies, joint international actions, etc.
- Publish the report within *three* months after the conference.

A further, almost natural requirement for 2nd generation TT exchanges would read to not only have TT *insiders* at the meeting, but also those who are potentially willing to employ TT as a tool for improving our living conditions and those who are supposed to benefit from all this work.

It would be quite likely to have a few fire chiefs, police department heads, sewage people and 'men of the street' at our meetings who would ask those questions which the insiders don't want to ask or can't ask, especially the nasty ones.

TRANSFER OF TECHNOLOGY THROUGH INTERNATIONAL LICENSING

Jacques Raffin

French Licensing Executives Society, Inc.

The economy and growth of industrial nations is founded on heavy investments by government and commercial enterprises in research and development which issues forth in a stream of new products, new processes, and new resources.

The trend today toward multinational business operations has speeded technology transfer from parent companies to foreign affiliates. Operating in a one planet milieu, the global company has provided for the efficient application of technology to the problem of economic development and higher standards of living for many people. But as we know, this is but one of the myriad ways technology transfer can be accomplished. Other ways include presentation of papers at professional meetings such as this, the publication of erudite articles in technical journals, the sale of products, joint ventures and last but not least, the must underrated and perhaps the most underappreciated, but in reality the most efficient - the international licensing of technology - the subject of my presentation today.

Licensing, a complex inter-relationship of technology, law, economics, management, and business is a sophisticated technique of transferring technical knowledge into economic reality. It requires a different type of skill and knowledge, it may invoke some changes of attitudes and values and it most certainly requires the manipulation of very different resources. These include patents, patent applications, designs, know-how and trademarks - otherwise designated as industrial property rights. Licensing has received an impetus in recent years due to many economic factors, of which three are most important: (1) Europe has become an exporter of inventions and know-how, and reciprocal licensing with U.S. enterprises has greatly increased; (2) New emerging countries aspiring to improve the standard of living of their people by speedy industrialization; and (3) U.S. investment activity thru subsidiaries, joint ventures and acquisitions of European enterprises was accompanied by licensing of U.S. owned industrial property rights.

International licensing is a multi-billion dollar business and it is a growing and dynamic area of technology transfer in most corporations. The U.S. has become a major exporter of industrial property (patent and technical know-how). The U.S. Commerce Department estimates that between 1965 and 1969 U.S. firms earned 6.4 billion dollars in royalty income. In 1970 the U.S. received 2.4 billion dollars in royalties while paying but 230 million dollars for the acquisitions of foreign technology. In 1971 the U.S. received 2.7 billion dollars in royalties.

What are the reasons for the international licensing of industrial property in the private sector:

1. To realize income from the money invested in the research and development of industrial property.

2. To convert idle or wasting patents and trademarks into cash assets.

3. To test market a product or to test a process.

4. To penetrate foreign markets blocked by high tariffs or sales barriers.

In the Government sector:

1. To provide the public with the immediate benefit of improved products and services flowing from government research.

2. To encourage private enterprise to seek profit-making opportunities in the betterment of the social sector (i.e., alleviation of water and air pollution, etc.).

3. To promote and increase the vigor of competitive forces within the economy.

4. To share the benefits of Government financed research and development with foreign nations.

Licensing is then, obviously, one of the most important forms of the international transfer of technology.

My purpose today is not to involve you with the specific issues of licensing such as the selection of a licensee and establishing a royalty base but to provide you with some fundamental concepts and a knowledge of the constraints involved in a successful licensing venture.

The industrial property rights granted in a licensing agreement are of three distinct types: patent rights, know-how rights and trademark rights. However, government licenses are confined to patents and with few exceptions the licenses are royalty-free and non-exclusive.

The combination of rights licensed in any particular agreement will depend on the respective resources and needs of the contracting parties and on the types of products or technology involved. (Patents are limited monopoly rights conferred by the government of an individual nation within its own area pursuant to its own local law.)

License agreements are primarily contracts and require careful preparation and drafting, as all contracts do, in defining the rights and obligations of the parties.

The Licensor's major obligations under a licensing agreement are of two kinds: (1) the granting of license rights with respect to owned or controlled patents, trademarks and/or know-how; and (2) the furnishing of information, technical assistance, components, and/or services of various kinds.

The licensee also incurs two major obligations: (1) to exploit the licensed rights and know-how in accordance with the terms of the contract; and (2) to pay agreed royalties and other remuneration to the licensor.

The primary purpose of the licensing negotiation is to settle and equate these reciprocal obligations. Virtually all of the other obligations and undertakings included in the licensing contract are for the purpose of implementing and safeguarding the bargain reached on these major points.

The following is a brief recital of the types of clauses commonly found in a typical license agreement between a French or EEC enterprise and a U.S. enterprise.

1. Identity of parties/address of each.
2. Whereas clauses:
   - representation licensor owns patents/know-how/trademark
3. Definition of key words such as "net sales", "field", "patent",etc.
4. Grant clauses:
   - scope of that which is given to licensee
   - patents/know-how/trademark
   - existing rights only or include future developments?
   - sub-license rights or not (affiliates?)
   - exclusivity or not (can be semi-exclusive)
   - territorial limits, if any

5. Royalty clauses:
   - minimums, down payment, running royalty
   - paid up total fee
   - some royalties are tied to a price index
6. Reciprocal grants:
   - license to licensor or outright grant of title
   - patents/know-how
7. Infringement clauses:
   - who sues and what happens if a third party infringes
   - some question re-defending licensee if he is sued
8. Confidentiality:
   - scope and term of years
   - if know-how only, should individual employees agree to secrecy
9. Warranty clauses
10. Admission of patent validity - for how long?
11. Term of license and conditions for termination-effective date
12. Accounting and right of licensor to inspect accounts - currency
13. Arbitration (sometimes provided)
14. Law of what country governs and language which governs if dual language
15. Force majeure
16. Assignability of license
17. Due diligence
18. Notices sent to whom at what address

It is important to note at this time that in reference to the sale or licensing of French industrial property rights including know-how, a French resident negotiating with a party abroad must file the prospective contract with the Ministry of Industry at least two months prior to its effective date according to law. (Decree No. 67-82, dated January, 27, 1967) The Ministry, however, cannot block the intended transaction.

This subject raises perhaps the biggest obstacles or constraints to international licensing agreements - Governmental Prohibitions.

The anti-trust laws of the Common Market and of the U.S. are perhaps the most important governmental constraints on domestic and international licensing within those jurisdictions. Article 85 and 86 of the Rome Treaty of March 25, 1957 which created the European Economic Community and Regulation 17/62 effective on March 13, 1962 and the U.S. Sherman and Clayton Anti-trust laws all have a common purpose - to prohibit:

1. Commercial or industrial combinations which prevent or limit the competition of the free enterprise system.

2. The creation of monopolies, that destroy or impede such competition.

3. Unfair competitive and business practices that hinder competition and contribute to monopoly.

   Licensing agreements cannot impede competition directly by including price, geographic, field of use and other restrictions or indirectly by royalty arrangements that discourage competition. By their terms they may not boycott or otherwise injure third person, or obstruct channels of distribution, and at the same time adversely affect innovation by means of closed pools, tie-in arrangements, discriminatory conditions as between different licensees and so on.

   They may not lessen the incentive to engage in competitive innovation by imposing limitations upon the use of new technology developed or acquired by the licensee or upon methods of distribution.

4. They may not cause competitive imbalance through excessive acquisition of technological property by purchase, merger or grant-back.

Other constraints arise out of the very nature of licensing transaction.

1. Inadequate market analysis by licensor
2. Product defects not known or understood by licensing executives.
3. Higher start-up costs than anticipated by licensee.
4. Insufficient attention, interest and support from licensee's top management, marketing, engineering and production executives.
5. Poor timing.
6. Competition (not just local or other American competition, but from third country competitors).
7. Insufficient marketing effort on the local scene (by licensor and/or licensee).
8. Inadequate licensee "after sales" effort, and
9. Weakness in licensee market research and competitive intelligence.

I have tried to briefly set forth the rationale to support my beginning statement that international licensing is in reality the most efficient way to transfer technology today. My goal in this presentation was to help you learn, feel, appreciate, and understand licensing.

Understanding as Alexander Pope might have put it, is the key to a drawer wherein lie other keys. When we come to appreciate and understand the transfer of technology through international licensing we can encourage beneficial change and the continued renewal of all our societies.

In closing, I recall the story of when Marshal Foch came to the United States after World War I. He had been the Supreme Allied Commander in 1918. He was assigned a young United States Lieutenant as an aide-de-camp and interpreter, because the old Marshal's English was poor, and the young lieutenant had served in France and spoke good French. The two had gotten to be fairly close friends. One day the old Marshal was looking down into the Grand Canyon. The young lieutenant was standing back a few paces, and kept thinking to himself "now there's the greatest man in my profession looking at this great feat of nature, the handiwork of the Almighty, and certainly that old gentleman is going to say somethin to pass on to my grandchildren." Well, the old Marshal finally turned to the lieutenant and said, "My what a lovely place in which to throw your mother-in-law."

Now nothing I've said today has been very profound. You may have been standing back waiting to hear something truly profound when the solution to the problem is in hand and quite commonplace. The simple facts are all I intend to give you. They are all I ask you to pass on to your collegues.

MAX. P. (Maximizing Profits) or How to Obtain Higher Financial Rewards from International Technology Transfer

Win Straube
Pegasus International - New York, Brussels, Sao Paulo, Tokyo

The exchange of technology can do more for a country than keep it competitive in world markets. It can add considerable prestige to both licensor and licensee. AND it can be profitable.

It is the latter feature which I would like to analyze, or more specifically: how to make international technology transfer MORE profitable.

I want to prove that higher than 5% royalty rates can be obtained - and why rates of up to 200% are being paid ... cheerfully.

I will present a case study to illustrate one method used to obtain higher than usual earnings for both licensor and licensee.

I hope that you agree with me that profits are desirable, for nobody wants to incur a loss in exchanging technology, nor does anybody want to give away the fruit of his labors for free. In international technology transfer profit is not a dirty word.

We at Pegasus International Corporation in New York - and in our branch offices around the world - consider it one of our main tasks not merely to accomplish technology transfer, but to make sure that both seller and buyer reap the highest financial rewards possible.

To do this we have no magic wand. On the contrary, we do a lot of painstaking detail work. We also use a computer. However, maybe the most decisive ingredient for the right outcome of our work is to combine a maximum of available facts with imagination. Therefore, our formula for maximizing profits from technology transfer is no secret, and I want to share it with you today.

But please do not expect a simple recipe. For us increasing financial rewards from technology transfer is more than a method; it is facts and information properly utilized and acted upon.

Technology transfer in the form of licensing a product or process to others is by itself a form of maximizing profits on existing know-how. For instance, if all your production capacity is used up, or if you don't want to invest in markets you are not ready for or unable to cover properly by your own efforts, technology transfer through licensing does two things:

(1) It makes somebody else in those international markets work with you instead of against you, and it may keep others from trying to copy or overcome your technological advantage - at least temporarily. In other words: It helps you maintain your international competitive position.

and

(2) It gives you additional revenue for which you didn't have to expend material or production labor.

I will disregard the obvious rewards of technology transfer here, such as being competitive or deriving a normal licensing income. I am dealing here strictly with the <u>PLUS</u> in profits which can be obtained. Anybody can give some technology away or obtain it and profit somehow by doing so. My concern, and the concern of my colleagues at Pegasus is: How much <u>MORE</u> can we obtain.

Let me stress at this point that successful and profitable technology transfer is <u>NEVER</u> a one-way street. Like in any good long lasting relationship of buyer and seller, both must obtain adequate benefits. The minute a licensor is so hungry to try and get all the profits for himself, leaving the licensee with little more than work and a nominal advantage, the relationship is not going to last very long.

Sharp deals are out. A steal or negotiated give-away is worthless. Apart of the fact that it would be unethical, since technology changes very fast nowadays, long term cooperation is more important than a short term financial gain. And only a clear understanding of who derives which benefits from what will ensure a healthy and lasting relationship.

The basis for <u>MORE</u> profits between parties offering and buying technology is solid ethical conduct and complete disclosure of intentions.

Coming to the HOW of maximizing profits from technology transfer, you are all familiar with the patent attorney or corporate licensing department, where when the question of technology transfer comes up, this issue is mainly a question of percentages and form. Standard contracts are used with blanks to be filled in.

> What is the percentage we charge on licenses of this type?
> 5%.
> Well, let's try for 8.

Or if he represents the other side:

> Let's try and get it for 3.

Or:

> Of course the agreement must be in our language and the jurisdiction in case of court action is to be that of our home town . . .

I am not saying that these questions are not important and should be neglected. Of course not. But what I am trying to focus attention on is this: If you want to increase financial rewards in technology transfer transactions, you cannot look at the process piecemeal, as a patent project, a licensing project, a marketing project, etc., but you have to look at it as a whole. You have to be very skeptical with boiler plate agreements and you cannot use a stereotyped approach to finding a market.

Research in depth is necessary. This is hard to do from your desk at home or by merely writing to consulates or chambers of commerce. You must get up close, very close. Your problem is always very specific - not general. If you want to get more for your technology or derive more benefits by obtaining it, it is obvious that you will have to do a little more than the ordinary.

Let me show you an example of what I am talking about. This case study is a true relationship as handled by our company, Pegasus International. It also is not a closed and shut case out of a dusty file, but one that is alive and producing healthy revenue every day.

I have brought with me a sample of the product which we were to sell or license for one of our American clients. Here it is. The device is very simple. It combines the compressed gas principle of the aerosol with an ordinary atomizer. There are three parts:

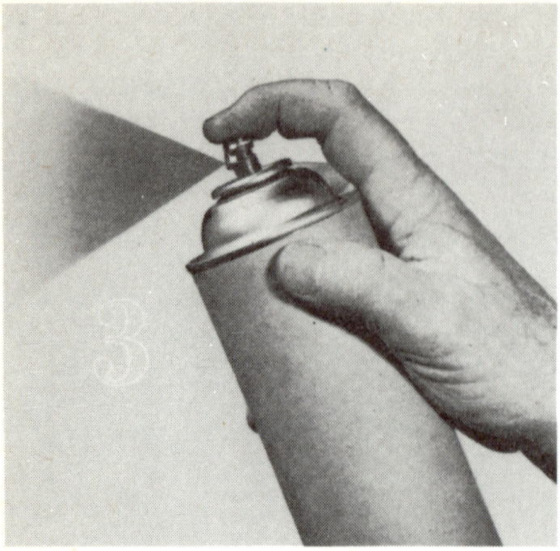

(1) the can holding the propellant gas closed off by a valve

(2) the glass or plastic reservoir container, which holds the material to be sprayed; and

(3) the bridge, which links the two together. Only the last, No. 3, is patented - a novel design of expansion chambers results in an increase of efficiency in the amount of gas being used to propel the material being sprayed - an increase of about 35%.

From a pricing point of view, the separate components cost

(1) 65¢ for the propellant container

(2) 10¢ for the jar with cap, and

(3) 15¢ for the plastic bridge.

In our initial thinking about this item for export sale, we had to consider that most customers might desire to produce the two non-patented components locally, and thereby avoid transport and import duty costs involved in purchase from the United States. This would mean that export sales would very quickly be limited to the bridges only, which at 15¢ each - and considering that they are reusable - would mean a very small dollar volume in return for the considerable promotional effort involved in introducing the product.

Since most license agreements provide for the manufacture, use and sale of the product being licensed, we felt that in this case we were not merely selling, but also conveying a right to USE the patented item, and that consequently, a royalty, as part of the sales price would be appropriate. Our market studies showed that five refills plus five jars were used - and I emphasize again the word USED - with each single bridge before the bridge wore out or became clogged or lost. It seemed appropriate to assess a royalty figure that approximated 5% of the total value of the USE of the product. Totalling up one bridge at 15¢, plus 5 jars at 10¢ each, plus five refills at 65¢ each gave us a total figure of $3.90. Applying a 5% royalty thereto gave us a royalty figure of 19.5¢. This would be a royalty higher than the cost of the bridge itself.

Therefore we compromised and added only 15¢ royalty to the price of the bridge, making its cost for export sale 30¢ each.

We were careful to point out in our negotiations with prospective licensees and customers that the royalty was calculated on the cost of not the bridge alone, but rather the cost of all the components used throughout the expected life of the bridge. At 15¢ royalty, this meant that the actual royalty percentage was 3.85%, which is generally considered an equitable percentage - although in fact this means a 100% royalty being paid on the item our client is supplying.

In the larger sense, thus, we established this pricing policy with the long term view that licensing the production of these units was inevitable, and we wanted to have our logic straight from the start to justify this figure.

License negotiations did in fact occur, with the result that a licensee was set up in one major industrial country with rights to sell there and in a few selected areas elsewhere. There was no objection to the 15¢ royalty rate, because of what we had done in the market first, which was to establish an export price based on the use of all components which was still competitive.

We now had a source outside the U.S. for the unpatented components of the unit, namely the refills and jars - as well, of course, as the bridges. With a growing international market for this device, we considered that one or more additional licensees might be required.

But before going to this stage we did another study to find out what it would cost to have the refills and jars made in the international market. It turned out that we could buy the refill fully packaged for 40¢ and the glass jars for 6¢ each. Rather than license someone else we went to our customers and told them that we would shortly be able to supply them with refills and glass jars at the prevailing U.S. factory price, but instead of their paying freight and import duty on products from the United States, the U.S. domestic price would now apply delivered duty-paid to their factory. They of course, were delighted, and immediately held forth about tripling and quadrupling sales in a few years.

We then, in association with our American principals, set up a separate company to develop this international market. With a nominal capitalization, this firm contracted with manufacturers to produce refills and jars and hold them in their ware-

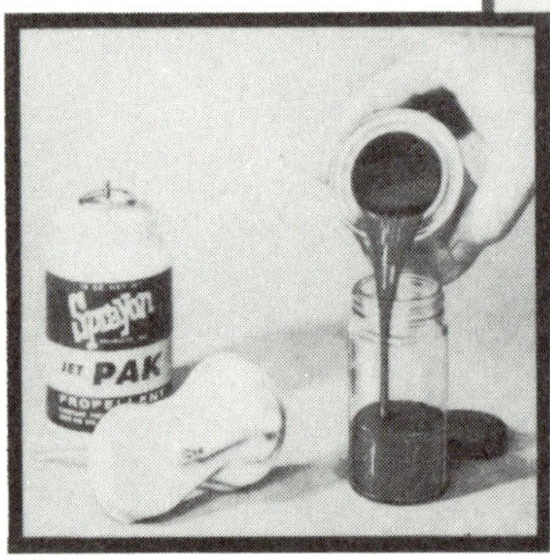

house until instructed to make deliveries in accordance with orders received by the joint venture company. Invoices were then rendered by that joint company to the buyers at the prices previously agreed upon. Eventually we arranged with the suppliers. to accept orders directly and invoice directly to the buyers, remitting to us only the price differential.

The net result of all these individual steps was: Instead of receiving a "normal" or "ordinary" royalty of 5% or 15¢ or 3/4¢ per bridge, the American licensor is now receiving 15¢ royalty per bridge plus the markup on the refills and jars, which amount to a total of 25¢ for the former and 4¢ for the latter, making a total per set of 29¢. Going back to our original premise that five refills and five jars are used with one bridge, his total profit now amounts to $1.45 plus the 15¢ from the bridge royalty, or $1.60.

Therefore instead of 3/4¢ royalty income per piece through "ordinary licensing" our American clients are receiving $1.60 via Max. P.

Mr. Max. P. as we call it is our short form for Maximum Profits.

In this case involving Max.P. represents a 200 fold increase in expected income through a combination of policies involving export sales, licensing, and joint or wholly owned subsidiary activities in the international market. All three activities cannot be divorced - one from the other - all these must be totally integrated. And, most importantly, a master plan has to be established BEFORE any commitments are made. Too many times isolated international activities are generated within a company with, in the end, a truly unworkable and frequently unprofitable conglomeration of separate and often conflicting operations being conducted.

So far with this example. I would like to add that techniques of this sort are not necessarily applicable to all types of technology transfer. I also would like to point out that all through this exercise, as you have seen, we did not merely bestow a benefit on our client, but also on the licensees by expanding their opportunities and giving them more value for less money. Naturally, under different circumstances the details also will differ.

To summarize: Max.P. can be introduced into international technology transfer by observing the following:

(1) Ethical and open dealings between the parties.

(2) Maximum knowledge of all facts concerning the technology, the use of the technology, the market, competition, prices, and alternatives.

(3) Ability to coordinate exports, service, support activities, licensing and cross licensing.

(4) It needs you.
You make the difference, the man or organization which puts these factors together and comes up with Mr. Max.P.

Too many times when one thinks of technology transfer or licensing one considers only the obvious straightforward agreement providing a royalty rate, minimums, and fees. And as well with joint ventures or wholly owned subsidiaries, one considers only that this requires a large scale investment in plant and facilities. There is a happy medium, where both risk and investment can be minimized and yet profit maximized. There are many other variations of this type which are outside the framework of this presentation. My brief appearance here today is merely to give a glance at some of the avenues which are open to benefit more from technology transfer.

In many places outside the United States, American technological progress reaching around the world has been felt as a challenge. However, as you can see from my small presentation here, maximizing profits on international technology transfer transactions need not be an American privilege. The same avenues and combinations of effort are available to everybody else, also to be practiced in the United States by non-Americans. Technology transfer is an international necessity. Whether it is profitable or not depends on the technology itself, the market conditions, research and sales efforts. But <u>HOW</u> profitable depends on you.

INTERNATIONAL TECHNOLOGY TRANSFER IN ONE INDUSTRY - AIRCRAFT

Marvin J. Cetron and James L. Duda

Forecasting International, Ltd.

The United States is rapidly approaching technical parity with its competitors in world markets. The closing of the technology gap is apparent in the balance of U. S. trade with the world, and in applications to the U. S. Patent Office where two out of every five are foreign as compared to one out of five twelve years ago. Although computers, electronics, aviation, and agriculture are still on the plus side of the balance of payments ledger, several of these industries are facing an increasing onslaught of technical and economic change. This change is due to competition with foreign innovation and the effects of productivity improvements, via automation on domestic employment levels. American firms have been losing their share of the market; and they are associating these losses with the technology transfer process.

A basic goal of any industry is to keep or improve its share of the market while maintaining employment levels. To accomplish this objective, U. S. based industries face many complex and diverse problems in the international market place. Most of these problems can be attributed to the fact that foreign firms have taken the technological lead, and the fact that, in some instances, certain foreign manufacturers can produce more cheaply. In the aviation industry these problems and their proposed solutions are far ranging, and exceedingly complex, involving such issues as capital and labor productivity, industry structure, private international investment position, and degree of government involvement. This paper identifies and discusses the role of the technology transfer process on the current status and direction of the U. S. aircraft industry.

Over the past eight years the U. S. aviation industry has exported over $20 billion more than it has imported; and has been the largest single contributor to U. S. export earnings. However, this position of strength shows signs of drastically weakening. Several foreign competitors have already developed their own commercial aircraft and have successfully entered the marketplace. For example, in selected aviation areas, such as V/STOL, European countries are starting to compete successfully. An LTV Aerospace Corp. subsidiary is now assembling helicopters designed by France's Aerospatiale. Nippon Aeroplane Mfg. Co. is taking aim at U. S. feeder airplanes with its US-11 short-hop turboprop. The European builders of the Mercure short-haul jet and the A-300B Airbus are also aiming at the U. S. market. To top it off, the U. S. is facing total loss of the supersonic transport aircraft market.

In the military sector the United States is also losing its share of the overseas market. Japan has decided to build its own fighter planes which might be a prelude to provision of Japanese made aircraft to Asiatic markets. A large share of the market is going to France's Marcel Dassault and his Mirage planes. The Netherlands, contemplating purchase of 200 planes, is considering Dassault's Mirage F-1 and the Viggen from Sweden's Saab. Northrop's Cobra is about the only U. S. airplane in contention for the Dutch order.[1] In addition, France has edged out the U. S. to supply modern weapons to Latin American nations. While the U. S. was hedging on the order to study the situation because U. S. arms sales reflect hemospheric policy and attempts to maintain some kind of parity, France simply provided the Latin American nations with what they wanted rather than attempt to limit, or to convince, these nations to invest in something else.[2] There can be little doubt that the U. S. aviation position is slipping and that this slippage may very well be irreversible.

R&D is the most critical factor in the growth and market position of the aviation industry in general. Industry leaders such as Sanford N. McDonnell, President of McDonnell Douglas; Robert Anderson, President of North American Rockwell Corp.; Allen E. Puckett, Executive Vice-President of Hughes Aircraft Co. and others have noted that domestic expenditures in R&D are declining and that successful R&D programs are crucial to future growth and survival.[3,4] The problem today, it is felt, is the declining government support of R&D in military research (and NASA for that matter). The investment in military research has paid off and is attributed to be the underlying cause for the current dominant U. S. position in the long-haul turbine aircraft market. The basis for this position is our effective use of the R&D data generated by the military during World War II and continuing government support of R&D since that time. Figure 1 presents the past and future of aviation. It also shows the relationship between military and commercial performance, as well as an indication of the timing of various programs.

However, the terms "R&D" and "technology gap" are meaningless without focusing on very specific technology at the "nuts and bolts" level. Successful programs are dependent upon the application and engineering refinement of demonstrated technology as reflected in the technical quality of equipment systems, superior production capabilities, and well-executed marketing plans. The success or failure of aviation programs is attributable more to these factors than to the technology transfer process as we know it. This is not to deny that aircraft developers the world over take advantage of major breakthroughs in technology whenever they occur. Variable sweep and vertical takeoff are obvious examples. But it may also be noted that in these cases, which may fairly be said to represent two of the more important recent advances in the aerodynamics sciences, Europeans have *not* been dependent on U. S. technology.

Both Britain and France, for instance, have flown fighter-type vertical takeoff aircraft appreciably more advanced in a technological sense than any American equivalent. Apart from a demonstration that the stubwing configuration was probably the most effective of several variable-sweep configurations, most European development has been independent of U. S. experience.[5] The Anglo-French SST, the Concorde, appears to rely in no important respect on American advances in either aerodynamics or structures. It stems instead from independent work conducted by Sud-Dassault and the British Aircraft Corporation.

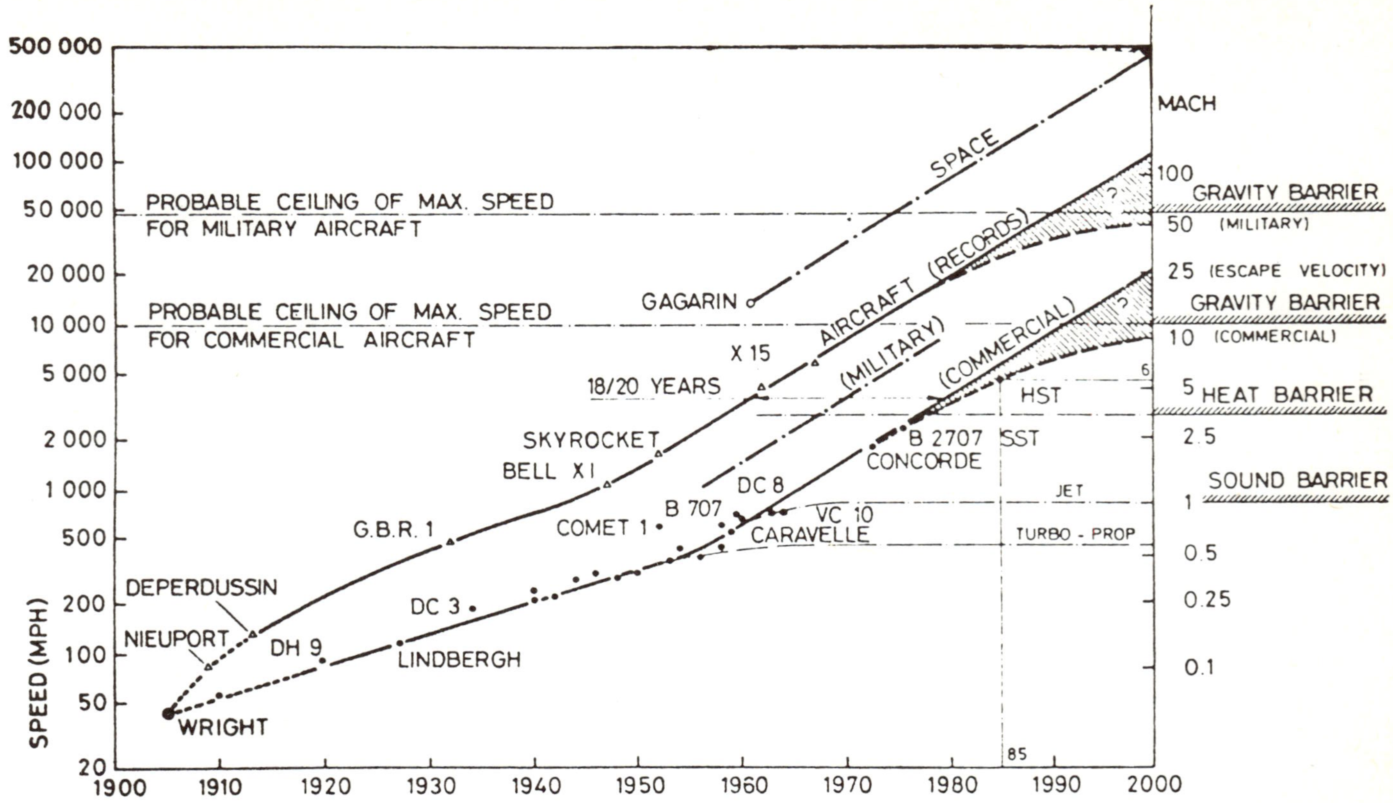

Source: Aviation's Role in Future Transportation *Technological Forecasting*[1] (1969), 5-16

Figure 1. The past and future of aviation.

Apart from the evidence considered above for and against technological transfusion, there are elements in the nature of European aircraft development processes that tend to impede heavy reliance on American technology. First, there is the matter of production rates and their effect on design and production factors.

Typically, European aircraft are built in smaller lots and are produced at an appreciably lower rate than their American equivalent. The tooling, and thus the production parts for production aircraft, are not, and cannot be highly similar. Thus, 561 Boeing 707/720 type aircraft had been built or were on the order books in 1971, while only 54 VC-10s were produced at Weybridge before production ceased in 1969.

Inputs are appreciably smaller too: deHavilland designed and tested the Trident aircraft using a design group that never exceeded 600 in total strength, while Boeing at one point had 1,600 designers working on development of the highly similar 727. (Both were maximum numbers: 450 to 500 was "normal" for deHavilland, and 1,100 to 1,200 for Boeing.)

There is no evidence whatever that either design influenced the other, save perhaps in initial concept. The Boeing airplane arrived somewhat earlier (but apparently had a later design inception), was somewhat more complex, and clearly was the more successful in the world market. There is no convincing evidence of technology transfer: both companies built on earlier experience in relatively successful transport aircraft (although the Comet had proved in the end to have a fatal design defect).

The commonly accepted explanation for larger American design staffs is that they presumably are required to plan and prepare for large production runs. Nevertheless, a threefold difference, or even more, in engineering labor inputs cannot be entirely explained in such terms. A more attractive explanation is that few European firms can afford the investment of resources that characterizes a major American aircraft development. They economize not merely by limiting staff size, but by limiting the scope of work. They do not, in fact, attempt to exploit the full range of nominally available technology. This is true even in the case of the Concorde. A case analysis of Concorde development and comparative domestic development follows.

The Concorde is the most ambitious commercial aircraft development yet attempted. However, the British and French were appreciably more conservative in SST design goals than their American counterparts -- and they relied for a technology base on the products of French and British research establishments, not American institutions. It would have been possible for the British-French design team to have chosen a variable-sweep design for the Concorde; the technology was nominally available to all who cared to read NASA reports on variable-sweep, and it was at least comprehensive enough to convince Boeing that variable-sweep should be used in a first-general supersonic transport. (That the concept overlooked some major faults of variable-sweep and that Boeing's SST later was changed to the sort of platform the British favored is somewhat incidental -- but it is interesting nonetheless.) In any case, the British proceeded from REA research on thin delta wings to an SST design, and the Americans proceeded from NASA research on variable-sweep wings to an SST design. Nor in this instance, at least, is there much evidence that the Anglo-French group drew at all heavily on the products of American turbine engine technology -- the Olympus engine is thoroughly British.

What is difficult in the development of an aircraft, what causes it to be unusually expensive, or sometimes unsuccessful, seems to be faulty engineering and faulty project management. *It is the detailed design of spars and landing gear, the fit of various parts, and the functioning of critical elements of the aircraft that present difficulty, not the technological base for the aircraft design.*

European engineering does not seem, on balance, to be inferior to its American equivalent. If the detailed engineering developed for one aircraft cannot be transferred to an aircraft even moderately different in design unless the original builder cooperates,[6] and if no European aircraft is clearly similar to any contemporary American aircraft, there is little support for the argument that the European aircraft industry is heavily dependent on the output of American technology.

Differences in process in a variety of efficiencies, in development and manufacturing practices and strategies, in tooling, in labor intensities, in labor costs, in program manning, and in materials costs appear to provide more reasonable explanations for observed differences in qualities and prices between comparable products of the aircraft industries of the United States and Western Europe.

At least, that is the indication of evidence drawn from this analysis of the Concorde and other general comparisons of a variety of American and European aircraft. There is, moreover, no convincing evidence that European aircraft manufacturers have drawn more extensively on the products of American technology -- except that *in some cases there may be more American technology to draw on because of the typically more rapid progress from design to production in the United States*. Finally, whenever levels of performance arising from technological advances are influential in affecting customer preferences, the United States seems to derive a significant advantage in being able to make a more rapid transition from design to production than can West European aircraft manufacturers.[7]

As far as domestic practice is concerned, there is a similar lack of transfer. Of the three principle turbojet-powered commercial transport aircraft built in this country in the late 1950's and early 1960's (the Boeing 707, the Douglas DC-8, and the Convair 880), the 707 appeared first, sold better, and generally set the pattern for its contemporaries.

Douglas had relatively good access to the engineering details of the Boeing airplane and full access to the aerodynamic and structural elements, or research technology, on which the Boeing airplane was based. Convair had somewhat less access to 707 details at the same stage, but clearly was able to obtain full access to the underlying technology, most of which was available in technical publications dating from 1954 and earlier.

Additionally, the well-known tendency of aircraft engineers to migrate from company to company and program to program, particularly pronounced in tight labor market conditions of 1958-1961, virtually insured that skilled personnel familiar with many details of one commercial program would sooner or later join a competing company. There was an interchange of personnel among Boeing, Douglas, and Convair after the 707 program began and before the DC-8 and 880 completed their design and development phases.

Nevertheless, neither the early DC-8 nor the 880 was as successful either technically or commercially as its predecessor. The 880 was probably the most economically disastrous commercial aircraft program in U. S. history (to that time), while the DC-8 proved to be appreciably more difficult to develop than the 707. Indeed, notwithstanding the earlier appearance of the 707 and the evidence that its design was highly acceptable to world airlines, Convair produced in the 880 an aircraft with higher seat-mile costs, higher maintenance and overhaul costs, lower passenger capacity, and generally less passenger attractiveness. The DC-8 encountered appreciable development difficulties, particularly in wing design, and proved more difficult and much more costly to produce than Douglas had anticipated. There is no indication, to say the least, that access to 707 data, or to the data on which the 707 design was based, gave Douglas or Convair an advantage. Both had to do their own engineering development.

Convair's engineers, and those who worked on the Douglas DC-8, appear to have been in no important way intellectually inferior to the engineers Boeing used in the 707 development program, and both Douglas and Convair had the apparent advantage of being able to proceed from a base of information developed by Boeing and from still more recent basic technology. Yet, neither the DC-8 nor the 880 was as successful as the 707.

A basic misunderstanding of the R&D process underlies much of the concern for inadvertent transfers of technology to Europe. First, the fundamental advances in technology are not necessarily American, as is often assumed. Application of new technology to utilitarian purposes is often done first in America, however. It is in the application, in the engineering refinement of demonstrated technology, that American aircraft development largely excelled. Engineering details are *not* readily transferrable without engineering drawings, which most aircraft companies tend to be reluctant to offer.

The main reason for the success of foreign or domestic aircraft programs is "aggressive" R&D coupled with the right combination of initiative, entrepreneurship, risk, capital, and timing. R&D program decisions must be made and public acceptance must be accurately gauged at a point in time which will give program proponents enough edge to assure them a firm grasp of the market before others get on the "bandwagon." Continued maintenance of this market share is also dependent upon aggressive R&D programs to improve the quality of the product, production techniques, or to develop new markets and products.

Since the curtailment of many military R&D programs and the subsequent realization of the consequences of that decision, technical and economic adjustment in the aviation sector has proceeded on several fronts. These activities are directed toward the maintenance of the technological lead through the forming of consortia, development of new markets, subsidies, more R&D for military/commercial programs, outright government sponsorship of commercial programs viz. U. S. SST, ease of antitrust laws, and other incentives. Some of these activities are discussed in the following paragraphs.

U. S. aerospace firms are forming joint ventures with foreign countries. Boeing will join with Japan on a $600 million venture to build a small (150-200 passenger) wide-bodied, low-noise, short takeoff airbus for use on domestic Japanese routes.[8] The General Electric Co. has joined forces with SNECMA, owned by the French government, to produce the CFM 56 aircraft engine for use in STOL aircraft. Pratt & Whitney will join forces with a German consortium, MTU, and an Italian group formed by Fiat and Alfa Romeo to produce the JTIOD, a competitive engine.[9] These engines will compete to power the next generation of commercial aircraft replacing the Boeing 727 and 737 and the McDonnell-Douglas DC-9. Other competition in this category is Britain's Rolls-Royce which is trying to put together an engine consortium with French, German, Swedish, Italian and Belgian manufacturers.

Joint ventures are convenient and effective ways of combining resources to share the risk of very costly development programs as well as to raise necessary capital. A beneficial side effect of these joint ventures was offered by William I. Spencer, president,

First National City Bank, New York, who believes that multinationals aid peace by increasing economic dependence among nations.[10] However, although consortia formation has been very beneficial to foreign countries, it may not be as beneficial to the U. S. where the lead is now being threatened. Joint ventures between the U. S. and foreign countries "sow the seeds" of future competition, especially when one of the countries involved presently has no share of the market.

An example of such a joint venture is the possible collaboration between Boeing and Italy's government-owned Aeritalia on a new transport. McDonnell-Douglas claims that Boeing is going to give the Italians a technology transfusion and thus produce a future independent competitor, especially because the Italians are producing no such aircraft at this time.[11] On the other hand, Aeritalia is probably contributing a substantial amount of capital, and there are reasons to believe that the net effect of such a venture could be positive. The point is that joint U. S./foreign ventures are taking place because domestic manufacturers do not want to risk total loss of their share of the market, or more drastically, loss of the whole company; and antitrust laws prevent joint ventures between U. S. firms. Corporations such as McDonnell-Douglas can not join with Lockheed or Boeing but can go in with Hawker Siddeley, Aerospatiale or Mitsubishi. Since new ventures these days easily risk two or three times the net worth of companies the size of Lockeeed or Boeing, there seems to be little choice.

Government subsidies could fill the gap. Many European countries are owned or heavily subsidized by their governments. Aerospatiale (SNIAS) of France is government-owned; the Italian government own 50% of Aeritalia; and the Nikon Aeroplane Manufacturing Company (NAMCO) of Japan is 53% government-owned![2,13] The French and British governments, subsidizing Concorde SST development, do not expect to recover any of the estimated $2.5 billion cost. Many other examples exist for the countries mentioned as well as for other countries.

An indirect form of subsidy would be more government investment in military needs which have potential civilian spin-off. A suggestion along these lines by Sanford N. McDonnell of McDonnell-Douglas is to develop a commercial airplane that has, for example, 70% commonality with a military airplane, which would ostensibly save 70% of the R&D cost already paid for by the government.

Total government sponsorship of commercial aviation is yet another alternative. This approach entails government participation in determining what the next generation commercial aircraft should look like, and then through competition and contract awards, paying for the R&D and supporting the production of the aircraft The now defunct U. S. SST program was proceeding along these lines before it was dropped.

The international market place offers its own possibilities. The U. S. share of the market can be enhanced by trade with the USSR and Red China as well as through independent increases in demand.[14] Production rates and employment levels can be maintained through increasing demand in passenger seat-miles. Increases are likely, owing to the escalating world population, growing affluence in both developed and developing countries, and the increasing availability of air travel through trade agreements and international travel arrangements between members of the Free World and Communist countries. It is recognized, however, that in the short run the market can be influenced by many factors, political events, economic perturbations, and spectacular crashes as evidenced by the in-flight break-up of the Tupolev TU-144 in front of a quarter-million audience at the Paris air show.[15]

All of the activities just briefly described basically deal with ways the U. S. can maintain the technological lead and market share in the aviation industry. Each has varying degrees of implications with regard to the international technology transfer process; from exposure to technological breakthroughs to direct technology transfusion at the most basic levels, as well as in national trade policy implications. Because each alternative has its own characteristics, good and bad, it is expected that ultimately there will be some mix of joint ventures and government aid in one form or another.

In any event, it is apparent that the days of total U. S. dominance in the aviation industry are over. Furthermore, it does not appear that the present U. S. position is being eroded by the international technology transfer process, per se, so much as by the results of effective and independent R&D programs by European consortia. In response, U. S. entrepreneurs, hampered by antitrust laws, are going for joint ventures with foreign firms in order to keep a share of the market. Although such ventures appear to be beneficial in the short term, both the short and long term effects of these technology transfusions bear special scrutiny.

Because of the complexities of the international economic environment, a thorough study is warranted in order to assure that full consideration is given to the future ramifications of any corrective action by the U. S. Government. In this manner the U. S. will best serve the public interest and espouse policies which are equally just to our foreign friends.

Problems now faced by U. S. aviation firms should be examined by determining the degree to which foreign firms have cut into the U. S. market, how U. S. firms have reacted, and how have domestic economic measures affected the market share. It is recommended that the current status and trends of the aviation industry be studied with respect to each country or countries involved. Although many excellent studies have been completed on this subject, they all point out the singular lack of a comprehensive data base. Therefore it is recommended that special measures be taken to assure a complete and accurate data base. The kinds of information required are listed below:

- Product Line
- Sales Volume
- R&D expenditures as percent value added in manufacturing
- Foreign sales as percent domestic sales
- Profit as percent sales
- Foreign earnings as percent total earnings
- Labor as percent manufacturing costs
- Length of production runs
- Civil aviation balance of trade
- Air carrier financial position: capital assets, overseas equity, etc.
- U. S. vs. foreign bank orders
- Other similar factors

Study of such indicators will yield response patterns to economic and technical change in the U. S. and world economy. Whether indicators, or combinations of such indicators demonstrate stable, increasing, or decreasing trends could very well explain certain interactions and, support or disprove stated hypotheses. International cooperation would be required to obtain current and specific data.

In addition to the lack of comprehensive data on a world-wide basis, decisions are made without taking advantage of the techniques of quantitative analysis. The observation that many phenomena are characterized by exponential growth provides the mathematical basis for quantitative analysis, as related to technological forecasting. The exponential provides the analytical tool for projecting performance, predicting the timing of technological events, and analyzing the rate of substitutions of one technology for another. By accurately fitting an exponential curve to preliminary data early in the process in a selected technology, a powerful predictive tool can be developed for planning and decision making. In this regard, another specific area worth stydying is the importance of timing; R&D programs, it seems, can occur too soon or too late. The Comet I, for instance, was ahead of its time as evidenced by its relationship to the curves depicted in Figure I presented earlier, and its early demise as a passenger carrier. On the other hand, neither the Douglas DC-8 nor the Convair 880 were as commercially successful as the Boeing 707, apparently because they appeared too late. The eventual economic fate of the Anglo-French Concorde, the Russian TU-144, and the suspended U. S. B-2707 SST may very well be a function of the timing. The techniques of technology forecasting and technology assessment can lend some insight on timing factors relating to technical feasibility, economic viability, and social and political acceptability of such concepts.[16]

If such a comprehensive study could be successfully accomplished, the results would constitute a general contribution toward the solution of the complex and serious problems of economic and technological adjustment now faced by many countries, and would benefit the affected firms, governments, and the quality of life in general.[17]

REFERENCES

[1] "Can the Pentagon Mix Maseratis and VWs?", *Forbes*, January 15, 1973, p. 19.

[2] "The French Arms Invasion," by Jean Ross-Skinner, *Dun's Review*, August 1971, p. 52.

[3] "Making U. S. Technology More Competitive," *Business Week*, January 15, 1972, pp. 44-45.

[4] Sanford N. McDonnell, "Another Export Industry is in Jeopardy," *Forbes*, May 15, 1973, p. 84.

[5] It is claimed, for instance, that the wing box of the F-14 being built by Grumman owes as much to Dassault-sponsored research as to Grumman's earlier exposure to the technology incorporated in the F-111.

[6] A direct copy is possible, of course, but except for some copying of German rockets, aircraft, and engines, a Russian copy of the American B-29, and one Pratt-Whitney engine reproduced by the Swedes (a-1 during or immediately after World War II), there has been no large scale unlicensed copying of aircraft and such for some 25 years.

[7] This aspect is developed in some detail by Forecasting International, Ltd. in *A Study of How Technology Transfer Affects the Competitive Position of the U. S. in the World Aviation Market*. Arlington, Virginia, Forecasting International, Ltd., 1972.

[8] "Planes for Japan," *Business Week*, Oct. 14, 1972, p. 35.

[9] "A Transatlantic Aircraft Venture," *Business Week*, June 9, 1973, p. 36.

[10] "Why Multinationals Are Under Fire at Home, Abroad," *U. S. News and World Report*, May 21, 1973, p. 66.

[11] Sanford N. McDonnell, "Another Export Industry is in Jeopardy," *Ibid*.

[12] "Europe -- The New Commercial Challenge," *Aviation Week and Space Technology*, Vol. 94, No. 22, McGraw-Hill, May 31, 1971, p. 56-209.

[13] International Air Transport Association, *World Air Transport Statistics*, No. 15, Montreal, Canada: 1970.

[14] "The Communists Like Those Boeing Jets," *Business Week,* Dec. 23, 1972, p. 26.

[15] Bill Gunston, "Concordski and Concorde," *New Scientist,* June 7, 1973, p. 595.

[16] See works by -- Marvin J. Cetron and Christine A. Ralph, *Industrial Applications of Technology Forecasting,* (New York: Gordon Breach Science Publishers, 1971); and Marvin J. Cetron and Bodo Bartocha, *The Methodology of Technology Assessment,* (New York: Gordon and Breach Science Publishers, 1972).

[17] Owing to space limitations, it was not possible to fully develop all of the background, and evidence presented in the two Forecasting International, Ltd. studies (total of 279 pp.) on which this paper was based. However, the following reports are available upon written request to:

Forecasting Intrrnational, Ltd.
1500 Wilson Boulevard
Arlington, Virginia]2209
U. S. A.

(a) Forecasting International, Ltd., *A Study of How Technology Transfer Affects the Competitive Position of the United States in the World Aviation Market.* Arlington, Virginia, Forecasting International, Ltd., 1972.

(b) Forecasting International, Ltd., *A Study of the Criteria to be Used in the Selection of Projects.* Arlington, Virginia: Forecasting International, Ltd., 1972.

TECHNOLOGY TRANSFER IN JAPAN

Keichi Oshima

University of Tokyo

At the time of the Meiji Restoration a hundred years ago, the Japanese government set a national goal to modernize industry by introducing Western patterns. There had been relatively high standards in indigenous handcraft technologies, but the importance of transfer of technologies from abroad became one of the most important bases of government policy. Even policies for education and science strongly emphasized the need to catch up with advanced Western countries. Such policies were oriented much more toward transfer of technology than were comparable policies in Europe. Faculties of engineering were set up in the universities from a very early period. Substantial contributions to industrial development were made by their determined efforts to learn foreign technologies and to train high-level engineers working in close collaboration with both industry and government.

The efforts mainly concentrated on transfer of scientific knowledge and technology from abroad. It should be emphasized that this transfer of technology was part of an integrated national economic and social policy meant to construct a modern industrial country. The past experience of transfer of technology in Japan can best be understood, then, in context of the economic and social development of the country.

Technology Transfer After the War

In the pre-war period, the official Japanese national goal was the combined one of "economic prosperity and military strength." After the war, the entire national objective was reconstruction of the economy. Japanese industry's target was set up with the goal of strengthening Japan's competitive capability in international trade and transforming the industrial structure towards advanced capabilities in "heavy chemical industries."

Except for a few fields such as shipbuilding and the optical industry, basic technological innovations chiefly originated abroad. In two cases the technologies transferred to industry were originally developed for military use during and before the war. In other cases Japanese technology after the war was far behind that of European countries or the United States, and there was little chance to develop competitive domestic technology in a short time.

The post-war success of Japan in introducing innovations and transforming the country's industrial structure can be seen in a review of the relative percentage of heavy and chemical industries--namely the percentage of machinery products (including electrical machinery), metals and chemicals--as part of the total of all industry. In 1950, these industries comprised about 50% of the total. By 1960 the proportion had reached 64%, which is equivalent to that in the industrialized countries of Europe and the U.S. The structure of products for export followed the same pattern of industrial production, with a delay of a few years. The relative share of heavy and chemical industries in the export trade was 38% in 1955, 43% in 1959 and 62% in 1965.

This shows that the transformation of Japan's industrial structure started in production for the domestic market, and then extended to the export trade. Demand in the international trade market has followed a similar trend--a decline in raw materials, food and natural fibers and an increase in advanced industrial goods such as machinery, vehicles, and chemicals. This pattern proved to be extremely favorable to Japanese industry's increasing competence in international markets.

Analysis of contributions made by imported technology to sale of products in several industries is shown in Table 1. In 1961, the contribution to sales of whole manufacturing industries was 33.6%, in 1967 it was 20.6%. Such major sectors as electrical, chemical, machinery, and iron and steel contributed substantially to overall sales, and were based on foreign technologies. While foreign technologies' impact seems to be decreasing in some industries, it is still large.

Payment by Japan for imported technology in 1967 amounted to 90,563 million yen ($250 million). This amount comes to about 2.8% of all sales based on imported technology. Of the total paid for imported technology, 32.4% was for items in electric machinery, 28.7% for items in chemical technology and 10.1% was expended in other machinery technology.

Table 1. Contribution of Technologies Imported From Abroad to Japanese Sales

(Million Dollars)

<table>
<tr><th rowspan="2"></th><th colspan="2">Sales (1967)</th><th colspan="2" rowspan="2">B/A (%)
(1961)</th></tr>
<tr><th>Total Sales
(A)</th><th>Sales Related to Imported Technology
(B)</th></tr>
<tr><td>Chemistry</td><td>7,850</td><td>2,111</td><td>26.8</td><td>(56.0)</td></tr>
<tr><td>Textile</td><td>1,550</td><td>258</td><td>16.9</td><td>(15.7)</td></tr>
<tr><td>Pulp, Paper</td><td>711</td><td>44</td><td>6.3</td><td>(29.5)</td></tr>
<tr><td>Rubber</td><td>706</td><td>402</td><td>56.9</td><td>(45.4)</td></tr>
<tr><td>Ceramics</td><td>878</td><td>160</td><td>18.3</td><td>(19.1)</td></tr>
<tr><td>Iron, Steel</td><td>5,508</td><td>440</td><td>8.0</td><td>(61.1)</td></tr>
<tr><td>Electric machinery</td><td>7,775</td><td>3,400</td><td>43.7</td><td>(30.8)</td></tr>
<tr><td>Precision machinery</td><td>372</td><td>87</td><td>23.5</td><td rowspan="2">(25.2)</td></tr>
<tr><td>Machinery</td><td>3,214</td><td>528</td><td>16.4</td></tr>
<tr><td>Transportation machinery</td><td>5,217</td><td>333</td><td>6.4</td><td>(5.9)</td></tr>
<tr><td>Total Manufacturing Industry</td><td>43,542</td><td>8,994</td><td>20.6</td><td>(33.6)</td></tr>
</table>

Innovations developed in Japan emphatically were not wholly derived from imported technology, but they have been combined extensively with indigenous Japanese R&D.

The fact of heavy Japanese industrial expenditures for R&D is emphasized by the 1969 figure, when they totalled 628,352 million yen ($175 million). Expenditures were mainly in electric machinery (26.9%), chemistry (21.9%), transportation machinery (11.1%), and machinery areas (8.5%). This indicates that heavy and chemical industries, with the largest share in technology importation also makes the major expenditures in domestic R&D. A survey by the Japanese Science and Technology Agency shows how indigenous R&D is related to the import of technology. An inquiry was made into the percentage of each level of R&D already in progress at the time the new technology was imported. Of 1,500 companies reporting on the period from 1950 to 1966, 83% were already doing some aspect of R&D related to the technology imported, and 31% were already carrying on some related development work. This shows that the import of technology was not isolated from, but was in fact parallel to efforts of indigenous R&D. This enabled industry to absorb and transfer the imported technology smoothly into integrated technological systems. In other words, most of the indigenous R&D was an integral part of the technological innovation process which had imported technology as a core. Important too is the fact that many imported technologies have been improved after their introduction into Japan, and some of these technologies were first commercialized by Japan.

In summary, technological innovations in Japan which have contributed so heavily to the growth of the economy have been achieved by a very efficient combination of imported foreign technology and a high standard of effort on domestic R&D.

Characteristics of R&D

Let us now consider some of the characteristics of R&D activity in Japan. First of all, although the present figure of 1,195 billion yen ($3.9 billion) in 1970 is comparable to expenditures in advanced European countries, Japan's past R&D expenditures have been relatively low compared to those of advanced countries. The R&D ratio to GNP is also low, remaining at about 1.6% up to 1966 and reaching only 2% in 1970.

The unique characteristic in Japan is that the proportion of the total R&D expenditure funded by industry is very high and that it is still increasing.

R&D expenditures funded by industry were always over 68% in the 1960's, and they rose to 74.7% in 1970. It is notable that the largest part of the R&D expenditure goes for economic growth purposes. Even for government expenditures, the OECD figures show that the Japanese government's R&D budget in 1967/8 was 75.4% for economic growth purposes and only 16.5% for such national projects as military, nuclear energy, and space. Expenditures for health and welfare were 5.7%. This indicates that the initiative for R&D activities was taken by industry, yet the main objective of R&D even in the government was to contribute to or at least be supportive of industrial efforts for economic growth.

Other notable characteristics of Japan's R&D is its high proportion of support to basic research and extensive research manpower. In many countries the proportion of basic research support decreases with the increase in total R&D expenditures, but Japan is an exception. In Japan not only is the proportion of basic research high (23.3%), but the total 1970 expenditure, 278 billion yen ($900 million) is larger than that in any Western European country. Only the U.S. figure is higher.

There have not been very large expenditures for development of national projects like those mentioned above, and so it is understandable that government expenditures for basic research is relatively large compared with that for development. Even in basic industry, expenditures classified as basic research are relatively larger than they are in other countries. For example, Japanese industry spent 9.7% of its 1970 R&D expenditures on basic research, which may be compared with figures of 3 to 5% for Western countries. This fact also indicates that most research in Japan is oriented to the general promotion of technical and scientific activities, and not so much to the development of large projects.

The number of researchers in Japan is very high. In 1970 there were 194,347 researchers in Japan. This is not only high in absolute terms, since Japan is third in the world--in proportion to total population, 1.6 researchers per thousand, this figure is higher than that of any European country.

The fact that the level of general education is high in Japan is of course relevant. The percentage of attendance at primary school is 99.9%, the highest in the world. The period of compulsory education is 9 years; attendance at high school is 70%, and at college and university level it is 20%.

Students of natural science and engineering in higher education are about 20% of the total, and this is somewhat lower than the number of enrollees in these fields in European countries. The proportion of engineering students in this group is extremely high, however--more than 70%. This trend reflects the Japanese governmental policy of meeting the industry's demand for high-quality technical manpower.

In Japan much greater effort is made to improve the scientific and technological base for technology transfer rather than to directly produce new technological development. This can be seen by studying the direct links between basic research in universities and public institutions and developments in industry. In most cases the links are not strong. Rather, basic research activities have been mainly to supply a scientific and technological soil of high standard to cultivate active technological innovations through transfer of technology from abroad.

It will also be useful to mention that the contribution of universities in Japan's basic research is high. Some 65% of the expenditures for basic research is spent at universities. This is weighted heavily toward engineering (42%) and medicine (32%). This shows that economic and social objectives have a much stronger influence on research in universities in Japan than in other countries, a situation which can be understood from the historical background of scientific developments in Japan.

Government and Industry

As may be surmised from the characteristics of Japanese R&D, there were no extensive national R&D projects to influence the growth of industry. Rather the development of technology was carried out as part of the all-encompassing effort to reconstruct and advance Japanese industry to internationally competitive standards. The consensus on the national goals of modernizing the industrial structure and increasing the economic standard of living was well accepted and the policies of government and the activities of industry have been in fairly good accord.

For example, the directions and targets for development of important technology-intensive industries such as petro-chemistry, electronics, and machinery, were decided after effective communication between government and industry, informally or officially in the form of advisory committees to government. Experts from industry, government and universities have all participated. In these cases, technology has been treated in the same context as other economic factors such as investment, labor and the market. Most of the measures related to the policy for industrial technology were handled and decided by the Ministry of International Trade and Industry, although responsibility for coordinating general policies for science and technology was allocated to the Science and Technology Agency of the Prime Minister's Office. For this reason there has been close coupling of technological and economic policies, and this has also been important in promoting successful technological innovations.

From the statistics, it can be deduced that the initiative for selecting specific technologies and actions for their development, including importation of key innovations and further improvement through domestic R&D for industrial adaptation, have been taken by industry on its own initiative. Matters of general framework--projections of future industrial development, setting priorities of fields to be promoted, the basic policy on the scale and allocation of investments and the principal direction of technology, have been the province of the consensus-forming committees. This work has been especially important in the earlier periods, because the government has controlled and allocated scarce resources according to the framework decided by the committees. The two major resources, of course, have been funds for investment and foreign currency. The system has functioned to coordinate importation of foreign technology in accord with the goal of transforming the industrial structure to "heavy and chemical industry." Additionally, it has assured investment for commercialization when the technology had been successfully developed.

In some important cases, the government has launched special temporary measures or laws to promote technological innovation and rationalization of specific, technology-intensive industries. A committee would be set up to decide specific measures and programs of promotion. The government has then allocated investment funds with favorable conditions, has granted favorable tax benefit and exemptions to the anti-cartel provisions of the Anti-Monopoly Law, if the industry will agree to follow the prescribed programs. Such measures clearly show that the transfer and development of technology has been treated as an integral part of economic policy in Japan.

Favorable governmental policies have encouraged enterprises to take a progressive attitude toward introduction of advanced foreign technology and have promoted efforts of domestic R&D to bring the technology into commercialization rapidly.

It should be stressed that even though there have been several measures and actions taken to promote development of technology for industry, the most important point has been that good consensus on a goal has brought about very clear guidelines toward development of economic and technological policy for the government, and towards industrial objectives and criteria for technological innovations and R&D.

Conclusions

It is clear that the main features of the past in Japan were what is called "absorptive" strategies. The core technology was transferred from abroad in most cases. However, indigenous efforts of R&D and the supply of high-quality research manpower have played a most important role.

R&D in the process of transfer has been essential in absorbing and integrating the transferred core technology into the total system of industrial production. It has been very important for new technology to contribute substantially to the growth of the economy, because the greatest economic impact of technology comes from the advancement of the entire technological structure of industry, and very much less from isolated improvement of some specific technology.

If there had not been favorable combination of R&D coupled with the transfer of technology, Japanese industry would not have been able to make such a rapid and self-sustaining transformation to an advanced technology-intensive industrial structure, which has been the basis of Japan's high rate of economic growth.

In the appendix following, a few typical cases in the chemical industry are described for reference.

APPENDIX

Case I, Synthetic Fiber

Vinylon and Nylon

The development of "vinylon" and "nylon" in Japan may offer a typical example of successful case of development of new technology, and transfer of technology. In this case there already existed a technological background of basic science and related technology.

In the 1930's, Japanese production and technology of rayon was at the highest level of the world and leading textile companies established their research laboratories. In the universities many research projects on rayon and basic macro-molecular chemistry were going on, and some study of new artificial fibers from protein. Announcement of the success of the first synthetic fiber, "nylon," by duPont in 1938 gave strong stimulation to the Japanese researchers to search for new synthetic fibers, especibecause nylon was anticipated to become a powerful competitor to silk, which was the most important indigenous natural fiber.

In 1939, Professor Sakurada of Kyoto University announced his success in the synthesis of polyvinylalcohol fiber, "synthetic fiber No. 1." This was followed by independent successes of the Kanebo Co. and Kurashiki Co. in polyvinylalcohol fibers the next year, and by Toyo Rayon Co., which announced its success in synthesizing nylon, "Amilen."

To promote commercialization of these synthetic fibers, the Japan Synthetic Fiber Research Institute was established in 1941 by a fund from the government and by contributions from 20 private enterprises. The research was carried out in several groups consisting of researchers from private enterprises and from universities. The government offered subsidies for five years. During this period test plants for production of less than one ton per day of polyvinylalcohol fibers, vinylon, and nylon six were constructed.

This was in the difficult period during World War II, but these developmental projects were able to establish a basis for commercialization of vinylon and nylon in Japan.

After the war, the government decided on a policy of promoting the synthetic fiber industry and it offered such favorable aids as privileges in taxation, financing and measures such as special allocation of electricity and other scarce materials for production and development of vinylon to Kurashiki Co., and for nylon to Toyo Rayon Co.

In 1949, both companies constructed pilot plants of one ton per day and started production. Though Rayon's process of production of nylon six was independent of the duPont patent, in 1951, Toyo Rayon acquired licenses of patents for nylon production and fabrication from duPont in order to improve its products. In the initial stage, both companies faced many difficulties in improving the quality of the fibers as well as the development of markets. Helped by the protective measures of the government, they gradually expanded their markets and eventually enjoyed the benefits of initial enterprise in the field.

The later demand for nylon far exceeded the government's projection. The program of production of nylon by Nihon Rayon, under the license of patents from Inventa Company of Switzerland, was approved in 1954, and several other companies entered the field later.

The government policy was to protect the enterprise in its infant stage of development of new technology. Free competition with other enterprises was introduced to avoid monopoly, when the economic standing of the enterprise was established.

By 1967, production of nylon and vinylon fibers had increased to the following:

nylon	188,000 tons
vinylon	61,000 tons

The case of vinylon and nylon is an example of how the Japanese government has given direct support in the establishment and transfer of technology from development to commercialization.

Case II, Petrochemistry

Ethylene

The petrochemical industry is an example of successful introduction of new technology to industry with direct support of the government. In contrast to synthetic fibers, all the technologies were imported from abroad and practically no technological development was performed in Japan at the start.

In the period after World War II, materials for the Japanese chemical industry were produced from coal and acetylene from calcium carbide. Many intermediate materials were imported from abroad.

With development of petrochemistry abroad, the low price of supplies and the growth of the market for petrochemical products in Japan, it became clear that a petrochemical industry should be introduced into Japan in order for the country's chemical industry to acquire international competence in international trade.

In 1954, the Ministry of International Trade and Industry (MITI) started a committee for petrochemical technology. This consisted of experts from industry, government and universities to study rational policy for a petrochemical industry. In the next year, on the advice of the committee, MITI decided on a policy meant to promote the petrochemical industry in Japan--to (1) secure a stable supply of raw materials for the rapidly growing synthetic fiber and plastic industry; (2) establish domestic production of raw materials such as ethylene and its derivatives (demand for these was expected to increase rapidly but the materials were imported in total because their production relies on a capability in petrochemistry); and (3) decrease in the domestic prices of the main raw materials for the chemical industry to the international standard to acquire international competence in chemical and related industries.

The main problem in implementing this policy was to determine how to allocate the ethylene production capacity. In the petrochemical industry--a typical process industry having the merit of large scale--the production capacity of ethylene should be larger than a minimum economical unit.

In Japan, industry was very progressive in entering into the petrochemistry field, and many enterprises expressed their willingness to start production.

Judging from the capacity of the minimum economical unit of ethylene production and the expected demand of products, the government decided to concentrate production of ethylene at four production centers. Industry was asked to coordinate the programs of petrochemical industries into groups at the four centers, and government offered favorable aids to these groups. All technologies for production were imported from abroad. Aid in financing was made through a loan from the Development Bank, for investments. Concessions in taxation were given.

This was the first phase in promoting the petrochemical industry, which lasted until 1959. All four ethylene centers then went into operation, with total production of 80,000 tons per year.

After the first phase the capacity of each ethylene center was increased and programs for additional centers were approved. General coordination of the expansion programs were carried out at the committee of MITI consisting of experts from industry and government. In most cases the programs for enterprises were very progressive. They tended to exceed the estimated demand of the market and the government had to regulate them. This was done by urging new investors to join or by adjusting the timing of construction and the enlargement of plants.

Most of the technology was imported. The government's influence was effected by regulation on approval of import of foreign technology. The government intervened in the conditions of agreements for licensing of patents.

In the case of approval of new production plants, priority was given to enterprises using their own technologies. In case of approval of licensing of foreign technology, priority was given to enterprises advanced in their own technological development in this area.

As in the case of synthetic fibers, import of products competitive with domestic production was restricted. Domestic manufacturers were asked to supply the products with prices comparable to imported products. The ethylene production capacity in 1964 was 920,000 tons at nine centers, and expansion to a 1970 capacity of four million tons per year was anticipated when this data was compiled.

Synthetic Rubber

Synthetic rubber, SBR, is a unique case of direct investment by the government for the promotion of industry.

In 1951, a study group for the synthetic rubber industry was set up by MITI, which studied the feasibility of SBR production in Japan. Industry started its own study group in 1952, and it became a preparatory committee for establishment of a synthetic rubber production company in Japan.

Two main points were to be studied: one was the raw material, butadiene, which had two possibilities (one from ethyl alcohol and the other from petrochemical products). The second was the production capacity whose minimum economical unit was 30,000 tons per year, much exceeding the expected market.

After much disagreement, in 1956 it was decided to establish a company with government contribution in investment under a special law. The raw material to be used was a series of petrochemical products. Ethyl alcohol was not economically competative even though it could be supplied by indigenous agricultural production. Though three companies wanted to start the business, it was decided that one company would be preferable if the market were to be secured.

The technology was introduced from the U.S. The Japan Synthetic Rubber Company was started in 1957 as a joint venture of rubber and petrochemical companies, with 40% capital participation by the government.

A special low-interest rate loan was given by the Japan Development Bank. Initial production capacity was 45,000 tons per year and it was decided that SBR must be sold with a lower price than imported SBR or natural rubber. No import duties were imposed on importation of synthetic rubber.

At the beginning, the sales of the company were much less than the initial estimate and there were difficulties in establishing the market. However, the market expanded. After increases in production capacity the company started making a profit in 1961. Now production has increased to 250,000 tons per year. In 1968 the government started to sell its shares to industry and in 1969 the company became entirely privately owned. The special law was then terminated.

TECHNOLOGY TRANSFER IN DEVELOPING COUNTRIES

Samuel N. Bar-Zakay

Manager, Center for Technological and Interdisciplinary Forecasting (CTIF), Tel Aviv University, Israel

"A poor beauty finds more lovers than husbands."
George Herbert, *Outlandish Proverbs*, 1640

INTRODUCTION

Any type of technology transfer can be of organized or incidental nature.* [1] Consequently governments or enterprises are faced with three policy alternatives to enhance technology transfer: to enhance the organized type, the incidental or both. I shall assume that the output in each case is the same.

The objective of this paper is to present an intuitive opinion of the author, that developing countries could more easily enhance organized transfers than incidental ones, and that this simple thesis has not yet been realized by the developing countries.

The examples, which will serve as background for the discussion, are taken from transfers between countries and between applications. This was done since in my opinion[2] the forces involved in the process of technology transfer are the same, whether it takes place within a country or between countries, although there are, of course, differences in the magnitude of forces and their relative importance in each individual case. Therefore, one should be able to apply experience in one type of transfer, say that occurring within one country, to transfer between countries.

---

* See definitions on the following page.

DEFINITIONS

1. When scientific or technological information generated and/or used in one context is re-evaluated and/or implemented in a different context, the process is called technology transfer. [1]
2. Incidental technology transfer- initiated by the donor. Organized technology transfer- initiated by the recipient. [1]
3. Transfer Agent: An individual or an organization who initiates a transfer process of technology.

   Transfer Agents:*

   1. Individuals
   2. Corporations
   3. Governments
   4. Universities
   5. Research Institutions

CASES OF TECHNOLOGY TRANSFER

In this section three basic tables are given, listing:
1. Cases of Technology Transfer between countries.
2. Cases of Technology Transfer between applications.
3. Missed opportunities of Technology Transfer.

It is clear from the introductory comment that I see <u>no</u> fundamental difference between the cases in Table 1, Table 2, and Table 3. These are separated mostly for illustration purposes.

The choice of cases represents the personal perception of the author and was based on the availability of literature. Hopefully it is a random sample.

Background information for each table is given in appendices 1,2,3.

The designation "organized" or "incidental" is based on the information available to the author at the time of writing.**

---

* Agents 2-5 always involve individuals. The opposite is not necessarily true.

** The author accepts full responsibility for misinterpretation of literature data and for any other mistakes.

TABLE 1

CASES OF TECHNOLOGY TRANSFER

(between countries)

No.	From: Context I	Capabilities in Context I	To: Context II	Needs in Context II	Transfer Agent	Type of Transfer
1	ITALY	Fiat manu-facturing and marketing	YUGOSLAVIA	Car produc-tion. Export.	Corporation	Organized
2	PHILIPPINES	Developed "miracle rice"	THAILAND	Grow more rice.	Research Institute	Incidental
3.	U.S.A.	Nuclear weapons technology	RUSSIA	Nuclear weapons technology	Individuals	Organized
4	GERMANY	Volkswagen manufacturing and marketing	BRAZIL	Car produc-tion. Export.	Corporation	Unknown
5	FRANCE	Color television technology	RUSSIA	Color television technology	Government	Organized
6	TURKEY	Training facilities	JORDAN	Higher Education	University	Incidental
7	U.S.A.	Soyabean drink technology	GUYANA	High protein content food.	Corporation	Incidental
8	ISRAEL	Biological pest control	U.S.A.	Fight Florida Red Scale	Research Institute	Incidental
9	BRITAIN	Textile mills technology	JAPAN	Industriali-zation	Individuals	Organized
10	U.S.A.	Advanced medical and welfare services	GABON	Medical techniques	Individuals	Incidental

TABLE 2

CASES OF TECHNOLOGY TRANSFER

(between applications)

No.	From: Context I	Capabilities in Context I	To: Context II	Needs in Context II	Transfer Agent	Type of Transfer
1	Nuclear irradiation	Destroys life	Fruits infection	Destroy insect and bacteria	University	Unknown
2	Space research	Satellite earth survey	Earth resources	Find new resources	Government	Incidental
3	Osmosis process	"Filter salt"	Desalination	Reverse Osmosis	Individuals	Incidental
4	Biological research	Enzymes technology	Detergent industry	Remove protein stains	Corporation	Organized
5	Polymer research	Polyox reduces turbulent flow	New York Fire Department	More water through hoses	Research Institute	Incidental
6	Teleoperators	Unmanned Moon exploration	Human augmentation	Help handicapped	Government	Incidental
7	Air Force	Helicopter development	Toy industry	New toys	Individuals	Organized
8	Nuclear research	Study of the atom structure	Leakage in pipes	Locate leakage	Research Institute	Incidental
9	Search for new structural materials	High modulus lightweight in composites	Soldiers armor	Protect life at light weight	Corpor ation	Organized
10	Computers	Fast storage and retrieval of information	Urban planning	City scape simulation	University	Organized

TABLE 3

CASES OF TECHNOLOGY TRANSFER

(missed opportunities)

No.	From: Context I	Capabilities in Context I	To: Context II	Needs in Context II	Reason for Missed Opportunity	State of Opportunity
1	Solar-energy technology	Low cost energy	INDIA	Low cost energy. Fertilizers.	Lack of analysis capabilities	Hopeful
2	JAPAN	Continuous welding process	U.S.A.	Modern shipping industry.	Lack of forecasting	Too late
3	ISRAEL	Modern agriculture technology	EGYPT	Modern agriculture technology	Political	Still Hopeful
4	U.S.A.	Composites research	TANZANIA	Sell more sisal	Lack of analysis capabilities	Still Hopeful
5	International sugar technology	Increase sugar production	CUBA	Increase productivity and export.	Lack of analysis capabilities	Still Hopeful
6	BRITAIN	Carbon fibers technology	U.S.A.	Carbon fibers	Inability to take high risks	Too late
7	Machine Bull	Computers technology	FRANCE	French owned computer technology	Lack of forecasting and planning	Too late
8	SIERRA Leone	Iron-ore deposits	World	Iron	Lack of analysis capabilities	Too late
9	BELGIUM	Land available near port	BELGIUM	Land for industry	Bureaucratic	Still Hopeful
10	World industry	Imports	BRAZIL	Export diversification	Lack of analysis capabilities and planning	Still Hopeful

DISCUSSION OF TABLES 1-3

Provided the author has not made too many mistakes in giving the designation "organized" and "incidental" to the various transfers, the following conclusions can be drawn:

1. Transfers to developing countries are primarily of the incidental type.
2. Transfers between applications are about equally distributed between incidental and organized cases.
3. The cases of missed opportunities of transfer could have been avoided had there been a stronger capability for organized transfer in the developing countries.

Unfortunately the sample size is too small to say with confidence that the conclusions are true in general. However, if we assume that the world picture is really represented in the tables then a question may be raised: Why is the ratio of organized to incidental in Table 1 and 3 different from that in Table 2?

My answer is that the Developing Countries (DC) lack the capabilities required for organized transfers. Moreover, in my opinion the DC have not yet come to realize that their control of organized transfer is greater than that of incidental cases.

Consider for example the long existing needs of Guyana and Gabon in Table 1, and the still unsatisfied needs of Cuba and Brazil in Table 3 as examples of the long time it takes to satisfy recipient needs by incidental technology transfer.

Furthermore, once the donor has finally decided to come to the DC (to the "seller's market" [3]) transactions tend to be off-the-shelf, not based on creative reformulations of problems. [3]

The result may be either neo-colonialism or failure of the technology transfer. The latter is in view of thoughts [2] and findings [4] on the importance of adaptation to the success of technology transfer.

Policies aimed at attracting donors to DC constitute a game known to all countries, both developed and developing. Unfortunately even those countries who play the game well, do not gain much: There is evidence to suggest that "capital is available in apparently abundant amounts within a relatively low rate of interest...skilled labor and reputable management are the shortage." [3]

As will be pointed out later on here, the situation is similar in Israel.

From the list of missed opportunities we can see that:

1. Missed technology transfer cases include both inter-country and inter-application transfer (e.g. solar energy to India).
2. In some cases the donor of technology lost an opportunity (e.g. carbon fibers in Britain) and in others the recipient is the loser (e.g. sugar in Cuba).
3. The reasons for lack of transfer range from political (agriculture-Egypt) through bureaucratic (land-Belgium) to lack of knowledge (Sisal-Tanzania).
4. Successful technology transfer can be achieved across thousands of miles and may constitute a failure across short distances.

Since the reasons for missed opportunities range from political through bureaucratic to lack of knowledge, it would be wrong to assume that policies aimed at enhancing the capability of organized transfers will bridge all gaps. Yet one may raise a terribly simple question: Why are there so many DC in which there is not even one group in charge of organized technology transfer?*

In the State of Israel a step in the right direction was taken in January 1971. The Center for Technological and Interdisciplinary Forecasting (CTIF) performs several tasks that can be categorized as organized technology transfer. One example will be given here.

The CTIF performed a ten-year forecast of computers in Israel (1971-1981) with an objective of translating the findings into short term decisions.

The work took six months and is covered in four volumes.

Among other things the CTIF came to realize the tremendous importance of LSI (Large Scale Integration) to the future of the Israeli electronic industry. As a matter of fact the CTIF came to the conclusion that Israel will not be able to compete on the world markets unless extensive use of LSI starts immediately.

---

* The name of the organization is of no importance as was pointed out before.[1]

The CTIF also weighed the risks and benefits of producing LSI chips in Israel, (both for local consumption and for export). The production costs, the size and structure of the local and world markets were analyzed in detail.

Subsequently, the CTIF invited a large American LSI producer to come to Israel, in order to update the Israeli industry on recent developments in LSI. As a result of the visit the CTIF learned that some of its numbers and estimates were inaccurate. Yet, two goals were achieved.

1. The Israeli electronics industry received "an early warning."
2. The CTIF focused the discussion by presenting data for deliberation.
3. Most important, the American LSI Company was approached by various Israeli electronics manufacturers and some contracts were signed (Since then 2 more companies have visited Israel for discussions on the same subject).

In 1972 the CTIF started several other projects, one of which is concerned with technology transfer from the defense to the civilian sector in Israel. [4]

Whatever the technology transfer output of this project will be, the CTIF will surely raise enough "naive" questions [5] to stimulate thinking in government and industry about the process under study.

The findings of this study and of others in the U.S. and elsewhere will probably not be applicable to many DC (the methodology may be applicable). There are marked differences in technology transfer in different countries.

For example, if the transfer of sophisticated technology necessitates the availability of R & D personnel at both donor and recipient organizations, then marked differences exist between the U.S. and Israel. Of the total U.S. R & D funds, about 50% come from the private sector. In Israel the figure is less than 5%.

We must therefore conclude that in order to enhance organized technology transfer, each developing country must develop tools and collect data applicable to its own environment.

A comprehensive search into alternative policies to enhance organized technology transfer in at least one developing country, seems urgently needed.

APPENDIX NO. 1 (to table 1)

1. Italy-Yugoslavia

The international corporation Fiat from Italy made a package deal in Yugoslavia. Fiat licenses production to Crvena Zastava. In 1966 the latter produced in Yugoslavia 40,000 units under Fiat License.[6] As part of the agreement the Yugoslavs could sell their products in certain other countries.

Thus Crvena Zastava has earned foreign exchange (e.g. in Greece) to pay for equipment it had to purchase for the plant. ("Yugoslav firms are required to cover imports requirements with exchange earnings from export sales."[6])

The Italians were given preferential treatment in selling Fiat in Yugoslavia, by reduced tariffs for example.

This initial transfer of technology was followed by the Yugoslavs building replacement parts factories and exporting some of the production to Italy. The initiative for the transfer came from Yugoslavia.[6]

2. Philippines-Thailand

The International Rice Research Institute (IRRI) in the Philippines is supported by several organizations.[7] IRRI was founded in 1960 for the purpose of conducting research on how to increase rice production in the tropics.

It has developed the "miracle rice" IR-8 which is now grown in several Asian countries.

A new strain designated IR-5 has more regular grains and is thus more popular with the Asian people. This particular strain is grown in Thailand which may have a comparative advantage over the Philippines: The IR-5 needs much solar radiation. Since the Philippines is a cloudy country compared to Thailand, it may grow better in the latter country.

The beneficial effect of the new strains of rice on the GNP is obvious.

3. U.S.A.-Russia

Many cases of espionage have been published since the beginning of the Atomic Era.[8] We have included these since they demonstrate dramatically how technology transfer can be initiated by individuals. It can be argued that individuals are always the transfer agent.

The example given in the table may also point out that a country can be a supplier of technology (not necessarily military) against its wish.

4. Germany-Brazil

In a similar case to the Fiat deal in Yugoslavia, Volkswagen has transferred its technology to Brazil.

We give this additional example in the automotive field to demonstrate that a common border for such a transfer is not a must.

Successful technology transfer can be achieved across thousands of miles, and may sometimes not be achieved across short distances.

An example of short distance transfer (across a river within a country), which is not taking place, will be discussed under "missed opportunities."

It is interesting to note that "in 1959, Volkswagen tried to market 300 passenger cars in the United States from its Brazilian subsidiary (in order to overcome a temporary production bottle-neck in Germany and to earn foreign exchange to pay for parts imported into Brazil)". [6]

The above is illuminating since technology may be transferred to one country with the purpose of penetrating a whole area. This marketing strategy will be evident in some of the other examples (e.g. Monsanto Company in Guyana).

> Volkswagen also makes available complete standardized ready-to-use management systems covering the whole gamut of operation, systems planning, procedure development, control patterns, inspection schemes, and so forth. [9, 10]

5. France-Russia

The French Government has signed an agreement with its Russian counterpart covering color television technology. The French are responsible for developing the color network in Russia, as well as for supplying the tubes. The French Compagnie Francaise de Television en Couleurs (CFT) is carrying out the deal, requested by Russia.

6. Turkey-Jordan

An outstanding Turkish training effort is given by the Middle East Technical University (METU) located in Ankara. Financed by the Turkish Government, but with additional support from the U.N., CENTO U.S. AID, and the Ford Foundation, METU began with a faculty of 3 and 49 students in 1956. By 1965 it had about 3,000 students including 235 from abroad. [11]

Of the 235 foreign students 45 were from Jordan.

7. U.S.A.-Guyana

The shortage of proteins in the Developing Countries is well known. [12]

In view of the shortage several American corporations have been interested in the technology of high-protein-content-food. [13]

For example, Monsanto Company has developed the technology of a soft drink made of soybean called Puma. (A similar drink called Saci is manufactured by Coca-Cola in Brazil.) Monsanto Company is producing Puma through a franchise in Georgetown City, Guyana. [14]

Although the local manufacturer may not be aware of some of the proprietary aspects of the process (due to the nature of agreements in the soft drink business) much technology has been transferred in the form of process conditions, quality control, etc. This technology transfer has enabled local marketing, probably with the aim of penetrating the whole area.

8. Israel-U.S.A.

The Florida Red Scale is a dangerous threat to every orange grove. Since the largest export item from Israel used to be oranges (currently it is second after diamonds), the Israelis were constantly looking for ways to fight this pest.

A comprehensive research effort at the Volcani Institute of Agricultural Research [15] led to a novel method of fighting the Florida Red Scale: Biological Control.

It was found that in Hong Kong the Scale has a natural enemy called Aphytis holoxanphus. The development of the conditions of reproducing the Aphytis was studied systematically with the support of the Israeli Citrus Marketing Board.

After the study was completed it was found that the release of the Aphytis in the orange groves of Israel practically eliminated the problem of the Florida Red Scale. [16]

The technology was transferred to several countries, among them South Africa, Mexico and Florida.

The initiative came from the Israeli scholars.

9. Britain-Japan

The economic development of Japan is unique in many respects and is far from understood. However, every scholar of Japan will probably agree that part of the growth rate is due to the ability and desire of the Japanese to absorb foreign technology.

The ability may exist in other nations as well, but the desire, sense of importance and pride in absorbing foreign technology is unique to Japan.

It is interesting to note that this process took place in the 1800's [17] as much as it is taking place today.

> The transfer of technology began right after the Meiji restoration in 1868. A technical program to absorb foreign industrial techniques was initiated by the Ministry of Industry established in 1870. In 1879, half the Ministry's budget was spent on foreign experts in railway buildings, workshops, and in mining. British experts were engaged in the textile mills. Experts in cement, fertilizer, paper and glass manufacture were also employed to set up factories, train technicians, and teach industrial techniques. [17, 6]

While the Japanese may be unique in their willingness to absorb foreign technology it is clear that the U.S. is receptive as well. Nelson pointed out [18] several examples of the U.S. debt to foreign technology, as the ball-point pen (Hungary by way of Argentina), cellophane (France), crease resisting fabrics (England), DDT (Switzerland) and insulin (Canada).

It would be instructive to compare the receptiveness, the results and the future trends of technology transfer in the U.S. versus Japan.

10. U.S.A.-Gabon

The transfer given in the table is one of many performed by the Peace-Corps.[19]

Ranging from hospital operations to surgical nursing, through pharmacology, health technology was transferred by the Peace-Corps. Many other technologies in a range of countries were transferred as well by the American Peace-Corps.

APPENDIX NO. 2 (to table 2)

1. Radiation

The deadly radiation of atomic reactors is being studied in the University of Baghdad, Iraq for food preservation.

> .....Work is being carried out on radiation of dates to keep them free from insect infection. The export of dates brings Iraq five million dinars (roughly $15 million) a year; and is an important industry.[20]

2. Satellite

Space research turned to Earth Research with the use of satellites to survey earth resources. NASA has reported:[21]

> A photograph of part of Iran supplied by TAC led to the identification of previously unknown salt domes which are of interest as potential sulphur sites.
> The application is notable for two reasons.
> First, the wide area of coverage of the photo, when tied in with existing geological mapping of adjacent areas, eliminates the need for extensive aircraft and ground field work saving hundreds of man years of work and substantial expense.
> Second, color gradations in photographs aid materially in identifying salt domes that don't quite pierce the surface. This kind of dome is most difficult to detect from photos taken at aircraft altitudes.

3. Desalination

Every student learns in school about the osmosis process. Through a semipermeable membrane water will pass freely tending to equalize salt concentration on both sides of the membrane.

The natural process was put to use by reversing it with the use of pressure. The Reverse Osmosis principle, a new method of desalination of sea water, was developed in a relatively short time to become an industrial technique.[22] The discovery did not originate in the study of desalination.

The idea has led to the development of hollow polymer fibers having inside diameters of 25 microns in order to increase the surface area of the "membrane."[23]

The idea was further combined with the centuries known fact that pressure increases towards the depth of the seas. Thus, a portable emergency unit was devised, which makes use of this pressure for purifying water.[24]

4. Enzymes

Biological research studies tried to unveil the structure and action of the biological catalysts: the enzymes.[25]

The detergent industry was always interested in a method to remove protein stains caused by blood, eggs, skin particles, etc. The idea of including a protein splitting enzyme in the washing powder originated in a small Swiss soap factory in the 1930's according to one source.[26]

The idea reached an explosion in 1967 both in Europe and the United States.

In the Netherlands, one of the two countries where the enzyme products were first introduced (Switzerland is the other) enzyme brands were estimated to account for $35 million in 1967.[25]

5. "Slippery Water"

In a press-release from the Office of the Mayor of New York, on May 13, 1969 it was noted:

> Mayor John V. Lindsay today participated in a demonstration of "slippery water", water containing minute quantities of a special chemical that enables it to flow through fire hoses with far less resistance, and thus with much greater speed than has previously been possible.
>
> "Slippery Water" will permit the use of a smaller, less bulky hose with the same water flow as in the present equipment. With this lighter hose fire fighters will be able to climb stairs and reach remote locations more

rapidly. It will also permit the delivery of large volumes of water over greater distances in conventional size hoses.

Initiated by The Rand Corporation, as part of its work with the New York Fire Department, the "slippery water" system is being developed and adapted by the Fire Department with assistance by the Union Carbide Corporation, which produces the special chemical.

Polyox was developed by Union Carbide for making extruded or calendered sheets in the plastics industry. [27]

6. Teleoperators

NASA is concerned with the development of teleoperators because many astronautical targets are so far away that they must be explored by proxy. Yet the amplification and extension of man via the teleoperator concept transcends space exploration. A survey of this fascinating technology embraces many advances made in the nuclear industry, in undersea exploration, in medicine, and in the engineering of "man amplifiers." [28]
.....The basic concept is that man can communicate across a barrier to mechanical actuators that can operate under loads too great for an unaided man, or in an environment too hostile or too far away for him to conquer in person. A teleoperator augments a normal man, or, in the case of prosthetics, helps a handicapped man become more nearly normal.

7. Toys

Toys based on modern technology are found around the world. A beautiful helicopter was introduced into the market recently: With the aid of a small electric motor it takes off and maneuvers like a real helicopter.

8. Isotopes

Isotopes are found in small quantities in nature and can be produced artificially by nuclear reactions. Isotopes possess properties which enable their detection. These properties are not found in more abundant atoms in nature.

The Nahal Sorek Nuclear Research Center in Israel [15] has put these facts to use in locating water leakage in pipes. By adding a small quantity of certain isotopes to a pipe at its source, the location of the leakage can be detected by monitoring the radiation along the pipe.

This technique was also used for determining the path of the pipeline from Lake Tiberias to the Negev which is the largest irrigation project in Israel. [29] It was initiated by the Nuclear scientists.

9. Composites

Research into new light-weight materials has led to the "reinvention" of composites. [30] The search was directed mainly for structural materials for aircraft and the like, where a saving in structural weight can be translated into larger payload. However, it was postulated that since the materials which were developed (called composites) had some of the properties of metals, but were much lighter, they could be used for armor applications. The heavy armor used in the Middle-Ages is obviously not practical for modern warfare.

Indeed several corporations have developed armor, which was used in Vietnam, [31] upon the request of the U.S. Army.

10. Computers

It has been noted [32] that in 1945 television, jet travel and digital computer industries were commercially non-existent. In 1965 these industries contributed more than $ 13 billion to the U.S. GNP and an estimated 900,000 jobs, and have affected the quality of life in the U.S.

Binary arithmetic computers were designed to perform rapid calculations of addition, subtraction, multiplication, division, etc. Analog computers were developed at about the same time as the digital counter of Pascal in the 17th-18th century. [33]

Since then computers have been developed for a variety of applications. [33,34] However, one novel application has been introduced recently. Using the memory of huge computers, plus their property of quick retrieval, Professor Peter Kamnitzer was able to develop visual simulation of a city for urban planning. [35] Through manipulation of the computer the user can "travel through the pre-registered details" in a visual manner.

This novel technique was initially used for aerospace simulation.

For the urban planner it provides an "interactive system" in which the designer becomes part of the environment and vice versa.

APPENDIX NO. 3 (to table 3)

1. Solar Energy - India

Cow dung is now burned in India in order to provide heat for cooking. If solar energy was used for this purpose, cow dung could be used as fertilizer. With 75% of the labor force in India engaged in agriculture, such an effort, if successful, would be very meaningful.

However, the mere mention of the above idea is not nearly enough to start such a transformation. What is required is about two years of non-experimental research into the feasibility, economics and technology of such a transformation. This, to the best of our knowledge, has not been done. [36]

Several technologies which are required for making solar energy at low cost source are developing rapidly. (See for example National Physical Laboratory in Israel). [15]

2. Continuous Welding - U.S.

Due to the Six-Day-War in the Middle-East, oil supply lines lengthened by thousands of miles, and the demand for tankers grew by millions of tons.

While U.S. and British yards were still laboriously constructing their ships, the Japanese poured large sums into automated equipment for continuous welding and thus reduced labor cost and construction time. [37]

All these savings brought down the cost of a tanker built in Japan from some $200 per deadweight ton for a 25,000 ton vessel to $104 for a 50,000 tonner and $75 for tankers of 150,000 tons and up. [37]

Powered by resurgent world trade and later by the Korean-War demands, shipping enjoyed an almost uninterrupted boom from 1945 to 1953. Three years later the first Suez crisis touched off another boom.

The Japanese were prepared. [37] In 1945 the largest supertanker afloat carried 23,000 deadweight tons. By 1956 the size of the supertank had grown to 45,000 tons. In 1965 the largest ship reached 210,000 tons. Now 800,000 tonners are under construction in Japan.

The U.S. missed the opportunity to become the world's largest ship builder.

3. Agriculture - Egypt

The capabilities of Israel in modern agriculture techniques are manifested in its aid to many Developing Countries in Africa, Asia and South America.[11]

For example: Three cooperative-type training villages were set up in Cameroon by Israelis to train about 500 young people for permanent cooperative settlement or as animators (catalysts) in their native villages.

In Nepal a 10,000 acre tract was developed with the help of Israelis for settlement by 1,100 families in new villages with some cooperative services.

In Venezuela a center has been established where Israeli advisers train professional agricultural personnel, economists and engineers involved in regional planning.

Since Egypt and Israel share a common border it is clear that agricultural technology from Israel could have been of value to Egypt. (GNP per capita $170 in 1968).

Capital does not seem to be scarce in the area. Loans by Kuwait (GNP per capita $3,462 in 1968, 70% illiteracy) to other Arab countries have totalled more than $400 million between 1962 and 1965.[11]

In this case the missed opportunity for technology transfer is due to political reasons.

4. Sisal - Tanzania

We have indicated[1] that it is in the best interests of Tanzania (GNP per capita $88 in 1971) to develop new applications and new technologies for Sisal. This obvious need was stressed also in the context of Sisal in the Yucatan region.[38]

However, as in the case of India and solar energy, it seems that the non-political, interdisciplinary, non-experimental local tools for developing new uses do not exist.

5. Sugar - Cuba

A review of the production yields of sugar cane reveals a strange phenomenon: In spite of the fact that Cuba depends on sugar cane production more than any country in the world, it does not have the highest yields. This can be seen in the following table. [39]

SUGAR CANE PRODUCTION AND YIELDS 1967-68

	Total Hectares* of Land	Total Production Metric tons	Yield Kg/Hectare
U.S.A.	196,000	13,293,000	67,800
Philippines	706,000	15,525,000	50,800
Cuba	1,000,000	40,000,000	40,000

* Hectare = 2.47 acres.

It was noted [40] that "the chief barometer of the nation's economy in Cuba is the sugar industry which represents about 85% of exports".

It seems then, that Cuba is missing the opportunity to increase the yields of its major product by acquiring the sugar technology available, say, to the Filipinos.

One may go one step further and question why have not the Cubans developed a Sugar Institute similar to the Rice Institute in the Philippines?

6. Carbon Fibers - Britain

During the last ten years the British have developed novel carbon fibers, and their technology. [41] This technology was partially responsible for a $2 billion sale of Rolls-Royce engines to Lockheed Aircraft Corporation in spite of pressures of the local U.S. jet engine makers: General Electric and Pratt & Whitney. [42]

The British sought mass production of the carbon fibers. [43] But the various companies which contributed to the development in England could not decide on a course of action, neither could the British Government. [44]

The British public opinion requested a considerable investment in the U.K. to ensure that the original British research results will not be "given away free."

After a $240,000 market survey ICI came to the conclusion [45] that a large U.K. carbon fiber plant would not be commercially attractive at present. It found that the U.K. carbon fiber process, using high cost acrylic fiber and complex engineering to produce continuous filaments, is too costly.

However, at the same time several American corporations [46] (as Union Carbide, Great Lakes Carbon, etc.) have announced their entry into the production of carbon fibers. In many cases this was done by small modifications of the British patents.

Thus, Britain lost the opportunity to transfer its technology to the U.S. and to earn foreign exchange and prestige in the process. Indeed, several licensing agreements did take place between U.K. and U.S. firms, [44] but these were of a fragmented nature and did not maximize the original British strengths.

7. Computers - France

Inasmuch as France yearned to have its own computer technology, the Machines Bull computer firm was sold to General Electric [47] (1963). Three years later the French Government had to start Plan Calcul [48] in order to ensure that not all the French computer technology will be transferred "cost free" to the United States, and that this technology will not be arbitrarily developed, neglecting the local transfer interests of France.

By selling Machine Bull at the time, the French missed the opportunity to have their own computer technology and to transfer it for proper compensations to the U.S. and to other countries.

They also missed the opportunity to transfer computer technology within France due to the same sale. The computer market in France today is controlled about 70% by IBM, 20-25% by General Electric and 5-15% by Control Data Corporation.

With new investments in the newly created Compagnie Internationale Pour e'Informatique the French hope to correct their mistake.

8. Iron Ore - Sierra Leone

The following has been noted with regard to iron deposits in Sierra Leone: [49]

When there are so many possibilities of finding a good mineral deposit elsewhere in the world, a government or company that sits on a resource too long may find in the end that it has no value. By the time it was finally decided to exploit the Tonkolili iron-ore deposit, in Sierra Leone, for example, it was no longer feasible -- other iron-ore deposits in Africa and Australia had been discovered in the meantime that were much more economic.

What held up the decision to exploit the iron ore deposit? Was the Government of Sierra Leone aware of the other deposits? Did it evaluate in time the technological and economic aspects of developing its iron ore deposits?

Continuing the discussion on the above resource it is suggested: [49]

With modern technology, it is always possible that some invention or process will alter the whole economics of an industry, and the kind and quality of raw materials it needs. If a resource is already being exploited, it may be able to continue to compete; if it is not, interest in it may disappear.

To what extent is the Government of Sierra Leone aware of new technologies with which it may interact?

What combination of technolgies should be transferred to Sierra Leone in order to make profitable use of the iron ore deposits?

9. Land - Belgium

The river Schelde separates the Province of East Flanders from the Province of Antwerp. While the Province of Antwerp is crowded with developing industries, and is in urgent need of more land, the other side of the river is used for agriculture.

Due to a debate between the bureaucracies of the two provinces, as to whose authority covers the agricultural side of the river (once industry moves there)*, Belgium industry cannot move across the river. The result: Industry moves to other countries, as the Netherlands for example.

Technology transfer cannot be achieved in the same country across a river. [50] An extreme case which demonstrates dramatically the possible bureaucratic barriers to technology transfer.

---

* There is no Port Authority in Antwerp as in New York for example, where it has authority over the parts of the port which are in New Jersey and New York.

10. Diversification - Brazil

The following paragraph speaks for itself: [51]

> The present Government realizes the need to increase and diversify exports. It has publicized this need and everyone now accepts the fact that special efforts must be made to generate the foreign exchange necessary to pay for the goods Brazil requires to achieve faster growth and still meet its foreign debt commitments.
>
> But so far an effective export mentality has not been fully developed. Manufacturers still think that it is the other fellow who must do the exporting. Most Brazilian businessmen also believe that coffee, iron, and cacao are the answer to the country's needs. They do not see that a much sounder economy would be created through the export of $1 million worth of 1,000 different products than through exporting the whole $1 billion in a few commodities.

Why is the diversification of industry in Brazil progressing so slowly? Is there enough knowledge of what technologies are most compatible with the local agricultural crops and existing industries? The answer to the latter question seems to be negative. [52]

REFERENCES

[1]Bar-Zakay, S.N., "Policymaking and Technology Transfer: The Need for National Thinking Laboratories," The Rand Corporation, P-4497, December 1970. *Policy Sciences*, 2, No. 3, September 1971.

[2]Bar-Zakay, S.N., "Technology Transfer Model," The Rand Corporation, P-4509, November 1970, *Technological Forecasting and Social Change*, 2, Nos. 3/4, 1971.

[3]Strassmann, Paul, "The Manufacturing Experience of Mexico and Puerto Rico," *Technological Change and Economic Development*, Cornell University Press, New York, 1968, pp. 34, 123.

[4]Bar-Zakay, S.N.,"Technology Transfer from the Defense to the Civilian Sector in Israel," NATO Advanced Study Institute, June 24-July 6, 1973, at the Novotel Paris-Every, France.

[5]Bar-Zakay, S.N., "Why Do Governments Employ 'Think Tanks'?," 20th International Meeting of the Institute of Management Sciences, Israel, June 1973.

[6]*Industrial Technologies for Developing Economies*, Jack Baranson, 1969, Praeger Publications.

[7]*The Ford Foundation Annual Report*, October 1, 1968 to September 30, 1969, Ford Foundation, New York, New York.

[8]*Soviet Atomic Spies* by Bernard Newman, 1952, Robert Hall, Ltd. and *Business Intelligence and Espionage*, Dow-Jones-Irwin, Inc., 1966.

[9]Heymann, Hans, Jr., "Promoting the 'D' in R&D Dubious Models and Relevant Strategies," P-4387, The Rand Corporation, 1970.

[10]Schmidt, Dr. Werner P., "The International Transfer of Management Skills: Volkwagen's Needs, Experiences, and Plans," ITOMS World Conference of AIESEC, Torina, November 1969.

[11]Laufer, Leopold, *Israel and the Developing Countries: New Approaches to Cooperation*, The Twentieth Century Fund, New York, 1967.

[12]*World Protein Resources*, 1966 Advances in Chemistry Series, American Chemical Society.

[13]Barber, Richard, *The American Corporation*, E. P. Dutton & Co., Inc., 1970.

[14]*American Soft Drink Journal*, July 1970.

[15]*Scientific Research in Israel, 1972*, National Council for Research and Development, Prime Minister's Office.

[16]I. Harpaz and D. Rosen, "Development of Integrated Control Programs for Crop Pest in Israel," Chapter 20 in C. B. Huffaker (ed.), *Biological Control*, 1970.

[17]*The Origins of Entrepreneurship in Meiji, Japan*, Johannes Hirschmeier, Harvard University Press, 1964.

[18]Nelson, Richard, et. al., *Technology, Economic Growth and Public Policy*, The Brookings Institute, 1967.

[19]Shriver, Sargent, et. al., "Ambassadors of Good Will, The Peace Corps," *National Geographic Magazine*, Vol. 126, No. 3, September 1964.

[20]"Science and Technology in Developing Countries," Proceedings of a Conference held at the American University, Beirut, Lebanon, December 1967, Cambridge at the University Press, 1969.

[21]Technology Transfer Examples, TU Division, NASA Headquarters, February 15, 1970.

[22]"Portable Water from the Sea at 10 to 15 Cents Per 1000 Gallons to be Feasible in the Future Using Reverse Osmosis," *Chemical Engineering News,* May 13, 1968, p. 57.

[23]"duPont Entered Reverse Osmosis Water Treatment Competition," *Chemical Engineering News,* May 29, 1967, p. 47

[24]"Emergency Sea Water Demineralizer," U.S. Patent 3, 156, 645, November 1964.

[25]Waley, S. G., "Mechanism of Enzyme Action," *Quarterly Reviews,* Vol. 21, No. 3, 1967.

[26]"The Enzyme Explosion," *Detergent Age,* May 1967, p. 47.

[27]*Modern Plastics Encyclopedia,* Vol. 43, September 1965, p. 1120.

[28]*Teleoperators and Human Augmentation,* NASA Technology Utilization Division, SP-5047, December 1967.

[29]Huschke, R.E., R.R. Rapp and C. Schutz, *Meteorological Aspects of Middle East Water Supply,* The Rand Corporation, March 1970, RM-6267-FF.

[30]Albert, G.H., "Dietz Composite Materials," 1965 Edgar Marburg Lecture, American Society for Testing and Materials, Philadelphia, Pennsylvania.

[31]*Ordnance,* May-June 1968, p. 582.

[32]*Technological Innovation--Its Environment and Managemement,* U.S. Department of Commerce, 1967.

[33]Bobrow, Davis B. & Judah L. Schwartz, eds., *Computer and the Policy-Making Community: Applications to International Relations,* Prentice Hall, 1968.

[34]Feigenbaum, E.A. and J. Feldman, ed., *Computers and Thought,* McGraw-Hill, 1963.

[35]Professor Peter Kamnitzer, UCLA School of Architecture, Director and Producer of the film, "City-Scape," also *Architectural Design* (Britain), September 1969; article by Professor Milnt, *Progressive Architecture,* June 1970; General Electric Moon Landing Visual Simulation.

[36]Peter E. Glasser, et.al., *Space Technology Transfer and Developing Nations*, Arthur D. Little, NASA CR-1222, October 1968, p. 79. U.S. Clearinghouse for Federal Scientific and Technological Information N68-37216.

[37]*Fortune*, September 1967. Also "Japan: Shipbuilding Goliath," *Marine Engineering/Log.*, Vol. 73, October 1968.

[38]McGuigan, William, *Some Possible New Directions for Henequen Bay*, Hudson Institute, HI-1252-DP, August 22, 1969.

[39]Data collected from the *FAO Production Yearbook, 1968*, Vol. 22.

[40]*The 1970 World Almanac*, Newspaper Enterprise Associates, Inc.

[41]"Advanced-fiber Composites Spark Materials Revolution," *Product Engineering*, October 23, 1967.

[42]*Fortune*, March 1969.

[43]*Aviation Week and Space Technology*, November 25, 1968.

[44]"Who Should Make Carbon Fibers?", *The Economist*, Business Section, December 7, 1968.

[45]*Chemical and Engineering News*, November 17, 1969.

[46]*Reinforced Plastics and Other Composites*, Skeist Labs, New Jersey, 1971.

[47]Servan-Schreiber, J.J., *The American Challenge*, Atheneum House, Inc., 1968.

[48]Plan Calcul for the Establishment of Compagnie Internationale Pour e'Informatique.

[49]Karmack, Andrew M., *The Economics of African Development*, Praeger Publications, 1967.

[50]Study Center for the Expansion of Antwerp, Private Research for the Chamber of Commerce, Antwerp.

[51]"The New Brazil--Prospects for Stability in Profits," *Business International*, 1965.

[52]Glasser, Peter E., et.al., *ibid.*

TECHNOLOGY TRANSFER IN "PARTIALLY DEVELOPED" COUNTRIES

Dr. Leslie G. Cook

Manager, Planning And Program Analysis
Corporate Research Staff
Esso Research & Engineering Company

"What do you mean by a Partially Developed Country?" I hear you ask.

I was requested to examine with you, as a part of the program for this conference, the Technology Transfer problems of a particular group of countries; not those generally spoken of as "underdeveloped," nor those generally spoken of as "developed," but those suffering from the peculiar problems of "one-sided" development.

By this was meant those countries with well-developed, internationally competitive raw materials industries, but without secondary manufacturing industries that enjoy similar outstanding international stature. This lack of international competitive stature in secondary industries can range, of course, from a total absence of secondary industry to a rather considerable secondary industry which however serves essentially only the domestic market behind a protective wall of tariffs or other controls--essentially a contrived and noncompetitive situation.

Probably "Partially Developed" is a confusing term for these countries, but anyway it is of these countries I will speak. And there are many of them. Nearly all of those countries with rich raw materials resources urgently needed by the industrialized nations come in this category.

The countries in this category have many common characteristics and problems, which are very different from those of the so-called "underdeveloped" countries.

First, they are relatively prosperous due to their raw materials export industries. Concern about the problems of the secondary manufacturing industries may be vocal, but this concern does not reach to the heart--or perhaps I should say to the pocketbook, since the economy is quite prosperous without them.

Second, these countries tend to acquire, along with other effects of prosperity, a well-developed and broadly based educational system. From this comes a steady flow of well-trained, skillful, capable technical people for whom adequately challenging jobs are not available. For much of this flow of people, trained in the newest areas of science and technology, the sophisticated, modern, highly competitive secondary industries are the natural and proper outlet and these industries are exactly those that are missing or maintained in an unsatisfying, local contrived situation. Such countries are plagued with an ever-growing group of frustrated and thwarted scientists and engineers. The most dynamic and creative of these try to migrate, if they can, to the leading industrialized countries--"the brain drain"--where modern, internationally competitive secondary industry does exist that can properly use their talents. The almost incredible infiltration of Canadian scientists and engineers into the upper echelons of U.S. technological industry (and universities also) is a classic example. However, for many "partially developed" nations this outward migration is impeded by language and culture barriers and sometimes by immigration restrictions, further aggravating the home situation.

Third, the subject of "Technology Transfer" generally becomes a major vocal issue, but with emphasis primarily on "jobs for the skilled." Since the technically skilled are in a minority and the country is prosperous without their help, this issue tends to remain vocal and tends not to be taken seriously as a national issue.

Fourth, because of the general prosperity resulting from the raw materials exports, wages tend to become high, nearly as high as in industrialized countries. This rules out the introduction of low wage, labor-intensive secondary industries. Populous "underdeveloped" countries are always searching for "labor-intensive" industries to provide jobs for the low paid, the unskilled. But to the "one-sided" developed countries these labor-intensive industries are unattractive. Nothing but capital-intensive, internationally competitive secondary industry will meet the already existing wage and skill structure.

Thus, whereas the "underdeveloped" country has avenues for entering job-rich secondary industry by taking in labor-intensive offshoot technologies (such as, for example, the threading of computer core memories), the already prosperous "partially developed" country finds no acceptable entry except at the most sophisticated, capital-intensive level. This puts it in direct competition at the embryonic stage with the fully developed industrial strength of the highly developed industrial countries. As a head-on competitive situation, it is almost impossible, as example after example of failure to "break in" to the international competition only too well attests.

But the "one side developed" country has nevertheless a very serious need to enter the secondary industry world competition, and also has unique assets to parlay into a successful entry. It has raw material strengths which "underdeveloped" countries only wish they had! Where then is the "hang up" in identifying and following a successful entry strategy?

I think it lies in two things.

First, it has become a sort of wishful doctrine that raw materials resources, whatever they may be--coal, oil, copper ore, iron ore, beef, pulpwood, tea, rubber, coconuts, wheat, rice, or what have you, are inexhaustible and essential--not subject to substitution. Both parts of this doctrine are essentially fallacious, as example after example in the last 50 years shows clearly. There are already "partially developed" countries in desperate economic straits today because they wanted to believe in this doctrine, and did until the moment of disillusion when it is, of course, too late. Indeed, one of the main thrusts of science in the last 50 years (it can be clearly seen in retrospect) has been the discovery and development of "alternate" raw material sources. This has been exceedingly successful, much to the dismay of one raw materials industry after another.

Second is the human habit of wishful thinking--it can't happen to us! "Our" raw materials will always be essential. The industrialized nations will always have to come to us. We have no need to worry about our prosperity!

The primary essential for successful Technology Transfer is a recognition by the recipient that he has an urgent need for it that takes priority over all else. We all know where this recognition has moved Japan in the last 20 years. In a way, Japan may have been fortunate that it had very little in the way of raw materials resources. It had no opiate of "raw materials prosperity" to dull the clarity of national understanding. There was no way the Japanese could talk about their "inexhaustible resources" and fool themselves into thinking that prosperity had to continue coming to them without self-sacrifice, perception, courage, and hard work. Unless and until this recognition of imperative and priority need develops as a national will, the more technical aspects of Technology Transfer must remain essentially irrelevant.

It is however to these technical aspects of modern, sophisticated Technology Transfer, appropriate to "one-sided developed" countries, that I now wish to speak.

The world has become one large shopping center today in a totally new sense--a sense that cuts across national boundaries and political ideologies. Wendell Willkie called it "one world" quite a few years ago, but many of us do not want to fully face it yet. There have, of course, been military conquerors throughout history who assayed to create "one world" and, temporarily, some did. But today's "one world" is totally different, deriving from totally new forces--the advances science has made in understanding man's material world and bending that understanding to his own use. The results of the discovery of the wireless, of the transistor, of uranium fission, of penicilins, of DDT, of synthetic elastomers, plastics and textile fibers, of hybrid corn--these and dozens more are world market forces that are not temporary and will not go away. And traditional methods of protecting and isolating local markets never again can have their historic effectiveness; they are relatively impotent against these new forces of technology. This is a totally new type of conquest. Historically conquests were imposed by the conqueror against the will of the conquered! This new technological conquest derives its force from the desire of the victims to enjoy all the benefits of being "conquered." Shades of that marvelous Peter Seller's film, "The Mouse That Roared"!

It is evidently vital to the economic welfare of the industrialized nations themselves to maintain their own technological leadership in this fast moving, competitive world market. The competition is sharp. There have been some rather notorious examples in the last 15 years of what happens when a developed country loses out in major technology competition--loses large investments, loses large orders or hoped-for orders, and of course loses skilled jobs and taxable income. And where does this leave the "one-sided developed" country, the would-be embryonic entry into this top league, high stakes technological poker game?

In this intense competition for technology leadership, it is a fact of life that Technology Transfer from someone who has the technology to someone who hasn't (but wants it) must be a strictly business transaction, one that must be negotiated till both parties feel satisfied with the bargain. This may seem obvious for technology transfers between industrially developed countries themselves; it is equally true however for transfers to "one-sided developed" countries--the subject of this talk.

It is sometimes said that the barriers to successful Technology Transfer are really the barriers to the transfer of knowledge, that these are one and the same. I do not believe this is correct at all. As an extreme illustration, consider how long it would take an electronic specialist marooned on a desert island to construct a wireless transmitter to ask for help when he had no copper, no glass, no vacuum pumps, no electric furnaces, no batteries, no single crystal silicon--nothing. When Daniel Defoe marooned Robinson Crusoe on an island in 1720, he contrived to provide him with basic hardware from the shipwreck; otherwise poor Robinson could never have survived except on wild berries, even in a novel! In a less extreme vein the lesson is equally clear. Successful Technology Transfer requires much more than just the transfer of knowledge--it requires at least the transfer of a full supply of the essential basic hardware and prepared materials.

It requires even more than that. As every toiling practitioner of Technology Transfer knows, at least six elements are absolutely essential for success--viz:

1. The transfer of knowledge and know-how.
2. The availability (by transfer if necessary) of all the needed equipment and starting materials.
3. A real felt need, a conviction that the transfer _must_ succeed.
4. An environment--political, legal, economic, cultural and social--of positive encouragement; a willingness to cut through manmade barriers.
5. A knack of taking full advantage of all beneficial local factors--raw materials, people, location, etc.
6. An assurance of access to the international market and of the credible capability to meet competitive conditions in that market--not only cost but quality, service, reliability, delivery, etc.

The first two are largely technical, and I want to say more about them. The last four are largely social, political and economic, and I am going to leave those for others to deal with. I will assume they are adequately taken care of (although, as was pointed out to me by a colleague who was kind enough to read and comment on my draft, they generally are _not_ properly taken care of; generally are far too cavalierly dealt with, especially by governments, and this is responsible for all too many economic fiascoes. Incidentally, this colleague is present today and may well wish to comment further during the question period.

My _first essential element_ is the adequate transfer of knowledge and know-how. Knowledge and know-how have several characteristics which have been known for hundreds of years--viz:

1. _Genuine_ technical knowledge can in principle always be purchased (if it exists), and usually pretty cheaply. There is of course a particularly dangerous hitch--the danger of fraud, of which more in a moment.

2. Technical knowledge can in principle easily be pirated if the selling price is set too high, and penalties against pirating are very difficult to enforce. This has the same dangerous hitch--however only more so. The would-be pirate must be an expert in the goods if he is not to make a fool of himself.

3. The would-be purchaser (or pirate) must have a technical sophistication and expertise at least as great as that of the seller if the transfer is to be followed by success. The deliberate selling or offering of inadequate or fraudulent knowledge or know-how is very easy and common--buyer beware! Probably more money-losing fiascoes result from failure to adequately heed this warning than from any other single cause.

My second essential element is the availability of needed equipment and starting materials. This has characteristics also known for hundreds of years--diametrically opposite to those of technical know-how--viz:

1. First-rate equipment and materials cannot always be bought, are rarely cheap, and in general are priced at what the international market will bear. The prices may not have much connection with intrinsic cost and none with the cost needs or desires of the would-be "transferee."

2. Equipment and materials of quality cannot easily be pirated or otherwise "got cheaply." Trying to dodge the world market is very likely to lead to faulty material or equipment and eventually to inordinate loss and expense.

3. Adaptation to take advantage of local raw materials is not easy--often because there are no advantages and the phrase is twisted to mean accepting local problems and disadvantages.

The "one-sided developed" nation, attempting to enter the world market in some secondary industry area by Technology Transfer, must recognize what the seller of the technology is up against. Any "seller" of technology, faced with these characteristics, must take elementary precautions against theft or piracy of knowledge and know-how by the customer, whoever he may be. Anyone who does not take steps to prevent outright theft of technical knowledge and know-how in today's technological world is as foolish as a man who leaves his worldly wealth lying in piles of dollar bills--or franc notes--on the sidewalk for any knowledgeable passer-by to pick up. Keeping your money in your pocket helps to develop healthy, successful business dealings in an atmosphere of mutual respect. And it is rather obvious what the seller must do. He must take the elementary precaution of tying the sale of knowledge and know-how to the transfer of materials and equipment. And the purchaser, if he is wise and prudent, will recognize that a tie-in is his best--perhaps his only--real insurance that a successful Technology Transfer will occur.

This whole subject of Technology Transfer is coming under increasingly intensive discussion for obvious reasons, but in two quite different professional circles.

On the one hand are the social professionals--politicians, economists, and sociologists who discuss Technology Transfer in terms of the broad generalities of social needs (the need for jobs--particularly for unskilled jobs; the need to use local raw materials--particularly if they are low grade; the need for capital--particularly if prospects for its eventual return are dubious) and how to encourage these things by influencing the economic trade-off with special tariffs or other regulatory tools.

On the other hand are the technical professionals--engineers and scientists who discuss Technology Transfer in terms of the practical problems of making a specific new enterprise a success right down at the technical and management level.

I have observed that these two professional groups, the economists and social scientists interested in broad generalities and economic policies and the engineer-scientists interested in the practical details of the success or failure of specific enterprises, interact almost not at all. They are separate and closed professional circles. A paper written on some aspect of Technology Transfer by an economist will invariably cite as references dozens of papers by other economists, but hardly ever the writings of engineers or scientists as if they did not exist; likewise, papers on Technology Transfer by engineers or scientists will cite other engineer-scientists' papers, but hardly ever papers by economists as if they too did not exist.

Now Technology Transfer--to be successful--must be successful from both points of view. It must fit in with sensible and desirable economic policies, and it must be realized successfully in technical practice. And these two requirements are not separable worlds--they are tightly interlinked. The economist, sociologist, or politician can easily specify an economic requirement--say a viable technology with a high labor-to-capital- ratio--which the technical man simply has not got and cannot produce because it does not, and cannot, exist. It has been my observation that most of the horror stories of Technology Transfer can be traced to one or the other of these groups having got the upper hand so to speak, selling a program of Technology Transfer which made sense to him but which foundered eventually on the shoals of the other specialty. It is hard for the economist, sociologist and politician to accept that economic, social and political theory today will in the end be strongly conditioned, perhaps even totally negated by the simple and obvious facts of the physical sciences; it is equally hard for the engineer-scientist to accept that the most beautifully sophisticated scientific technology with obvious technical advantages may in fact founder on the shoals of competitive marketplace reality. But Technology Transfer is no respecter of persons, or of professions. And until we learn how to get these two professions to work together instead of in separate circles, we will continue to have Technology Transfer fiascoes on a totally unnecessary scale.

And this is a very difficult challenge. The reason it is so difficult, I believe, does not lie in human orneriness, but rather in the nature of our education process. On the one hand most social scientists and economists can only be described as technically illiterate; they took little or no science or engineering at school. On the other hand most scientists and many engineers can only be described as social sciences illiterate; they took little or no training in economics or sociology at school. These two professional groups with rare individual exceptions--in my observation--find it difficult to even converse; they have no common language of terms, no common concepts, no common ideas. And those who need to benefit from their joint advice may be pardoned for occasionally becoming confused. Yet this is one of the most crucial problems in successful Technology Transfer.

This problem cannot be solved by getting them together on committees--that mainstay solution for everything to which governments are so prone to turn. I have seen a committee of sincere economists and scientists manage to sit together a whole day in dedicated effort--and communicate absolutely nothing but total frustration.

I think the most useful advice one might give to the "one-sided developed" nation attempting to enter the competitive international world of modern secondary manufacturing is to engage the professional services of at least one man who is reasonably literate in science and engineering on the one hand, and economic, social and political sciences on the other. These are rare animals, and the genuine article is not always easy to identify from the sham; but in the highly competitive technological world economy of today and tomorrow, their wisdom is absolutely essential to avoid failure and to ensure success.

It is a tough game, launching more or less from scratch into the world's #1 league of technology competition. It requires a cold, objective perception, a unity, a dedication, a commitment and a degree of self-denial which most prosperous "one-sided developed" countries find hard to muster. They are peculiarly handicapped by their own temporary and evanescent prosperity.

OBSTACLES TO THE TRANSFER AND ADAPTATION OF IMPORTED TECHNOLOGY IN THE AFRICAN COUNTRIES

'D. Babatunde Thomas

Economist, Lagos, Nigeria
and Florida International University

INTRODUCTION

The co-terminal relationship between technological advances and steady economic progress in the advanced countries (ACs) is explained largely by the factor endowments, and more importantly, by the systematic development and application of scientific knowledge. The rapid growth of capital and skill formation served to broaden the capacity to develop indigenous technology and, when necessary, advantageously adapt technologies developed elsewhere. This important foundation greatly facilitated the self-perpetuation of technological advances and provided the broad base of science and technology (S & T) which spurred the economic and social development of these countries.

In stark contrast, the developing countries (DCs) generally lack the capacity to develop the required capital and skill formation for a replication of the experiences of the ACs. Serious shortages of scientific and technological manpower keeps industrial science at a rudimentary state and greatly constrains the development of indigenous technology. This in turn has limited the extent of the diversification in economic activity for the advancement of traditional agriculture and development of viable manufacturing activities. In the past two decades, pressures for rapid economic development have made reliance on imported technology more and more inevitable. But similar obstacles threaten the effectiveness of technology transfer (embodied in imported capital equipment for example) and its adaptation to the conditions and needs of the DCs. Two important obstacles viz.: (i) the existing mechanisms of international technology transfer (TT); and (ii) the weak base of S & T in the DCs; encompass most of the technical, socio-economic, and political constraints against successful importation, assimilation and advantageous use of imported technology in DCs.

This paper focuses on the economic implications of these obstacles to the development efforts in the African countries. Existing S & T policies based on the UNESCO (Lagos and Yaoundé) Conferences on the Organization of Research and Training in Africa are discussed and possible arrangements for effective TT are then examined. [5 & 7] Most of the considerations examined are not however peculiar to Africa.

FACTORS AFFECTING THE TRANSFER OF TECHNOLOGY TO AFRICAN COUNTRIES

African countries remain traditionally agricultural in spite of the dynamic role of technological advances in international trade. The contribution of R & D to the 'product cycle' and changing comparative advantage in world trade has been limited mostly to the ACs in Europe, North America, the Soviet Union, and Japan.

All bilateral and multilateral trade arrangements between African countries and the ACs through GATT, UNCTAD, EEC, etc., are primarily to improve terms of trade. Technology considerations have yet to be integrated into these trade arrangements when, in effect, imported products that are protected by monopoly privileges through patents and other modes of capital and TT tend to be over-priced and hence have an adverse effect on the terms of trade of the DCs in general.

Although the suitability of modern technology for Africa has been questioned based on differences in factor endowments, climatic conditions, and the total ecosystems, there is no conclusive evidence that modern technology and know-how cannot be adapted advantageously to the tropical ecosystems. Trade policies designed to encourage domestic and foreign investment in industry have increased the African countries' rate of capital formation and, indirectly, the transfer of modern technology. But, the latter neither subsumes assimilation nor diffusion. Recent rapidly increasing factor costs and the depletion of key resources in the ACs have also necessitated an increase in the overseas operations of multinational corporations. Part of their investments are undertaken to protect export markets in response to African countries' import substitution programs. This suggests that climatic conditions and political factors are now playing a lesser role than previously attributed to them and the comparative advantage of the DCs in labor cost has become a dominant factor in the multinationals' search for high profits.

The transfer of technology to the African countries is usually in the 'embodied' form in imported machinery. The lack of appropriate manpower to exploit these technologies to meet local conditions and needs, and the obstacles posed by the existing modes of transfer have fostered complete dependence on foreign technology and limited indigenous technological initiative. For example, agricultural technology (for food production and processing) 'embodied' in imported tractors, irrigation systems, fertilization processes, new cropping, canning, and refrigeration, have not been advantageously used due to improper adaptation to tropical agriculture, lack of effective organizational know-how, and requisite marketing techniques. Similarly, industrial technology 'embodied' in imported machinery and other capital equipment has had limited effects on industrial development. This can be attributed largely to the weak base of S & T, lack of industrial discipline and adequate entrepreneurial and managerial know-how, and finally, inadequate technical training to facilitate the adaptive capability of the labor-force to use modern technology in a way that is most beneficial. One economic implication of this relative technological stagnation in the DCs has been a widening in the income gap between the ACs and the DCs.

Mechanisms of Technology Transfer

The most commonly used mechanisms of TT are licensing and patent agreements, coproduction arrangements, direct enterprise-to-enterprise transfers, industrial consultancy services, and the importation of machinery and other capital equipment. The proprietary characteristic of most of these mechanisms is illustrative of the serious built-in hindrances to complete and effective TT to the African countries. The abstract view that technical knowledge is freely transmittable between nations is therefore erroneous.

In comparison to the ease of transfer and the advantageous use of imported technology in ACs such as the Soviet Union, Germany, and Japan, the narrow base of S & T and the incipient state of industrial science in the African countries puts them in a weak recipient position in technology trade and makes the advantageous use of the instructive and adaptive potential of the technology 'embodied' in imported capital equipment very difficult. Consequently, the movement of capital and technology by means of the existing modes of transfer to the African countries has failed to provide any stimulus for immitational initiatives. The long duration of the monopoly privileges conferred on the modes of transfer such as patents and licenses tends to hinder the ability to fully assimilate and adapt imported technology. The proprietary and monopolistic features of the existing modes of transfer have therefore rendered them inoperative between unequals.

TABLE I

CUMMULATIVE FINANCIAL AND HUMAN RESOURCES IN SCIENCE AND TECHNOLOGY

Country	in Year	Estimated number of: Scientists & engineers 1 Total	2 engaged on R & D	2 as % of 1	Technicians 3 Total	4 engaged on R & D	4 as % of 3	National expenditure on research and development
BOTSWANA	1967	180	10	5.6	606	6	.9	317,500 Rand(a)
CAMEROON	1967	800	80	10.0	3,000	140	4.6	350,000,000 francs CFA
CONGO (Brazzaville)	1966	231	17	7.4	415	3	.7	248,300,000 francs CFA
ETHIOPIA	1967	4,406	...	...	7,753	...	...	...
GHANA	1966	5,137	167	3.3	27,893	551	2.0	...
IVORY COAST	1967	...	204	...	...	174	...	1,661,000,000 francs CFA(b)
KENYA	1964	2,536	...	...	6,893	...	...	...
MALAGASY	1966	...	195	...	...	230	...	997,000,000 francs MG
MALAWI	1967	1,253	14	1.1	5,028	...	...	... (c)
NIGERIA	1966	3,970	1,723	43.4	6,997	...	...	£N 9,770,000
RWANDA	1967	207	19	9.2	454	7	1.5	24,400,000 francs Rwanda
SOMALIA	1965	175	4	2.3	938	10	1.1	770,000 Somali shillings(d)
TOGO	1967	215	...	...	388	...	...	18,108,000 francs CFA
UPPER VOLTA	1967	160	48	3.0	102	32	31.4	...

Source: The Promotion of Scientific Activity in Tropical Africa, UNESCO, Document NS/SPS/11, SC/SP. 68.XIII.II/A. Paris, 1969, p. 21.

(a) Provisional.
(b) Of which 806 million francs CFA is "aide étrangère."
(c) Total expenditure by the government on research is 0.3%-0.4% of GNP.
(d) For 1967.

Although it is still difficult to show conclusively that the marginal cost of transferring technology to the DCs in general is relatively low, there is evidence that the price of imported technology (in the form of royalties, patents and consultancy fees, purchases of machinery, etc.) has been unduly high. [3] To stimulate the flow of new technology to the DCs, the mechanisms of TT have to be reconsidered and the fees sharply reduced. Arrangements for effective TT and assimilation are better achieved through direct transfer, free of monopoly privileges and direct in providing training for instructive and adaptive services to importing industries.

The Base of Science and Technology

The capacity to develop and take advantage of new technology depends not only on facile mechanisms of TT but more importantly on a broadly based scientific and technological manpower with scientists, engineers, technologists, technicians, etc. Table I shows most of the African countries that presently have a national science policy, their stock of scientific manpower, and the financial resources devoted to research. The shortage of this type of manpower and consequently the weak base of S & T in African countries is one of the primary impediments to their development efforts.

The initial and wide recognition of the need for national science policies and technological advances in African countries were a result of the Lagos Conference in 1964. [5] Prior to the Conference, the majority of the countries had neither a science policy nor a national institution to coordinate research activities. The 'Lagos Plan' as the outcome of the Conference was later known, sought to encourage and promote a long-term plan for the development of S & T. The Conference noted that African countries are handicapped by: (i) absence of adequate national scientific policy; (ii) lack of any machinery for coordinating and preparing such policy; (iii) suffering from extreme shortages of senior scientific and technical personnel; and (iv) not possessing enough training and research institutes. [7;p. 11] It recommended a minimum allotment of 0.5 percent of GNP to research and 200 scientific workers per million of the population as the primary targets all African countries should strive for by 1980. See Table II.

A few countries, notably Nigeria, Ghana, and Senegal have already surpassed these targets. Nigeria presently spends the equivalent of 1 percent of its GNP as recurrent expenditures on scientific research. One of the primary aims of the 1973/74 budget is a review of the existing procedures for remitting

TABLE II

Target number of scientists needed as a function of projected population increases and interrelationship between gross national product and research expenditure targets

	1970	1975	1980
Population (thousands)	277,882	312,333	353,243
Aggregate number of scientists at 200 per million population (thousands)	55.6	62.4	70.6
Gross national product (millions of U.S.$)	39,696	49,895	65,238
Expenditure for education (millions of U.S.$):			
Total expenditure	2,865.0	3,492.6	4,160.4
Higher education	499.6	722.6	1,029.7
Expenditure for research (millions of U.S.$):			
Total expenditure	198.5	280.6	326.2
Fundamental research	39.7	56.1	65.3
Other research	158.8	224.5	260.9

Source: <u>Outline for a Plan for Scientific Research and Training in Africa</u>, UNESCO in association with UN-ECA, Paris-7e, 1964, p. 21.

royalties, patents, management and technical fees so as to improve international transfer of technology. Senegal currently spends over 0.7 percent of GNP (or 6 percent of gross fixed capital) on scientific research annually, and Ghana has over 300 scientific workers per million population. Most of the other countries are well on their way to meeting both targets.

Between the 1964 Lagos Conference and the taking stock at the Yaoundé Conference in 1967, scientific and technical institutions in tropical Africa had proliferated to over 2,000. Between the 1963/64 and the 1969/70 UNESCO surveys, [6 & 4] research institutions on food and agricultural sciences rose from 581 to 1163 and those engaged in industrial research rose from 36 to 347. In 19 of the 34 countries, national science policy institutions had been established with plans for their integration into national economic and social development. The resultant coordination of resource allocation for technical development is *prima facie*, a significant achievement. Although the surveys are preliminary and incomplete, they are informative. They show that these institutions are concentrated in only a handful of countries. The dismal results from the efforts of these countries to develop indigenous technology and to adapt imported ones also suggest that the increase in the number of research institutions is purely a numbers game.

A fortiori investment in scientific research is undertaken primarily by government or quasi-government institutions mostly for pure research. In specific countries - Nigeria, Ivory Coast, Ghana, and Senegal, the effect of R & D on the national economy suggests that the output of research is less than commensurate with the consumption of resources in research activities. Lack of the appropriate entrepreneurial and organizational links between applied research institutions and industries has tended to impede the adoption and diffusion of new technical know-how, and the maximization of benefits from investment in research. The remoteness of research to the direct needs of industry is inherent in the isolation of research institutions (usually in universities) from the practical problems of society. This lack of unison between research institutions and industry is one of the most important missing links for advantageous use of modern technology. This link can best be provided by the development and mobilization of proper entrepreneurial and managerial manpower.

Factor Proportions, Choice of Techniques, and Choice of Products

An important extension of the question of how suitable modern technology is in the African countries is the compatibility or incompatibility of local conditions with factor proportions and the scale of operation dictated by specific imported technologies. Clearly, a necessary condition for compatibility is a choice of technique suitable for the attainment of high employment in an unskilled labor surplus economy with small market size. However most, if not all, of the world stock of technologies require high wage capital-intensive techniques and a large market size for economies of scale.[1] Hence, a serious consideration in the choice of technology is its adaptive design for an appropriate choice of technique. This requires knowledge of the technical possibilities of the available alternative sources of technology.

---

[1] The common conceptualization of the relationship between a change in the technique of production and a change in technology is the two-factor isoquant representation where changes in techniques are shown by a movement along an isoquant and a shift in the isoquant represents a change in technology. The restrictive aspects of this presentation are familiar and

(continued)

An important criterion for rational choice of technique is the maximization of the rate of return to the imported capital given existing and possible changes in skill formation, or the minimization of capital utilization per unit of output. Apart from other handicaps in the pursuance of such an objective, capital price distortions, for example through foreign exchange controls may cause capital and imported technology to be valued below their true cost and in turn cause a distorted rate of return. Such distortions invariably lead to wasteful capital deepening when control-free relative factor prices suggest labor-using bias in the choice of technique.

The range of such a choice and the scope for factor substitution may be limited in some cases by the choice of products. Import substitution programs requiring choice of products intended for direct competition with imports and patterned after tastes in the ACs will likely require a choice of technique similar to that used in high income markets. The markets for these products and hence the choice of technique depend largely on imported inputs and cater to only a small segment of the domestic population. The interdependence of choice of products and choice of techniques suggests that the success of the continuing search for labor-intensive techniques in African countries is dependent on the proper choice of products. This is an important consideration in deciding on the importation of foreign technology and the development of indigenous ones.

TECHNOLOGY TRANSFER, INDIGENOUS TECHNOLOGY DEVELOPMENT AND LONG-RUN INCOME GROWTH

Indigenous technology development in African countries has so far been a slow secular process. Clearly, neither TT by itself nor indigenous technology development offers complete hope for the long-term development of these countries. The need for a simultaneous use of both approaches is underscored by the complimentarity of effective TT and a broad base of S & T. The effectiveness of TT depends on the base of S & T and the more effective and complete the transfer, the greater the chances of its stimulating indigenous technological initiative. Such an integrated approach requires the cooperation and support of the ACs who are major trading partners of the African countries, namely the EEC, the U.S., and Japan. Initially, this could be addressed by the integration of TT into commodity trade arrange-

[1](cont.) it would be perfunctory to discuss them here. For our purpose, it is more illuminating to simply describe a technology as a configuration of techniques of production where a technique relates to the factor combination in a particular production process at a particular time.

ments and the liberalization of the existing modes of acquisition and use of modern technology.

A rise in the capacity of the DCs to develop indigenous technology and effectively assimilate and utilize imported technology is a potential source of expansion in world trade, and an important precondition for their long-run income growth. Such technical capacity expansion requires, in addition to international support, a reorientation of government funding of research to practical applications in industry.

A variant of the Chinese experience and the early Japanese practice of redesigning imported capital equipment so as to make the best use of available local resources are potentially viable directions for the new S & T policy in African countries and therefore merit serious attention. A possible direction is the promotion of pilot projects in agricultural and manufacturing technology intended to effect transfers which involve demonstrations and training in the technological possibilities of imported technology; assisting indigenous entrepreneurs in perceiving the potential 'learning' effect in the use of imported technology; and encouraging indigenous technical innovation and their diffusion.

The establishment of a systematic relationship for the flow of the needs of industry (agricultural and manufacturing) to applied research institutions and the flow of results from research to industry, especially to small scale firms, through an institution specializing in the management of applied research and the marketing of technical innovations is a viable functional approach for the establishment of the missing link between research and their beneficial use in industry.

The problems of balkanization limits resource availability and market size in a large number of African countries thus restricting national technology development policy to a narrow range of R & D projects. By establishing an operational basis for cooperation in research projects, exchange of documents and designs and for other information sharing, costly duplication of efforts could be greatly reduced. Another obvious imperative is that national import substitution programs aimed at increasing the use of domestic raw materials, the conservation of foreign exchange, and export promotion in regional markets require cooperation and active coordination of development programs. The primary obstacle to any form of regional cooperation is the design of the appropriate formula for benefit-sharing. The appropriate methodology for benefit-sharing is therefore a matter of urgent concern. This is particularly important for the OCAM's Science Committee recommendations on the regionalization of scientific research. More importantly, regional research cooperation should be preceeded by a well-defined national science and technology policy.

There is no doubt that the road to successful technological development is rough but with a reliable vehicle and cooperation among its operators, the completion of the journey in a reasonable time is assured.

CONCLUDING REMARKS

The vast reservoir of the world's technical knowledge concentrated in the ACs has much to offer the development of African countries and other DCs provided the internal and external obstacles to effective and complete TT are greatly minimized. In this respect, the ACs have important roles to play in trade arrangements by giving due consideration to the technological needs of the African countries. This is of great importance in the present and future trade arrangements between the EEC and the African associated states.

It is also of vital importance that African countries develop their base of S & T in order to be able to take full economic advantage of scientific knowledge. Without significant improvements in their scientific and technical know-how all the objectives of short- and long-term development are likely to be more difficult to achieve.

REFERENCES

[1] Baranson, J., *Industrial Technologies for Developing Countries*, Praeger, N.Y., 1969.

[2] Thomas, 'D. Babatunde, 'International Transfer of Industrial Technology and the Emerging Nations', Paper presented at the 13th World Conference of the Society for International Development, San Jose, Costa Rica, Feb. 1973.

[3] U.N. Economic Commission for Latin America, 'The Transfer of Technology and Its Relation to Trade Policy and Export Promotion in Latin America,' *Economic Bulletin for Latin America*, Vol. XIV, No. 1, 1969.

[4] UNESCO, *Enquete Sur le potential Scientifique et Technique des Pays d'Afrique*, UNESCO, Paris-7e, 1971.

[5] UNESCO, International Conference on the Organization of Research and Training in Africa in Relation to the Study, Conservation and Utilization of National Resources, Lagos, Nigeria, July 28 - August 6, 1964. See *Outline for a Plan for Scientific Research and Training in Africa*, UNESCO in association with UN-ECA, Paris-7e, 1964.

[6] UNESCO, *Scientific Research in Africa: National Policies and Research Institutions*, UNESCO, Paris-7e, 1966.

[7] UNESCO, The Yaoundé Symposium on the *Promotion of Scientific Activity in Tropical Africa*, Yaoundé, Cammeroon, 1967, UNESCO, NC/SPS/11,SC/SP.68.XIII.II/A, Paris-7e, 1969.

[8] West Africa, 'Making Research Pay', Dec. 31, 1971, p. 1548.

PART V: Industry Experiences

The papers included in this section were all presented by individuals actively participating in corporate technology transfer efforts. Ken Craver leads off with an overview of the corporate environment for technology transfer emphasizing the role of the individual entrepreneur in new product innovation. The same subject is treated by Fernand Debreyre, who discusses planned and unplanned innovation based upon the experiences of several Belgian firms. In a similar view, the importance of unplanned innovation resulting from "curiousity motivated research" is emphasized by Derek Medford. Dieter Altenpohl discusses the acquisition mechanism as a means for transferring technology into the firm while the paper by David Ben Daniel describes the General Electric program to transfer technology out of the firm through the creation of spin-off's.

A method for third-party transfer assistance to firm-to-firm transfer is described by Quentin Hartwig based upon the experience of the Pennsylvania Technical Assistance program. The section concludes with a paper by Donald Cunningham discussing experimental designs for gathering more data on the technology transfer process and a report of the Institute Workshop on Industrial Application of Technology Transfer prepared by Dieter Altenpohl and Philip Wilmot.

TECHNOLOGICAL INNOVATION IN THE CORPORATE ENVIRONMENT

J. Kenneth Craver

Monsanto Company

My discussion of the difficult task of technological innovation within a large multinational corporation will be based on my own experiences as well as those of others.

The most successful commercial innovations have been those where a specific market need was brought to the attention of a knowledgeable and creative individual who was also an entrepreneur or was closely associated with an entrepreneur.

The history of an innovation in plasticizer technology which led to further innovations in other related fields will be discussed. Some unsuccessful innovations will be described to illustrate how new projects can fail.

I define technology transfer as the process of transforming knowledge into the action that produces a desired change. In this sense, this process is an activity recognized as Commercial Development in the growth and innovative industries - chemicals, synthetic fibers, computers, electronics, and aerospace - but the discipline is not restricted to any particular industry.

Industrial commercial development is expected to perform two very important functions. One, to bring market needs to the attention of research, and two, to transfer the products of research out into the market.

It is these two functions of commercial development or technology transfer that I want to discuss in light of my own almost 30 years' experience in the field.

Although the art of discerning market needs and translating research results into salable products has been practiced informally in industry since the Phoenecians, commercial development emerged as a formal discipline shortly after World War II when the Commercial Chemical Development Association was organized in the U.S.A. This professional group (which subsequently dropped the name "Chemical" to become the Commercial Development Association) has become a nucleus for crystallizing the various elements that go to make up commercial development.

The first step in the process of industrial commercial development is generally a market research study, whose aim is to identify the potential markets for a new product or service, or to define a need for which a new product or service might be developed, and to obtain as much information related to market size, economics, and product specifications as it is possible to obtain. Most importantly, market research must summarize its findings and communicate its conclusions to research and engineering and sales for action.

The next step in the process is that of market development. This is the introduction of a new product into an existing market or an existing product into a new market as expeditiously as possible. You might say that this is the innovative act of taking a research or engineering achievement and putting a foundation for future sales under it. It is an entrepreneurial function and, as such, requires that the development man must interact both with his colleagues and his superiors within his organization and with his potential customers outside his organization.

Success in technology transfer or commercial development is highly dependent upon the creative and innovative skills of the individual developer or transfer agent. He must be creative in that he can use or encourage others to use unconventional techniques to arrive at original solutions. He must be able to see beyond the obvious or to persuade others to do so. He must be able to recognize the basic technological, economic, and societal relationships involved in a problem.

When a new concept is generated by the creative act, the process of innovation must take over. This is a highly individual skill of managing the political problem of introducing change (in the sense of a new product or a new market or a new service) and maintaining the momentum and enthusiasm that carries the project through in spite of reversals and setbacks.

Machiavelli described the central problem of technology transfer in his astringent fashion as, "It must be remembered that there is nothing more difficult to plan, more doubtful of success, or more dangerous to manage than the creation of a new system, for the initiator has the enmity of all who would profit by the preservation of the old institutions and merely lukewarm defenders in those who would gain by the new ones. The hesitation of the latter arises in part from the fear of their adversaries who have the laws on their side and in part from the general skepticism of mankind, which does not really believe in an innovation until experience proves its value. So it happens that whenever his enemies have occasion to attack the innovator, they do so with the passion of partisans, while the others defend him sluggishly so that the innovator and his party are alike vulnerable." ("The Prince" - Chapter VI.)

In spite of this sage advice, there are many of us who find the challenge of development and innovation an irrestible one.

Turning now from the abstract to the concrete, I would like to describe to you several case histories, all of which sprang from one original research discovery. Two of these cases are of highly successful developments; while another, pursued by the same people and with the same enthusiasm, can only be described as a failure.

In the early 1940's, polyvinyl chloride resins were emerging as versatile and economic replacements for many of the established products when on the U.S. market - rubber, linoleum and leather. It was then and still is necessary to compound these resins with sizable amounts of plasticizers in order to fabricate them into their final forms and to impart various degrees of flexibility required. Plasticizers had been used for other plastics - cellulose nitrate, cellulose acetate and to some extent rubber - but these applications were less demanding in terms of quantity of plasticizers required and in overall performance than was the new market for polyvinyl chloride. Monsanto was then and still remains a major supplier of plasticizers, but most of our 1940 products could not perform as well as the market demanded.

This problem was first defined by a market research study which established that a large potential market existed for a product which had the right combination of viscosity, volatility, solubility and cost.

Some highly creative and imaginative organic synthesis in our research laboratories led to an entirely new class of plasticizers, the alkyl diaryl phosphates. Application research established that these products had all of the desirable characteristics needed for a polyvinyl chloride plasticizer including some desirable extras - nonflammability and low toxicity. A specific member of this class, octyl diphenyl phosphate, was introduced into the market as Santicizer 141 in about 1945. Intensive development work quickly established commercial acceptability of this entirely new product. By 1948, a commercial plant had been built and was operating and, since then, over 100,000,000 pounds of this plasticizer have been produced and used throughout the world.

At about the time that Santicizer 141 was going into commercial production, the Douglas Aircraft Company came to Monsanto with a request for a nonflammable aircraft hydraulic fluid. The hydraulic fluids in use at that time were low viscosity hydrocarbons, which were highly flammable and which had been responsible for a number of disastrous fires in commercial aircraft. As a result of these accidents, the Douglas DC-6 airplanes had been grounded until this problem could be corrected.

As described to us, what was needed was a nonflammable fluid with good low temperature viscosity, nonvolatility, good lubricating qualities, and compatibility with the rubber hoses and sealing compounds used in aircraft.

We recognized the "fit" between these requirements and our class of alkyl diaryl phosphates. We submitted a small sample of a pilot plant run of an alkyl diaryl phosphate to Douglas and, within a matter of days, we had a request from them for one 55-gallon drum to be shipped air express as soon as possible. By running our small pilot plant around the clock, we succeeded in making the product within a week. Douglas found the material not only to meet their initial requirements but to exceed many of them, particularly in lubricity and low temperature viscosity. On the other hand, it was completely incompatible with existing rubber hose and sealing compounds, swelling them or dissolving them. Suitable replacement rubber formulations were identified in the laboratory and a choice was made to replace all of the elastomeric components of the Douglas hydraulic systems with materials compatible with the new fluid, which became known as Skydrol. The performance of this original product and its modifications and improvements were such that it is now used in all commercial aircraft in the western world. Stimulated by these two successful developments, our research and application people looked for other uses and reasoned that these products would make excellent paint removers. They were tested in this application and found to be quite effective: nonvolatile, nontoxic, nonflammable, and easy to use. An application patent was taken out and commercial development began to try to find a market. In spite of extensive effort by the same people who had successfully brought off the plasticizer and Skydrol project, the paint remover effort was a complete failure. No one wanted the product, no one bought it, and the project was dropped.

Based on my experience and these case studies, I would propose the following dictum. <u>Success</u> in technology transfer requires the following combination: A new technology or a new concept, plus a transfer agent, plus a <u>real</u> market need. All three of these were present in the successful cases. <u>Failure</u> in technology transfer occurs if any one of them is absent. In the case of the paint remover project, there <u>was</u> new technology, we <u>did</u> have skilled transfer agents (or development men, if you will) but we lacked an <u>actual</u> market need. We thought one existed but no one in the market felt any need to change.

FACTORS MAKING PRODUCT INNOVATION SUCCESSFUL
Findings of a field study in Belgium

Fernand Debreyne

Centre E. Bernheim pour l'Etude des Affaires, University of Brussels

GENERAL PURPOSE OF THE STUDY

Barriers and stimulants to technology transfer were at the core of research that I had privilege to conduct some years ago, from 1967 to 1971, at the University of Brussels, Belgium.

The specific purpose of the project, however, was not to study technology transfer per se, but, more largely, the main factors that make product innovation successful -- I mean here: a success on the market. Of course, many factors that make innovation successful are not basically different from the factors that make product innovation successful -- I mean here: these factors are alike to those required to launch on a market, with a reasonable return on investment, a product that has nothing to do with innovation (for example: selling Coca-Cola to Zulu tribes, or manufacturing and distributing something under license). We rather concentrated our attention on factors that are specific to innovation, and doing so, we naturally encountered technology transfer.

Let us thus keep in mind, when discussing technology transfer matters on the basis of our findings, that the problem was not faced by us in its general aspects but only insofar as it appeared to be related to a commercial success. For instance, brilliant technology transfers that had no happy destiny on any market until now were not taken into account, even when one can presume that they may possibly have one in the future. In other terms, a single invention, something discovered in a laboratory, or something that some starving pioneer richer in ideas than in banknotes tries to sell to persons who are richer in banknotes than in ideas, with the hope they will make money with the idea or prototype. Thus, such inventions are not what we called innovations. Nor were innovations, to our eyes, items that were once developed and sold to some Government Agency or Army Department and that have no other possible use. Nor, *a fortiori*, items that were developed by a firm for internal use -- for instance some machine, or genuine system, that are not innovations, according to our criteria. Let us underline, however, as we will mention later once more, these items can often give way to innovations: i.e., they represent a germ.

METHODS

In the field, we made use essentially of historical methods, Preliminary desk studies had mainly shown that rather contradictory results or comments were published on the matter. It seemed to us it was useful to take again of the root the problem, and thus to conduct first, thoroughly, some case investigations of firms having innovated. We took into account all possible elements of the situation and environment that were brought to the firms' managers attention at the time and that could have influenced their decisions and behaviors in innovating. We worked by multiple interviews, searched for proofs in old records of the firm, tried to reconstitute the historical truth when testimonies were contradictory or vague.

On these bases, we could build in each firm a model of the innovating process, as it once functioned in the past. It soon appeared that structural resemblances existed between those cases. We extended our sample to test and refine our hypotheses, visiting some 60 Belgian firms that were known to have succeeded in at least one product innovation in the last 20 years. Those firms were of different sizes and belonged to various sectors of economic activity. As a first step, we interviewed the managing team of these firms in order to determine what they considered themselves as innovation, how they felt about the advantage or disadvantage of having innovated, and how they were led to do so.

At this stage, we eliminated a great number of these firms, for it appeared that the so called innovations were in many cases not consistent with our criteria of successful product innovation. So we carried on the inquiry only in 24 firms, trying to make clear how the managers were led to innovate, how they proceeded, and to what factors their successes were due. An average of 5 days of inquiry per firm appeared to be sufficient to test the preceding findings.

Aside from the historical method, we could test the operationality of our models and gain some more ideas on the practical way by which barriers can be pulled down: indeed the research team seized the opportunity to follow the innovating process in real time during two years in a firm that had just encountered problems in properly defining the product on which it would concentrate its technological research capabilities in the coming years. We intervened in that process, using what many authors now call action research methods. We were lucky enough to clarify the situation.

We have now seen roughly which methods were used. The rest of my talk is devoted to the main findings related to our purpose of today.

MAIN CHARACTERISTICS OF SUCCESSFUL PRODUCT INNOVATION

In absolutely all cases of successful innovation, one finds the conjunction of three elements. Our visits to the firms gave us many examples, too, of attempted innovation that led to failure -- in such cases, at least one of these elements was found lacking. We may consider these elements as necessary and, I should say, nearly sufficient, conditions to succeed in innovation. I say "nearly sufficient", instead of "sufficient", to let room for chance and good luck, and also room for the particular talents that are necessary to set up these elements and permit them to merge.

These basic elements are:

1. At the highest decision levels of the firm, a motivation, coupled with a strong managerial authority.

2. An anticipated definition of the new product, comprising: a) a clear view of the decisive advantage the product shall represent for its users, that ought to be purchasers of an expanding market; b) the conviction that it is feasible; c) the will to discourage a too easy imitation by competitors, which assumes that advanced technology or at least a genuine know-how are to apply to its production.

3. A reasonably well organized management system.

MOTIVATION AND MANAGERIAL AUTHORITY

Spirit of enterprise and availability.

Adequate motivation doesn't depend exclusively on the fact that top management thinks only in terms of business success. Naturally, it helps. But among the barriers that can prevent firms from innovating -- even firms having a fairly good spirit of enterprise -- we frequently found existing projects that were in the course of execution and on which everybody's energy and all available resources are focused for some years. A possible technology transfer is then delayed or even ignored.

Another barrier is the simple fact that the business is good, the market of traditional products continuously expanding without being threatened by anyone. After all, one may think it is wise not to take a risk when everything is going well. As a matter of fact, many of our innovating firms were motivated to take risks and to develop their new products precisely because they had a problem or felt they could have one in the future.

Problems and fears

Many psychologists have a bad opinion of fear, and consider it a motivation of doubtful quality. Anyway, most people are subject to fear (especially the clients of the psychologists), and managers do not escape the rule -- for the sake of innovation, fears about the future are the number one stimulant.

Consideration of internal and external statistics, reflection on the trends of evolution of the business practiced by the firm, information on projects of competitors, are factors that make the stimulation more effective. Even when such figures and information do not justify any concern about the future, they may often suggest that something should be done to give some economic value to various scraps, waste materials and residues of the current production, which can stimulate research, and, in the best cases, lead to product innovation.

Pragmatism of the goal

Also stimulating, and capable of giving a firm a definite orientation toward innovation rather than toward other ways of making money, is the clear view of the product on which it can base its expectations. In other terms, our second element, that we previously called "anticipated definition of the new product", enforces the necessary motivation. As long as a general intention to innovate has not found a concrete base in the form of a definite product toward the realization of which everybody can collaborate, that intention remains highly vulnerable: any other project, any other kind of solution to the strategic problem of the firm, can still therefore, be preferred to innovation.

We particularly experienced that during what we called our action research. It happened in a middle sized firm (about 900 employees) that comprised two production departments. Having some fears about a possible reduction of his market shares in the long range, the manager intended to create a third department in order to diversify his activities, this without knowing what kind of products that department would ever make. He just wanted a given return on investment, and that with his new branch he could take a leading place among its still unknown competitors thanks to the use of some advanced technology. The range was ten to fifteen years, before that department should be fully functional.

As you see, the motivation to innovate was present, at least at the upper level. When we intervened in that firm, they had already, since one year, held about twenty internal meetings at executive level, and carried out as many surveys on the possible developments of recent technology, without having retained any worthwhile product idea. Actually, as we could soon see, nobody in these committees agreed with the ideas of anyone else, most of all not with the ideas of the boss. Most of the time of the 20 meetings had been devoted to examining, discussing, criticizing expansion ideas that had nothing to do with innovation.

This story contains both an example of the vulnerability of a too general intention to innovate and an example of lack of managerial authority.

Authority

Contrasting with the situation we have just described, the power, the interest, the convincing and control pressure of top management were present in all cases of successful innovation we could study. Those stimulants appear to be most necessary. However, they can best be effective in cases in which the boss knows himself what he wants, this, in terms of a definite product, instead of in terms of some odd-structured task, offering a too open system of thought, like: "We must have a third branch within 15 years".

That remark emphasizes the importance of the second basic factor of successful innovation, that we will now explain and comment on.

ANTICIPATED DEFINITION OF THE NEW PRODUCT

Issuing such a definition requires basically that somebody would have had the synthesis in mind, on the one hand, of the needs of certain possible purchasers for certain as yet non-existant products and, on the other hand, of technical possibilities of realizing these products by means that the firm could develop and master within a reasonable span of time.

Making such a synthesis appears to be one of the most difficult tasks of managerial activity. Some authors have proposed various techniques that could guide the manager's way of thinking in such circumstances -- among authors, those who have written on creativity, brain storming, synectics, and especially here Abraham Moles (University of Strasbourg) who developed what he called discovery matrices.

Anyway, none of our sample-firms had made use of these rather sophisticated systems to make the necessary synthesis. They all did it empirically, and most of them without even being aware they had done it. From the way they proceeded, lessons are to be learned.

Two different kinds of synthesis processes were observed.

In the first kind, the product was invented or discovered without having any previous idea of its value for potential purchasers. That value was revealed later on to the managers, and sometimes much later. It's only at this stage that one can speak of an existing project for a new product. All that happens before is a latent process.

In the second kind of innovating process we could observe, the market and general characteristics to be given to the product were defined in advance. These are specific cases of planned innovation. Technological transfer then comes later, it's an engineer's job. The laboratories and development departments go to work to find out ways of realizing top management's aims. These are expressed, in the best cases, in the concrete form we have already encountered : a) mention of the decisive advantage in using the product that must be guaranteed ; b) mention of tentative research directions to follow as to ascertain feasibility ; c) mention of a will to invest much brain work and specialized means of technology when implementing the project, so as to avoid a too easy imitation by possible competitors.

We will now give and discuss various examples of the first kind, then examples of the second kind of the necessary synthesis.

WHEN TECHNOLOGY TRANSFER PRECEDES THE MARKET DISCOVERY

The first kind of processing appears to be the most frequent one, which doesn't mean it is the best one. Because the success relies here on a great deal of chance. Here are some examples.

Paedogenesis

The laboratory of a firm constructed some new instruments to make lighting measures for its own use. Purchasers of lighting sources produced and sold by the firm asked for appropriate measuring means. So the firm decided to open a new production department and to increase the variety of these new instruments, which required further technology transfers.

A similar process can be observed in other examples, like that of a firm producing mechanical kneaders. This firm had developed, to use them in its maintenance workshop, a new set of genuine machines to sharpen used gears. The casual visit of a high-ranking engineer of a car building corporation opened to that firm the totally unexpected market of the automobile industry, interested in sharpening gears.

Such a process makes us think of what geneticists call "paedogenesis", meaning that some steps in the evolution of species consist in the birth of beings that have characteristics proper to the foetus or to very young individuals of a pre-existing species. As in this case the innovation has its origin in the work done, or in the tools used at an intermediate stage of the production process of an existing item.

The trouble is that, like in the evolution of species, natural selection condemns to death a large amount of such attempts to make something new. For the few that succeed, there are many that fail. So that kind of innovation process appears to be an enormous waste. One can think that wasting is a law of nature breaking down routine and introducing change. Anyway, it is a matter of fact that some firms consider that kind of waste as normal, and even, let their laboratories make "research for the sake of research".

Research for the sake of research.

These research laboratories discover new properties, new technological processes, new products, that have no economic value, at least in the near future. It happens too, rather frequently, that, searching to obtain a certain effect, they discover something that was not expected and that represents a greater economic value than what they were searching for. For instance, something that happened in the labs of a Belgian firm manufacturing photographic film : While working on a colour-radiography (which was only a gadget or prestige project in the eyes of the marketing responsibles of that firm), they discovered that an X-Ray photograph in blue and white had properties that an ordinary radiogram in black and white has not. Note that we are facing here a typical paedogenesis : the decisive advantage for the users of radiograms became clear, at an intermediate stage of the development of full colour radiography.

Such an opportunity, and the developments they are giving to it, consoles the managers of that firm for having accepted to let their laboratories play with colour.

Main lessons

There are, we think, two main lessons to learn from these observations :

a) Changes, in the broad sense, even when they are not introduced with a direct profit motivation, have a virtue per se : by paedogenesis, the economic value of a part of a known process, product or instrument may be revealed. However, that process has the fault of being uncontrolled. Success relies on chance.

b) Second lesson: such a chance can be assisted. In this sense, any firm desiring to find some practicable diversification would have the advantage of paying systematic attention to the possible transfer of improvements realized by its own means, in its own workshops or laboratories. The transfer seems to be the most practicable toward markets constituted by people who have to solve problems of the same nature. It still requires imagination and also the will to go further than the immediate transfer.

Going beyond the immediate transfer

We have already seen an example of further transfers when that firm, led by paedogenesis, produced genuine instruments to measure lighting. Let us bear in mind that once a market was revealed by the demands of clients, the firm developed other instruments along the same line, trying to put still more brain work and advanced technology into it, so as to satisfy the condition of success we have already underlined : discourage imitation. A similar case was seen with radiograms.

When the primary technology precedes the market discovery, such behavior cannot always be observed : we saw many cases during our inquiries, of half-achieved transfers, that sadly lead the innovation to failure.

Contrasting with such half-achieved transfers are the cases where the market discovery preceded the technical one : technology transfers are then always actively sought, and brought to fruition.

WHEN MARKET DISCOVERY PRECEDES TECHNOLOGY TRANSFERS

Here are cases of fully conscious and voluntary planned innovation. It is obviously the surest way to succeed. Strangely enough, also a way that is seldom used by firms, at least in Belgium.

Perhaps an explanation of this rarity should be that to proceed to a synthesis between human needs or problems, and possible developments of technologies, assumes a broad and diversified culture, that must here be on a par with other talents : that of exerting a sane authority, that of obtaining necessary means by persuasion and negotiation. The union in one's personality of all of these talents can be rare too, even at the highest level of middlesized firms. Let us add that when the firm is bigger and has reached the size where a man alone can no longer make the desirable synthesis, lack of communication, filter, staff against line and committee problems are there to make the task more difficult and the fully planned innovation relatively rare. Perhaps that in a country like Belgium, where the culture of the engineer is often still preferred to the business mind at the top of traditional firms, that kind of innovation has less chance to flourish than in some other parts of the world.

Anyway, even in Belgium, we observed cases of the kind in question. They had a common characteristic : a deep consideration for problems encountered by clients was always the beginning of the process.

Consideration for client's problems

We should better describe that process by means of a full example. Just before World War II, in times where public lighting was often realized by suspending electric lighting points to wires above streets and roads, a Belgian firm had, among many other productions, a small production of half-globes made of glazed sheet iron that were classicially used as reflectors at the time. The manager dreamed of a leader position in the market of public lighting accessories, and chose to realize such progress in the conception of the product that clients all over the world would not hesitate to give him preference on the market, his competitors being left behind for some many years.

To specify the new product, he answered a series of questions, such as : "What are the human needs beyond today's use of those primitive reflectors?" In other terms, "to the resolution of what problem do these reflectors serve?" ; "considering that problem, what are the main annoying constraints and defaults of the solution so far offered to it?" ; "what are the trends of the evolution of the needs, according to recent developments of new lighting sources and considering the modern appeal for more and more efficiency and comfort?".

Any product on the market, he thought, can be seen as a tool that people use to solve problems. Successful innovation consists in inventing and industrializing better tools. Insofar as the advantage offered by these tools is great, and insofar as a sufficient number of people would be responsive to it, the market is guaranteed, even when prices are higher.

The new solution. Its technological consequences.

In the case studied, it was decided that the main advantages of the new tool for its users had to be : to provide an equal distribution of light on road surfaces, without areas of darkness ; to avoid glare and waste of light ; to be reliable without requiring any maintenance work (let us bear in mind that the early existing reflectors called for regular maintenance because of the corrosion of iron and the fragility of glass parts).

The decision implied many consequences : the foundation of laboratories and testing benches to make research on the distribution of light, development of new measure instruments, research on the reflecting properties of aluminum, on the chemistry of aluminum alloys at the micro-cristal level, on injection machines able to inject aluminum at adequate rates into moulds of the proper size on the properties of plexiglass on heating effects, etc., this, as far as technology is concerned. Other consequences affected the marketing, finance and general management of the firm : the project led in particular to mergers and to joint ventures between firms, as to a total reorganization of distribution channels.

We have given a rather detailed example, because we felt it could help to gain an insight into the philosophy of the process.

Executive mobility as a stimulant

In another of our cases, where the innovation also proceeded from the consideration of client's problems and needs, the clients in question were not people whose problems were familar to the firm : the idea came from an executive who just arrived in the firm, coming from another firm where he had had experience of that kind of consumer. He rapidly made a bridge between their problems and the germs of solution contained in the particular technology of his new firm. The story I have in mind is the one where a product destined for building industries was developed starting from textile technology. This case leads us to presume that a certain executive mobility stimulates innovation, as long as other conditions are present. Among these conditions we cited before : a reasonably well organized management system.

A REASONABLE WELL ORGANIZED MANAGEMENT SYSTEM

On this point I will not be long. Colleagues of other Belgian universities have particularly born their research on this type of factor.

Let us only point out here that the first of the elements we have noticed as important to succeed in product innovation, the managers' motivation, depends on the information system that can show them there is a problem looming at the horizon.

Innovation usually demands time. Management systems where the managers cannot become aware, with a fairly long anticipation, of coming economic difficulties, are systems that we may consider as constituting a solid barrier to genuine technology transfers of great magnitude. I say : of great magnitude, because even day-to-day improvements given to existing products assume that some transfers are made. But such transfers are not what we specifically called innovations.

Now to carry out a project, once it has been formulated in terms of specifications of the future product, no doubt that a reasonably good organization is required, the major points being on the one hand competence at the head of the various functional divisions of the firm, and on the other hand, of course, authority and power to unify the work of these divisions.

CONCLUSION

We have described as most secure the process in which market discovery precedes and inspires technological transfer. This is particularly valid in cases of great innovation projects, mobilizing many energies in the firm.

However, the other process, where market possibilities are discovered only after that novelties appeared in the technological field, is not incompatible with the first one. On the contrary, we observed that research, even strictly oriented, often gives unexpected results, others than those which are looked for. These unexpected results, with naturally a high percentage of failures, may give way to interesting and marketable developments, following thus the second process we described.

When success comes out of the second process, it often takes a relatively modest place among the other activities of the firm : we are no longer in presence of great innovation projects mobilizing nearly everyone. It has also a characteristic other than its lack of importance : that of being nearly unpredictable.

For instance a fairly large group of glassmaking factories discovered successively two things, some years ago : first, in its laboratories, by chance, the way of making perfectly round balls of glass of microscopic size ; and secondly, this no longer in its laboratories but in the field, that an American firm making abrasive paste to clean some parts of jet airplanes has a need to include some other material than sand in its paste, because of the too rough action of sand on the surfaces. You can guess what happened, now you know the elements of the story. But obviously not any manager could have decided in advance that that Belgian firm would try to find out a technological way to improve abrasive paste. The process was unpredictable. The success depended on the chance of a happy encounter. Let us add that microscopic glass balls are not its primary market : it remains small, as one can imagine.

As firms generally prefer security to the unpredictable, one can wonder that the observed innovation processes so often obey the pattern that appears to be the most insecure and uncertain. Perhaps a general trend to face rather immediate problems than long range ones is here in question. But investigating that point would be another research.

ACKNOWLEDGEMENTS

The research was supported by funds provided by the Belgian National Council for Scientific Policy and administrated by the Belgian Productivity Office. It was a part of a large research program, associating five Belgian universities on the general theme of "Technology changes and business administration".

The team of the University of Brussels (Department "Centre E. Bernheim pour l'Etude des Affaires"), who researched the factors of successful innovation was composed of Pierre Hertay and Guy Leroy, industrial psychologists ; Tony Reekmans and Jacqueline Desmedt-Roelandts, commercial engineers ; Serge Mihailov, economist ; Fernand Debreyne, industrial psychologist, project leader.

TECHNOLOGY TRANSFER OF CURIOSITY-ORIENTATED RESEARCH

R. Derek Medford

"There are some who feel dismay that science, pure science, should have become so costly, that its cost should transcend what one individual nation, even the wealthiest, could hope to afford. I am afraid I do not share this dismay. On the contrary, if one considers the future of humanity, confined on a small globe, interdependent as never before, living off a limited biosphere, which until recently we have exploited with no thought either for our neighbors or for our children, if we consider that we also possess technological competence capable of providing plenty on a planetary scale if we were not hampered by the dead-weight of our preatomic and preglobal traditions, if you remember all this, one should rejoice in costly science, rather than feel dismay."

Prof. Abdus Salam (1971)
Nobel Laureate

Introduction

If technology assessment is ever to be an effective tool, its practitioners must be prepared to analyse some of the barely predictable outcomes of potential innovations stemming from pure, basic or curiosity-orientated research (COR).

At present, positive recommendations on constructive methodology for the assessment of COR would obviously be premature because the multiple impacts of this kind of research have only recently come under extensive scrutiny. The purpose of this chapter is to fuel the debate on organization and control of COR and discuss the extent to which technology assessment techniques, "proven" on mission-orientated and applied research can be usefully applied to "purer" research activities.

Difficulties of Evaluating COR

Prognostication on COR compared with prognostication on applied research is as an intelligence exercise is to a market survey. Curiosity-motivated researchers are justifiably sensitive to outside surveillance, and much more effort must be spent by the analyst in familiarizing himself with the science behind COR than is customary in the assessment of applied research and much of the science is new and the analyst, in the COR context, cannot disruptively appear as an assessor of potentially new technology, benefit-cost accountant, or a manager. Empathy in discussion of COR is best generated by demonstrating more than a shallow knowledge of the scientific principles involved.

Some aspects of technology assessment methodology as applied to mission-orientated research are obviously valuable in an assessment of COR. However, the detailed intricacies of formal benefit-cost analysis are not, in this stage, seen by most experts to be deeply relevant to an understanding of COR. Conversely, the intelligence aspect of assessment work (information collection and collation) is universally acknowledged to be *very* relevant, since COR has inherently much dependence upon synergy and serendipity.

At the time of writing, open publication on the problems of assessment of COR has tended to be excessively biased towards ex post examination of the impact old COR has had on currently successful, and unsuccessful, innovations in industry. Good examples of this kind of work are TRACES (1968), Gruber and Marquis (1969), SAPPHO (1972). On the other hand, structured analyses of the possible innovations yet to come from current COR are, in general, only contained in confidential commercial documents which make entrepreneurial bids for high risk capital, documents supporting claims for research funds, and industrial and military documentation which examines "outre" ideas for signs of narrow sectorial benefit.

One way around the obvious restrictions on circulation of technology assessments of current COR would be to circulate information on the methodology used, or intended to be used, in assessment so that the general points of principle could at least be discussed. A brave attempt by Byatt and Cohen (1969) was made in order to improve the methodology which could be applied to assessment of COR. Predictably, their proposed methodology would not even stand up against the evidence of ex post case studies although it did have a considerable catalytic effect on analysts who would assay one step further in the creation of method.

To this author's knowledge no generally acclaimed improvement has been made on Byatt and Cohen's rejected, heuristic, attempt at methodology. Therefore, in this chapter a cursory examination of the Byatt-Cohen Model will be made against one of the excellent case studies from TRACES (1968) and a speculative, but original, methodology will be proposed for calculating the notional benefits of COR.

Retrospective Study

An early worthwhile objective of Science Policy Studies, made under the auspices of the UK Department of Education and Science, was to find means of identifying the economic benefits stemming from previously completed COR and use the findings to help make policy decisions on current COR. The Byatt and Cohen (1969) paper proposed a conceptual means of achieving this objective by means of a mathematical model of the innovation process. Their model depended upon constituent equations for the incremental change in discounted net economic benefit of a science-based industry which could be brought about by notional delays or advancements in COR and/or information transfer.

Before considering general comments on work done in an attempt to verify the Byatt-Cohen Model, it is useful to obtain some numerate understanding of what Byatt and Cohen were about. One of the theoretical "pivots" of the original Byatt-Cohen Model in algebraic form, as follows:

Change in discounted net benefit $=\{\Delta\ V_i \exp[-(n+T)r]\}$

$$=\{\Delta(V_i)-rV_i\Delta(n)-rV_i\Delta(T)\exp[-(n+T)r]\}$$

where Δ's mean "change in" (thus $\Delta(n)$ is the change in n), V_i is the ultimate benefit of the science-based service or industry, r is the discount rate, and n and T the innovatory and information time lags. The longer representation of the right-hand side of the equation is not vital to the present discussion, it simply shows how the change in discounted net benefits is comprised of components dependent upon the changes in information and innovation times and the change in the magnitude of the undiscounted net benefit. For example, suppose the time lags in innovation and communication of information are reduced by two years each, i.e. $\Delta(n) = \Delta(T) = 2(-)$ (minus sign for a reduction), and the ultimate benefit is consequentially improved by 10% (through earlier access to information and marketing "know how"). Then we have

$$\frac{\text{Change in discounted net benefit}}{V_i \exp[-(n+T)]} = \text{Fractional change}$$
$$= [0.1 + r(2+2)]$$
$$= 0.5$$

if, say, discount rate = 10%.

It will be noticed that the time lags are additive; thus it has been assumed that the process leading from COR is simply sequential. That is to say that a sequence of separate steps COR, applied research, invention and development, prototype construction, market research, and sales leads to benefit. Extensions of formulas similar to the last above were recommended by Byatt and Cohen, the extensions to be made by multiplying the equations with "impressive" factors which purported to allow for statistical chances that the COR would eventually lead to industrial innovation, and relate true benefit to idealized commercial sales of the graphic form shown in Fig. 1.

Questions arising from the Byatt-Cohen formulation of their model are

(1) How is benefit apportioned, even in a simple chronologically sequential chain, when more than one piece of OCR is involved?

(2) If the innovatory process is not simply sequential but multiple connected and sometimes parallelled in time, how does one

Fig. 1

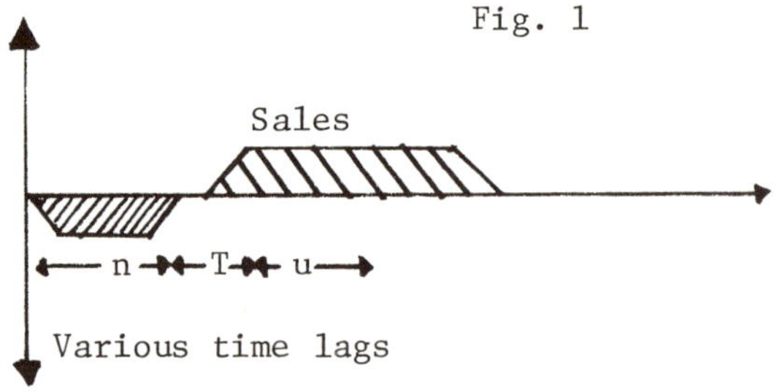

then calculate the likely contribution of the COR inputs and check on the sensitivity of discounted benefit to various time delays?

(3) Can simple statistical factors for the a priori probability of COR leading to industrial innovation be obtained for such a class of disparate events as that thrown up by ex post studies?

(4) For such a complicated thing as a market, can idealized commercial sales functions be accepted?

The author examined the ex post studies from TRACES (1968) for some evidence to corroborate Byatt and Cohen's assumptions. In TRACES, key events in COR, applied or mission-orientated research, development, and application are denoted as differently shaped nodules on a multi-coloured, prolific, dendritic growth map which depicts the historic cross fertilization of apparently unrelated projects and disciplines.

Only on one of the five innovations studied in TRACES (Electron Microscope) does the dendritic growth ultimately converge to a sequence of steps from COR to profit. Starting from the data of the successful innovation and working backwards through time, it was found that the first mention of "basic" research occurs five years before the first commercial electron microscope was produced. This research was in direct line of development of the microscope and so does not appear to require consideration of parallel events at that time.

Looking at the definition of this event it is seen, however, that the "basic research", or COR, was carried out in order to investigate the phenomenon of emission. The important feature so far as it effects the innovation was the designing of an electron microscope to study this phenomenon in the laboratory. How much of the cost of this COR should be considered as cost to be offset against the later benefits gained from the electron microscopy?

Going further back in time, the position outlined in the TRACES case study becomes even more complicated. TRACES considers the derivation of the Energy Distribution Law by Boltzman to be an important event in the chain of events leading to the innovation of the electron microscope. *But* this work has had important effects in almost every branch of physical science. Would the innovation of the electron microscope have taken place without Boltzman's Law? The simple answer is that we do not know.

Going back to the original problem of decision making on COR, "could we have known in 1870 that the work of Boltzman would have helped to make possible the development of a new instrument with an enormous range of use in metallurgy, physics, chemistry, biology, medicine, etc?". We think not.

An examination of any of the other COR events in the TRACES case studies reveals the same story. All that can be said is that each of the innovations examined has depended to a very great extent on a fund of knowledge accumulated as a result of curiosity-motivated research. Many of the COR projects have no easily identifiable connection with even the general field of the innovation. The well-known truism is that the value of the research result has occurred, very often, long after the COR discovery has been made.[1] This has required that there has been a successful, but labyrinthine, transfer of knowledge between the discoverer of the COR phenomena and the more practical inventor or innovator. It is not possible, in the general case, to take account of the cost of this transfer which may have been made across national boundaries, disciplines, and even decades of time: it is a tribute to human ingenuity that the transfer actually takes place. An additional factor which must be taken into account is that, in TRACES, only successful events have been considered. It is not known how many people at what cost were trying to solve similar problems. Should the costs of unsuccessful attempts be included in a model? Perhaps the labyrinthine process of useful information transfer takes a minimal time which cannot be reduced beyond a certain lower limit (although with present communication efficiency we are nowhere near such a limit now and great improvements can be made) and if we do not do the COR now perhaps we will not have the innovations we need in the "discounted" future.

No discussion of the pioneering methodology of Byatt and Cohen would be complete without some consideration of the following conceptual problems.

Is it fair to all other contributors to an innovation to attribute the net improvement in economic, or social, benefit to the originating COR? In order to calculate the notional importance of COR, some notional approtionment of benefit to the different contributors must be attempted.

---

[1] Because COR information may have been obtained long ago in another place, it does not follow that society has not paid a collective price for it, or that society should not lay down good COR for future generations. An entrepreneur can certainly obtain benefit from COR without contributing fairly to its cost but this does not mean that COR is not important to innovation or that society should, say, reduce expenditure on COR. It can always be asserted that COR is likely to produce benefit, even in the extreme case when the outcome of research is the production of "malignant" knowledge. At least then one knows what one should avoid.

Assuming that the answer to the last question is "yes" and that the last sentence is accepted, how, then, do we apportion economic and social benefits to the contributing events?

In the author's opinion it is manifestly unfair to apportion the benefit in the ratio of the discounted cost of each stage of the innovation process. Some allowance must be made for the investment of intellectual and entrepreneurial skill. Contributors who invest capital on low a priori probability of success and then rely largely on intellectual and entrepreneurial effort should be correctly acknowledged.

In the next section a hypothetical method is developed for the notional apportionment of benefit which appears intuitively to be fair because magnitude of capital, entrepreneurial and intellectual, investment are all allowed for and later contributors are given credit for carrying on the success of their predecessors. Suggestions are also put at the end of the next section for the extension of this notional approtionment technique to parallel and multiple connected processes of innovation, similar to the realistic development profiles shown in the TRACES study. It is hoped that these tentative suggestions are encouraging to future methodology. They are not intended to stimulate the construction of large computer-based decision trees for the allocation of funds to COR. Very much more careful thought needs to be given to the effect of COR culturally, for future survival, and economically before methodology can be safely accepted into decision making in COR.

Notional Apportionment of Benefit

As an indication of difficulties to be overcome in any project starting with COR and ending with a commercial profit or a worthwhile service or amenity, the quotient of expectance of total benefit to expectance of total cost is adequate. With a simple binary, success or failure, chronological sequence this quotient is,

$$\frac{\text{Expectance of total benefit}}{\text{Expectance of total cost}} = \frac{p_R p_D p_M \overline{B(r)}}{\overline{R(r)} + p_R \overline{D(r)} + p_R p_D \overline{M(r)}}$$

where the discounted costs of the various chronological processes are $R(r)$ for research, $D(r)$ for development, $M(r)$ for marketing and manufacture, and $B(r)$ is the discounted benefit, all calculated for a discount rate of r. The "probabilities" of success at the conclusion of each phase are taken to be p_R, p_D, and p_M.

If the assumption is that it is acceptable to apportion benefit and other credits to the contributing events of research, development, and marketing according to discounted costs of each stage, then notional shares of benefit would be apportioned in the following ratio.

$$\overline{R(r)} : \overline{D(r)} : \overline{M(r)}$$

But such an apportionment would be grossly unfair to those who have contributed and risked capital and intellectual effort on low a priori probability of success.

How, then, do we, allow for research finesse or intellectual effort when the results of innovatory research are so uncertain? An attempt to answer this question can only be made if subjective estimates of a priori difficulty are acceptable as probabilities (and these should ideally allow for serendipity and synergy). One method of notional apportionment would be according to the contribution made by each phase in reducing the appropriate uncertainty.

For example, the contribution of increments in the probability of success at each stage produces marginal enhancements in the benefit-cost quotient in the following ratio[1].

$$\overline{R(r)}\,\frac{\Delta p_R}{p_R} : \left(\overline{R(r)} + p_R\overline{D(r)}\right)\frac{\Delta p_D}{p_D} : \left(\overline{R(r)} + p_R\overline{D(r)} + p_R p_D\overline{M(r)}\right)\frac{\Delta p_M}{p_M}$$

At any time during the chronological sequence of probability, Δp can be obtained by subtracting the initial estimate of probability of success from the current probability. Thus on successful completion of the overall sequence, the values of Δp would equal 1-(initial value of p). (Common sense needs to be applied if a notional apportionment of expectance of benefit is attempted before completion of the overall task *and* the a priori estimate of probability has turned out to be unduly optimistic because this would mean at some intermediate time that the value of Δp would be negative!)

The main concern here is the apportionment of achieved benefit (not expected benefit) so that the values of Δp are necessarily positive and the above ratios with $\Delta p = (1-p)$ give the notional

---

[1] This result is easily obtained by differential calculus; alternatively it can be derived by subtracting the quotient determined by probabilities (p) from the quotient determined by probabilities ($p + \Delta p$).

shares of the achieved benefit. It is my opinion that such an apportionment is essentially fair because it allows for magnitude of both capital and intellectual investment (the latter is likely to be proportional to (1-*p*) and it also gives credit for carrying on the success of previous contributors.

To consider what this means in quantitative terms, consider a "typical" ratio of research to development to marketing and manufacture expenditures common to industry, i.e.

$$\overline{R(r)} : \overline{D(r)} : \overline{M(r)} = 1 : 10 : 100$$

and consider that a priori probabilities of success at each stage become larger as one moves forward chronologically through the processes

$$p_R : p_D : p_M = 0.5 : 0.8 : 0.9$$

Then, according to the hypothetical apportionment, the ratio of notional shares of benefit would be

$$(0.5/0.5):(1+0.5.10)(0.2/0.8) : (1+0.5.10+0.5.0.8.100)(0.1/0.9)$$
$$=1 : 1.5 : 5.1$$

which is quite a different result from the initial ratio of capital investment!

In the limit of all a priori probabilities equal to unity (i.e. every stage is certain of success), the ratio of notional apportionment of benefit reduces to the ratio of the spends incurred up to the completion of each stage, which is

$$\overline{R(r)} : (\overline{D(r)} + \overline{R(r)}) : (\overline{M(r)} + \overline{D(r)} + \overline{R(r)})$$

and this ratio bears some relationship to the actual pressure put upon the management of each stage since an essential phase (linking what has gone before to the achievement of benefit) is under pressure because of what has irredeemably been spent and what is going to be the benefit after total success.

Return on Capital Investment in COR

Although, in real life, the suggested notional apportionment of benefit has no real substance as payment, it is salutary to calculate the notional return on capital that one could fairly expect on COR investments. Before doing this it is important to realize that a programme containing a COR phase may commit capital which could be more safely spent in a conventional operation where a good of unconventional kind is manufactured or a conventional service is to be provided. It is instructive, therefore, to use the above notional shares of benefit and calculate the smallest acceptable (assuming "notional" shares become "real" shares) threshold benefit for the overall process.

Suppose that the smallest acceptable return on marketing and manufacture, set by the market is represented by the rate of return, m, and that the notional rate of return on COR capital is c. Then the simple situation of sequential contributions yields the following relationship between *sufficient* returns to COR and marketing and manufacturing capital.[1]

$$\overline{R}(1+c) + \overline{M}(1+m) \frac{\overline{R}\Delta p_R}{[(\overline{R}+p_R\overline{D}+p_R p_D\overline{M})\Delta p_M]} \frac{p_M}{p_R}$$

Substitution of the numerical values used earlier into the above relationship yields

$$(1+c) = (1+m)(900/46) \sim 20(1+m)$$

Thus we see that a moderate chance of success in COR of 0.5, given the acceptance of the notional share-out procedure, demands a return on COR capital greater than twenty times the lowest acceptable return on capital used for manufacture and marketing. If the same organization funds both the COR and marketing and manufacture, this is not a startling conclusion. However, the conclusion may be somewhat startling to companies solely performing contract R&D with pay-off linked to ultimate profits or to organizations buying R&D from without.

Naturally, with different ratios of $(\overline{R} : \overline{D} : \overline{M})$ and $(p_R : p_D : p_M)$, the ratio of the returns to the individual contributors obviously change. In general, $\overline{R}<\overline{D}<\overline{M}$ and the more curiosity-motivated

---

[1] These are returns *after* allowing for discounting, a comparison of internal rates of return is algebraically more difficult but the above mode lends itself to an approach which yields, the percentage return on the present value of investment.

the research the smaller p_R becomes. The qualitative change is not surprising *but* the quantitative variation can be larger than is expected.

Threshold of Benefit for an Acceptable Venture involving COR

At some time during an assessment involving COR, before rigorous application of full analysis, it is prudent to examine the largest possible benefit of the overall venture. When the foreseeable benefit is small, prudence obviously dictates an end to speculation unless the inferred social benefits balance the deficit. But how small is "small" in this context? It is possible to develop an empirical rule-of-thumb based on the foregoing analysis of a simple sequential process.

$$\overline{B(r)}\ \frac{S_M(p_R\; p_D\; p_M)}{S_R + S_D + S_M}$$

is the notional share of expected benefit accruing to the marketing and manufacturing process,where the symbols S are functions of the discounted expenditure and a priori probabilities of success (see above). Since it is reasonable to assume that the return to marketing and manufacturing must at least equal to the lowest acceptable return determined by market forces, we must have

$$\overline{B(r)} > \frac{(1+m)\overline{M(r)}}{p_R\; p_D\; p_M}\left(\frac{S_R + S_D}{S_M} + 1\right)$$

Usually the ratio $(S_R + S_D)/S_M$ is less than unity and the following empirical rule emerges.

$$\overline{B(r)} \gtrsim \frac{(\overline{M(r)}(1+m))}{p_R\; p_D\; p_M} \text{ (a number between 1 and 2)}$$

If the multiplying number is close to unity, the result is a familiar one from ordinary commercial transactions but it is of some interest to note that one rarely need to insist on more than twice the customary commercial benefit.

Where the probability of success in research is less than 0.5 the reader is warned that the above empirical rule of thumb should not be used. Furthermore, in *all cases* of excessively high expenditure on COR or development, or when a priori probabilities of either are very small, the penultimate formula should be used.

Further emphasis must also be placed on the fact that all the calculations of this section have been made on the assumption that management is firm in handling the project and that an unsuccessful phase will not be allowed to repeat itself.

COR which does not Sequentially Fit into a Development

Apart from assuming control by good managers, the above analysis is built up for a very simple sequential chronological process, viz.

COR → development → manufacture and marketing → benefit

Students of real innovations will, as mentioned earlier, realize that hardly ever has COR been successful in such a straightforward way. Research programmes are, for example, often duplicated or necessarily complemented by other COR programmes and a real chronological flow diagram is likely to have the appearance of a "spider's web" or the "Gordian knot". In theory such intricate choronological patterns are amenable to analysis and a few relatively simple patterns are analysed below:

(a) In the case of COR which is comprised of more than one necessary piece of research, each with its own expenditure and probability of success, the following simplex pattern could emerge.

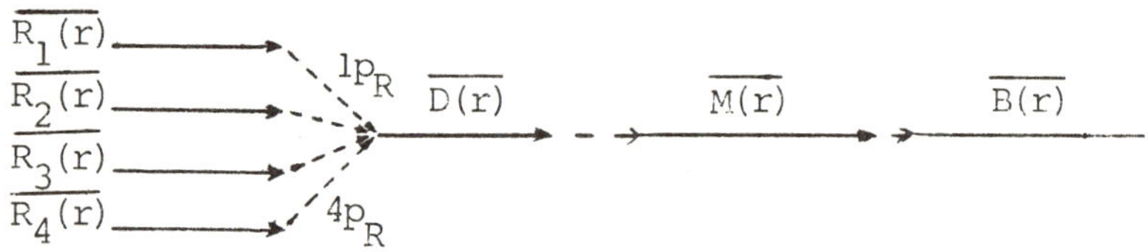

If all pieces of COR are essential to success in the research stage, then the overall probability of success in research is

$$p_{COR} = ({}_1p_R\ {}_2p_R\ {}_3p_R\ \ {}_4p_R) = \prod_{n=1}^{4} ({}_np_R)$$

and the associated research cost is:

$$\overline{R(r)}_{COR} = \sum_{n=1}^{4} \overline{R\ (r)}$$

If the quantities P_{COR} and $\overline{R(r)}_{COR}$ defined above are substituted into the previous formula for notional shares of benefit instead of p_R and $\overline{R(r)}$, the shares between *total* COR, development, and marketing and manufacture are found.

Surprisingly, the individual notional shares of the total COR apportionment to the individual COR pieces of research contain no terms which allow for expenditure and are as follows.

$$\frac{\Delta_1 p_R}{_1 p_R} : \frac{\Delta_2 p_R}{_2 p_R} : \frac{\Delta_3 p_R}{_3 p_R} : \frac{\Delta_4 p_R}{_4 p_R}$$

Note that in sharing the total apportionment to COR between individual contributors, there is no allowance for the individual COR research costs! This does not mean that the hypothetical method of apportionment of benefit has broken down: it simply means that, although one piece of COR may be very much more expensive than another, one cannot claim consideration for this spend if the expensive research is useless without the contribution from the cheaper piece of research.

(b) By way of comparison, consider the case in which COR consists of two independent pieces of research with success in either sufficient. Then the following chronological pattern would emerge.

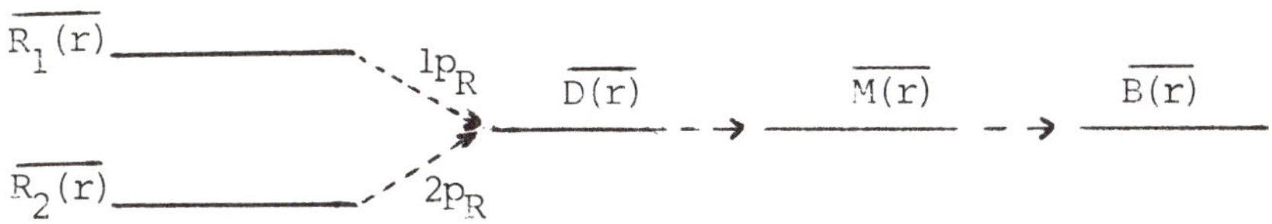

If either piece of research is adequate, the success can be obtained with successful completion of either piece of COR or simultaneous success in both pieces. Thus the overall probability of success is

$$P_{COR} = {_1 p_R} + {_2 p_R} + {_1 p_R}\,{_2 p_R}$$

and total expenditure committed to both pieces at the onset is

$$\overline{R(r)}_{COR} = \overline{R_1(r)} + \overline{R_2(r)}$$

Substitution of these quantities into the first formula for notional shares of benefit will once again yield the shares between total COR, development, marketing and manufacture. The notional shares of the total apportionment to COR, made to the individual pieces of COR are given by

$$\Delta_1p_R(1+{}_2p_R) : \Delta_2p_R(1+{}_1p_R)$$

In this case at the completion of the total effort (through to benefit) the values of both Δp's need not necessarily be equal to $(1-p)$. If only one piece of COR is totally successful, the other piece can have a share attributed to it (the logic is that it has provided insurance against COR failure) but only if there has been some enhancement over the a priori probability. Naturally, if the value of Δp for the less successful case is zero, it has provided no insurance against failure, and its allocated share of benefit is zero. This last consideration does not arise in the last case because if any of the Δp's is zero in *(a)*, when all contributary COR is essential, there is no benefit to share out.

(c) The final simple case is purely sequential but in this case the COR consists of two successive pieces of research with the second dependent upon the success of the first.

$$\begin{array}{ccccccccc} \overline{R_1(t)} & & \overline{R_2(r)} & & \overline{D(r)} & & \overline{M(r)} & & \overline{B(r)} \\ \text{———} & \cdots & \text{———} & \cdots & \text{———} & \cdots & \text{———} & \cdots & \text{———} \\ & {}_1p_R & & {}_2p_R & & p_D & & p_M & \end{array}$$

and the quotient of benefit-cost is

$$\frac{\text{Expectance of total benefit}}{\text{Expectance of total cost}}$$

$$= \frac{{}_1p_R\,{}_2p_R\,p_D\,p_M\,\overline{B(r)}}{\overline{R_1(r)}+{}_1p_R\overline{R(r)}+\,{}_1p_R\,{}_2p_R\overline{D(r)}+{}_1p_R\,{}_2p_R\,p_D\overline{M(r)}}$$

which according to the original hypothesis yields notional shares of benefit in the following ratio

$$\{\overline{R_1(t)}\}\ \frac{\Delta_1p_R}{{}_1p_R}$$

$$\{R_1(r) + {}_1p_R\overline{R_2(r)}\}\ \frac{\Delta_2p_R}{{}_2p_R} :$$

$$\{\overline{R_1(r)} + {}_1p_R\,\overline{R_2(r)} + {}_1p_{R2}p_R\overline{D(r)}\}\ \frac{\Delta p_D}{p_D} :$$

$$\{\overline{R_1(r)} + {}_1p_R\,\overline{R_2(r)} + {}_1p_{R2}p_R\,\overline{D(r)} + {}_1p_{R2}p_Rp_D\,\overline{M(r)}\}\ \frac{\Delta p_M}{p_M}$$

In the above apportionment, the relative shares to the separate pieces of COR are made according to a ratio which is dependent upon the relative expenditures. The reason for this is (as opposed to the two previous cases) that the second piece of COR has the responsibility for carrying on the success of the first. (In the previous cases it was assumed that the separate pieces of COR were running concurrently in time and the expenditure on each piece was approved at the start of the whole process.

[In order not to be hoist by his own petard the author has made an inadequate attempt at technology assessment of specific COR which would conceivably lead to the production of an X-ray (three dimensional) holographic microscope.

Space does not permit a description of this effort here but the attempt is masochistically revealed in *Environmental Harassment or Technology Assessment*. Elsevier, Amsterdam, 1973.]

Concluding Comments on the Prospective Evaluation of COR

Difficulties of attempting an evaluation of COR have been described early on in this paper. My own reaction to the study is that two principal questions arise.

> "Is there any point in trying to assess the future general outcome of COR when it can be argued that all attempts to acquire knowledge are a justifiable end in themselves?, and
>
> "If there is a need for this kind of technology assessment, and if we accept the inherent difficulties outlined earlier, how can the process be efficiently conducted?".

There are two main answers to the first question. Some respondents would say "no" because they claim that the direction of COR will be determined by the propensities of the researchers and the fashion of the moment (which may be as unpredictable as the fashion in apparel). Others would say "yes" because, in their blacker moments, they see the COR scene as mainly comprised of researchers galloping "like hell along the wrong racetrack" and they would wish to see some societal control effectively determining the general direction of the COR activity. Whatever answer is preferred, there is a need for *more* information about COR activities and, if one believes (as I do) that the second answer is correct, the need is based upon wider consideration than that of simply enhancing the efficiency of the COR that just happens to be currently in vogue.

References

Byatt, J.R.C. and Cohen, A.V. (1969) "An Attempt to Quantify the Economic Benefits of Scientific Research", *Department of Education and Science, Science Policy Studies No. 4*, London: Her Majesty's Stationary Office.

Gruber, W.H. and Marquis, D.G. (1969) *Factors in the Transfer of Technology*, pp. 155-176, Cambridge, Mass.: MIT Press.

SAPPHO (1972) *A Study of Patterns of Innovation in the Chemical and Scientific Instrument Industries*, a report delivered to the Science Research Council by the Science Policy Research Unit (SPRU), University of Sussex.

ACQUISITION OF TECHNOLOGY AS ONE SPECIFIC WAY TO ACHIEVE TECHNOLOGY TRANSFER

Dr. D. Altenpohl

Swiss Aluminium Ltd., Zurich

First, let me remind you about the two main alternative routes in the innovation process of industrial firms. Number one is in-house research with product--or process--development as a result of this research. Number two is acquisition or licensing of research results or technology from a third party leading to innovation.

The main question I will deal with is: What are the incentives and what are the limiting factors for each of these approaches to achieve innovation?

Technology acquisition is a specific case of technology transfer where you pay for the acquired technology either by taking a license or by purchasing another company or by making a joint venture. There exists another way of technology transfer which is free of charge: this is, for instance, "NASA-fallout" or other available new technology or research results--perhaps from a different type of industry being transferred into the new R&D or investment program. For the text of my paper, I will use the term "technology acquisition," which includes both versions of technology transfer.

Acquisition of technology as a very fair and legal method of keeping up in the front line of modern technology, can best be illustrated by the success of Japanese industry. Technology acquisition has even been described as a so-called "Japanese method." From 1950-1970 Japan took 13,000 licenses of foreign products or processes, but in recent years Japan has granted many licenses to other countries. This illustrates the volume and importance of trade in new technology in our days.

I will try to explain a few of the rules of technology acquisition. The concept I shall present is not a complete one, but will serve as a basis for discussion and for further consideration.

Although I have briefly mentioned the example of Japan, I will not present a macro-survey for a whole country, which may be quite different from a conclusion for a single firm. I will describe how from the viewpoint of one specific company, my own research may compare with the alternative of acquisition of new technology.

I shall choose the example of an unidentified company wanting to enter into new activities by diversification. In Figure 1, we have plotted schematically the turnover achieved by diversification in percent of the total turnover of the company. We assume that at the beginning the company has diversified only about 2.5 percent of its turnover. Now we describe schematically the next 10 years, and we distinguish between three sectors.

Sector I is turnover resulting from in-house research, which this company started in the first year with the purpose of diversification. After two or three years we can expect a small influence on the turnover since it is well known, from existing research laboratories, that about seven to ten years is required before the real impact comes out of research as new turnover.

Sector II is acquisition of technology, and here we assume that this brings the company into diversification much faster than their own R & D (Sector I). Sector III is turnover from investments, based on well-known technologies.

Why did we distinguish between Sectors II and III? The main difference between them is that Sector II needs a certain research effort before investments can be made and turnover can be achieved, whereas Sector III does not. Sector III is just plain investment using well known technology with very little research. In Sector II, the so-called leverage system is applied, where a certain approach element is needed to make the investment successful. Sector I comes entirely from the company's own research.

In the next figure (Figure 2) we have chosen the example of Sector I for new turnover based on company-sponsored research. We assume 10 units each year for research. These units could be 10 million dollars or another currency. After about 6 - 10 years, you can reach a sizeable amount of new turnover generated by your own research. The advantages are obtaining proprietary know-how for future investments, and training of specialists, which applies as well for Section II. The disadvantages are the 6 - 10 years of large expenditures for R & D until large new turnover is possible. And another disadvantage is relatively high risk.

In the next figure (Figure 3) we look at Section II, which shows new turnover based on acquisition of technologies, new to the acquiring company. This technology should have passed the prototype or pilot stage successfully. The advantages of this procedure would be that the engagement only happens if the risk can already be calculated, which means after this technology has reached a certain maturity. The second advantage is the quick availability of new technology with relatively low R & D expenditures.

There are also disadvantages, of course, such as payments of license fees, determining the value of a company providing the technology, and some problems with searching for available opportunities. Still there will be in many cases quite a difference in favor of Section II,compared with Section I.

The next figure (Figure 4) describes the increase of turnover by investing in well-established technology (Sector III).

The advantages of putting a lot of money into a conventional plant with very little R&D involved are low expenditures in R&D, and that timing and cash flow of the project can be rather accurately predicted. This is a typical project which large industries carry out all the time to erect another steel mill, or another concrete plant. Very little R&D is typical in this sort of investment.

The disadvantage is that there is no technological advantage over competitors and investment is endangered by new developments. This leads to conclusions in favor of Sector II, which is acquisition of technology (Figure 5). Low probability of failure and cooperation with or financing by third parties are easier to obtain. R & D expenditures are rather low, by reason of the leverage.

Now we will demonstrate, how TA leads to TT within a large company. This results in an effective way to arrive at short or medium-range diversification, and easy adaptation of the supply to the demand in different markets, which means flexible marketing strategies. This will be described in Figure 6. It means that if a company wants to penetrate into new markets, for instance, into countries of the third world, acquisition of technology might be a very flexible instrument for doing so. You might be well off in acquiring technology and in then implanting it in a chosen new market.

Figure 6 explains methods of exploiting an invention in different areas by adapting it to specific needs of the local market. This activity is well suited for a multi-national company with its specific resources in R & D and marketing. Figure 6 leads us to a specific group of TT activity, providing technological know-how to developing countries, a hot issue at this very moment.

As it turns out, multinational companies are best suited to transfer technology into developing countries. Certainly, they have nowadays and in the future to avoid becoming a dominating power.

Technology acquisition has to be interrelated with the company's organization and present activities. Basically, each large company can acquire any existing technology. But a lot of failures have shown that some requirements have to be met for a successful acquisition.

The next figure (Figure 7) shows one of the possible organizational forms to acquire new technologies in a large company. In our example, a "new ventures group" exists which evaluates various technologies. There is a selection committee, which then makes proposals to general management. In this committee there is always a member of the division responsible for realization, and the so-called "new ventures group " --- professionals who can compare 20 or even 50 different items according to standard criteria. Technology acquisition in this case might then be a real tool for diversification.

The next figure (Figure 8) gives a brief explanation of a case of technology transfer resulting from interaction between basic research and new technology. We have chosen the case of NASA-fallout. Plotting technological progress, or "technological sophistication" against time, we went back to the ten years between 1960 and 1970 when most of the NASA basic research was carried out. In our example, the highest sophistication is represented in the basic research of NASA.

The technology of aerospace construction lags behind theoretical designs. It is well known, for instance, that basic research has already provided the design of much further improved propulsion systems but technology could not follow (for instance: Lack of heat resistant material for prolonged use at 1400°C). This means that the problem was solved theoretically in the laboratory but could not be realized in actual practice. The civilian technology, which also keeps improving all the time, is much further below in sophistication and progress than the NASA technology.

It has often been predicted that basic research and technology of NASA and other government-financed research centers (two upper curves in Figure 8) will lift civilian technology. Therefore in Figure 8, curve "d" was expected. If there would be no influence of NASA-fallout, we would have curve "f." What happened so far corresponds to about curve "e." The main problem that our American friends have described to us is that NASA and other advanced technology centers move ahead mush faster than the civilian technology. Thus, the gap between the two groups of technology is widening, as it is shown in Figure 8.

Therefore, the taxpayer and even leading men in industry say that the government in U. S. and other highly developed countries should slow down the fast advancing basic research and technology activity because of the widening gap to civilian nonaerospace applications.

We have illustrated in the next figure (Figure 9) the number of projects which a research institute might investigate, versus the total expenditure for one new product going into true production. As we can see, about 50% of the expenditure is only for selecting projects, coming down to three prototypes or two so-called "zero-series." The timing for technology acquisition is when perhaps between 40 and 80% of the money has been spent.

Looking at Figure 9, the question comes up whether it is possible to find someone who spends a lot of money sorting out many ideas to make the results available to a third party. But there are such organizations, not only NASA and other Government supported research teams, who are trying and sorting out virtually hundreds of ideas, materials and processes. Also, large private R & D institutions like GE allow "fallout" of their work available to third parties.

In the next figure (Figure 10) we have plotted against time the maturity of a product or process, beginning with the original idea. In phase "a" this invention or development, so often occurs, and makes good progress at first. Afterwards it is for some time difficult to predict whether it will succeed, stagnate or collapse. In our example it disappears from view, perhaps for years. I sometimes compare projects in stage "b" with a little mountain stream that goes underground and disappears between rocks, so you believe it is gone forever. But further down it comes out again. Perhaps not much has been lost of the original input. This comeback might take, in our time scale, a few years or even more than ten years. After this comeback the project might make good progress as we have shown in Section "c" of the curve.

We have explained in Figure 10 that there are three different possible timings for technology acquisition. Some courageous people might acquire during time interval "a." I can give you an example of this.

A master of the art of technology acquisition is Mr. Hutzenlaub, who many years ago acquired a rather unfeasible development called the "Wankel" engine and recently sold it jointly with the original inventor for a large sum to General Motors and to six Japanese companies on a license basis. The Wankel engine was acquired at phase "a" when leading experts did not believe in success and while prototypes were still under test, it went through phase "b" but came back again. Although phase "b" is also a time for acquisition, the best timing is certainly "c," where the probability of success is already easier to establish. Most large companies would acquire here.

What are the reasons for the developing group giving up here at "c" or needing someone else as promoter? Some of the possible reasons are lack of capital or lack of experts or no feasible application or they just cannot solve a detail problem.

For approximately two years, many large companies in the U.S. have a large part of their technology made available for acquisition or licensing. According to findings of the Industrial Research Inst. acquisition in chemical industry in the U.S. has been rather small in the last ten years relative to total sales. But recently licensing has increased at a remarkable rate. Consultants and employees of chemical companies are touring Japan and Europe constantly. They have the following main reasons for doing so:

1. Present economic climate:
   licensing is surer, there is less risk

2. To get technology which is already is existence

3. Many companies that would not have licensed several years ago will do it now as one way to get additional funds.

A survey has shown that two-thirds of the IRI companies present at a specific seminar had formal license-seeking groups. One-third of them were created in the last two years. These licensing groups report either directly to the president, to the patent department or to the corporate marketing vice-president of their company.

My exploration was mainly directed at new developments of such industries as metallurgy, metals transformation, machinery, electrical equipment, cars and heavy chemicals, and not so much to industries that are very research-dependent like pharmaceuticals. The industries which I have named are, in the coming years of the seventies, more development than research oriented. If a company of this group wants to diversify or to establish activity in a new advanced technology, it can, to a certain extent, rely on acquisition of technology as a supplement to its own research. But remember that the Sector II, which is mentioned in Figures 1 and 3, needs trained research people because of the "leverage." You always need experienced personnel in the research department to judge and to further develop the acquired "new" technology. Technology acquisition has to be well balanced against in-house research.

Technology acquisition can very well provide additional strength and success to a number of industrial R & D groups.

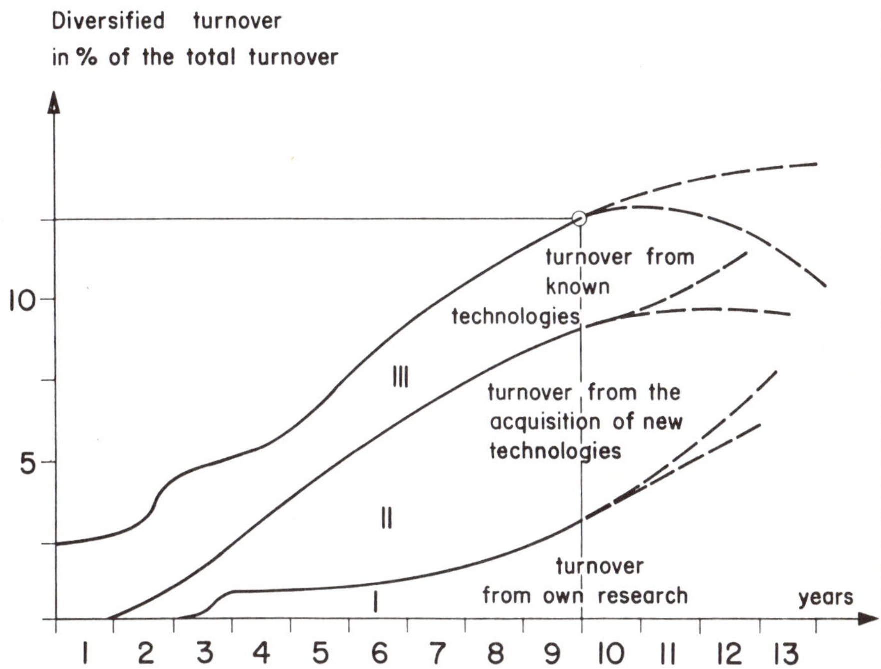

Achievement of new turnover by diversification (schematic)

Figure 1

Ratio: Expenses for R and D to corresponding turnover

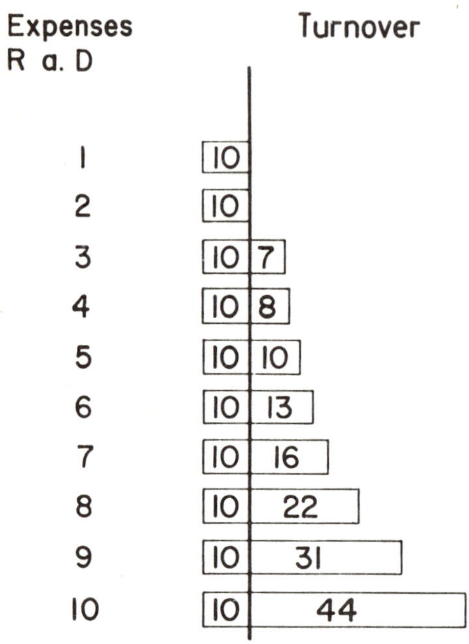

Advantages: 1. Obtaining of own basic knowledge for future investments.
2. Training of specialists, also for section II.

Disadvantages: 1. 6-10 years expenses for R and D until larger turnover is possible.
2. Relatively high risk.

Section I – Turnover based on own research

Figure 2

Ratio: Expenses for R and D to corresponding turnover

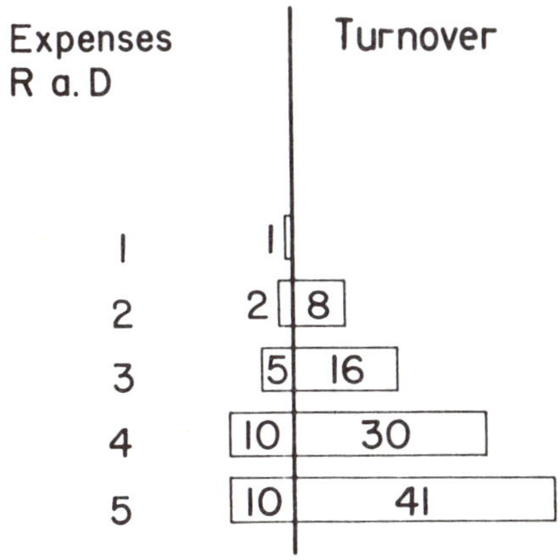

Advantages:

1. Engagement only if risk can be calculated.
2. Quick availability of new technologies with relatively low R and D expenses.

Disadvantages:

1. Payment of license fees, company value, etc.
2. Difficulties with the search for suitable objects

Section II – Turnover, based on acquisition of technologies

Figure 3

Ratio: Expenses for R and D to corresponding turnover

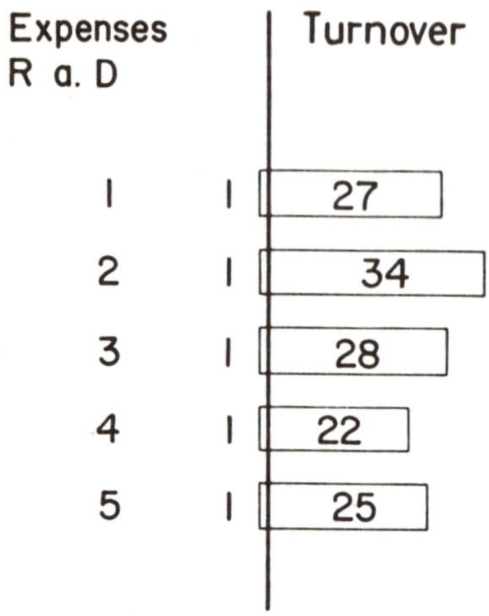

Advantages:

1. Low expenses for R and D.
2. Consumption of time and money clearly definable.

Disadvantages:

1. No technological advantage over competitors.
2. Endangering of the investments by new developments.

Section III – Turnover, based on known technologies

Figure 4

- Low probability of failure
- Cooperation with, or financing by third parties easier to obtain
- R & D expenses rather low
- Effective way for short- to medium range diversification
- Easy adaptation of the supply to the demand in different markets = flexible marketing strategy

Advantages in the Acquisition of Technology

Figure 5

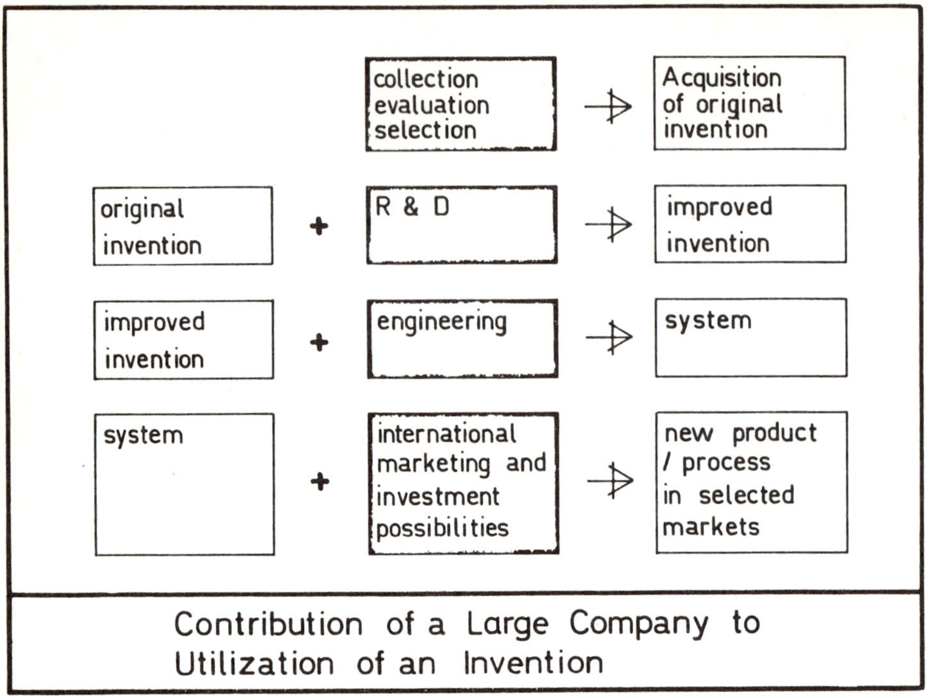
collection
evaluation
selection
Acquisition
of original
invention
original
invention
+
R & D
improved
invention
improved
invention
+
engineering
system
system
+
international
marketing and
investment
possibilities
new product
/ process
in selected
markets
Contribution of a Large Company to
Utilization of an Invention

Figure 6

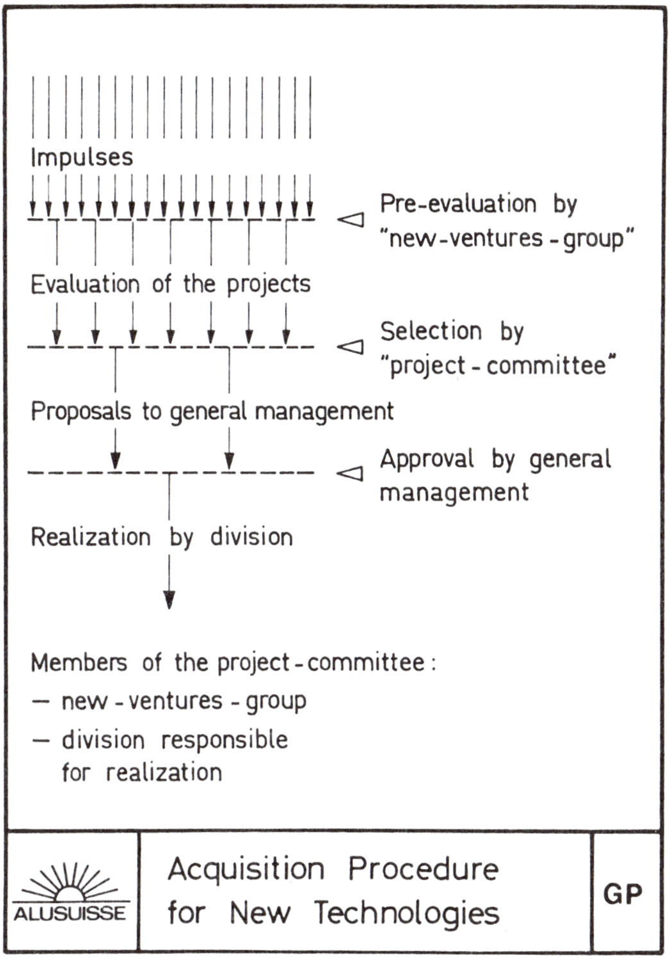
Impulses
Pre-evaluation by "new-ventures-group"
Evaluation of the projects
Selection by "project-committee"
Proposals to general management
Approval by general management
Realization by division
Members of the project-committee:
— new-ventures-group
— division responsible for realization
ALUSUISSE
Acquisition Procedure for New Technologies
GP

Figure 7

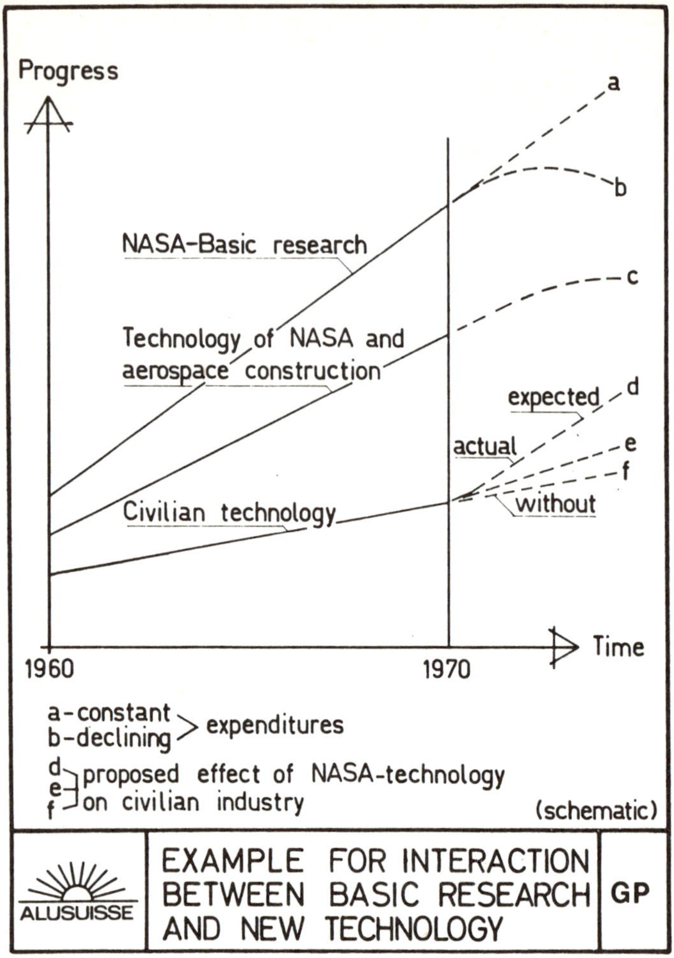
Progress
a
b
NASA-Basic research
c
Technology of NASA and
aerospace construction
expected
d
actual
e
f
Civilian technology
without
1960
1970
Time
a-constant
b-declining
expenditures
d
e
f
proposed effect of NASA-technology
on civilian industry
(schematic)
ALUSUISSE
EXAMPLE FOR INTERACTION
BETWEEN BASIC RESEARCH
AND NEW TECHNOLOGY
GP

Figure 8

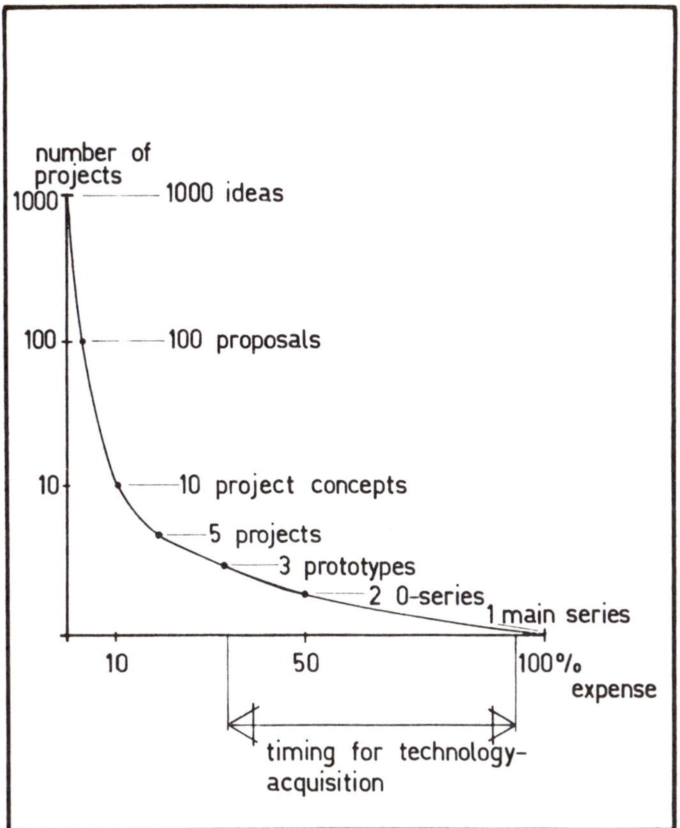

Timing for Technology Acquisition During R & D - Process

Figure 9

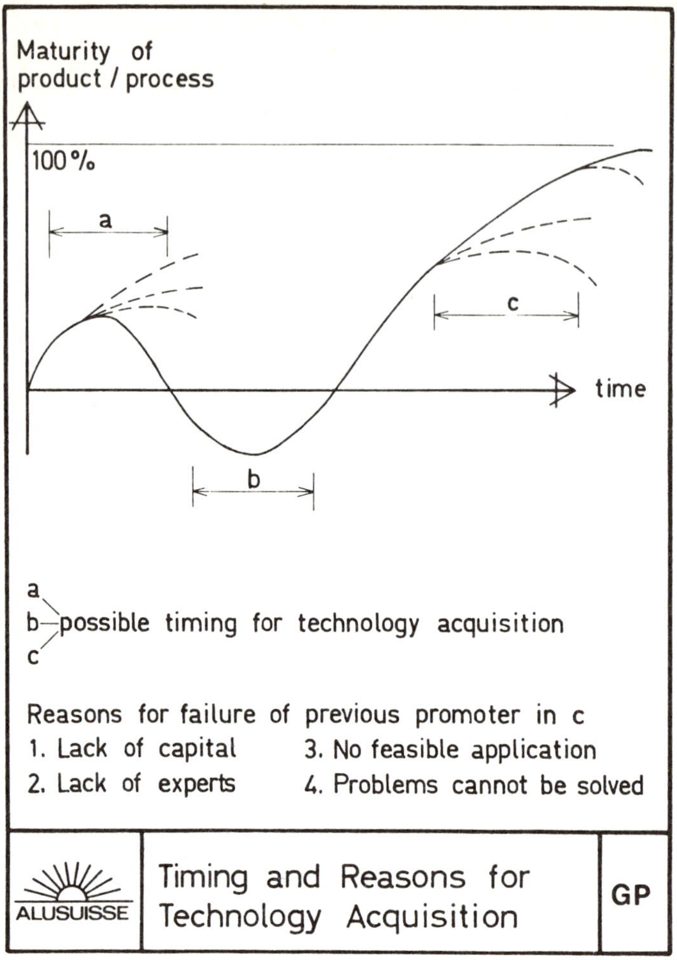

Maturity of product / process
100 %
a
b
c
time
a
b possible timing for technology acquisition
c
Reasons for failure of previous promoter in c
1. Lack of capital
2. Lack of experts
3. No feasible application
4. Problems cannot be solved
ALUSUISSE
Timing and Reasons for Technology Acquisition
GP

Figure 10

THE TECHNICAL VENTURES OPERATION

Dr. David J. BenDaniel

General Electric Company
Schenectady, New York, U.S.A.

The General Electric Company has begun an interesting experiment which is, perhaps, a significant innovation in American industry. The Technical Ventures Operation of General Electric has been established to aid in the start-up of independent spin-off businesses from products and ideas which were developed within General Electric but which the corporation itself does not wish to pursue. In this paper, I shall first describe some of the background associated with the need for corporate spin-offs and then provide a discussion of the actual organization and performance to date of the Technical Ventures Operation.

In an effort to prosper and to satisfy consumer demands, massive expenditures are annually made by American corporations in research, development, and applications. These new product development expenditures have indeed enabled growth. But from the point of view of effective use of resources, the R&D process can be deemed relatively inefficient, if one measures the process solely by the attrition rate of new products and ideas. An often-quoted rule-of-thumb is that, on the average in American industry, only one new commercial product success is attained for every sixty new product ideas initially screened.

There are numerous reasons for this high mortality rate of new product ideas, among which are:

- internal corporate review procedures do not enable the company to grasp and verify the full dimensions of future need for the product.
- "timing" for the technology may be wrong;

- the company has a narrow market orientation and the product falls outside that base;

- there may exist an organizational "stalemate" as to which internal component should carry the work forward;

- there may be lack of proper incentives to carry the work on;

- the company, especially a large firm, may be generating internal barriers against the successful growth of small business opportunities;

- OR, the product may, in fact, not be a good one.

The General Electric Company has long been aware that the really successful companies of the future will be those which can increase the effectiveness of their R&D expenditures and, at the same time, exploit those new product ideas which have not been carried to the final stage of commercialization in order to maximize the return on investment from these unused ideas. The primary processes of technical change may depend not only on research and development but also on ingenious applications of old techniques in response to market demands.

The Technical Ventures Operation (TVO) has the purpose of exploiting General Electric technology through the old technique of the small business. In return for an equity position in the small business and other securities, GE will aid in the establishment of a "spin-off" company which has as its goal the development and marketing of the new product idea or technology. GE retains a minority interest in the "spin-off" and the entrepreneur team members invest substantial amounts of their own funds. The majority of operating capital comes from outside sources: Venture capital or other industrial organizations.

The General Electric approach, as carried out by TVO, consists primarily of matching available new product ideas, technology and assets which otherwise would not be exploited in-house with entrepreneurially-oriented individuals, either from inside or outside GE, who have the expertise to start, to nurture, and to grow the new business opportunity, and with outside venture capital. TVO's functions include:

- Location of entrepreneurs and entrepreneurial organizations and the possible matching of products, processes and ideas with individuals or organizations presently within General Electric whose management approves their consideration of outside activities.

- Exploratory financial and technical business appraisals.

- Contact with investment bankers and other financial sources to acquaint them with business opportunities inherent in GE technology and with selected individuals.

- Administration of these outside capital sources in support of venture interests.

- Participation in the new ventures to the extent of advising on initial business plans and arranging for GE management participation on their Boards of Directors.

General Electric's corporate spin-offs through TVO now number six and include (1) Intermagnetics General Corporation (IGC), a firm set up to handle the production, distribution, and sales of all superconducting magnets, materials and systems formerly handled by GE's Superconductive Products Operation; (2) Ferrodyne Corporation, set up in May, 1972, with GE (3) Community Information Systems, a Chaska, Minnesota firm putting together two-way cable television communications linking homes with business and industry and educational facilities; (4) Instructional Industries, Inc., set up to augment educational and instructional processes through the use of new electronic technologies; (5) Nuclepore Corporation, a filtration products company exploiting a unique microfilter developed by GE's Nuclear Energy Products Department; and (6) Electro-Petroleum, Inc., a firm set up to develop electrical methods of secondary oil recovery.

GE's initial equity position in all six firms comes to several million dollars. In all cases, the company holds a minority interest and in several cases, additional securities as well. IGC, the oldest of the group, is now finishing its second full year of operation, and is profitable with annual sales in excess of $1 million. The other businesses are newer and are, at this point, all struggling for survival. Each has prospects which may amount to something but must first succeed in their initial formation phases.

These businesses have, by their nature, an interesting balance. General Electric does not control them. Nor are they controlled by venture capital or the employees. What has been constructed is a "three-legged stool", in which the interests of all three parties, employees, General Electric and venture capital, are all served.

The question is frequently asked as to what are the true motivations for each of the three parties in this new approach.

First of all, it is reasonably clear why the entrepreneurs want to do it. Even though it means leaving the womb of a large company, the venture process allows them to get a piece of the action to run with a business in which they believe. I would reiterate here that GE requires significant personal financial commitment of the entrepreneurs to the venture. It is not difficult to see why entrepreneurs want to form the venture and why they are willing to accept significant personal risks. The process is, in fact, an old American technique for becoming wealthy.

Why should GE do it? In situations which are ripe for the venture alternative, GE is faced with the following options:

1. Keep the business going.
2. Liquidate the business.
3. Sell the product, patents and equipment outright.
4. Transfer the business scope internally from one Division where it is not doing well to a more conducive home in another division within the corporation.
5. Permit the employees to start a venture.

Option 1 represents a status-quo solution. The whole process of producing a venture is never initiated if Option 1 looks viable. General Electric has first rights.

Option 2 has built into it significant liquidation costs (typically \$300,000 for a small project) in the write-down of inventories and equipment and the potential termination costs of employees. This is quite unattractive to a Division which would have to book these losses all in one year. It is, however, often chosen by default.

The direct sale, Option 3, is frequently not available. If the project is hanging in the balance, there are quite likely other companies with similar projects, all struggling. Therefore, the business in question may be difficult to sell and such a sale would barely cover book values; probably less than that and, therefore, the General Management is faced with booking a loss even if he were able to consummate a sale.

Incidently, sale of equipment and patents does not usually transfer the true worth of the organization; its committed employees and the "lowered entropy" associated with a going business are of significant value.

The fourth option - transfer within the corporation - is in all respects the best option available to a given General Manager faced with a troublesome project. He would like to find some other General Manager within the corporation to take over the problem from him. However, it takes two to tango; transfer within a corporation of a troublesome project requires a pretty enthusiastic General Manager in some other Division to take on the responsibilities.

The fifth option, the new mode of permitting the employees to form a venture, with GE retaining equity, is attractive to GE for two reasons; First, it enables GE to retain technological and market contact with the project. Recall, the business was, at least in the past, viewed as an area of corporate interest and so may still be in the future. Second, it permits GE to obtain a potential capital gain in the event of the substantial success of the venture. The investment advantages of a sponsored "spin-off" can be outlined in great detail; however, a single statement encapsulates the advantages:

<u>The venture provides a hedge to the General Electric Company</u>

In the event that the business fails, GE can write down the value of the assets (now on the books of TVO) and has merely postponed a previously contemplated liquidation. In the event the business survives, there are three possibilities: First, to sell stock and obtain capital gains; second, to hold its equity in anticipation of future capital gains; and third, to buy more stock and obtain control of the venture. (One must recognize that the future cost of such stock will be at fair prices consistent with the growth of the organization.)

The detailed mechanisms within GE are as follows:

Just previous to the formation of the venture, the assets are transferred from the books of the Division to the books of the Technical Ventures Operation. It is, therefore, the Technical Ventures Operation which bears the subsequent risk, (and, optimistically, might reap the subsequent rewards). The Division contributing the project is off the hook in exactly the same way as it would have been if the transfer to the books of some other General Manager had taken place.

Finally, why should venture capital do it? It is an interesting fact that all of the businesses formed by the Technical Ventures Operation have received venture capital, despite a recent difficult venture capital market. So successful has the Technical Ventures Operation been in attracting venture capital for the entrepreneurs of the ventures that the Technical Ventures Operation to venture capital is illustrated in the accompanying Figure. On curve 1 of the Figure is portrayed a normal product life cycle in terms of annual cash flow. In the early phases, money has to be spent to get the thing off the ground. With good luck and good management sooner or later the business begins producing a profit. On the other hand, certain projects are illustrated by curve 2. They have deviated from expected performance (for a variety of reasons often time quite special to the situation) and point "x" is the point of decision. The power of the TVO approach is demonstrated most adequately by recognizing that the full reorganization and re-incentivization of the project and formation of the TVO venture results in a translation of the origin to point "x" on curve 2. The subsequent performance of the venture carries with it beneficial effects of previous development and organization expenditures in past years, otherwise unrecoverable.

As a result of the powerful advantages to venture capital, ventures are formed at significantly different rates per share for the three parties involved. Typical of the ventures formed by the TVO is to have the GE assets transferred such that its price for stock is, say, $5.00 a share. Venture capital comes in at $10 or $15 a share. This is a good deal from the point of view of the General Electric Company based on book values, but it also is a good deal from the point of view of venture capital which recognizes that GE's true contribution is not the $5 a share but perhaps $25.00 a share, when one takes into account the development and organizational expenses incurred in previous years to get the point "x". The entrepreneurs are permitted to come in at preferred founders' rates lower than GE's to permit them to have a reasonable piece of the equity at what, for them, is a deep financial commitment.

Entrepreneurs, GE and venture capital, all get a reasonable deal. This is essential for the formulation of a small business. For what, in truth, is a small business? It is *potential* rather than *actual*. It can only achieve its potential if all parties feel they have a reasonable deal so that their collective incentives are made parallel and they can thereafter work together to achieve the success of the business. This fundamental principal of fairness to all parties is essential in the formation of the new business and is itself an old technique for business success.

As a result of the Technical Ventures Operation, new businesses are in existence that might otherwise never have had the opportunity to test the marketplace. Within the corporation itself an additional alternative has been invented which permits decisions to be expedited. The Technical Ventures Operation is, therefore, also a pioneering effort on the part of the General Electric Company to provide both a new alternative for Corporate decisions and, perhaps, a new alternative for Corporate growth.

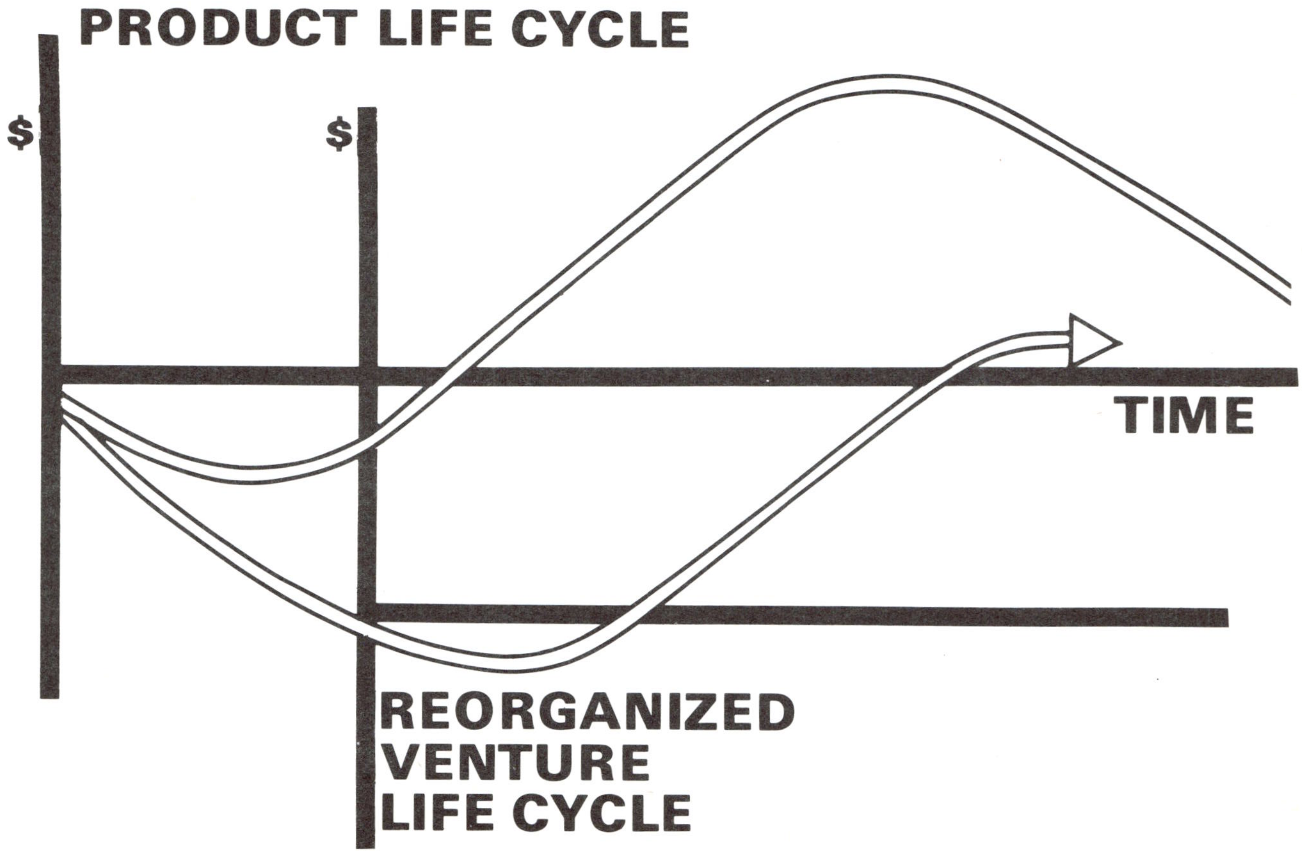

Forming a new venture from a reorganized project carries with it significant values, as illustrated graphically in terms of the translation of the start-up time shown above.

THE TECHNOLOGY WORKSHOP: A KNOWLEDGE APPLICATION FACILITATOR

Quentin L. Hartwig

Saint Francis College

INTRODUCTION

Combined forces of competitive world trade and public concern for practical applications of basic research has intensified efforts to accelerate the process of technology transfer. The transfer process involves the flow of knowledge across discrete interfaces which exist between the source of an idea and its point of application. The system-user interface between the technology transfer system and the terminal user is still poorly understood, and deserves considerable attention.

One type of "facilitator" to the system-user interface is the technology workshop. The Pennsylvania Technical Assistance Program (PENNTAP), a state knowledge dissemination effort based at Penn State University, funded 45 technology workshops from 1967 through 1972 with the purpose of acquainting societal appliers in industry of technology relevant to their needs. After a period of some five years of experience with the workshop concept, it seemed timely to conduct a study* of the workshops thus sponsored by PENNTAP with the purposes of (a) identifying specific transfers or applications of technology disseminated at the workshops; (b) assessing the impact of the workshop on the attitudes of the participants; and (c) providing direction for further dissemination efforts. This report covers these three areas.

* This study was funded by PENNTAP.

METHODS

PENNTAP provided the names and addresses of 2,700 participants in 45 workshops. The workshops were conducted by six institutions: Pennsylvania State University held five; Bucknell University, 4; Philadelphia College of Textiles and Science, 2; the Governor's Science Advisory Committee, 30; Duquesne University, 1; and Institute of Man and Science, 1. Examples of workshop titles can be found in Table 2.

The primary means of identifying specific transfers and assessing the impact of the workshops on the participants was through the use of a questionnaire. The questionnaire package contained a cover letter, a single-paged questionnaire, and a return-addressed, postage-paid envelope.

The brief cover letter stressed the importance of the survey by pointing out that the completed response might well have an effect upon future directions of PENNTAP activities. The questionnaire (Table 1) contained eight yes-no questions which covered the specific intent of the study and an invitation to write comments at the bottom of the sheet.

The percent response to the questionnaire was 32%. The yes-no responses were punched on the computer cards and the results were tabulated with an IBM 1130. Total numbers and percentages were computed for each of the 45 workshops. A computation was also made of the total responses to each question. The written comments were sorted by hand.

RESULTS

The yes-no response to the questionnaire is given on Table 1. A third of the respondents believed that their company did use specific technology or realized a savings as a result of information brought to their attention at a PENNTAP workshop. Double the number of respondents (70%) indicated that the workshops had a positive impact on their attitude toward the potential of new technologies in their line of business. However only a third, again, seemed to feel that their techniques for searching out information relative to their work needs were modified by the workshop.

Response to question 4 gives insight into directions for future technology workshops. A little less than half (43%) of the respondents would be interested in a follow-up to a previous workshop. Most participants (89%) would be interested in a workshop describing a specific technology in their field of interest. A slightly smaller percentage (83%) would be interested

TABLE 1

Questionnaire and Its Response (%)

	Yes	No	Checked Neither
1. Did you or your company use any specific knowledge or technology, or did you realize any savings as a result of information brought to your attention at the PENNTAP workshop? (If yes, kindly describe the benefits below.)	31	64	5
2. Did the workshop have a positive impact on your attitude toward the potential of new technologies in your business?	70	27	3
3. Have your techniques for searching out information relative to your work needs been modified by the workshop?	31	65	4
4. Would you attend another workshop if it was:			
a) a follow-up to a previous workshop?	43	37	20
b) a description of new technology in your field of interest?	89	5	6
c) a description of new technology of general interest to you or your firm?	83	9	8
d) to help you to cope with Federal and State regulatory legislation in such areas as safety standards and pollution control?	53	33	14

Comments (be assured your comments will not be made public without your prior approval):

---

in a workshop describing new technology of general interest to the participant of their firm. About half of the respondents (53%) would be interested in a workshop that would assist them in coping with Federal and State regulatory legislation in such areas as safety standards and pollution control.

Evaluation of the individual workshops organized by responsible institutions provided further insights. Penn State University scored a considerably higher percentage of affirmative

TABLE 2

Five Workshops with Highest Percentage Yes to Question One

Workshop #	% Yes to Question 1	4a	4b	Workshop Title
18	71	57	100	Industrial Corrosion
38	70	70	90	Metal Cutting Processes
30	69	50	69	Air Pollution Control in the Cement Industry
34	63	36	91	Chemical Instrumentation
9	60	40	90	Characterization of Powder and Aggregates

Five Workshops with the Lowest Percentage Yes to Question One

Workshop #	% Yes to Question 1	4a	4b	Workshop Title
17	8	47	97	New Technology and Materials in the Building Industry
32	10	10	70	Materials Evaluation and Testing
36	16	16	83	Computer Simulation Techniques
3	18	34	87	Modern Coatings Technology
6	20	13	80	Biomaterials in Medicine and Dentistry

responses to question 3, "Have your techniques for searching out information relative to your work needs been modified by the workshop?", than did other responsible institutions.

The five workshops with the highest percentage of yes response to question 1 (Did you use a specific technology?) and the five workshops with the lowest percentage of yes response to question 1 are shown above in Table 2.

Percent response yes to question 4a (Are you interested in a follow-up workshop?) was consistently less in those five workshops with lowest percentage yes to question 1 (Did technology transfer . . .?). On the other hand, percentage response yes to question 4a was consistently high in those five workshops with the highest percentage yes to question 1.

However, it should be noted that enthusiasm for other new technologies was not diminished in the minds of the participants of the five workshops with the lowest percentage yes to question 1. The percentage yes to question 4b (Are you interested in a workshop of other technologies?) was always high in these workshops.

Fifty questionnaires contained statements describing applications of technology or savings as a result of information brought to the attention of the attendee at a PENNTAP workshop. The number would be even higher if every written hint of transfer were counted.

Specific types of applications are wide ranging. Many relate to improved methods of manufacturing resulting in either improved products and/or savings in the manufacturing process. Financial considerations range from introduction of new laboratories to reduction of tool wear. More detailed descriptions will require a visit to the plant and discussions with individuals involved in the transfer process.

Several comments indicated that socio-economic conditions at the plant prevented early adoption of the new idea disseminated at the workshop. The most graphic statement was, "Your program supplied me with a million possibilities. There was no limit to what I could do. Then I went back to the panic situations and personality complexes and politics." It is impossible to estimate the number of participants of the workshops that are faced with this problem.

A small number of comments suggested that the material presented at the workshops was presented on too high a level, much too vague, not timely, or not directly related to everyday problems.

Most of the replies were received during the 10 day period following the March 10 deadline. Not all the replies were usable for a variety of reasons. For example, lists of participants at the workshops contained names of individuals who claimed that they either had not attended or had never heard of the workshops. A few stated that because too long a time had elapsed between the workshop and receipt of the questionnaire, they could not intelligently fill out the form and therefore sent it back blank.

DISCUSSION

PENNTAP sponsored a wide variety of technology workshops from 1967 through 1972 to disseminate the latest technological information to the industrial and public sectors of the Commonwealth. A study has been made of this five-year experience in the workshop concept, primarily through the use of a questionnaire. The purpose of the brief questionnaire was to identify transfers, and ascertain the impact of the workshop on participants through specific yes-no questions and an invitation to write in comments. The results favor the workshop concept.

A significant number of transfers have been identified. Furthermore, a majority (70%) of the respondents to the questionnaire indicated that the workshop had a positive impact on their attitude toward the potential of new technologies in their business. The participants were generally impressed with the prospect that their institution could be assisted through the introduction of technology not currently in existence at their firm. In terms of a long-range dissemination program, this judgement is important. These participants are most likely to return to home base and become involved in a transfer activity. If not a transfer activity, the individual will likely remain receptive for some time to a new stimulus which might be an announcement of a new workshop, a follow-up that a new technology described in some other media might be applied in the firm. This continued receptivity is probably of significant proportions since some 89% of the respondents indicated that they would attend another workshop.

The identification of numerous transfers and this created sense of receptivity supports the conclusion that workshops can be a productive method of disseminating information into the industrial and public sector.

Attention should be given to ways of encouraging a ripple effect from the workshop. For example, if the workshop participant returns to the plant with an idea of a transfer and adds a resource that makes the transfer possible, he is responsible for amplifying the effect of the workshop.

If, in addition, this receptive individual organizes or improves those other activities that identify and apply additional new technologies at the plant, this is yet another step of amplification. A further step of amplification would be if his actions caused accelerated flow of information into firms, plants, and organizations other than his own.

Certain clues on the returned questionnaires disclose barriers that can impede or prevent amplification at any one of the above steps.

Several respondents indicated that one or several socio-economic factors at the plant were a hindrance to the application of the technology. It could be financial, poor inter-personal relationships, or ultra-conservative management. These comments are not unusual. Utilization of new curricular material following attendance of an educational demonstration center was said to be hindered by a lack of money and facilities, and a shortage of staff. (1)

Other studies have shown that a major problem can lie in the communication between the researcher and the administrator. It was reported that, "Unfortunately, both groups suffer from the wearing of professional blinders. The researcher, in his involvement with the world of chi squares and factor analysis, is not frequently cognizant of the pressing problems and demands made of administrators. Conversely, administrators know little about the potential and limitations of research. To counteract this possible communication strain it is recommended that a flow of written and oral contacts be made to key decision makers within the plant for the purpose of reinforcing the probability of actual utilization." (2) These contacts might well be accomplished by the workshop director.

At the next step of amplification, encouraging the incorporation of other technologies into the plant operations may hinge on the workshop participant's ability to seek out relevant information. The fact that only a third of the respondents indicated that their techniques for searching out information relative to their work needs was modified by the workshop suggests that more attention should be given to educating the participant in the use of the many sources of information relative to stated problems. It is not uncommon in a plant conducting some form of R & D to lack even the most rudimentary library of journals and reference books. The engineer is often expected to purchase his own journals or visit a local library. Beyond that the whole image of ways in which data utilization is poorly handled in the educational institutions and within science itself. Therefore, workshops should not only be considered as a means of disseminating specific packages of information, but also as a classroom in the science of information searching.

At the amplification step beyond the plant, it is difficult for an employee at one plant to encourage more efficient use of technology at another nearby plant even though their individual successes can be mutually beneficial in terms of the overall economy. Yet a number of respondents commented that the workshops gave them the opportunity to visit with other participants and share ideas or gather information informally that collectively will provide an adequate fund of information necessary to sell the idea to management.

Experience elsewhere suggests one approach to facilitating the amplification effect outside the plant. Participants in a dissemination workshop conducted by the Oregon State System of Higher Education were under obligation to provide a workshop at their home institution for a minimum of twelve fellow members from that region. It was reported that participants in the second order workshop gained some knowledge of the material available. In addition, potentially valuable lines of communication were set up within the region among persons interested in similar topics. (3)

Other respondents attributed a sense of adult education to the PENNTAP workshops. This brings to mind that there are vocational-technical schools in nearly every sector of Pennsylvania. They, along with colleges and universities, might be encouraged to play a greater role in providing an arena where receptive appliers could gather further information or learn how better to gather information.

The following suggestions are specific to PENNTAP, but apply to any state or federal dissemination effort.

Two recent reports highlight resources that could be coordinated into the PENNTAP concept. It is suggested that they be incorporated into the PENNTAP framework rather than the reverse since PENNTAP has a successful framework for actively disseminating technology into the state economy and there seems little advantage to creating yet another.

The report (4), "Action Now; Partnerships: Putting Technology To Work," contained the resolutions of the National Action Conference on Intergovernmental Science and Technology recommending specific actions putting technology to work at a pace necessary to resolve social ills of housing, unemployment, etc.

Examples of such actions are the use of Federal laboratories in domestic programs and new arrangements among institutions in public service.

A similar recommendation (5) was reported in a study entitled, "Intergovernmental Uses of Federal R & D Centers and Laboratories," which was conducted by Arthur D. Little, Inc., for the Council of State Governments. The Arthur D. Little consultants stated that Federal R & D laboratories have a great deal to offer state governments in areas of technical assistance, technology transfer, and management support.

These recommendations suggest that numerous resources could multiply the PENNTAP impact significantly if they were extensively coordinated into the PENNTAP network.

SUMMARY

A questionnaire was mailed to 2,700 participants in PENNTAP-sponsored technology workshops held from 1967 through 1972. Analysis of the response to the questionnaire has resulted in the following observations and recommendations.

1. The PENNTAP technology workshops have been a successful means of transferring useful technology into the state economy. Significant applications of knowledge described at the workshops have resulted in better products, improved methods of manufacturing and new research endeavors.

2. A high percentage of the respondents to the questionnaire found the workshops useful and would be interested in attending future programs. This suggests the "Adult Education" or "Continuing Education" concept of a state-wide dimension.

3. Conclusions derived from the questionnaire results and recommendations found in the literature give guidance to the structure and follow-up to future workshops. Somehow, consideration must be made of the "conditions" at the home base of the participant that can impede application.

4. The workshop technique presents a "captured, live" segment of the applier population who can potentially, in turn, cause a ripple effect with their colleagues.

5. PENNTAP performs a critical function within the Commonwealth. Every effort should be made to coordinate other actual or potential "technology utilization facilitator" resources into the PENNTAP mission.

REFERENCES

(1) Lapan, S.D., et. al., The Effectiveness of Demonstration as a Means of Dissemination, Illinois Gifted Program Evaluation, University of Illinois, Urbana, Illinois, pg. 15, February 5, 1971.

(2) Goldin, G.J., K.Margolin, and B.A. Stotsky. "The Utilization of Rehabilitation Research, Concepts, Principles, and Research," New England Rehabilitation Research Institute, Northeastern University, Monograph No. 6, pp. 39 and 41, February, 1969.

(3) Karr, C., and B.C. Porter. "A Prospect for Research, Development and Evaluation Training," Oregon State System of Higher Education, Monmouth, Oregon, pg. 19, January, 1971.

(4) Anonymous. "Action Now; Partnerships: Putting Technology To Work," National Action Conference on Intergovernmental Science and Technology Policy. Published by Pennsylvania Office of Science and Technology, Harrisburg, Pennsylvania, August, 1972.

(5) Anonymous. "Intergovernmental Uses of Federal R & D Centers and Laboratories," Arthur D. Little, Inc., Washington, D.C., April, 1973.

SOCIAL RESPONSIBILITY, AN EXPERIMENTAL APPROACH

Donald E. Cunningham

National Science Foundation

Of late, much attention has been directed to the problems which result, in unanticipated ways, from economic growth. People are becoming more aware that the unforeseen consequences of the unassessed introduction of technologies into the marketplace can at times be enormous; that a cure for a specific ill can turn to a curse in an area apparently unrelated to the specific ill attacked. A commonly cited example involves the widespread use of DDT. DDT reduces disease and increases agricultural production but at the same time may contain the seeds of destruction for certain animal species. Another and more critical example involves modern civilization's need for energy and the attendant social, economic and environmental price of obtaining that energy. There are also pressing and, at present, unresolved problems of recycling and the responsible utilization of unrenewable natural resources.

Civilization has come to a point where individual goals have to be subjected to scrutiny in terms of societal collective goals. Aluminum is widely used for the manufacture of beer and soda cans despite its being difficult to reclaim. Because it is cheaper for the manufacturer to use aluminum in cans and because <u>his</u> profit is greater, does not imply that society as a whole should absolutely restrict the manufacturer from using aluminum. Some means must be found for determining the overall societal cost--if a determination cannot be made then there is no reasonable way of adjudicating who takes liability for a share of the costs. The method for determining overall costs has proven elusive. The question deals with a host of ponderables and imponderables. Physical costs, those to which dollars can be attached, are relatively easy to consider. (Even here, a just solution to the apportionment of the costs is by no means a trivial task.) But when the notion of overall societal good enters the arena, even the ground rules are difficult to set, never mind quantity.

Many have written about why technology assessment is now a must. The arguments range from those of scale of individual use of a technology (e.g. the auto) to the synergistic effects of high technology industries pushing boundaries so fast and so far that impact analysis is impossible since baseline data never has been acquired. The pace of development has simply overwhelmed attempts at rational prediction. The need for prospective rather than retrospective technology assessments has been made, and I think convincingly, by many. Having accepted those arguments as reasonable and believable, I would like to point out that there are still some difficulties inherent in implementing and validating their suggested courses of action.

Social Cost Forecasting

Basically the arena that must be explored is that of social interactions. What the social responsibility of any firm or interest group is depends directly on how society conceives itself at that time. And society will take a distribution of views. The interactions of the varied views might, if analyzed, lead to a path down which we might obtain a consensus of the value system society is using at that time and what it is likely to be in the future. Unless the value system can be determined, the term social cost has no meaning--and if social cost has no meaning, neither does social responsibility. Several approaches can be made in the determination of social cost. In one, available data can be analyzed and categorized historically. [1/] With a certain number of data points arranged in time sequence we could, hopefully, project to a future time with some degree of assurance. Two problems immediately arise. First, the indicators by which we measure social phenomena are not quantitative and therefore to a large extent a data base does not exist in usable form. This means that if we are to use historical data for social trend projection, a sizable analysis and categorizing effort is a necessary preliminary. In addition to the size of the task, the work of compilation itself is uncertain with no real degree of assurance that the effort would result in a usable output. Yet another major problem exists. This concerns the quality of social data

1/ Technology Forecasting and Assessment, I. Mitroff and M. Turoff, IEEE Spectrum, March 1973. This article offices an interesting review of forecasting methods and their philosophical bases.

as a function of time. Clearly as social consciousness develops, the realm of social concern multiplies and expands. The interaction of many socially concerned individuals undoubtedly has a greater effect than the efforts of a few. We have seen this occur time after time. When social concern becomes accepted it becomes institutionalized--or vice versa. There were many more environmentally concerned people when they realized that there was an environmental movement. (I may be faulted on cause-effect relationships but the message is clear.) In the context under discussion, though, this means that more recent data is more valuable and more important than earlier data. The historical collection referred to earlier may have as its principal output categorization of, or means to define, the social data which we are seeking but the categorization may indeed limit the utility of data collected because the categories prove to be not correct or useful in the context of projection. The projection, too, is likely to be necessarily a non-linear one. The conclusion we reach is that current data collection is more valuable than historic, that "current" data must be collected with a fair degree of repetition and that selection rules for what data to collect (and how) must be developed.

Are there other ways to derive information about the future social environment? There are many. Consensus mechanisms of various types exist--Delphi, cross-impact, cross support--to name a few. In general they differ in how the data is processed rather than conceptual framework. The result is a consensus scenario--individual scenarios also exist. The critical point though is that these techniques always are based on consensus projections. In a sense they use the personal experimental measures of social phenomena in place of the more categorized and defined approach alluded to above. They also use personal judgements on the non-linearity of social interactions rather than an attempt to measure recent social phenomena in quantified terms. Some attempts of this type are quite good--some are not. They all have the connotation of "winging it" and there is some doubt about their generalizability or transferability to a broader range of issues.

Where does this leave us? To categorize, we know that we as a society have come to a point where individual or corporate actions cause spillovers into areas not initially the target of the action. We know further that the spillovers of various actions add--that the spillovers form a "noise" background that reaches closer and closer to the critical level. While we know that one more window air conditioner will not bring Con Edison, and with it New York City, to its knees, we do know that human aspirations toward coolness, if left to themselves, can have that result. When they do, a cataclysmic result can occur. We know, too, that no individual or group of individuals would calculatingly cause that result. Thus social costs rise, people become aware, and the onus of social responsibility is heaped upon the manufacturer of window air conditioners. We understand the dimensions of the problem, but we don't understand how to make better approaches to the solution.

The tacit assertion of this paper is that until the machinery is available to evaluate in a quantitative way social costs as a function of time, it will be impossible to determine what anyone's share of social responsibility is.

I have not mentioned technology transfer specifically to this point. It has been sort of an underpinning to the discussion. Few would maintain that economic growth is not desirable and necessary for a nation or for the world. New ways of doing things, new products, and new services are all desirable goals. But any "new" activity in the context of social responsibility implies "new" social costs--costs that may be markedly different for alternate ways of satisfying the same functional need. The question is, "Can an estimate of those costs be made before a large scale technology acquisition takes place?" And then, "Can an apportionment of the social responsibility be made?" Obviously the second question cannot be answered until the first one is.

Socio-Economic Experimentation

With this backdrop, I think that a workable approach which can lead to a solution of the problems posed above involves the use of socio-economic experimentation. Having stated that, though, not much has been stated. Questions are left hanging concerning scale, measurement, instrumentation and design.

An Experimental Approach

When impacts of a serious nature accompany introduction of new technologies, and when it is difficult to predict in total the consequences, then it makes sense to attempt to determine the nature and extent of the impacts by small scale experimentation. This approach can provide better estimates of possible unanticipated effects accompanying the introduction.

To indicate that experimentation is necessary does nothing but state the obvious. In the area of social experimentation much must be done to clarify what is to be experimented upon, and how (and if) cause-effect relationships can be derived.

It is best to examine this type of experimentation by reviewing certain conditions which must be fulfilled if an experiment is to be meaningful and certain techniques which may be used to interpret the results of the experiment. The general considerations which lead to the choice of an experimental situation can be listed as follows:

- that a measurable output be obtained
- that the measurement be reproducible
- that the results obtained are transferable (conceptually) to other like situations
- that the results be useful in choosing further areas for experiments

This is not a unique listing but it sums up several salient points. First, it indicates a measurement of something must take place. Measurement implies a choice of variables--a definition and categorization of experimental variables. This goes to the heart of the discussion of the preceding section and represents no trivial problem.

The issue of transferability further complicates the problem of experimentation. Variables measured in one situation must have a direct comparability with those measured in another and apparently different situation. This sort of consideration leads to very strict requirements on structuring the problems to be addressed. In each application a choice of analytic description of the situation must be made and the description must be in terms of potentially measurable and generalizable variables. This point will be returned to later when the question of a search strategy is discussed.

How to interpret the effect of an experimental treatment is the next and related question. Three ways are considered though each present difficulties:

- by comparison between control groups
- by testing a hypothesis
- by comparison with a forecast

Control groups are fine in principle but their number offers perplexing problems if we are to establish statistical reliability. Hypothesis testing is more tractable but if experimentation is not to be prohibitively expensive then we must build webs and chains of hypotheses as we shall see later. The forecast technique would be attractive if the forecast could be made for times short enough to allow acquisition of data short enough for comparison with experiments. This implies forecasts within the next two to five years rather than to 1984 or the year 2000.

Despite the shortcomings, attempts must be made to use all three techniques with suitable concern for their weak points. More knowledge of the techniques will be developed when used if the experimentation also has as a purpose the upgrading of the observation technique through methodology validation and iteration.

The Experimental Situation

This leads us to the choice of experimental situations. Naturally we would like to perform experiments in critical problem areas--and that should be a long-term aim. This will probably be impossible to accomplish initially because of the nature of the universe for the experiments and our own state of understanding of the process. The situation is akin to that of the physicists who can solve problems involving two bodies or involving an infinite number of bodies but cannot precisely solve anything in between. The social experimental situation may really be worse for there are no guidelines indicating for sure which is the two body or infinite body problems. Suffice it to say, where there are critical problem areas there is extensive activity and where there is extensive activity there are, most likely, interactions leading to the non-linearities referred to earlier. The choice of experiment depends on ability to understand and structure as well as problem area importance. A compromise will be necessary.

In planning the experiment, due attention must be paid to why the experiment is to be performed. One goal is to estimate total social costs of the treatment applied to the experimental situation. Instrumentation and design must include means for picking up this information during the lifetime of the experiment since the information cannot be retrofitted to the projects undertaken. Note, too, that it involves a notion of what items contribute to overall social cost raising another series of questions about what is to be included and what is not. Unfortunately that detailed concern is worth a whole other paper's discussion. What is important here is that conceptually such considerations can be built into the projects undertaken.

Finally there is the necessity for making experimentation efficient--meaning obtaining reliable, important data at appropriate cost. This involves choices as to where experimentation may be carried out effectively and introduces the requirement that the results of experiments be noticeable (i.e. measurable as in the previous discussion). One way to approach the choice problem is to select areas where something is obviously delaying, or preventing totally, progress in the application of some class of product, process, or service. The experiment could be set up in such a way that the impediment or impediments to application be identified and attempts made to do something about the impediments. At this point two different types of experimental treatment emerge. In the first the impediment, once identified, might be overcome in particular cases, and the result of the treatment observed. In the second, the impediment might be removed and results observed. Here the first treatment is selective, and the second is general. Either treatment might make sense in given cases. The first would be appropriate where potential social costs of the action might be high and/or unpredictable. The second would be appropriate where the process was clearly incremental, where slight but widespread changes might occur, and where the impediment could be reintroduced either in its original form or in a simulation of its original form. In concrete examples, the first treatment might occur where agovernment has made a choice to stimulate a certain type of R&D activity, leading to successful export of a certain and specific type of technology. The second would occur if a government decided to spur R&D activity in general--by subsidy, tax rebate or other across-the-board means. There are many examples of this type.

Independent of the example chosen, though, concern must be paid to the cost of obtaining reliable data in the experimentation. This is an important parameter when dealing with a real situation of experimentation, and enters consideration in much the same way as that which led to choosing not the most important experiment, but rather the most important experiment that could be reasonably understood. As the methodology and analysis of social cost experimentation develops, more and more important experiments can be chosen since more and more background understanding will be available.

At this point it is useful to introduce the idea of cost per bit of information. If we were to choose an experimental situation where the result of the experiment would be only a "go/no go" (i.e. a yes or no answer as to success or failure) the cost per bit of information derived will be the total cost of the experiment divided by two. If the experiment were structured so that not only a "go/no go" answer were obtained, but also if "go" then another "go/no go" grouping of hypotheses could be tested. Diagrammatically, the situation is presented in Figure 1. Here we would construct a map of data points which could be analyzed through "sub-treatment" as labeled, but which may really be thought of as a carefully structured data collection analysis of the situation brought about by the primary experimental treatment. Each separate "go" or "no go" situation might be thought of as a test of a specific hypothesis--in the case of the

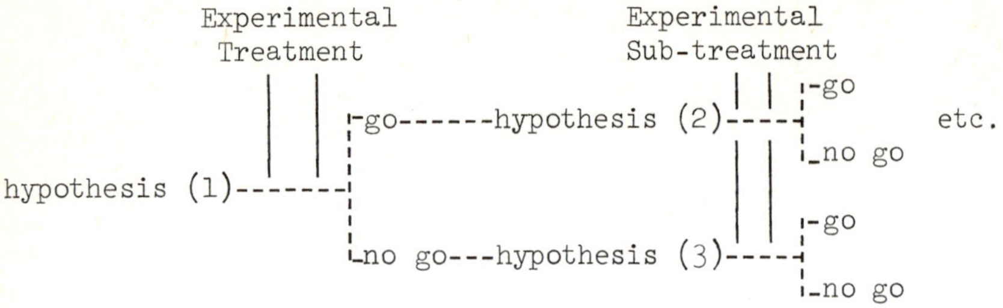

FIGURE 1

subtreatments a hypothesis connected to an upstream previous hypothesis. This structuring is extremely difficult in practice, but again can be conceptually proposed.

With n hypotheses of this type, then the number of bits would be:

$$\text{number of bits} = 2^n$$

provided all sub-hypotheses were independent, which they most probably will not be. A more realistic estimate of the number of bits might be:

$$\text{number of bits} = 2^n - f(n)$$

where f(n) is a function of n which indicates degree of cross linkage of hypotheses.

This type of approach it should be noted is more appropriate for the first case above where a specific impediment has been identified and is to be overcome rather than eliminated. A cost curve for this type of experimentation is presented in Figure 2. Curve (1) corresponds to one type of impediment, curve (2) to another, etc. The curves begin to rise in cost per bit at some point because of data collection problems. Also shown on the curve are estimates of reliability of data, which indicate the statistical nature of the problem. Again, without analysis, where little data is present, data is likely to be reliable provided the structuring of hypotheses has been carried out in a well controlled and analyzable way.

A cursory examination would indicate experimentation carried out as closely as possible to points A, B, and C as being the most efficient. Not appearing in Figure 2 explicity is the <u>total</u> cost of the experiment. Even when the cost per bit of experimental data decreases, the total cost of the experiment may rise --although this is not necessarily true, the cost to a certain

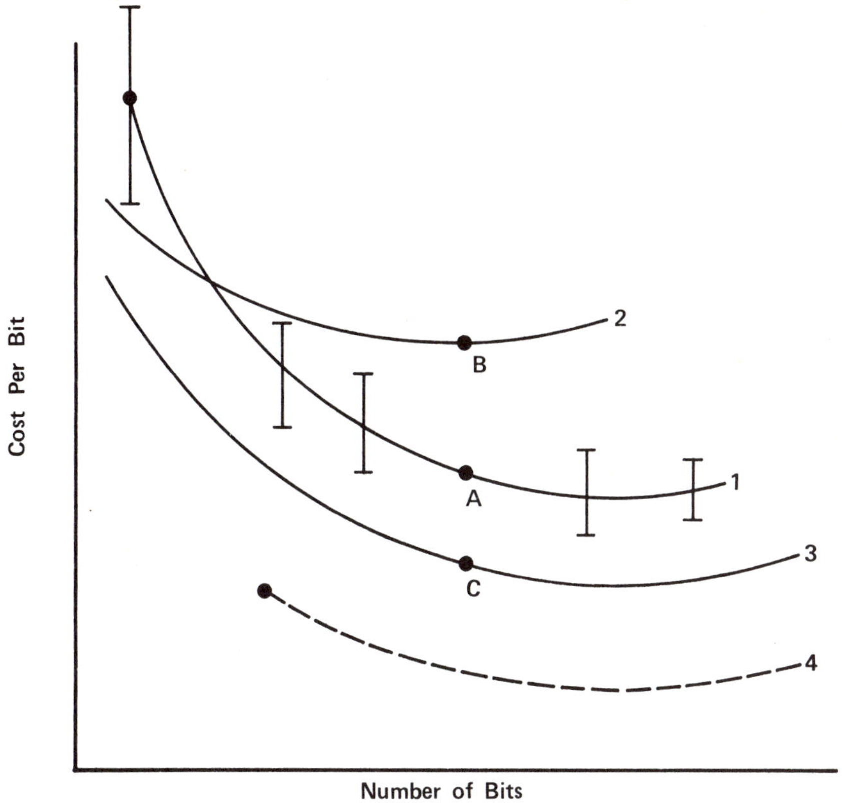

Figure 2

point may be effectively constant if the structuring can be carried out with more precision.

It is evident that this price structure may be considerably different for experimental situations of the second kind, those where an impediment is removed. There, there is a cost to study possible effects, design strategies, and data collection, but that cost is in a sense in the form of entry to the "game." Data then becomes available at the rate and degree to which it can be collected. A typical cost curve might be that shown by (4) in Figure 2. Reliability is, of course, related once again to the statistics of the collection and should be quite high.

Summary

While the preceding discussion has been concerned more with the general problems of structuring and analyzing situations where data might be obtained on the relationships between growth and social costs, it is suggestive of strategies which might be pursued. In particular, for economic reasons related directly to the degree of understanding of the conduct of socio-technological experimentation, it would seem that the primary mode of experimentation would involve removal of identified impediments in an across-the-board fashion. This does not necessarily imply that this action should be taken on a major scale. As more knowledge is gained specific targets could then be addressed on an individual basis.

What is evident is that a promising pathway to understanding of social impacts is through controlled experimentation. What is equally evident is that the pathway is ill defined. In such a case we should proceed with enlightened caution.

The previous discussion has concerned itself with the conceptual framework in which experimentation might take place. It has not dealt at all with a real situation. There are many such situations existing. To name but a few, all sorts of government (national, state and local) regulations exist ranging from safety to clean air. Laws exist, for example antitrust and finally, perioding the whole framework of our economic society--taxes.

It is within the power of the governments involved to modify on a lesser or greater scale these (necessary) constraints in one fashion or another--they do not have to be specifically removed. If the constraints are selectively removed or modified in an atmosphere where data of various types, including that pertaining to factors relating to social impacts, can be carefully measured inferences can be drawn regarding the potential across-the-board impact of more general modified constraints. Let us consider just one example.

As R. E. Slitor[2] has eloquently stated, "The tax laws are pervasive in their influence on decisions to spend, to invest, to work, to assume risk, and to retain, modify, or abandon products, production processes, and sales promotion policies. There is virtually no aspect of private industrial decision-making which is not now touched by the tax laws (whether or not the decision-maker is fully and rationally aware of the fact) and could not be

2/ Tax Incentives for Research and Development, R. E. Slitor--a report prepared for the National Science Foundation

influenced (possibly with the aid of some informational and educational support) by a change in the tax law to remove disincentives or to provide positive incentives for the accomplishment of research." Starting from this point, we could consider the effect of some type of tax rebate (either current, or delayed to some future time. Tax rebates are not an effective incentive, obviously, if the company is not paying taxes. Many technologically oriented firms do not pay taxes for 5-7 years, since they have no profits until then.) on either one of two bases. The first might be for some particular type of company, the description to be developed in such a way that historic and current data would be available, and further that the description would be meaningful to the Internal Revenue Service in either collecting taxes, or rebating taxes. In short the modification of law must be for an accurately described industrial sector. Another approach might be to make the potential rebate be related not to the company descriptors, but to end use. The end use might be one clearly related to one having obvious social worth--for example, products having primary application in particular areas of the public service sector.

The problem of coming up with suitable descriptors is by no means trivial in either case. It implies on the one hand a structuring of the nature of activity of a given industrial firm (or sector), or class of firms (or sectors). Further the structuring must relate quantitatively to the law, in this case the experimental variable which we are attempting to modify. This problem has been dealt with by many groups (see for example "Technological Innovation--Its Environment and Management."[3/]) This particular study abandoned this general idea because of perceived impending difficulties. At the present time I would recommend further analysis of the problem to pinpoint the difficulties in greater depth. Perhaps even more difficult is obtaining descriptors in terms of responsible end use. Here, clearly, the problem must be approached on an experimental basis. We will not know what the effect will be (even the primary effect let alone potential second order effects) until various well planned experiments take place. Alternate approaches should be tried and correlations of effects analyzed and catalogued.

One further consideration is cost. Because of the very pervasive nature of taxes, changes can have potentially enormous effects on government revenues. This fact would tend to indicate that any changes to be experimented with probably be of the "added activity" type. By this is meant that some baseline data be derived for perhaps the previous three to five years. No

3/ Technological Innovation--Its Environment and Management, U.S. Department of Commerce, January 1967

rebates would be available unless there is an increase in describable activity above that baseline--or some such assurance that no taxes presently collected will not be collected in the future due to experimental alterations in tax law.

As noted above taxes are only one of many areas, and tax rebates are only a subset of taxes. However, if significant experiments are to be performed this type of narrowing of the general field must be performed. In some ways "convential wisdom" must be used to pick the initial types of narrowing. With performance of experiments and development of models to describe the process better targeting can take place. The problems are not simple either technically or psychologically. Laws and regulations generally are on the books for good reasons. The situation may have changed sufficiently to partially alleviate the necessity for certain parts of the regulations, but almost always not all.

Underlying this paper is the concept of how innovation may be stimulated. Technology transfer, or acquisition lies within the total spectrum of the innovation process, and the discussion above, particularly as regards development of descriptors of end use applications impacts directly on the functional process of transfer. Technology is acquired in an evolutionary way--and with the help of an experimental framework in which to test the process an understanding of the mechanisms causing change will become understood and hopefully come more under our control.

INDUSTRIAL APPLICATIONS AND TECHNOLOGY TRANSFER -- REPORT ON THE WORKSHOP

Dr. Dieter Altenpohl and Philip D. Wilmot

Chairmen

The central point in the thinking of the Industrial Workshop was the significance of technological transfer to industrial strategy.

There is a strong indication that the major managerial decisions will be increasingly complex due to the necessity to take greater account of such factors as:

- environmental and social aspects
- increasing cost of R&D
- compression of product life cycle curves
- scarcer resources

The questions-- where can I obtain technological forecasts; where are the technological gaps; what business should we be in; which direction should R&D take--are taking on new significance.

A better understanding of technological transfer and its related environment is becoming vital to efficient management.

The work reported at this conference appears to be focussed at both the macro and micro level. The presentations and discussions have permitted a wider understanding of the applicability of such work to other problems. It is the view of the workshop that further work in the directions of putting individual studies into full context is needed. In this connection a conceptual framework which would relate such work to the forces internal and external to the company, industry or country, including the constraints upon them, would need to be considered. Only by putting the individual case into its appropriate context can improved and more generally applicable methodology be evolved. Due regard will have to be paid to personal, company and national proprietary rights.

The questions the Industrial Workshop posed were:

-- Is our present system for acquiring information on studies of this type sufficient?

-- If not, what new channels to process and to retrieve such information will be required?

The development of an effective means of communication is vital in order to provide the working material for industrial management decisions and the feedback mechanism between the practical user and the conceptual thinkers.

The discussions pointed out the necessity to define and to identify the need for and collection of specific cases supported by adequate data, in a proper framework of analytical research. Comprehensive conceptual frameworks for technology transfer must be created and developed.

Four needs were foreseen:

1. Data banks on technology transfer information

2. Conceptual frameworks

3. Analytical frameworks

4. Procedural, functional and methodological frameworks

The questions must be asked--what new institutional linkages are needed to diffuse knowledge of technology transfer processes and how can they be created?

The nature of the technology demand and the transfer process varies greatly between one type of industry and another. Consequently, a failure to consider such background can lead to errors in applying the experience of technology transfer in one field to another. Several frameworks have been suggested, but the following was suggested as a good working basis:

1. Primary industry (extraction of raw materials). Low demand for basic R&D, high demand for applied R&D.

2. Sales maximizing (commodity markets). Low demand for basic and applied R&D. High demand for logistic skills.

3. Performance maximizing (high technology markets). High demand for basic and applied R&D.

4. Market maximizing (retail trade). Low demand for basic and applied R&D. High demand on marketing skills.

5. Personal services.

6. Security.

7. Overhead (administration).

Technology transfer is particularly important in the strategic planning process in industry and the subject should be considered from the standpoint of the industrial planner or manager.

Technology transfer has always been a matter for such managers but new factors are making such planning more complex:

- environmental and social problems
- cost of R&D which is increasing
- a compression of the latter part of the product life cycle curve
- source resources (in some cases caused by geographical distribution)

Increasingly planners are asking such questions as:

- What business should we be in?
- Where are the "technological gaps"?
- Which direction should company R&D take?
- Where can I obtain technological forecasts?

What information about the process of technology transfer do managers need as input to this decision making process?

For some time the universities and others have been studying the managerial decision making process and a better understanding of the principles of technology transfer should be an essential part of these studies.

The work that has been reported over the course of this conference appears to be focussed at either the macro-level or micro-level of technology transfer. Although the presentations and following discussions have certainly extended the understanding of the process. From the point of view of this workshop, there has not yet been achieved a sufficient understanding of mechanism by which one can postulate on a sound basis a methodology as to how to transfer technology in the most cost effective way in the future areas of interest to the industrial participants.

This lack of a sound methodology appears to the workshop to be largely caused by insufficient attention to linking the macro-level work with the micro-level investigations. It was the consensus of opinion that this lack of coupling was due to the failure to relate the various technology transfer processes to a conceptual framework for the industrial sector. Such a conceptual framework must address in some depth the internal and external forces which influence the transfer process industry and how they differ from industry to industry. This conceptual framework is recommended for a research focus in order that a sound methodology of "how to" be evolved.

If technology transfer is to be of maximum benefit, the conceptual, the procedural and the functional frameworks must be considered within the constraints of individual subsets of activities which have individual goals with different motivating forces. These subsets can be industrial firms, governmental agencies and frequently the combination of these subsets. It is important to consider the area in which transfer of technology can take place and, in the ideal cases, ignore many areas where influences will preclude the transfer.

In order to provide the motivation, rewards must be offered which will stimulate the transfer of technology. These rewards must be a synthesis of some type as the creation of profit, growth and other individual factors may individually be in conflict.

The following members of the Conference contributed to the Workshop: Dr. Dieter Altenpohl, Mr. Justin A. Bereny, Dr. Richard Block, Mr. J. Kenneth Craver, Dr. Quentin L. Hartwig, Mr. John Holmgren, Dr. G. O. Krause, Dr. LeRoy Pagano, Dr. M. Reinel, Mr. C. A. Stone, Dr. A. Vanterpool, Mr. Philip D. Wilmot. As the Chairman, I should like to acknowledge the help of my co-chairman, Dieter Altenpohl, and the other members of the Workshop. I have tried to report the conclusions of this Workshop without attribution and hope that I am doing justice to all of my colleagues' ideas. Naturally, I will accept the blame for any errors or omissions.

PART VI: Keynote Address

TECHNOLOGY: MODERN MAN'S MAGIC

J. EDWARD ROUSH

MEMBER HOUSE OF REPRESENTATIVES UNITED STATES CONGRESS

Several years ago at a hearing conducted by the House Science and Astronautics Committee, one of the witnesses noted that today technology is modern man's magic; that through technology man plans and hopes to secure most of what he desires.

I would go further to say that modern man uses technology as primitive man used magic: to control a hostile environment, to subdue it to his own needs and interests; to control and prevent illnesses; to increase crops and regulate the weather; to ward off disasters; to lighten his physical and emotional load, eliminating some work and much fear, anxiety, uncertainty concerning the forces that surround him.

Technology utilization experts are the medicine men of modern society.

I would even go so far as to say that through technology and technology utilization man becomes not just an adapter, but a creator, fashioning a new universe for himself and his fellowmen. It is an exciting and a rewarding adventure to be a part of the 20th Century technological revolution, which is really what you have been discussing over the past week in this beautiful setting outside Paris.

As I appear before you tonight, I am thinking how Copernicus would envy us the climate of opinion in which we operate today, the support we are given by our governments, the results that we enjoy.

I say this well aware of the fact that some countries support technology efforts much less than others and that none supports technology utilization enough, or as much as we would wish. Nevertheless, at least in the United States, I have seen, over the years, a distinct and favorable change in attitude on the part of my government and the American people. I'd like to recount this for you and then indicate the immediate pressures inadvertently supporting increased public and private efforts at technology utilization.

Technology utilization, or technology enhancement as it is now being called, has meant to me the process by which knowledge acquired by government research dollars has not only served the original purposes intended, but has been adapted, revised, applied in other areas affecting the public interest.

Let me give you a little history of how the concept of technology utilization has developed over the years in the United States, particularly vis a vis the Federal Government.

I first became aware of the potentialities of technology utilization as a member of the House of Representatives Committee on Science and Astronautics, that committee which has been responsible for the Apollo shots. I was a member of that Committee for 10 years while the initial steps toward increasing U.S. scientific knowledge and landing men on the moon were underway.

As a representative of people from a Congressional district in Indiana, I was frequently questioned, before the success of the moon shots in particular, about the uses the Federal Government found for the billions spent for research and development over the last decade. Between 1960 and 1974, the toal amounted to approximately $194 billion (taking into account that the figures for 1973 and 1974 are estimates).

The funds were aimed at specific needs or goals. Research and development efforts by the Federal Government have been mission-oriented serving either the broad needs of the Department of Defense, the Atomic Energy Commission, the National Aeronautics and Space Administration, the Department of Health, Education and Welfare, or the Department of Agriculture. Until recently, the government had not gone beyond this kind of support into subsidizing private enterprise efforts or district economic goals, as have some other countries. Some new tentative moves in that direction seem to be in the offing at this time, as I will endeavor to show.

In the 1960's, the U.S. taxpayer, like my constituents in the heartland of the United States, could not be easily persuaded that such a gigantic budget for research and development was worth the cost, without some assurances that in his or her daily life, these expenses would make a difference, contribute to the improved quality and convenience of their lives.

So, in a sense, technology utilization, while intrinsically and independently valuable, has also been a means of proving to the American people that the R&D dollars, which ultimately came out of their pockets as taxpayers, were paying off in improved communications, new fuel sources, better health equipment, weather modification programs, even new and better cooking ware and household paint and clothing materials.

The fact is, however, that not all Americans are scholars who relish the exploration of the unknown simply because it is there. The great Frenchman, Alexis de Tocqueville, explaining why early 19th Century Americans were more concerned with the "applications than with the theory of science" distinguished between the aristocratic and democratic temperaments. "In aristocratic ages," he wrote, "the chief function of science is to give pleasure to the mind, in democratic ages to the body."[1]

Well, I believe there is an Aristotelian resolution of this seeming dichotomy between concerns of the body and the soul, matter, and spirit; a golden mean harmonizing the two. While I would in no way disparage basic research -- I see it as the source of all subsequent technological development and an essential ingredient in man's expansion of his knowledge frontiers--I have also become a kind of salesman for the practical and almost immediate values of R&D dollars when translated by means of technology utilization into everyday comforts, luxuries, necessities--gratifications for both the mind and the spirit.

Besides expanding our own dim view of the universe, I believe, and I have so stated many times, that the improved technology (products and processes) which can result from such increases in knowledge, can make it possible for us to: prognosticate the weather, control pollution, provide new and better production methods, improve health care, move toward a systems approach in handling large scale problems, such as mass transportation and housing, and create many now unknown creative comforts.

Because of these convictions, I introduced a bill in Congress in 1971 setting up an independent Office for Technology Transfer. I wanted to coordinate the present efforts at technology transfer by the government, for I found them disparate and unorganized. I wanted to create a single point of contact for government agencies and those interested in technology utilization outside the government.

The Office for Federal Technology Transfer would be at once a storage bank for all Federal technology utilization reports and activities; a source of grants and contracts for research; an organizer of regional science and technology centers staffed with people who would seek out potential technology uses and users. This bill is pending in Committee at this time.

I have also co-sponsored another bill which aims at some of these same goals by means of a different legislative vehicle. Congressman John Davis' bill, "The National Science Policy and Priorities Act of 1973" would, among other things, establish a Civil Science Systems Administration within the National Science Foundation to design and demonstrate technical systems which could provide improved public services.

The Civil Science Systems Administration would function through the award of contracts to industry, universities and other research organizations to channel technical talent and resources toward problem solving pointing to a "better quality of life for all Americans," as one of the co-sponsors noted.

On the Senate side, Senator Edward Kennedy, introducing this legislation, pointed out that our engineers can propel nuclear ships around the globe with the energy from a bucketful of fuel, yet we suffer from power blackouts at home; that we can cruise on the moon's surface; but we cannot comfortably and conveniently travel from suburbs to city; that we can design high speed computers, yet we cannot teach some children to read. The possibilities of technology utilization, he concluded, are incalculable, the fault lies with ourselves. I agree whole-heartedly.[2] This legislation is pending in Committee also.

Then last year, the Congress passed a bill providing for an Office of Technology Assessment so that we might have a research arm to help us explore the pluses and minuses, on a broad scale, of legislation that appears before us which has technology components. Indeed technology utilization is Janus-faced, with potentialities for good and ill, which we must carefully calculate, not to impede progress, but to prevent undesirable effects.

Perhaps these legislative efforts reflect some of the numerous studies and reports that have been conducted over the past decade by the House Science and Astronautics Committee, the Senate Subcommittee on Science and Technology of the Select Committee on Business, Departments and Agencies of the Executive Branch, of the Congressional Research Service in the Library of Congress, to say nothing of private research efforts in this regard. As a result the relationship between national policy, national goals, government R&D efforts, and economic growth have been fully explored. I think we are now beginning to act on what we have found out. We know that science and technology have a strong influence on economic growth and we have begun to realize that conversely economic growth or stagnation at this time can strongly influence our science and technology policies. It has accounted for a new note in federal policies as we shall see.

Several years ago, the President's own Task Force on Science Policy recognized that technology development is a tool for progress for highly industrialized countries. The Task Force described economic growth of the United States in the past half century as determined as much by "new technology" as it has been by the continuous investment of capital. "If a major national goal is increasing the quality of life for the mass of our population," the Task Force advised, "it becomes essential that continued technological development also be a high priority national goal."[3]

And Peter G. Peterson, then an Assistant for International Economic Affairs, responded to the President's request for a review and analysis of the changing world economy with another Report. This Report concluded that industry for both domestic and international consumption purposes must do more to develop new products and processes and advised that to achieve this goal "it is clearly appropriate that the Federal Government consider new methods for bringing forth the new technologies upon which productivity and competitiveness depend."[4]

Such statements were hardly imaginable five years ago and it is doubtful that they would have been listened to. They reflect a changed climate of opinion. Economic growth and technological progress might well have to be stimulated by government efforts to insure their continuance. The winds of change were blowing in the early 1970's.

This was further revealed in March of last year when the President came out with a Message on Science and Technology. What followed was not as grandiose as we had expected but certainly it was encouraging for the Executive Branch created three programs that 1) show a commitment to increased Federal efforts in technology utilization and, 2) indicate a knowledge of the kind of assistance required. These were a part of a "new Federal strategy" of the President's for refocusing scientific research and development on domestic needs.

One of the programs is the EXPERIMENTAL RESEARCH AND DEVELOPMENT INCENTIVES PROGRAM in the National Science Foundation. The objectives of this program are to find ways to increase the efficiency and speed of conversion of R&D results into new and improved products, processes, and services. The EXPERIMENTAL R&D PROGRAM will explore a range of possible incentives to innovations as a basis for later decisions on the best incentives to use to stimulate technology utilization in the civil sector and to overcome barriers and blockages preventing such conversion.

I am now a member of the House Appropriations Subcommittee which provides funding for this program. With $15,000,000 for fiscal 1973--which just ended June 30th--the EXPERIMENTAL R&D people have been concentrating on incentives that can be applied by the Federal government; in future years they expect to look at other types of incentives and to extend their interest outside technology utilization *per se*, to areas that have an impact on innovation, such as tax structure, patent policy, anti-trust legislation. The next fiscal year, which we have just begun, the program will focus on the "start-up" problem, emphasizing overcoming barriers to the creating of new technologically based businesses and products.

Meanwhile NSF through the EXPERIMENTAL R&D Program has started some promising experiments, four of them.

Briefly, the first of these is the FEDERAL LABORATORY EXPERIMENT. Often small companies cannot afford the testing necessary to prove use and saleability of new processes. They can neither find risk capital nor facilities for such experimentation. The National Science Foundation will use Federal laboratories to do "proof-of-concept" and "performance-verification" testing with the idea that if designs are certified as feasible, venture capital will be easier to come by.

The second project couples the R&D capacity of universities with the need for advanced technology and product development by industry, organizing research centers at three universities to support traditionally nonresearch industries which will then have a research arm where updating and improvement of products will be analyzed and researched. This is a Cooperative Research Experiment.

The third major effort under the EXPERIMENTAL R&D Program is the INNVOATION CENTER EXPERIMENT in which instruction programs and internships are being organized on three university campuses. Scientists and engineers are not necessarily entrepreneurs, so this program would marry Business Management school knowledge with technical and scientific knowledge to produce engineers and scientists who do know how to design a product and make it ready for the market. Inventions originating within the university, or taken from Federal files or submitted by individual inventors will be used in this program to learn product marketing by doing.

The fourth trial project is the TECHNOLOGY EXTENSION SYSTEM EXPERIMENT in which an integrated network of probably 30 technology agents will be established to work for selected local communities. Public Technology Incorporated, a non-profit corporation organized by non-Federal public sector officials, will coordinate the program with federal laboratories and local universities as back-up information service for these technology agents who are trying to match needs with unknown capabilities. This program especially interests me because something similar is included in my Technology bill in recognition of the importance of personal agents in matching needs and potentialities.

Unlike most government programs, these experiments are expected to phase themselves out with success, being picked up by businesses which become the beneficiaries of the advances made.

These four experiments fall under the EXPERIMENTAL R&D PROGRAM. Now there are two more programs, one in NSF; the other in the Department of Commerce.

The second new program is an EXPERIMENTAL TECHNOLOGY INCENTIVES PROGRAM (ETIP) within the Department of Commerce which is somewhat similar to the Experimental R&D, although the emphasis is different because NSF is science-oriented and the Department of Commerce is mission-oriented. ETIP has set as its goal to establish experiments to answer the following questions:

What alternatives in federal or federally-assisted procurement policies and practices would be cost-effective in stimulating technological invention and innovation and thereby increasing productivity and improving the U.S. competitive position?

Can the regulations issued by the government to protect public health, safety, and well-being, which control goods and services offered to public, be used to provide a positive stimulant to technological innovation?

And what cost-effective federal actions can be taken to improve the productivity of U.S. industry and its international competitiveness through facilitating the commercialization of the technical inventions of inventors and small R&D firms?

Working with $7 million for fiscal 1973 the Department of Commerce is planning to let contracts for experiments to be performed in cooperation with technologically-oriented private organizations and other federal agencies. The answers to these questions should be most helpful to TU purposes.

The last of the new programs devised by the Executive Branch is the NATIONAL R&D ASSESSMENT PROGRAM (distinct from the Office established by Congress) within the National Science Foundation. With $2 million in fiscal 1973, the NATIONAL R&D ASSESSMENT PROGRAM is charged with providing an objective study and assessment of how science and technology contribute to the achievement of national goals and objectives. Operating through research contracts with social scientists and non-profit research firms, the R&D ASSESSMENT people expect to analyze patterns of R&D and technological innovation; the incentives and decisions that underlie the existing patterns; and the implications that our choice of operations will have in shaping future patterns of R&D and technological innovation. This is a research operation aiming to discover how government assists or deters technology innovation and to illuminate the socio-economic effects of investing in R&D and the resultant technological innovations. I think it is a good idea. We are moving fast in uncharted areas and we need to know where we are going.

Obviously there is something new here. The Federal Government has gone beyond publication of R&D materials for potential users to actually helping out with risks, providing agents to match information and need, and research assistance to innovate. We have moved into the field of rather directly stimulating technology utilization for a broad range of human concerns and interests. Although the economic appears predominant, the products and processes will cover a number of consumer interests. We are into perhaps a major growth area in Federal relations for the future: supporting technological processes as well as basic R&D. Belatedly, we are arriving at a policy of technology enhancement, where others have already been.[5]

Back in the summer of 1971 when I was introducing my technology utilization bill yet another very significant Report was being put into the hands of the President. It was a Report on the subject of International Trade and Investment Policy. Endeavoring to assess both U.S. investment policies and trade for the 1970's and the relation between the two, the Report noted an important fact: that technology-based products are the United States principal asset in increasing its exports and meeting import competition. This "Williams" Report warned that Americans had to improve the "technological capacity which largely supports our export performance" and recommends "a much higher level of government support for research and development directed specifically to industrial objectives, as found in other countries." A new element in the growing interest in technology utilization was revealed in the next sentence of that recommendation, namely that: "Such assistance should be based on priorities reflecting careful analysis of where U.S. competitive advantages lie."[6]

That new element is the result of the circumstance of the recent United States trade imbalance. I hardly need remind you, although for most of you this is not your area of prime interest, that the United States over the past couple of years has suffered a trade deficit of, to us, startling proportions.

In July of 1971, the Secretary of Commerce, appearing before the Subcommittee on Science, Research and Development of the House Science and Astronautics Committee to discuss science, technology and the economy, warned that the United States might suffer a trade deficit that year.[7] Unfortunately, his prognostication was correct. The first U.S. trade deficit in the 20th century occurred in 1971. It was $2 billion. It was followed in 1972 by a trade deficit of $6.3 billion.

The reasons for this deplorable condition are multiple and I will not try to analyze our trade picture in any depth. I am here concerned with the impact this deficit has had, is having, and will have on governmental technology utilization efforts, as well as popular support for the same.

Just as there are many causes for the trade deficit, so there may be many solutions. And we are not short of these. Historically, the United States has advoated a "free trade" policy for a number of years now while other countries have progressively stimulated exports by direct and indirect means. We were not unaware of this difference in export policy, but until very recently there was no concern over the effects this would have on international trade and competition.

Now organized labor in the United States is demanding that the Congress establish quotas and discourage the operations of multinational corporations abroad.

In testimony before the Subcommittee on Science, Research, and Development of the House Science and Astronautics Committee, Andrew J. Biemiller, representing the AFL-CIO, succinctly summarized their position.

> We recognize the need to encourage research, science, and technology. But we also believe that the U.S. Government should not spend the taxpayer's dollars to export the taxpayers' jobs. That's self-defeating.
>
> The practical realities of today are that other governments encourage transfer of production, technology, and jobs into their countries, and discourage the transfer of production, technology, and jobs out of their countries.[8]

The United States, he thought, should do likewise and multinationals were the chief offenders in exporting jobs. The argument is well taken.

Multinational corporations are not unique to the United States and they are important in consideration of technology utilization efforts in several ways. They are the prime users of technology, prime developers both at home and abroad. A Report for the Senate encapsulated their relationship to technology thus:

> Multinational corporations based in the United States dominate the development of new domestic technology. They also are the principal institutions through which technology in its various forms is exported and imported.[9]

It is this dual role which creates the problem. While technological innovation on products for exportation from the United States contributes to a favorable trade balance, exportation of licenses and patents and production facilities by multinationalists to foreign countries reduces the need in those countries for items manufactured in the United States and therefore adversely affects trade.

Perhaps these corporations are protecting themselves in a market that would be filled by someone else. Yet, we have as the trade deficit has worsened, come to rely upon exports of "technology intensive" or "high-technology" items, such as chemicals, scientific and professional instruments, transporation equipment over "low-technology" products, such as textiles, apparel, petroleum products, and footwear.[10] It is the former that kept our foreign trade solvent and where we look now to redress the balance. As the major exporters of technology, multinationals are a source of concern.

I can therefore predict that the next legislation affecting technology utilization in the United States may well come in the form of a trade bill. Organized labor has asked for Burke-Hartke Bill, which by means of quotas would restrict imports; restrict the transfer of patents and technology; discourage foreign investments through changes in the tax laws.

The multinationals argue that they do not on the whole adversely affect jobs and trade. The Tariff Commission Report concluded that MNCs actually out-paced the non-MNCs in the high technology industries as "generators of net new exports" in the period 1966-1970 and that "MNCs appear on balance to have helped rather than hindered the expansion of U.S. trade in high technology goods."[11] Which should mean that over the long haul they would help our trade balance, but this does nothing for the immediate situation in which plants are forced to close or cut back due to competition from foreign markets.

The Tariff Commission's conclusion seems to generally agree with that of a Commerce Department analysis last year of multinationals in which the Department admitted that to the extent that multinational firms had reduced opportunities for export by diffusing technology abroad the "United States might conceivably derive a short-term benefit from the regulation of such transfer." The Commerce Report went on to warn that such a tactic could not be looked to as "a significant response to U.S. balance of payments or domestic employment problems" because the "net effect would most probably be only to encourage additional foreign efforts to develop their own new technologies and to independently imitate those of the United States."[12]

The issue is joined. I do not know how it will be resolved. The House Ways and Means Committee has been looking at trade proposals in addition to the Burke-Hartke Bill. In particular the President has made a trade proposal that would give him broad powers to raise and lower trading barriers in bargaining with the major United States trading partners, establish an improved program to speed anti-dumping complaints, and make some very limited tax changes concerning earnings from overseas investments. The problems raised by labor are not satisfactorily handled by this legislation as it appears now. It may undergo vast changes in Committee.

Technology utilization is further implicated in this picture as an ingredient, perhaps the essential ingredient of any solution we might devise to our present trade problems, if we would at the same time avoid drastic changes in prevailing policies. At the time, the Secretary of Commerce told that subcommittee of the House Science and Astronautics Committee that we might suffer a trade deficit in 1971, he also detailed the factors affecting our trade balance and steps we might take to improve the same and then concluded:

> The major element which we can influence decisively for the long run is the level of technological development. It may be our only hope of maintaining a future trade position adequate to support our balance of payments in the years to come.[13]

It will take continued intensive technological development to keep the United States competitive in the world market.

Space spin-offs are a long way from the main value attributable to technology utilization today. If we would look for solutions to our trade problem (and that is critical to the whole economy), and for that matter to our transportation, our health problems, our pollution concerns, our inflation and productivity worries, technology utilization must be one, if not THE ingredient in such remedies. Belatedly, I think we are realizing that we have finally recognized the value, the magic of technological utilization.

Perhaps Senator Kennedy's hope that we will use our vast technological potentialities to improve the power supply in our homes and offices, to reduce the headaches of inter-city travel, and to teach all children to read and to learn will now be realized. Recognizing that the fault does lie within ourselves because we have not achieved these goals, we can and are setting a course for our nation aimed at solving the practical problems of mankind.

FOOTNOTES

1. Alexis De Tocqueville. DEMOCRACY IN AMERICA. Edited by J. P. Mayer and Max Lerner. New York: Harper & Row, Publisher. C 1966. p. 429.

2. Senator Edward Kennedy. CONGRESSIONAL RECORD. January 4, 1973. p. S103.

3. Task Force on Science Policy. SCIENCE AND TECHNOLOGY: TOOLS FOR PROGRESS. Report to the President. Washington, D.C.: The Government Printing Office, 1970. pp. 11-12.

4. Peter G. Peterson. THE UNITED STATES IN THE CHANGING WORLD ECONOMY. (2 Vols.) Volume 1: A Foreign Economic Perspective. Washington, D.C.: U.S. Government Printing Office, 1971. p. 36.

5. See U.S. Department of Commerce. TECHNOLOGY ENHANCEMENT PROGRAMS IN FIVE FOREIGN COUNTRIES: A report distributed by National Technical Information Service, U.S. Department of Commerce, December, 1972.

6. Commission on International Trade and Investment Policy. U.S. INTERNATIONAL ECONOMIC POLICY IN AN INTERDEPENDENT WORLD. Report to the President. Washington, D.C. U.S. Government Printing Office, 1971. p. 120.

7. The Honorable Maurice Stans, Secretary of Commerce, Hearings Before the Subcommittee on Science, Research, and Development of the Committee on Science and Astronautics: SCIENCE, TECHNOLOGY, AND THE ECONOMY. U.S. House of Representatives 92nd Congress, First Session. Washington, D.C., U.S. Government Printing Office, 1971. pp. 3-32. Hereafter referred to as SCIENCE, TECHNOLOGY, AND THE ECONOMY HEARINGS.

8. Andrew J. Beimiller, Director, Department of Legislation, AFL-CIO, SCIENCE, TECHNOLOGY AND THE ECONOMY HEARINGS, p. 51.

9. U.S. Tariff Commission. IMPLICATIONS OF MULTINATIONAL FIRMS FOR WORLD TRADE AND INVESTMENT AND FOR U.S. TRADE AND LABOR: REPORT TO THE COMMITTEE ON FINANCE OF THE UNITED STATES SENATE AND ITS SUBCOMMITTEE ON INTERNATIONAL TRADE ON INVESTIGATION NO. 332-69, UNDER SECTION 332 OF THE TARIFF ACT OF 1930, Washington, D.C., U.S. Government Printing Office, 1973. p. 10. Hereafter referred to as *IMPLICATIONS OF MULTINATIONAL FIRMS*.

10. For a full description of the two kinds of technology see statement of Hon. Peter G. Peterson, Secretary of Commerce in *SCIENCE, TECHNOLOGY, AND THE ECONOMY HEARINGS*, 92nd Congress, Second Session, 1972, pp. 3-33; See also, "U.S. Leadership in High Technology Products: A U.S. View," Remarks by James H. Wakelin, Jr., Ass't. Secretary of Commerce for Science and Technology, Before the American Institute of Physics, Annual Meeting of Society Officers and Corporate Associates, New York City, September, 1971. p. 1-7.

11. *IMPLICATIONS OF MULTINATIONAL FIRMS*. p. 11.

12. U. S. Department of Commerce. *THE MULTINATIONAL CORPORATION : STUDIES ON U.S. FOREIGN INVESTMENT*. (2 Vols.) Vol. I: *POLICY ASPECTS OF FOREIGN INVESTMENT BY U.S. MULTINATIONAL CORPORATIONS*. Washington, D.C.: U.S. Government Printing Office, 1972. pp. 38-40.

13. *SCIENCE, TECHNOLOGY AND THE ECONOMY HEARINGS*, 92nd Congress, First Session, pp. 3-4.